Water Pollution: Control and Treatment

Water Pollution: Control and Treatment

Editor: Ezra Duncan

STATES
ACADEMIC PRESS
www.statesacademicpress.com

States Academic Press,
109 South 5th Street,
Brooklyn, NY 11249, USA

Visit us on the World Wide Web at:
www.statesacademicpress.com

ISBN: 978-1-63989-566-3 (Hardback)

Cataloging-in-Publication Data

Water pollution : control and treatment / edited by Ezra Duncan.
p. cm.
Includes bibliographical references and index.
ISBN 978-1-63989-566-3
1. Water--Pollution. 2. Pollution prevention. 3. Sewage--Purification.
4. Waste disposal in rivers, lakes, etc. I. Duncan, Ezra.
TD420 .W38 2021
628.168--dc23

Table of Contents

Preface

I am honored to present to you this unique book which encompasses the most up-to-date data in the field. I was extremely pleased to get this opportunity of editing the work of experts from across the globe. I have also written papers in this field and researched the various aspects revolving around the progress of the discipline. I have tried to unify my knowledge along with that of stalwarts from every corner of the world, to produce a text which not only benefits the readers but also facilitates the growth of the field.

Water pollution refers to the contamination of water due to the release of substances into water bodies. The pollution sources that pollute water are broadly categorized into point sources and non-point sources. The contaminants that enter the water bodies from a single and identifiable source are referred to as point sources of water pollution. These include discharges from factory and sewage treatment plant. Non-point sources of water pollution are defined as the contamination caused by the cumulative effect of small amounts of contaminants that accumulate from a large area. The various contaminants that cause water pollution encompass pathogens and a wide spectrum of chemicals. Water pollution can be divided on the basis of its location into surface water pollution and groundwater pollution. Most of the topics introduced in this book cover new techniques to control water pollution. It consists of contributions made by international experts. This book will prove to be immensely beneficial to students and researchers in this field.

Finally, I would like to thank all the contributing authors for their valuable time and contributions. This book would not have been possible without their efforts. I would also like to thank my friends and family for their constant support.

Editor

Evaluation of removal efficiency of 2-chlorophenol in aquatic environments by modified fly ash

Mohammad Malakootian[1*], Alireza Mesdaghinia[2], Shima Rezaei[3]

[1]Professor, Environmental Health Engineering Research Center, Department of Environmental Health, School of Public Health, Kerman University of Medical Sciences, Kerman, Iran
[2]Professor, Department of Environmental Health Engineering, School of Public Health, Water Quality Research Centre, Institute for Environmental Research, Tehran University of Medical Sciences, Tehran, Iran.
[3]MSc, Department of Environmental Health, School of Public Health, Kurdistan University of Medical Sciences, Sanandaj, Iran

Abstract

Background: Chlorophenols are classified as priority toxic pollutants. These acidic organic compounds present a serious potential hazard for human health and aquatic life. Chlorophenols accumulate in water, soil and air due to high stability, and impart an unpleasant taste and odor to drinking water and can exert negative effects on different biological processes. Among the different methods of removal, adsorption process by low price adsorbents, such as fly ash (FA) is common. Therefore, in this study, the effects of oxidation of FA as modified adsorbent were investigated when the adsorption of 2-chlorophenol (2-CP) was increased.

Methods: This experimental study was conducted from March to September of 2013. FA obtained from Zarand power plant (located in Kerman province) was oxidized with potassium permanganate. Effective factors on the oxidation of FA, such as temperature, oxidation time and concentrations of oxidizers were optimized. Raw sewage of Zarand coal washing plant was tested under optimal conditions. All tests were carried out according to the standard methods book for the examination of water and wastewater.

Results: Optimal condition for the preparation of oxidized FA was obtained at 70°C, 1 hour, and 1 mM of potassium permanganate concentration. The absorber obtained was able to remove 96.22% of 2-CP under optimized conditions (pH=3, 2 hours, adsorbent dose 0.8 g and room temperature). The removal efficiency of the real wastewater under optimal conditions was 82.1%.

Conclusion: Oxidized FA can be used for the removal of this pollutant from industry wastewater due to its high efficiency of removal in real wastewater, it is easy and inexpensive to prepare and could modify the sorbent.

Keywords: Chlorophenol, Adsorption process, Modified adsorbent.

***Correspondence to:**
Mohammad Malakootian
Email: m.malakootian@yahoo.com

Introduction

The presence of chlorophenol in the environment is one major environmental concern due to its potential toxicity, carcinogenicity, mutagenicity, poor biodegradation, high stability and widespread pollution (1,2). Chlorophenol has 19 derivatives. The Environmental Protection Agency (EPA) classified 9 compositions of chlorophenol in priority categories. One of the most important is the 2-chlorophenol (2-CP) (3,4). Short-term exposure with high amount of 2-CP in laboratory animals leads to increasing respiratory rate, weakness, tremors, seizures, coma and finally death (5). The maximum allowable concentration of 2-CP in drinking water is 10 µg/L (6). This contaminant has high solubility in aquatic environment (7). This combination has been reported usually in petrochemical wastewater refineries, leather and textile industries, steel plants, manufacturing of pharmaceuticals, plastic, paint, paper industry, pesticides, herbicides and wood preservative (8-10). Commonly used techniques for the removal of phenol and its derivatives are catalytic and photocatalytic decompositions, electrochemical conversion, coagulation, flocculation, resins, microbial decomposition and adsorption (11-14). Recycling of waste materials as an economical solution to the purification of aquatic environment, in addition to environmental pollution control, is consistent with the policy of recycling waste materials management at its source . The application of low cost and the easy availability of materials in wastewater treatment have recently attracted great interest (15). Several studies have been performed using industrial waste materials, such as fly ash

(FA), to absorb NO_X, SO_X, organic compounds and mercury in air, heavy metals, color and organic compounds in water (14,16). FA, obtained from fossil fuels, can be considered as an alternative to activated carbon, due to the presence of high amount of unburned carbon particles (3,14,17). FA is a by-product of coal burning power plants, its characteristics are dependent on several factors, such as the method of burning coal, type of burn, fuel composition, combustion zone temperature, time of residence particle in the combustion zone, etc (18-22). Several studies have been carried out to enhance adsorption capacity, in which FA was modified and activated. Luo et al (23) in China, used absorbent oxidation with potassium permanganate to increase the efficiency of adsorption. In this study, the oxidation effect of FA on 2-CP adsorption efficiency in aquatic environments was investigated in order to increase the adsorption capacity of the absorbent.

Methods

This experimental study was conducted in the laboratory from March to September of 2013 at the Environmental Health Engineering Research Center of Kerman University of Medical Sciences. Raw FA of Zarand thermal power plant was obtained from cyclone dust collector. In total, 100 to 200 mesh particles were selected. Sieved FA was washed for 48 hours, using a distilled water to FA ratio equal to 10, in order to remove impurities and dissolution of metal oxides in water. Electro-conductivity (EC), pH and total dissolved solids (TDS) of the solution were measured before and after washing (24). The washed FA (WFA) was dried in an oven at 105°C to attain a constant weight (by trial and error method). Thereafter, FA was oxidized with potassium permanganate (PC) oxidizers to improve the efficiency of adsorption (23) and thus, oxidized FA (OFA) was obtained. So at first, the oxidizer solution was prepared with the desired concentration using potassium permanganate powder. The solution pH was adjusted in acidic range using sulfuric acid (Purity of acid: 98% and molarity =1) to increase the power of the oxidation.

Preparation of OFA

Optimal amounts of the effective factors (oxidizers concentration, time and temperature of oxidation) on absorbent oxidation were then determined:

1. Determination of PC concentration: To determine the optimal concentration of oxidizers, FA was oxidized at different concentrations (0, 0.5, 1, 5, 10, 30 mM) with potassium permanganate. This process was performed at room temperature for 30 minutes. Then, the adsorption efficiency of 2-CP (at concentration 50 mg/L) was determined at room temperature, contact time: 1.5 hours with 0.5 g of OFA obtained at different PC concentrations. The optimum PC concentration was selected by comparing the results.

2. Determination of temperatures of oxidation: In order to determine the optimal temperature of modification, FA was oxidized at different temperatures (room temperature (28 ± 2), 50, 60, 70 and 80°C). This process was performed for 30 minutes at the optimal concentration of oxidizers. The temperature was set at the desired range using paraffin bath. Thereafter, the optimum temperature was selected by comparing the 2-CP removal efficiency with OFA obtained at different temperatures.

3. Determination of time of oxidation: FA was oxidized at different times (0.5, 1, 2 and 3 hours). This experiment was set at the optimum temperature and concentration of oxidizers. The removal efficiency of 2-CP was then determined with OFA obtained at different times.

At all stages of this experiment, to separate the OFA from the oxidizers solution, the samples were centrifuged for 10 minutes at 3500 rpm. The OFA was then washed with distilled water on high speed stirrer for 4 hours and filtered using 0.45 micron filter. The samples were dried in an oven at 105°C to attain a constant weight. The scanning electron microscope (SEM) of the adsorbents was prepared before and after modification.

Sorption studies

The effective factors on adsorption process were optimized after optimization of all effective parameters on FA oxidation.

In order to investigate the effect of pH on absorption, 100 mL of 2-CP solution with different pH (2-1) at room temperature (2 ± 29°C) was added to 0.5 g of OFA. The pH was adjusted using 0.1 molar NaOH and HCl.

Different doses of OFA (from 0.3 to 1 g) with 100 mL of 2-CP solution (50 mg/L) were added at different times (for 30 to 180 minutes) in order to evaluate the effect of two factors (adsorbent dose and contact time) on the removal efficiency of 2-CP.

2-CP with different concentrations ranging from 20 to 250 mg/L were placed in contact with OFA at optimal conditions and pseudo-first order and pseudo-second order kinetic equations were tested, Freundlich and Langmuir isotherms were investigated also to adsorption of 2-CP on OFA (25, 26).

Wastewater from Zarand coal washing plant located adjacent to the power plant was sampled and transported to the laboratory under standard conditions. After determining the physico-chemical quality of real raw sewage, the 2-CP removal efficiency was determined by OFA under optimal conditions.

To determine the reusability of the adsorbent, the used FA in 2-CP removal was separated from the reaction solution and dried after washing with distilled water. Finally, the 2-CP removal efficiency was determined again.

The adsorption solution was exposed to the environment for 24 hours and then the concentration of 2-CP was measured to determine the desorption of contaminants from the surface of the absorber.

In order to obtain accurate results, all experiments were

repeated three times. The determination of 2-CP was done spectrophotometrically on a spectrophotometer UV-1800 SHIMADZU model at wavelength of 274 nm. All tests were conducted at the same time with distilled water and with a synthetic sample containing 2-CP in order to avoid interference of adsorption of residual permanganate on the adsorbent surface. Afterwards, the test solution with distilled water was placed in a spectrophotometer as a blank sample. Stirrer speed of 3001 Heidolph MR Model in adsorption process tests was set at 400 rpm. The required chemical substances were produced by Merck Company. Statistical analysis was done using SPSS version 19.

Results

The EC and TDS of FA solution increased after and before being washed for 48 hours. As EC and TDS increased from 2.3 to 4.07 (ms/cm) and 0.08 to 1.71 g/L, respectively. But, pH was fixed and equal to 6.2 after washing for 48 hours. The SEM images of the absorber before and after oxidation are shown in Figure 1.

Figure 1 confirms an increase in porosity of the adsorbent surface after oxidation process. As shown in the figure, streaks surface was more open and the surface uniformity of the FA was reduced after oxidation. This effect was intensified by increasing the time. The weight percent of elements and compounds in WFA and OFA samples is shown in Table 1.

Oxidizers concentration

Figure 2 shows changes in 2-CP removal efficiency by changing the PC concentration in the absorber oxidation process.

When PC concentration is zero (WFA), 2-CP adsorption efficiency is less (67.3%) than the situation that absorber is oxide (OFA). But the removal efficiency decreased by increasing the concentration of the oxidizers. The highest adsorption efficiency obtained at lower PC concentration (1 mM) was 81.9%.

Effect of oxidation temperature

The effect of changes in oxidation temperature of FA on removal efficiency is shown in in Figure 3.

Increasing the temperature of FA oxidation (from 28 to 80°C) resulted in increased 2-CP removal efficiency (from 81.9 to 89.5%). Hence, 80°C was selected as optimum oxidation temperature of FA.

Oxidation time of FA

The effect of oxidation time on removal efficiency is shown in Figure 4.

As the time of absorber oxidation increases, the removal efficiency increased as well. But there was no significant difference in adsorption efficiency of OFA obtained at different times. Therefore, 1 hour was selected as the optimal time by considering the appropriate efficiency.

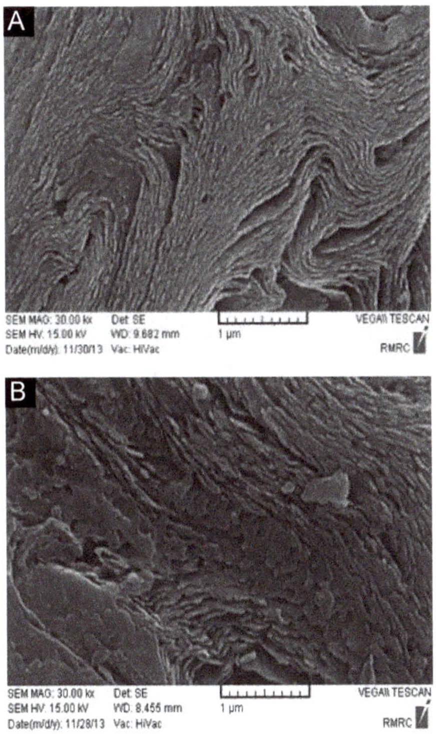

Figure 1. SEM image of a sample fly ash (FA) . (A) Raw FA, (B) oxidized FA.

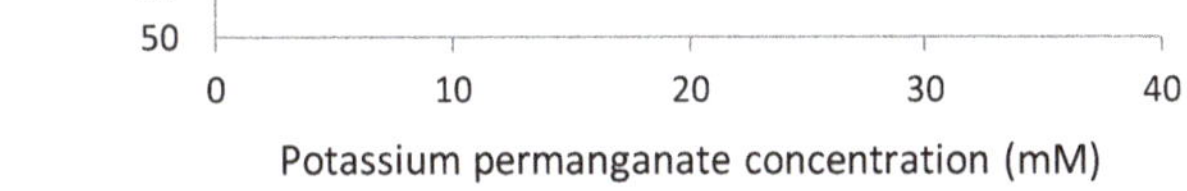

Figure 2. Effect of potassium permanganate concentration in the removal of 2-chlorophenol (2-CP).

Table 1. Weight percent of elements and compounds in WFA samples and OFA

Present elements in samples (wt.-%)					Present compounds in samples (wt.-%)						
CO_2	SO_3	V_2O_5	Fe_2O_3	TiO_2	Carbon	Sulfur	Oxygen	Iron	Vanadium	Titanium	
91.97	6.22	1.2	0.61	0	77.94	7.73	10.91	1.32	2.09	0	WFA
94.92	5.08	0	0	0	86.52	6.8	6.68	0	0	0	AFA

Abbreviations: WFA, washed fly ash; OFA, oxidize fly ash.

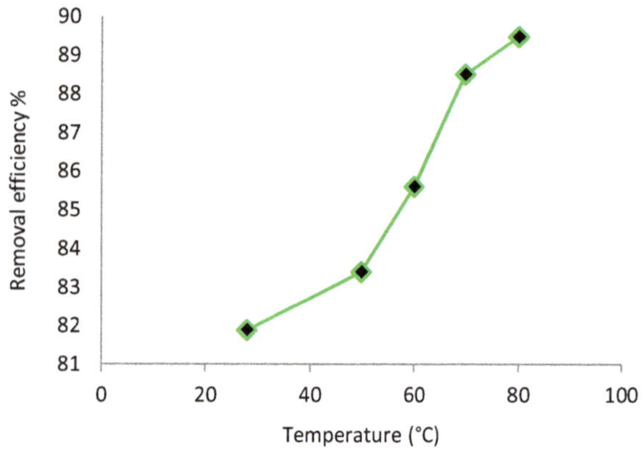

Figure 3. Effect of oxidation temperature on the removal efficiency of 2-chlorophenol (2-CP).

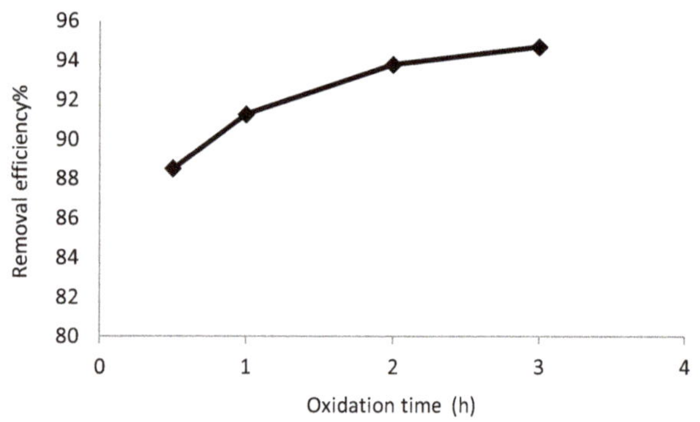

Figure 4. Effect of oxidation time on the removal efficiency of 2-chlorophenol (2-CP).

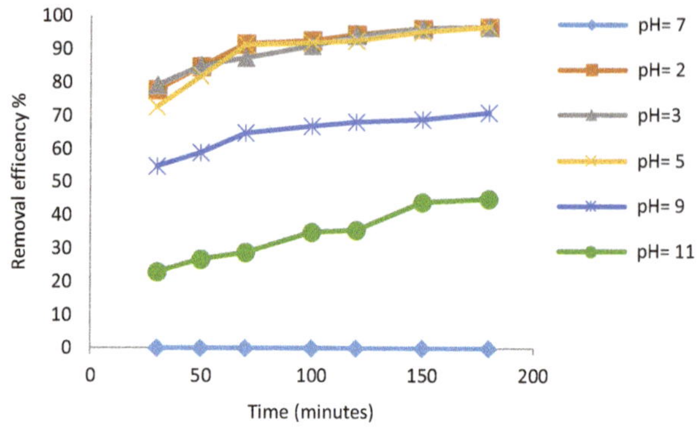

Figure 5. Effect of pH on the removal efficiency of 2-chlorophenol (2-CP).

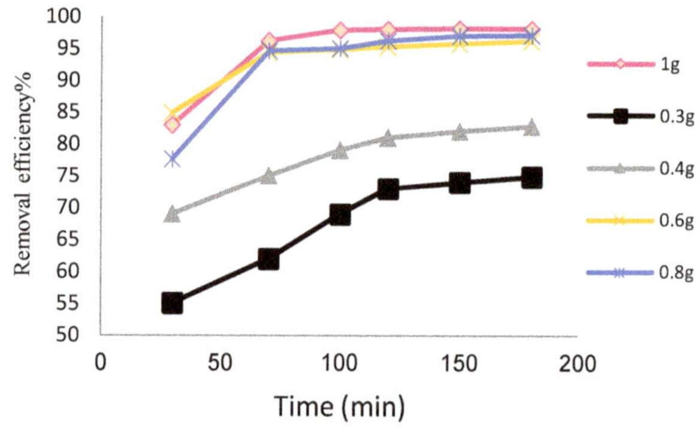

Figure 6. Effect of adsorbent dose and contact time on the removal efficiency of 2-chlorophenol (2-CP).

Effect of pH

In this study, the effect of pH is shown in Figure 5. pH reduction resulted in an increase in the removal efficiency, therefore, pH = 3 was obtained as the highest efficiency. The removal efficiency decreased greatly at neutral pH and increased again under alkaline conditions.

Effect of adsorbent dose and contact time

Changes in removal efficiency by the differing adsorbent dose and contact time are shown in Figure 6.

As the adsorbent dose and contact time increases, the removal efficiency and adsorption capacity increases as well. 0.8 g in 120 min was selected as optimum adsorbent dose.

Kinetics and isotherm of adsorption

An increase in the initial concentration of 2-CP resulted in an increase in the rate of adsorption and reduced the removal efficiency. Batch adsorption studies were conducted in varying concentration of 2-CP. The amount of sorption at time t, q_t (mg/g), was calculated using the following equation:

$q_t = (C_0 - C_t)\, V/M$ (1)

where C_0 is the initial concentration of 2-CP, C_t is the concentration at a given time t, V is the volume of the 2-CP solution in L and M is the weight of the activated carbon in g. For the prediction of isotherm parameter, the Freundlich and Langmuir isotherms were used. The Freundlich isotherm was used based on the assumption that the sorption process is involved in a heterogeneous layer. The nonlinear form of the Freundlich equation is as follows:

$qe = K_F C_e^{1/nF}$ (2)

The linearized form of the Freundlich equation:

$\log q_e = \log(K_F) + 1/\, nF \log(Ce)$ (3)

where q_e is the amount of adsorbate adsorbed at equilibrium (mg/ g), C_e represents an equilibrium concentration of the adsorbate (mg/L), K_F is the Freundlich adsorption constant related to the adsorption capacity of the adsorbent ((mg/g) (L/mg)1/n) and 1/n is the adsorption intensity. The Langmuir isotherm assumes that adsorption proceeds in the homogeneous sites, which are involved in the intramolecular binding on the adsorbent surface and attains a specific site occupation. Thus, no extension of adsorption was found on the occupied site of the adsorbent material. This states that the adsorption process is a monolayer process (26). Table 2 shows the parameters of isotherm and kinetic equations in 2-CP adsorption by oxidized FA.

The Freundlich isotherm was found to be the best expositor for equilibrium conditions of 2-CP absorption. Based on the determination coefficient (R^2) and the comparison

Table 2. Isotherm and kinetics parameters for the adsorption of 2-CP (adsorption values are in mg/L)

Adsorption isotherms						Adsorption kinetics							
Freundlich			Langmuir-type 1			Second degree- type 1					First degree		
1/n	K_f (m/g)	R^2	q_m(m/g	K_a	R^2	$q_{e(exp)}$	$q_{e(cal)}$	*k	h	R^2	R^2	k_1	q_e
0.362	4.07	0.906	0.042	0.074	0.8	6.91	7.57	0.011	0.635	0.996	0.81	0.023	3.24

between calculated equilibrium adsorption ($q_{e(cal)}$) and the obtained equilibrium adsorption amount through testing ($q_{e(exp)}$), the 2-CP adsorption of oxidized FA was followed by second degree kinetics.

Recyclability of adsorption and desorption rate

Studies have shown that reusability of the adsorbent (OFA) is possible. The removal efficiency of 2-CP under optimal conditions during the two times of adsorbent recycling were 78.4% and 63%, respectively. By evaluating the desorption rate, it was shown that after 24 hours, there was no desorption through the absorbent by contaminant. The removal efficiency of 2-CP in the actual wastewater of Zarand coal washing (82.1%) was less than the synthetic samples (96.2%) due to the presence of available interference in the actual samples.

Discussion

The effect of absorbent oxidation

The increased EC and TDS after washing the raw FAs for 48 hours, indicates that the soluble metal oxides, in the presence of water absorbent, are soluble. The risk of leakage of dangerous metals from the absorbent to the aqueous solution reduced when these compounds were removed from the absorbent during the pickling process (27,28). The increasing removal efficiency of oxidized FA is induced by the oxidation of carbon present in the FA with MnO_4^-. During the oxidation process, this carbon is converted to other forms of carbon (C=O, C-OH, OH-C=O, C-O-C). Thus, the ability of oxidized carbon is higher than raw carbon. Also, the observed increase in oxidized adsorption could be attributed to the presence of C=O double bonds in the oxidized FA, and absorbing interactions between the absorbent and pollutants is always the π-π type (29). Jeon et al in North Korea reported that the oxidation process of alginate acid to alginate carboxyl acid by potassium permanganate leads to an increase in carboxyl groups in the oxidized sample rather than the non-oxidized samples, indicating that the carbon type was changed (30). These results are in line with the results obtained in the present study. The observed reduction in adsorption efficiency with increasing concentration could be attributed to the deposition of potassium permanganate oxidizer on the surface of the absorbent. The empty spaces were filled with the oxidizing agent and the adsorption capability was reduced by reducing porosity on the adsorbent surface. The increasing removal efficiency of the oxidized absorbent at high temperatures was due to the effect of heat on absorbent oxidation.

Effect of pH

At acidic pH, chlorophenols are neutralized, whereas the absorbent surface is surrounded by hydronium ion. This results in an increase in surface tension between chlorophenols and binding sites on the adsorbent surface (31). At alkaline pH (greater than pH_{ZPC}), along with increase in the percentage of ionized species, the number of negative charged sites on the adsorbent surface increased. Thus, adsorption will decrease due to electrostatic desorption between the negative surface of the absorbent and Chlorophenol anions (2,8,12,32). The result of this study is in line with the result obtained by Baker and Ghanem in Japan on the removal of 2-CP using natural Zeolites (33). The reduction of adsorption efficiency could be attributed to the weakening of the interaction forces between Chlorophenol and the adsorbent surface (34).

Effect of adsorbent dose

Increasing removal efficiency and adsorption capacity by increasing the amount of adsorbent, could be attributed to the availability of empty binding sites. Nadavala et al (13) in America using calcium alginate and chitosan beads and Bayramoglu et al (35) in Turkey, using yeast bio-absorbent, reported that as time and adsorbent dose increased, the adsorption capacity of 2-CP increased as well. Their obtained results are consistent with the findings of the present study.

Effect of 2-CP original concentration

The number of adsorbed component molecules increased, when the concentration of 2-CP in a fixed number of adsorption sites increased. The generated competition resulted in reduced efficiency and increased q_t. The initial concentration provides a large driving force to overcome the mass transfer resistance of monochlorophenol between solid and aqueous phases. Kamble et al (10) in India, using FA-modified zeolite by detergent; and Mangrulkar et al (12) in India, using MCM-41 observed that initial concentration increases resulted to a decrease in efficiency and increasing q_t. The results of this study are consistent with the mentioned studies.

Conclusion

The oxidation process removes impurities, increases porosity and results in an increase in adsorption efficiency. During oxidation process, the carbon present in the absorber is oxidized and converted to carbon species containing greater affinity interaction with the contaminants. Finally, the adsorption rate improves. In this study, an adsorbent with high adsorption capacity to wastewater treatment of coal washing plant adjacent to the power plant was obtained using a sample and inexpensive modified method, which requires no transportation cost and sophisticated technology. In addition to environmental pollution control, reuse of waste FA is consistent with the policy of recycling at source and waste management discussion.

Acknowledgments
This study is one of the approved projects for students theses in the Environmental Health Research Center and was sponsored by the Vice-Chancellor for Research and Technology of Kerman University of Medical Sciences.

Competing interests
The authors declare that they have no competing interests.

Authors' contributions
All authors participated in the design of the study, performed the literature search and wrote the manuscript, data acquisition, analysis, and interpretation. All authors critically reviewed, refined, and approved the manuscript.

References

1. Altunlu M, Yapar S. Effect of OH^-/Al^{3+} and Al^{3+}/clay ratios on the adsorption properties of Al-pillared bentonites. Colloids Surf A Physicochem Eng Asp 2007; 306(1–3): 88-94.
2. Namasivayam C, Kavitha D. Adsorptive removal of 2-chlorophenol by low-cost coir pith carbon. J Hazard Mater 2003; 98(1–3): 257-74.
3. Andini S, Cioffi R, Colangelo F, Montagnaro F, Santoro L. Adsorption of chlorophenol, chloroaniline and methylene blue on fuel oil fly ash. J Hazard Mate 2008; 157(2–3): 599-604.
4. Sun D, Zhang H. Electrochemical determination of 2-chlorophenol using an acetylene black film modified glassy carbon electrode. Water Res 2006; 40(16): 3069-74.
5. Xia S, Zhanga Z, Zhong F, Zhang J. High efficiency removal of 2-chlorophenol from drinking water by a hydrogen-based polyvinyl chloride membrane biofilm reactor. J Hazard Mate 2011; 186(2-3): 1367-73.
6. Tu Y, Xiong Y, Tian S, Kong L, Descorme C. Catalytic wet air oxidation of 2-chlorophenol over sewage sludge-derived carbon-based catalysts. J Hazard Mate 2014; 276: 88-96.
7. Yang B, Zhang J, Zhang Y, Deng S, Yu G, Wu J, et al. Promoting effect of EDTA on catalytic activity of highly stable Al–Ni bimetal alloy for dechlorination of 2-chlorophenol. Chem Eng J 2014; 250: 222-9.
8. Anbia M, Amirmahmoodi S. Adsorption of phenolic compounds from aqueous solutions using functionalized SBA-15 as a nano-sorbent. Scientia Iranica 2011; 18(3): 446-52.
9. Fattahi N, Assadi Y, Hosseini MR, Jahromi EZ. Determination of chlorophenols in water samples using simultaneous dispersive liquid–liquid microextraction and derivatization followed by gas chromatography-electron-capture detection. J Chromatogr A 2007; 1157(1–2): 23-9.
10. Kamble SP, Mangrulkar PA, Bansiwal AK, Rayalu SS. Adsorption of phenol and o-chlorophenol on surface altered fly ash based molecular sieves. Chem Eng J 2008; 138(1–3): 73-83.
11. Lin SH, Pan CL, Leu HG. Equilibrium and mass transfer characteristics of 2-chlorophenol removal from aqueous solution by liquid membrane. Chem Eng J 2002; 87(2): 163-9.
12. Mangrulkar PA, Kamble SP, Meshram J, Rayalu SS. Adsorption of phenol and o-chlorophenol by mesoporous MCM-41. J Hazard Mate 2008; 160(2-3): 414-21.
13. Nadavala SK, Swayampakula K, Boddu VM, Abburi K. Biosorption of phenol and o-chlorophenol from aqueous solutions on to chitosan–calcium alginate blended beads. J Hazard Mate 2009; 162(1): 482-9.
14. Sarkar M, Acharya PK. Use of fly ash for the removal of phenol and its analogues from contaminated water. Waste Manag 2006; 26(6): 559-70.
15. Visa M, Andronic L, Lucaci D, Duta A. Concurrent dyes adsorption and photo-degradation on fly ash based substrates. Adsorption 2011; 17(1): 101-8.
16. Wang S, Wu H. Environmental-benign utilisation of fly ash as low-cost adsorbents. J Hazard Mate 2006; 136(3): 482-501.
17. Malakootian M, Almasi A, Hossaini H. Pb and Co removal from paint industries effluent using wood ash. International Journal of Environmental Science & Technology 2008; 5(2): 217-22.
18. Lin JX, Zhan SL, Fang MH, Qian XQ, Yang H. Adsorption of basic dye from aqueous solution onto fly ash. J Environ Manage 2008; 87(1): 193-200.
19. Malakootian M, Mesdaghinia AR, Rezaei S. Efficiency of ortho-chlorophenol removal from aqueo us solutions using activated Fly Ash of Zarand Fossil Fuel Power Plant. Journal of School of Public Health and Institute of Public Health Research 2014; 12(2): 81-92.
20. Tiwari M, Sahu S, Bhangare R, Ajmal P, Pandit G. Elemental characterization of coal, fly ash, and bottom ash using an energy dispersive X-ray fluorescence technique. Appl Radiat Isot 2014; 90: 53-7.
21. Zhang A, Wang N, Zhou J, Jiang P, Liu G. Heterogeneous Fenton-like catalytic removal of

p-nitrophenol in water using acid-activated fly ash. J Hazard Mate 2012; 201–202: 68-73.
22. Zhou L, Chen YL, Zhang XH, Tian FM, Zu ZN. Zeolites developed from mixed alkali modified coal fly ash for adsorption of volatile organic compounds. Mater Lett 2014; 119: 140-2.
23. Luo F, Liu Y, Li X, Xuan Z, Ma J. Biosorption of lead ion by chemically-modified biomass of marine brown algae Laminaria japonica. Chemosphere 2006; 64(7): 1122-7.
24. Lu SG, Bai SQ, Zhu L, Shan HD. Removal mechanism of phosphate from aqueous solution by fly ash. J Hazard Mate 2009; 161(1): 95-101.
25. Doğan M, Alkan M, Türkyilmaz A, Özdemir Y. Kinetics and mechanism of removal of methylene blue by adsorption onto perlite. J Hazard Mate 2004; 109(1–3): 141-8.
26. Kumar M, Tamilarasan R. Modeling studies for the removal of methylene blue from aqueous solution using Acacia fumosa seed shell activated carbon. Journal of Environmental Chemical Engineering 2013; 1(4): 1108-16.
27. Kashiwakura S, Ohno H, Kumagai Y, Kubo H, Matsubae K, Nagasaka T. Dissolution behavior of selenium from coal fly ash particles for the development of an acid-washing process. Chemosphere 2011; 85(4): 598-602.
28. Kashiwakura S, Ohno H, Matsubae-Yokoyama K, Kumagai Y, Kubo H, Nagasaka T. Removal of arsenic in coal fly ash by acid washing process using dilute H2SO4 solvent. J Hazard Mate 2010; 181(1–3): 419-25.
29. Rodríguez-Reinoso F. The role of carbon materials in heterogeneous catalysis. Carbon 1998; 36(3): 159-75.
30. Jeon C, Park JY, Yoo YJ. Characteristics of metal removal using carboxylated alginic acid. Water Res 2002; 36(7): 1814-24.
31. Ali I, Asim M, Khan TA. Low cost adsorbents for the removal of organic pollutants from wastewater. J Environ Manage 2012; 113: 170-83.
32. Anbia M, Lashgari M. Synthesis of amino-modified ordered mesoporous silica as a new nano sorbent for the removal of chlorophenols from aqueous media. Chem Eng J 2009; 150(2–3): 555-60.
33. Baker HM, Ghanem R. Evaluation of treated natural zeolite for the removal of o-chlorophenol from aqueous solution. Desalination 2009; 249(3): 1265-72.
34. Şenel S, Kara A, Alsancak G, Denizli A. Removal of phenol and chlorophenols from water with reusable dye-affinity hollow fibers. J Hazard Mate 2006; 138(2): 317-24.
35. Bayramoglu G, Gursel I, Tunali Y, Arica MY. Biosorption of phenol and 2-chlorophenol by Funalia trogii pellets. Bioresour Technol 2009; 100(10): 2685-91.

2

Assessment of water supply system and water quality of Lighvan village using water safety plan

Mojtaba Pourakbar[1], Mohammad Mosaferi[2*], Saeed Lak[3]

[1]MSc of environmental health engineering, Department of Environmental Health Engineering, Faculty of Health, Tabriz University of Medical Sciences, Tabriz, Iran
[2]Professor, Tabriz Health Services Management Research Center, Tabriz University of Medical Sciences, Tabriz, Iran
[3]MSc Student of Environmental Health Engineering, Department of Environmental Health Engineering, Faculty of Health, Ardabil University of Medical Sciences, Ardabil, Iran

Abstract

Background: Continuous expansion of potable water pollution sources is one of the main concerns of water suppliers, therefore measures such as water safety plan (WSP), have been taken into account to control these sources of pollution. The aim of this study was to identify probable risks and threatening hazards to drinking water quality in Lighvan village along with assessment of bank filtration of the village.

Methods: In the present study all risks and probable hazards were identified and ranked. For each of these cases, practical suggestions for removing or controlling them were given. To assess potable water quality in Lighvan village, sampling was done from different parts of the village and physicochemical parameters were measured. To assess the efficiency of bank filtration system of the village, independent *t* test was used to compare average values of parameters in river and treated water.

Results: One of the probable sources of pollution in this study was domestic wastewater which threatens water quality. The results of this study show that bank filtration efficiency in water supply of the village is acceptable.

Conclusion: Although Bank filtration imposes fewer expenses on governments, it provides suitable water for drinking and other uses. However, it should be noted that application of these systems should be done after a thorough study of water pollution level, types of water pollutants, soil properties of the area, soil percolation and system distance from pollutant sources.

Keywords: Bank filtration, Lighvan, Risk ranking, Water safety plan, Water pollution

***Correspondence to:**
Mohammad Mosaferi
Email: mmosaferi@yahoo.com

Introduction

Safe drinking water is one of the necessities to sustain life and a satisfactory (adequate, safe and accessible) supply of water must be available for all peoples. Qualitative assessment of water leads to improvement in water quality and is an important factor in water treatment (1).

Although two-thirds of the earth is covered with water, available fresh water is rare. Along with water scarcity, water pollution is the other issue threatening fresh water resources (2,3).

Industrialization without considering sustainable development principles, population increase, modern and mechanized agriculture, etc have contributed to the pollution of water resources in recent years. This is much more significant especially in the case of developing countries (4,5).

In some villages in the country, domestic sewage flows through streets and open canals. This kind of wastewater collection leads to the contamination of surface and groundwater. Also, leachate with lots of nitrogen and microorganisms are the other source of pollution. Typically, these sources of pollutants are source of pathogenic contamination for drinking water.

Over time and with increased attention to the quality of drinking water during the last few years, different methods have been used to assess the quality of drinking water resources. Such measures over the last few years in relation to drinking water in the world is the implementation of water safety plans (WSPs). Given the importance of access to safe water, WSP since 2004 has been on the agenda of the World Health Organization (WHO) (4,6-8). Monitoring and evaluation of water from catchment to the

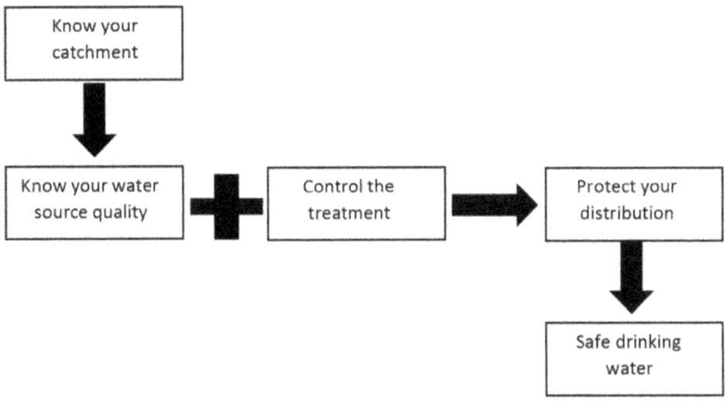

Figure 1. "Catchment to consumer" approach to risk management of the safety of the drinking water.

consumption point is taken into consideration, and study programming is reached as shown in Figure 1 (7).

The primary objectives of a WSP in ensuring good drinking-water supply practice are the prevention or minimization of contamination of source waters, the reduction or removal of contamination through treatment processes and the prevention of contamination during storage, distribution, and handling of drinking water. A WSP has three key components which are; system assessment, operational monitoring and management and communication plans (7).

Lighvan is a village in Meydan Chay rural district, in the central district of Tabriz county, East Azerbaijan province, Iran. At the 2011 census, its population was 5524, in 1553 families (9).

It is famous for its slight weather during summer time and cold weather in winter. The main occupation of the people is agriculture, dairy and gardening. Lighvan's traditional feta cheese is the most famous cheese all around Iran.

To our knowledge, there is no study about the quality of drinking water supply in Lighvan village and assessment of the village system performance. Therefore, the present study was aimed to identify risks and hazards threatening water quality and health of the villagers and to prioritize and monitor the risks and hazards. The performance of the river bank filtration (RBF) of the village was also investigated.

Methods

Before 2004, drinking water in Lighvan village was provided only from the spring. This spring was sanitized 20 years ago. The source of the water to this spring is provided from RBF. In recent years, the RBF system could not solely provide water for the village due to population increase in the village. In 2004, a well was drilled to a depth of 160 meters on the south side of the village. Well water after disinfection with chlorine along with RBF water are pumped to a storage tank on top of a mountain overlooking the village from where the water is injected to the water distribution network of the village.

In this study (conducted in 2014), the first step was to identify the hazards and potential risks in different parts of the water distribution network and prioritize them. A variety of different semi-quantitative and qualitative approaches have been proposed for ranking risks. In this study the ranking proposed by the WHO was used. The semi-quantitative approach used in the present study is shown in Table 1 (9).

The study identifies and prioritize the risks in accordance with the above matrix and then strategies to modify or reduce the risks were also presented.

Lighvan river originates from the northern slopes of Sahand mountain. Several springs and rivers join to form the Lighvan river. The river flows from south to north and irrigates downstream lands. This river flows in relatively deep valley in the downside of the village.

In this study, 3 water samples from different parts of the river flowing through the village were taken. The samples were taken from entrance, middle (next to RBF system), and the lower course of the river in village. Parameters including pH, electric conductivity (EC), hardness, major cations (Ca^{2+}, Mg^{2+}, Na^{+} and K^{+}) and major anions (HCO_3^-, SO_4^{2-}, Cl^-, F^- and NO_3^{2-}) were measured. All the analyses were done according to standard methods for examination of water and wastewater (10).

Since the water to the village is supplied from two separate sources (well and RBF system), sampling was also done from these points. The samples were collected in acid-washed polyethylene terephthalate (PET) bottles after 5 minutes discharge of the current and 3 times after washing of the bottles. To evaluate the water quality at the point of use, 3 samples were taken from different parts of the network (Figure 2).

Table 1. Scoring matrix for ranking risks[a]

Likelihood	Severity of consequences				
	Insignificant	Minor	Moderate	Major	Catastrophic
Almost certain	5	10	15	20	25
Likely	4	8	12	16	20
Moderately likely	3	6	9	12	15
Unlikely	2	4	6	8	10
Rare	1	2	3	4	5

[a]Risk rating: Low, <6; Medium, 6-9; High, 10-15; Very high, >15.

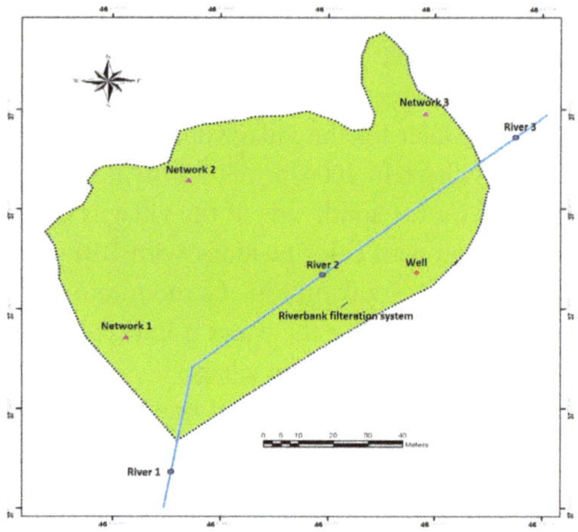

Figure 2. Sampling points and RBF system of the village.

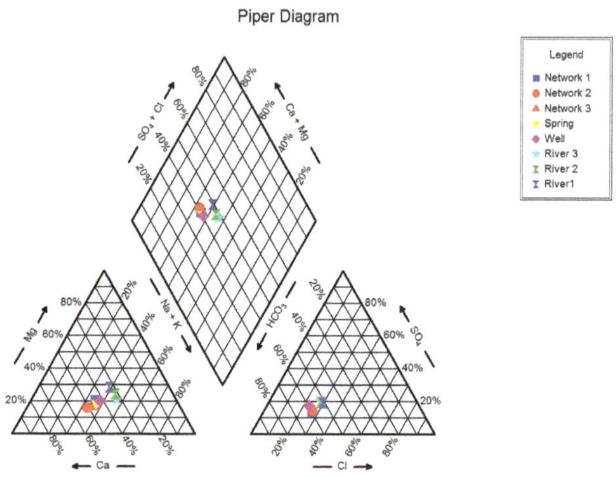

Figure 3. Piper diagram for the analyzed samples.

In order to study the performance of the filtration system of the village and impact of the river water on the treated water by RBF, statistical analysis was used. To assess the normality of the variables in spring and summer seasons, Kolmogorov–Smirnov test was used and confirmed. For the comparison of mean values of the indices, independent *t* test was used. $P<0.05$ was considered significant.

Results

Potential hazards

All potential risks of water quality of Lighvan village are shown in Table 2. Likelihood and severity of each of the risks was determined by 3 environmental health experts and finally the extent of these risks was ranked.

As indicated in the table, the sanitary wastewater of the village is one of the main potential source of pollution. This is particularly important in the case of RBF.

Physicochemical analysis of water samples

Results of the physicochemical analysis are presented in Table 3. In terms of hardness, water is grouped as soft water (<75 mg/L $CaCO_3$), medium hard (75–150 mg/L $CaCO_3$), hard water (150-300 mg/L $CaCO_3$) and very hard water (>300 mg/L $CaCO_3$). As specified in the table, the range of total hardness in the network water samples was 124-136 mg/L $CaCO_3$, indicating that the studied wa-

Table 2. Potential pollution risks ranking

Event	Likelihood	Severity	Risk ranking
Contamination of drinking water by sanitary wastewater discharged to river	5	4	20
Drinking water contamination by sewage from dairy products canters, especially in the months of January through the end of July	3	3	9
Drinking water contamination by animal waste	4	2	8
Drinking water contamination by agricultural drainage	3	2	6
The risk of secondary contamination due to lack of residual chlorine in the network	2	2	4
Contamination of drinking water by the wastewater from nearby school	2	1	2

Table 3. Mean values of physicochemical analysis of water samples

Sample	pH	EC μs/cm	TDS mg/l	Turbidity NTU	F mg/l	Cl mg/l	SO_4 mg/l	NO_3 mg/l	Ca mg/l	Mg mg/l	Na mg/l	K mg/l	Hardness mg/l	Alkalinity mg/l
Network 1	7.20	320	208.0	0.21	0.6	36	23.8	9.4	35.2	9.7	30.0	4.2	128	112
Network 2	7.02	315	204.8	0.34	0.6	36	25.5	7.4	41.6	7.8	28.9	4.1	136	112
Network 3	7.12	318	206.7	0.39	0.6	36	26.1	7.2	36.8	7.8	31.2	4.2	124	112
RBF	7.03	310	201.5	0.26	0.6	32	32.5	8.5	35.2	8.7	31.2	3.9	124	116
Well	7.02	340	221.0	1.16	0.6	36	35.2	7.3	36.8	10.7	35.8	4.1	136	128
River 3	7.35	360	234.0	20.40	0.6	44	41.1	6.9	30.4	11.7	47.3	4.2	124	112
River 2	7.60	335	217.8	12.60	0.6	42	38.9	4.2	27.2	12.6	41.5	3.5	120	108
River 1	7.90	326	211.9	4.10	0.6	40	33.9	3.1	24.0	12.6	31.2	3.3	112	96

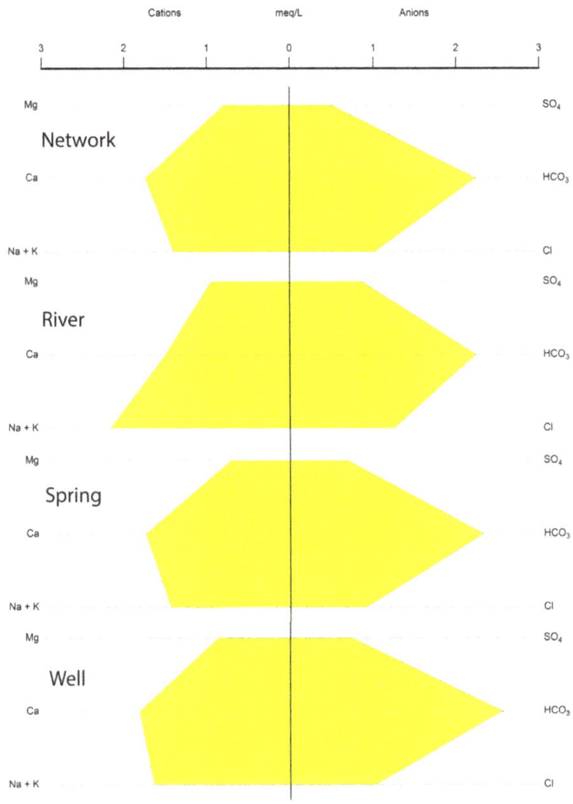

Figure 4. Stiff diagram for the analyzed samples.

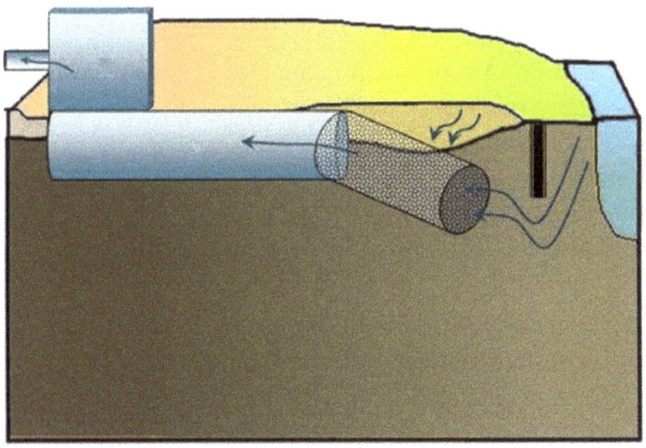

Figure 5. Schematic of RBF system.

ter samples could be grouped as medium hard water. Electrical conductivity was almost same in different parts of the water distribution network, indicating that there is no change in dissolved solids.

As shown in Table 3, the amount of dissolved solids in the RBF is less compared to the well.

Figure 3 shows the piper diagram for samples taken from the Lighvan village. As shown, samples taken from drinking water contains higher bicarbonate ions compared to samples taken from the river.

Figure 4 shows the stiff diagram of the water samples for network, river, spring and well. As shown in the figure, sodium and chloride in the river water is higher than in drinking water. The type of river water is sodium-bicarbonate, while it is calcium-bicarbonate for drinking water.

Evaluation of river bank filtration system performance

RBF is a type of filtration that works by passing water to be purified for use as drinking water through the banks of a river. Schematic diagram of Lighvan village RBF is illustrated in Figure 5. As shown in the figure, water passes through river bank and then, the purified water is taken for drinking purpose. After passing the river water through the RBF, reduction was observed in pH, turbidity, total dissolved solids (TDS), magnesium, sodium, sulfate and chloride. However, there was a slight increase in bicarbonate, potassium, nitrate, calcium and hardness.

The results of P value between the mean values of the physicochemical parameters of the river and RBF system shows that there is only significant correlation between chloride (P=0.049) and magnesium (P=0.032).

Discussion

One of the main problems that threaten the safety of the village water is sanitary wastewater and the wastewater discharged to the river from dairy production centers. Considering that a large population inhabits the village, thus a large amount of wastewater is produced. The intrusion of wastewater to the river and passage through the RBF could threaten public health of the villagers. It is strongly suggested to modify the wastewater collection network of the village and to channel the collected wastewater downstream of the village. Then it could be treated by natural wastewater treatment systems, and after being purified up to an accepted level it could be discharged to the river. Minimal corrective action which could be given should be that discharge of wastewater to the river from the upstream of the RBF should be stopped.

About the case of animal wastes, which could be the main reason for an increase in nitrate concentration of the RBF effluent, it is suggested to allocate some sites out of the village for waste storage.

It is suggested to construct special canals for collecting agricultural drainage particularly near the well and RBF system.

About the case of river water quality, due to the intrusion of wastewater, the amount of dissolved solid increases in the river water on passing through the village. As previously mentioned, there are lots of diary production centers in the village, these centers also discharge their wastewater, which contains high amount of salt (salt is used for preservation of cheese) into the river.

According to the lithology maps of the studied area, geological structure of the village is made of volcanic and sedimentary rocks. Based on the results of the present study, bicarbonate and calcium of the water increase after passing through the RBF. This could be attributed to the geological structure of the study area which is made up of volcanic and sedimentary rocks (Eq. 1) (11).

$CaCO_3 + H_2CO_3 \rightarrow Ca^{2+} + 2HCO_3$ (1)

As shown in Table 3, potassium was also increased in the water on passing through the RBF system, which was also observed in groundwater. This could be due to the presence of potassium-feldspar minerals in the geological structure of the area. Potassium-feldspar can react with carbonic acid and generate potassium and increase alkalinity as shown in Eq. 2 (12).

$$H_2CO_3 + KAlSi_3O_8 + H_2O \rightarrow K^+ + HCO_3^- + 2H_4SiO_4 \qquad (2)$$

Subsurface water storage systems have the advantages of natural capacity of water and soil therefore, in addition to little cost to governments, it can generate high quality water (13-15). In this study some chemical parameters such as hardness, alkalinity, nitrate, potassium and calcium increased after treatment which is similar to the findings of a study conducted in Egypt on the Nile River filtration system. However, the amounts of magnesium, sodium, sulfate and chloride decreased which contrasts the findings of the Nile River study. In addition, similar to the study in Egypt, turbidity and amount of suspended solids reduced significantly (16).

Although nitrate was low, it increased after passing through the filtration system. This could be due to accumulation of animal waste near the RBF system which leads to the filtration of nitrogen from animal waste into treated water.

In general, in the construction of these kinds of systems, there should be much emphasis on water pollution, type of pollutant, soil type, soil permeability and the distance of the system from pollution sources. While this method is recommended for use in different parts of Iran, comprehensive studies in this field is also proposed for accurate and continuous monitoring of water for chemical and biological parameters before and after construction of the system and during its operation.

Conclusion

Many risks and hazards of pollution could be controlled by implementation of rural WSP. Therefore, implementation of these kinds of programs, especially in rural communities is recommended.

Although bank filtration imposes fewer expenses on governments, it provides suitable water for drinking and other uses. However, it should be noted that application of these systems should be done after a thorough study of water pollution level, types of water pollutants, soil properties of the area, soil percolation and system distance from pollutant sources.

Acknowledgments

This article was a report from the data obtained during a MSc thesis, registered in Tabriz University of Medical Sciences (TBZMED). We also appreciate the Environmental Chemistry Laboratory of the Faculty of Health, especially Mr. Mohammad Abedpour.

Competing interests

The authors declare that they have no competing interests.

Authors' contributions

All authors contributed equally and participated in the data collection, analysis and interpretation. All authors critically reviewed, refined and approved the manuscript.

References

1. Noshadi M, Azarpeikan A, Nohegar A. Investigation of water quality in Bandar Abbas drinking water using cluster and factor analysis. Water Res J 2009; 3(5): 79-82. [In Persian].
2. Quevauviller PP, Fouillac AM, Grath J, Ward R. Groundwater Monitoring. Wiltshire: Wiley; 2009. Water quality measurement series.
3. Samadi MT, Sadeghi S, Rahmani A, Saghi MH. Survey of water quality in Moradbeik river basis on WQI index by GIS. Environmental Health Engineering and Management Journal 2015; 2(1): 7-11.
4. Kathuria V. Controlling water pollution in developing and transition countries—lessons from three successful cases. J Environ Manage 2006; 78(4): 405-26.
5. Kathuria V. Informal regulation of pollution in a developing country: evidence from India. Ecol Econ 2007; 63(2–3): 403-17.
6. Summerill C, Pollard SJ, Smith JA. The role of organizational culture and leadership in water safety plan implementation for improved risk management. Sci Total Environ 2010; 408(20): 4319-27.
7. World Health Organization (WHO). Guidelines for Drinking Water Quality. 4th ed. Geneva: WHO; 2011.
8. Baum R, Amjad U, Luh J, Bartram J. An examination of the potential added value of water safety plans to the United States national drinking water legislation. Int J Hyg Environ Health 2015; 218(8): 677-85.
9. Census of the Islamic Republic of Iran, 2011. http://www.amar.org.ir/
10. APHA, AWWA, WPCF. Standard Methods for Examination of Water and Wastewater. 21st ed. https://www.standardmethods.org/. Published 2005.
11. Dehnavi AG, Sarikhani R, Nagaraju D. Hydro geochemical and rock water interaction studies in East of Kurdistan, N-W of Iran. International Journal of Environmental Sciences and Research 2011; 1(1): 16-22.

12. Tesoriero AJ, Spruill TB, Eimers JL. Geochemistry of shallow ground water in coastal plain environments in the southeastern United States: implications for aquifer susceptibility. Appl Geochem 2004; 19(9): 1471-82.
13. Hubbes SA. Riverbank Filtration Hydrology. Kluwer Academic Publishers; 2002. Nato Science Series - Earth and Environmental Sciences; vol 4.
14. Ray C. Riverbank Filtration: Understanding Contaminant Biogeochemistry and Pathogen Removal. Netherlands: Springer; 2002. Nato Science Series; vol 4.
15. de Vet WW, van Genuchten CC, van Loosdrecht MC, van Dijk JC. Water quality and treatment of river bank filtrate. Drinking Water Engineering and Science 2010; 3(1): 79-90.
16. Shamrukh M, Abdel-Wahab A. Riverbank filtration for sustainable water supply: application to alarge-scale facility on the Nile River. Clean Techn Environ Policy 2008; 10(4): 351-8.

3

Survey of bacterial contamination of environment of swimming pools in Yazd city, in 2013

Hossein Jafari Mansoorian[1], **Salman Zarei**[2*], **Narges Khanjani**[3]

[1]Lecturer, Environmental Health Engineering Research Center, Department of Environmental Health, Kerman University of Medical Sciences, Kerman, Iran.
[2]MSc of Environmental Health Engineering, Chief of Environmental Protection, Ilam Gas Company, Ilam, Iran
[3]Department of Environmental Health Engineering, School of Public Health, Kerman University of Medical Sciences, Kerman, Iran

Abstract
Background: Infections are readily transmitted as a result of bacterial contamination of swimming pools. Therefore, hygiene and preventing the contamination of swimming pools is of particular importance. The objective of this study was to determine the amount of bacterial contamination in indoor pools of Yazd in 2013.
Methods: In this descriptive and analytical study, all indoor swimming pools of Yazd (12 pools) were evaluated during the spring and summer of 2013, in terms of bacterial contamination. In order to determine contamination, a sterile cotton swab was used for sampling. On average, 45 samples were taken from different surfaces in each pool (shower, dressing room, sitting places in sauna, platforms and around the pool). In total, about 540 samples from all pools were tested for bacterial contamination.
Results: The results show that from 540 samples, bacterial contamination was observed in about 93 samples (17.22%); and was seen more in showers, edges of the pool and jacuzzis, and the slippers used in swimming pools. The most important isolated bacteria types were *E. coli, Actinobacteria, Pseudomonas alcaligenes, Pseudomonas aeruginosa* and *Klebsiella pneumonia*.
Conclusion: The results indicate the presence of bacterial contamination on the surface of these places. It is recommended that health authorities should pay more attention to cleaning and disinfecting surfaces around the pool, showers, dressing rooms etc, to prevent infectious disease transfer as a result of contact with contaminated swimming pool surfaces.
Keywords: Bacterial contamination, Health indices, Swimming pools, Yazd

***Correspondence to:**
Salman Zarei
Email: : ab.salmanzarei@gmail.com

Introduction

Swimming is a sport that all ages enjoy. Swimming pools attract a lot of people, especially during warm seasons. However, due to mismanagement of these swimming pools and the poor quality of swimming pool water, the transmission of water borne diseases has become a serious problem worldwide (1). People who swim in these pools are in danger of acquiring various gastrointestinal, dermatologic, ear and upper respiratory tract infections. Also, eye complications can occur in people who use contact lenses (2-6). Studies show that in addition to water, some free living bacteria and amoebas, can live in swimming pool facilities (including warming systems, ventilation and air conditioning) or other humid surfaces and some of them may lead to various respiratory, dermatologic or neurologic diseases (7). Research shows that different infections are transmitted by direct or indirect contact with surfaces and facilities (8).

In swimming pools, there is a possibility of direct or indirect transmission of infections from one swimmer to another by eye, ear, upper airway or skin contact (9-11). Also, the dirty surfaces, toilet, wardrobe and personal devices of swimmers are capable of transmitting diseases (12). Generally, infections are transmitted by inanimate objects such as door handles, showers, toilet seats, cupboards and chairs. When these objects are in continuous contact with humans, they are the main source of transmitting infectious diseases and pathogenic organisms (13).
In order to prevent the transmission of bacteria, swimming pool operators should keep the inner surface of swimming pools and their surroundings clean. These surfaces increase the probability of transmission and also facilitate human contact with pathogens (14). In many circumstances, swimmers use the platform and chairs around the swimming pool for changing their clothes, and small particles may remain on the surface of their clothes and

the bacteria in these particles are easily transferred when these devices are used (15). Also, studies have shown that 35% of people do not take a shower before jumping into the pool and this can increase the probability of infectious disease transfer by several folds (16).

If recreational equipment inside and around pools are not properly cleaned, they can transmit microbes through unwashed hands (15). In these pools, dermal infections and rushes due to *Pseudomonas aeruginosa* have been reported (11). Certain microbes are capable of producing a slimy thin layer or biofilm, to protect them from disinfectants (15). When the biofilm is cleaned from the surface, the microorganisms are scattered in water.

Increase in the prevalence of water borne diseases and diseases related to public swimming facilities in the past decades, has shown that bacterial analysis of local swimming facilities is necessary for the health and safety of people (17). Therefore, to safeguard the health of swimming pool staff and swimmers, it is necessary to manage the health risks surrounding swimming pools (18). Yazd has a warm and dry climate. The swimming pools of this city can become a suitable place for the growth of fungi and bacteria. Therefore, this study was conducted to evaluate the bacterial contamination of surfaces in the swimming pools of Yazd, during the spring and summer of 2013. By determining the contamination rate of swimming pools in this city and using the results, appropriate action can be taken to promote health.

Methods

In this analytic and descriptive cross sectional study, all the active indoor swimming pools of Yazd, including 7 public (governmental) and 5 private pools, were evaluated through the spring and summer of 2013 in regard to bacterial contamination. The swimming pools studied were all indoor pools used by both men and women in different hours. Pools owned by specific organizations and those in residential complexes were excluded.

In order to determine bacterial contamination, sampling was done using a sterile cotton swap. All swimming pools were visited and different surfaces of the pool (including shower, changing rooms, chairs, sitting places in saunas, baths, platforms and around the swimming pools) were sampled and cultured. Sampling was done monthly and in two rounds in the morning and afternoon. Researchers decided to take 8 samples from each swimming pool and 576 samples in total. However, because parts of some swimming pools like the sauna and jacuzzi were closed due to repairs, in average 45 samples were randomly taken from different surfaces and totally, 540 samples from all swimming pools were examined for the presence of harmful bacteria. Researchers also attempted to find out the time of pool disinfection and conducted sampling at a time that is in compliance with sampling standards (19). This time was when the pool was crowded and the duration between disinfection time and washing of pool surfaces. Therefore, the sampling time was the same for all swimming pools.

Sampling was done using a sterile cotton swap from different surfaces of swimming pools. In order to do this, the swap was rubbed on surfaces with 20×20 cm area (19,20) and was then cultured separately on Blood Agar and Eosin Methylene Blue, manufactured by the Micromedia Company in Germany.

The media were then transferred to the microbiology laboratory at the Medical School and kept in the incubator for 48 hours at 37°C. Colonies were determined based on their appearance, gram color, differential tests such as coagulase, DNAse, sensitivity to novobiocin MR-VP, mannitol fermentation test, motivity and hemolysis test and culture on specific media, based on techniques mentioned in the book for standard techniques of water and wastewater experiment (19). Data analysis was done using frequency tables, descriptive statistics, proportions (percents) and also graphs from Excel 2013. Statistical analysis was done using SPSS version 16 and chi-square test.

Results

In this study, bacterial contamination of surfaces in indoor swimming pools of Yazd was evaluated and the relation between contamination and type of swimming pool (private or public), years of operation, location and site of sampling were recorded. From the 540 samples collected during spring and summer, 93 cases (17.22%) were contaminated and 447 (82.78%) were negative or without bacterial contamination. Figure 1 shows the contamination of swimming pools in different months. The positive samples contained at least one or several types of bacteria. The most important isolated types were *E. coli*, *Actinobacteria*, *Pseudomonas alcaligenes*, *Pseudomonas aeruginosa* and *Klebsiella pneumonia* (93 samples).

The positive samples were mainly from the changing cubicles, the edges of swimming pools and jacuzzis, diving platforms, cold and warm water tabs and slippers. The most contaminated samples were taken from 2 public and one private pool. Tables 1 and 2 show the frequency and type of bacterial contamination, according to type of

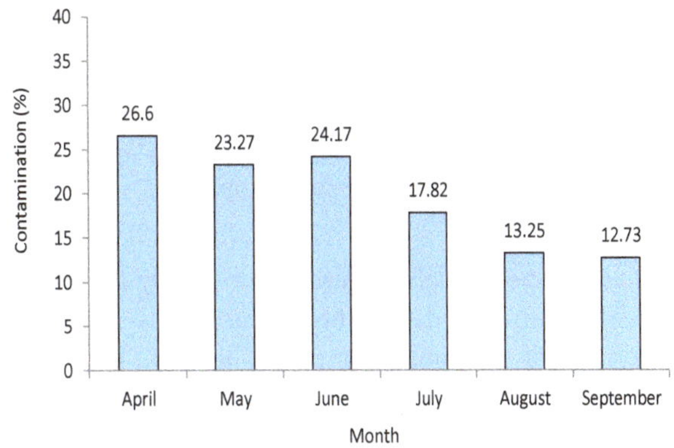

Figure 1. The distribution of bacterial contamination of the swimming pool surface by sampling month.

Table 1. The distribution of bacteria isolated from private and public (governmental) swimming pools

Swimming pools	Bacteria											
	E. coli		*Proteus*		*Pseudomonas*		*Actinobacteria*		*Klebsiella*		Total	
	No.	%	No.	%	No.	%	No.	%	No.	%	No.	%
Private pools	24	25.8	1	1.07	5	5.38	0	0	3	3.23	33	35.48
Governmental pools	35	37.63	2	2.15	13	13.98	5	5.38	5	5.38	60	64.52

Table 2. The distribution of bacteria isolated from different parts of the swimming pools under study

Sampling site	Bacteria											
	E. coli		*Proteus*		*Pseudomonas*		*Acinetobacter*		*Klebsiella*		Total	
	No.	%	No.	%	No.	%	No.	%	No.	%	No.	%
Changing rooms	5	5.38	0	0	0	0	0	0	0	0	5	5.38
Changing cubicles	5	5.38	0	0	2	2.15	0	0	1	1.07	8	8.6
Shower	13	13.98	0	0	4	4.3	1	1.07	2	2.15	20	21.5
Slippers	11	11.83	1	1.07	4	4.3	1	1.07	3	3.23	20	21.5
Platforms	1	1.07	1	1.07	1	1.07	0	0	1	1.07	4	4.3
Pool edges	8	8.6	0	0	1	1.07	0	0	0	0	9	9.68
Jacuzzi edges	6	6.45	1	1.07	2	2.15	2	2.15	1	1.07	12	12.9
Diving platforms	3	3.23	0	0	4	4.3	1	1.07	0	0	8	8.6
Sauna	3	3.23	0	0	0	0	0	0	0	0	3	3.23
Slide	1	1.07	0	0	0	0	0	0	0	0	1	1.07
Pool ladder	3	3.23	0	0	0	0	0	0	0	0	3	3.23
Total	59	63.44	3	3.23	18	19.35	5	5.38	8	8.6	93	100

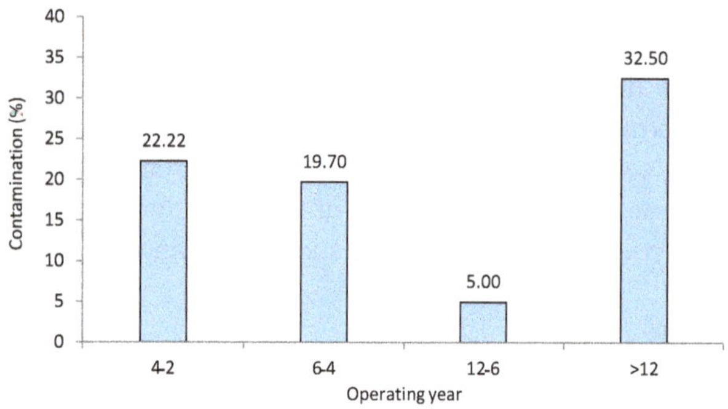

Figure 2. The distribution of bacterial contamination of the swimming pool surfaces by operating years.

pool and site of sampling. Also, in this study the rate of contamination was determined according to the operating years of the pool (Figure 2). The result of statistical analysis showed that there was a significant association between bacterial contamination and type of pool ($P<0.01$), operating years of pool ($P<0.01$) and site of sampling ($P<0.01$). But there was no relation between the rate of bacterial contamination and the sampling months ($P=0.067$).

Discussion

Studies show that the incidence of infectious diseases has increased in recent years. This may be due to increase in the population of vulnerable people or those with immunodeficiency and also increased contact of people with contaminated environments (20). This problem has been highlighted by different studies performed in Iran (21-26) and other areas of the world (10,12,27,28) about bacterial contamination of pools. In the current study, 17.22% of the samples were positive for bacteria, and different types of bacteria including *E. coli*, *Proteus*, *Acinetobacter*, *P. alcaligenes*, *P. aeruginosa* and *K. pneumonia* were isolated.

Garrido Mata et al (29) in 2013 in Portugal, sampled different areas (such as surfaces, showers, etc.) as well as the pool and Jacuzzi water of a hotel, and observed Coliform bacteria such as *E. coli*, *Pseudomonas* and *Staphylococcus* in their samples. In Hutcheson et al (30) study, in Atlanta, 42% of the samples contained *E. coli* and Pseudomonas.

In another study in Greece, 32.9% of the samples were contaminated and resistant bacteria such as *P. alcaligenes*, *Staphylococcus aureus*, *P. aeruginosa* and *K. pneumonia* were isolated (7). Also, in a study conducted by Mansoorian et al (31) about health indices of water in Kerman swimming pools, showed that the total coliform, heterotrophic plate count (HPC), and *Pseudomonas aeruginosa* in 100%, 14.6% and 31.4% of the samples were more than standard, respectively. The results of this study are in line with the results of other studies; however, contamination of *S. aureus* was not observed in our study and the percentage of bacterial contamination (17.22%) was less than other studies. This result indicates better hygiene control in the swimming pools of Yazd.

Based on the results, the highest contamination was seen in April and it decreased in the other months. Generally, bacterial contamination in spring (24.9%) was more than summer (15.76%). This increase maybe due to increase in the number of people visiting swimming pools after February (the winter season) and the unpreparedness of swimming pools to accept this extra load of visitors. As the weather gets warmer, the number of hygiene inspectors performing surveillance on swimming pools increases, resulting in a decrease in contamination. In Sohrabi et al (32) study in Zahedan, the highest bacterial contamination was seen in summer and was 26.5%, which is different from what was recorded in this study. This difference maybe because of the increase in the number of visitors in summer in comparison to spring, in Sohrabi et al study.

Also, in this study, the relation between bacterial contamination with pool type (private or public) and operation years of the pool was evaluated. The results showed that bacterial contamination was higher in public pools (64.52%) than private pools (35.48%), this is in line with the studies of Papadopoulou et al (7) and Tesauro et al (33). The reason might be that nowadays, the cost of running swimming pools has increased due to increase in the cost of detergents and disinfectants, cost of water, electricity, gas etc. This leads to increases in customer fees and decrease in service quality. The low level of environmental hygiene, detergents, disinfectants, lighting etc are due to financial problems and is seen more obviously in public pools.

The results of this study show that the percent of bacterial contamination in pools up to 12 years old, decreased with increase in operating years. The reason for the decrease in pool contamination may be due to increased management experience. But increase after 12 years is probably due to aging of the building and facilities of the pools. The results showed the most bacterial contamination in the changing cubicles, edges of jacuzzis, showers and slippers. This can cause disease in vulnerable people or those with immunodeficiency. It is necessary to observe hygienic regulations and inform people about the consequences of not obeying hygienic regulations, by hanging posters at the entrance of swimming pools and changing rooms. This will make swimmers more responsible and prevent the transmission of diseases by showering before swimming.

Conclusion

The aim of this study was to determine the amount of bacterial contamination in indoor pools of Yazd, in 2013. The results show that from the 540 samples collected during spring and summer, 93 cases (17.22 %) were contaminated and 447 (82.78%) were negative or without bacterial contamination. The most important isolated types were *E. coli*, *Actinobacteria*, *P. alcaligenes*, *P. aeruginosa* and *K. pneumonia* (93 samples). Based on the result of this study, contamination is higher in old swimming pools and is related to the weariness of the building and facilities. Therefore it is necessary to do critical repairs in these swimming pools and replace old and worn out equipment with new and appropriate ones. Also, the walls and surface should be covered with tiles for better cleaning. Also, it is recommended that the swimming pool directors should observe hygienic regulations completely and in addition to disinfecting pool water, pay more attention to the cleanliness and disinfection of pool surfaces, showers, changing rooms etc, to prevent the spread of infections.

Acknowledgements

The authors wish to thank the managers of the swimming pools of Yazd and also the members of staff of the microbiology labs, who cooperated with this study.

Competing interests

The authors declare that they have no competing interests.

Authors' contributions

SZ designed the study. HJM performed the literature search and wrote the manuscript. NK carried out statistical analyses and significant improvement in English. All authors participated in data acquisition, analysis and interpretation. Also, all authors critically reviewed, refined and approved the manuscript.

References

1. Amburgey JE, Bowen JD, Fielding RR, Lu P, Arrowood MJ. Cryptosporidium reduction and outbreak risk: intelligently balancing filtration, disinfection, and recirculation subsystem efficiencies. Fifth International Conference Swimming Pool & Spa; Italy: Istituto Superiore di Sanità; 2013.
2. Rasti S, Assadi M, Iranshahi L, Saffari M, Gilasi H, Pourbabaee M. Assessment of microbial

contamination and physicochemical condition of public swimming pools in Kashan, Iran. Jundishapur Journal of Microbiology 2012; 5(3): 450-5.
3. Wu YT, Tran J, Truong M, Harmis N, Zhu H, Stapleton F. Do swimming goggles limit microbial contamination of contact lenses? Optom Vis Sci 2011; 88(4): 456-60.
4. Rabi A, Khader Y, Alkafajei A, Aqoulah AA. Sanitary conditions of public swimming pools in Amman, Jordan. Int J Environ Res Public Health 2007; 4(4): 301-6.
5. Sule IO, Oyeyiola GP. Bacteriological assessment of some swimming pools within Ilorin Metropolis, Kwara, Nigeria. Bioresearch Bulletin 2010; 1: 13-7.
6. Barna Z, Kádár M. The risk of contracting infectious diseases in public swimming pools: a review. Ann Ist Super Sanita 2012; 48(4): 374-86.
7. Papadopoulou C, Economou V, Sakkas H, Gousia P, Giannakopoulos X, Dontorou C, et al. Microbiological quality of indoor and outdoor swimming pools in Greece: investigation of the antibiotic resistance of the bacterial isolates. Intl J Hygiene Environ Health 2008; 211(3-4): 385-97.
8. Ayan S. The hidden danger for the users of playgrounds and sport complexes. Life Sci J 2013; 10(1): 78-84.
9. Casanovas-Massana A, Blanch AR. Characterization of microbial populations associated with natural swimming pools. Intl J Hygiene Environ Health 2013; 216(2): 132-7.
10. Thorolfsdottir BO, Marteinsson VT. Microbiological analysis in three diverse natural geothermal bathing pools in Iceland. Int J Environ Res Public Health 2013; 10(3): 1085-99.
11. Salvato JA, Nemerow NL, Agardy FJ. Environmental Engineering. John Wiley & Sons Inc; 2003. p. 1196.
12. Lika M, Dako A, Meçe O. The microbial pollution in pools and diseases conected with them. Natura Montenegrina, Podgorica 2010; 9(3): 859-66.
13. Nworie A, Ayeni J, Eze U, Azi S. Bacterial contamination of door handles/knobs in selected public conveniences in abuja metropolis, nigeria: a public health threat. C J Med Res 2012; 6(1): 7-11.
14. World Health Organization (WHO). International Health Regulations Guide to Ship Sanitation Third Edition. Geneva: WHO; 2007. p. 39-60.
15. Lutz J K, Lee J. Prevalence and antimicrobial-resistance of pseudomonas aeruginosa in swimming pools and hot tubs. Int J Environ Res Public Health 2011; 8(2): 554-64.
16. Agbagwa O, Young–Harry W. Health Implications of some Public Swimming Pools located in Port Harcourt, Nigeria. Public Health Research 2012; 2(6): 190-6.
17. Cappello MA. Assessing bacteriological contamination in public swimming facilities within a Colorado metropolitan community. J Environ Health 2011; 73(7): 19-25.
18. Guthrie B. Conference report: health and hygiene in pools: a snapshot of a conference by the Royal Society for Public Health. Perspect Public Health 2013; 133: 84.
19. American Public Health Association, American Water Works Association, Water Environment Federation. Standard methods for the examination of water and wastewater, part 9000 microbiological examination: 9213B. Swimming Pools. 21th ed. Washington: APHA; 2005. p. 133-9.
20. Leoni E, Legnani P, Mucci M, Pirani R. Prevalence of mycobacteria in a swimming pool environment. J Appl Microbiol 1999; 87(5): 683-8.
21. Shahriari A, Nafez AH, Norouzi S, Heidari M. Investigation of common microbial indicators in swimming pool of gorgan city. Journal Safety & Health of Ardabil 2011; 2(2): 17-26. [In Persion].
22. Neghab M, Gorgi H, Baghapour M, Rajaeefard A, Moemenbellah-Fard M. Bacterial contamination of the swimming pools in Shiraz, Iran; relationship to residual chlorine and other determinants. Pak J Biol Sci 2006; 9(13): 2473-7.
23. Dindarlu K, SoleimaniAhmadi M, Zare S, Abdi H, Heidari M. Hygiene condition of bandar abbas swimming pools, 2003. Medical Journal of Hormozgan 2005; 9(1): 41-6. [In Persian].
24. Nikaeen M, Hatamzadeh M, Vahid Dastjerdi M, Hassanzadeh A, Mosavi Z, Rafiei M. An investigation on physical, chemical and microbial quality of Isfahan swimming pool waters on standard indicators. Journal of Isfahan Medical School 2010; 28(108): 346-56. [In Persian].
25. Yousefi Z. Study of the contamination condition of water swimming pools in Sari city for the Staphylococcus Aureus. Iran J Health Environ 2009; 2(3): 178-87.
26. Nikaeen M, Hatamzadeh M, Vahid DM, Hassanzadeh A. Predictive indicators of the safety of swimming pool waters. Water Sci Technol 2009; 60(12): 3101-7.
27. Abd El-Salam MM. Assessment of water quality of some swimming pools: a case study in Alexandria, Egypt. Environ Monit Assess 2012; 184(12): 7395-406.
28. Agbabiaka TO, Oyeyiola G. Microbial and physicochemical assessment of Foma river, Ita-nmo, Ilorin, Nigeria: an important source of domestic water in Ilorin metropolis. International Journal of Plant, Animal and Environmental Sciences 2012; 2(1): 209-16.
29. Garrido Mata MS, Rocha Nogueira JM, Heitor AM. Identification and characterization of microbiologic critical points in swimming-pool surfaces. Fifth

International Conference Swimming Pool & Spa; 2013 April 9-12; Roam, Italy. Istituto Superiore di Sanità; 2013.

30. Hutcheson C, Cira R, Gaines SL, Jones KR, Howard W, Hornsby D, et al. Microbes in pool filter backwash as evidence of the need for improved swimmer Hygiene-Metro-Atlanta, Georgia, 2012. Morbidity and Mortality Weekly Report (MMWR) 2013; 62(19): 385-8.
31. Mansoorian HJ, Rajabizadeh A, Jafari Modrek M, Doulatshahi S, Hatami B. Water health indices in Kerman swimming pools, in 2011. J Health Dev 2013; 2(2): 128-37. [In Persian].
32. Sohrabi A, Qureshi MI, Dehdar M, Rakhsh Khorshid A. Study of fungal and bacterial contamination of Sheltered swimming pools of Zahedan, in 2002. 6th National Conference on Environmental Health; 2003.
33. Tesauro M, Bianchi A, Consonni M, Bollani M, Cesaria M, Trolli F, et al. Hygienic profile of the water in Milan swimming pools: a three-year comparative study. Ann Ig 2010; 22(4): 345-55. [In Italian]

4

Study of Co(II) adsorption from aqueous solution using protein granules produced from chicken feather

Mohammad Hossien Salmani Nodoushan[1], Mohammad Hassan Ehrampoush[2], Lida Rafati[3], Maryam Farzi[4*], Kazem Barzegar[5]

[1]PhD in Analytical Chemistry, Department of Environment Health, Shahid Sadoughi University of Medical Sciences, Yazd, Iran
[2]Professor, Department of Environment Health, Shahid Sadoughi University of Medical Sciences, Yazd, Iran
[3]PhD Student, Department of Environment Health, Shahid Sadoughi University of Medical Sciences, Yazd, Iran
[4]MS, Department of Environment Health, Shahid Sadoughi University of Medical Sciences, Yazd, Iran
[5]PhD Candidate in TEFL, Head of the English Language Department, School of Medicine, Shahid Sadoughi University of Medical Science, Yazd, Iran

Abstract
Background: Co(II) constitutes a part of vitamin B_{12}, hence, it is necessary for human health. However, at concentrations higher than the permissible limits, humans and animals suffer adverse chronic effects. It is necessary to reduce the concentration to a permissible level. In the present study, pre-purification and thermal modification of chicken's feather increased their porosity and they were used to adsorb Co(II) ions from aqueous solutions.
Methods: Chicken feather was procured from a slaughter house, washed with detergent and dried at laboratory temperature. To increase the porosity and remove the organic pollutants attached to the feather, feathers were heated in the oven for 1 hour at 165°C. Some batch experiments were conducted to optimize the parameters affecting the adsorption process, such as solution pH, initial concentration of Co(II) and contact time, at a constant agitating speed and temperature.
Results: The pH_{zpc} of chicken's feather granules obtained was 5.3. The results showed that a time of 60 minutes, pH value of 9 and initial concentration of 10 mg/l were the optimum conditions for Co(II) removal by the adsorption process. Increasing the initial concentration of Co(II) from 10 to 50 decreased the removal efficiency from 52% to 26%. The pseudo-first order kinetic model provided the best correlation ($R^2 = 0.998$) for adsorption of Co(II) on chicken's feather granules.
Conclusion: The increase in electrostatic repulsion forces between positive charges of Co(II) ions decreased the removal efficiency at higher concentrations of Co(II) ions. The maximum efficiency of Co(II) adsorption was obtained at pH= 9, which is higher than the pH_{zpc} of the adsorbent.
Keywords: Adsorption, Chicken feather, Co(II), Protein.

***Correspondence to:**
Maryam Farzi
Email: : 2mfarzi@gmail.com

Introduction

Heavy metals are produced from different industrial activities and regarded as the main cause of water pollution (1,2). They accumulate in microorganisms, aquatic plants and animals, and finally enter the environment and the human food chain, resulting in health problems (3). Cobalt is a common toxic metal which contaminates the environment remarkably (4). It constitutes a part of vitamin B_{12} and consequently, its presence in the food chain is necessary for human health. However, at concentrations above the permissible limits, humans and animals suffer chronic adverse effects (5). The Environment Preservation Agency (EPA) has determined 5 ppb as the maximum allowable amount of Co(II) in drinking water (6). Different technologies, such as electro-dialysis, ion exchange, reversed osmosis, ultrafiltration, chemical precipitation and adsorption have been used for the removal of metal ion, which have low cost in performance (7-9). Among the different technologies for purification, absorption is regarded as a fast, cheap and efficient method. Recently, the use of industrial and agriculture wastes have increased, as reported by some researches (10). Chicken feather is a natural and cheap adsorbent. It was previously used to provide creatinine nanoparticles which had a good efficiency in the removal of pollutants. In addition, using the waste of feather as an adsorbent, is an environment – friendly method and cleans subsequent contamination from the environment. It has been proven that the redundancy of chicken's feather which is insoluble has non-toxic creatinine and has great ability to adsorb inorganic pollutants (11,12). In previous studies, chicken's feather was used to remove aquatic pollutants such as heavy metals including As, Cd, Cr, Cu, Zn, etc (13,14). Zhou et al (15) conducted a study to adsorb oil from aquatic solutions using a creatinine porous foam made of chicken's feather. Their results showed that the modified adsorbent had a great adsorption capacity for the removal of oil products. Eslahi et al (16) estimated the feasibility of producing nanoparticles from feather wastes after modification with ultrasonic and enzymatic hydrolysis. Their results showed that the particle size decreased after ultrasonic hydrolysis. In this research, the thermal stability of feather nanoparticles increased with enzyme hydrolysis during ultrasonic hydrolysis. In 2013, Pandima Devi and Muthukumaran (17) used chicken's feather as an adsorbent in the removal of color from water. In their study, adsorption capacity was determined with regard to conditions of pH, contact time, temperature, adsorbent amount and adsorbate concentration. The data were consistent with Freundlich and Langmuir isotherm. Yang et al (18) modified feather with tannic acid 5% and used it to remove metal cations from aqueous solutions. Their results indicated that the fiber of the feather modified with tannic acid had a high adsorption for metal cations. They concluded that it can be used as metal ion absorbent from aqueous solutions (18). In the present study, pre-purification and thermal modification of chicken's feather increased its porosity and were used to adsorb Co(II) from aqueous solutions.

Methods

This descriptive study was carried out in 2015 in the Chemistry Laboratory of the Health School, in Shahid Sadoughi University of Yazd. All chemicals used for the tests were of analytical purity grade, and purchased from the Merck Company. The following equipment were used for this study namely a distillation device, analytic sensitive scale of 0.0001 g accuracy model HR200 of A & D Japan, pH-meter model MI 151 of Wagtech, orbital shaker (GFL 137) and atomic absorption spectrophotometer model 20 AA Varian.

Preparation of adsorbent

Chicken's feather was procured from the slaughter house, washed with detergent and dried under the laboratory temperature. To increase their porosity and to remove the organic pollutants attached to the feather, the dried feathers were heated in the oven for 1 hour at 165°C. The protein granules were obtained after grounding the feather and then, the ground feathers were sized with standard sieves of 60 and 100 meshes (250 and 149 μ diameter). To remove organic materials, the separated granules were rinsed with ethanol 20% and distilled water. After drying, the granules were kept in desiccators for the next tests.

Adsorption experiments

To investigate pH_{ZPC}, 50 ml of the 0.1M NaCl solution was introduced to 8 flasks of 250 ml and the pH was adjusted in the range of 2-12 by 0.1M NaOH or HCl solutions, 0.1 g adsorbent was added to each flask. The suspension was shaken for 24 hours and the final pH was recorded after filtration.

The parameters affecting the process, such as pH of solution, initial concentration of Co(II) and contact time for Co(II) adsorption, were optimized by batch adsorption. For adsorption tests, first, a stock solution of 50 mg/L Co(II) was prepared from cobalt nitrate salt and the required concentrations were provided with dilution of stock solution by distilled water. To determine the effect of contact time, the experiments were carried out at a contact time of 30, 60, 90, 120 and 180 minutes, while all other parameters were kept constant. To determine the effect of pH on the adsorption efficiency of Co(II) using protein granules of the feather, all the parameters were kept constant and pH was varied in four levels of 3, 5, 7 and 9 by a buffer solution. Then, 0.1 g of adsorbent was added and was in contact with the pollutant for a fixed time, so that the process of adsorption could take place. In the next stage, to examine the effects of initial concentration on the adsorption efficiency, all the parameters were kept constant and the initial Co(II) concentration was varied by 10, 20, 30, 40 and 50 mg/L.

For all the tests, to have a good contact of adsorbent with Co(II) in the suspension and to control the temperature, each sample was agitated by an orbital shaker at 150 rpm in a 250 ml flask. After a desired contact time, the flasks were removed from the shaker and the absorbent was separated from the suspension by a Whatman filter. The clear solution was analyzed to get the remaining Co (II) concentration by using an atomic absorption spectrophotometer. The following formula was used to calculate the adsorption efficiency and capacity:

$$\%RE=\frac{(C_0-C_e)}{C_0}\times 100 \qquad (1)$$

$$q=\frac{(C_0-C_e)}{m}\times V \qquad (2)$$

Where C_0 and C_e, are the initial and final concentrations of Co(II) (mg/l) respectively, *V* is volume of suspension (L), *m* is mass of adsorbent, *q* is adsorption capacity and *RE* is removal efficiency (3).

Results

The pH_{zpc}

pH_{zpc} is the pH at which adsorbents have a net zero surface charge. Figure 1 shows the plot between pH_{final} and $pH_{initial}$ for determining pH_{zpc}. The results indicated that the pH_{zpc} for chicken's feather granules was 5.3. Chicken's feather granules becomes more positively charged when pH is less than pH_{zpc} and more negatively charged when pH is above pH_{zpc}.

The adsorption experiments were conducted in a stepwise manner. In each stage of the test, regarding a variable, other conditions were constant and the next stages were examined with optimum parameters being constant from previous stages.

The effect of contact time on adsorption efficiency

Figure 2 shows the effect of contact time on the removal efficiency of Co(II) in aqueous solution. The efficiency of granules of chicken's feather to remove Co(II) increased with an increase in contact time. The removal efficiency increased from 52% to 59% with the increase of contact time from 30 to 180 minutes.

Regarding the obtained results, the equilibrium time of 60 minutes for the Co(II) adsorption on the modified chicken feathers was chosen as the optimum contact time and it was constant in the next experiments.

In order to investigate the adsorption characteristic, kinetic constants of adsorption were determined by pseudo-first order and pseudo-second order equations based on the solid phase adsorption. Table 1 shows equations and parameter values of the linearized forms of the pseudo-first order and pseudo-second order model for adsorption of Co(II) onto granules of chicken's feather.

According to the coefficient regression of kinetic investigation, the high correlation coefficient (R^2=0.998) was shown in the pseudo-first order model.

The Effect of pH on adsorption efficiency

Figure 3 shows the effect of solution pH on Co(II) removal with chicken's feather by a batch adsorption process. To investigate this, the adsorption equilibrium was studied under four pH levels of 2, 5, 7 and 9 at a constant time of 60 minutes.

As can be seen in Figure 3, increase in pH had a remarkable effect on the removal efficiency. Hence, increasing the pH from 3 to 9 enhanced the absorption efficiency from 0% to 52%.

The effect of initial Co(II) concentration on adsorption efficiency

The effect of initial concentration on the removal efficacy shown in Figure 4 indicates that an increase in the initial concentration of Co(II) caused a decrease in the removal rate. Also, an increase in initial concentration from 10 to 50 mg/L, decreased the removal efficiency from 52% to 26% at contact time of 60 minutes.

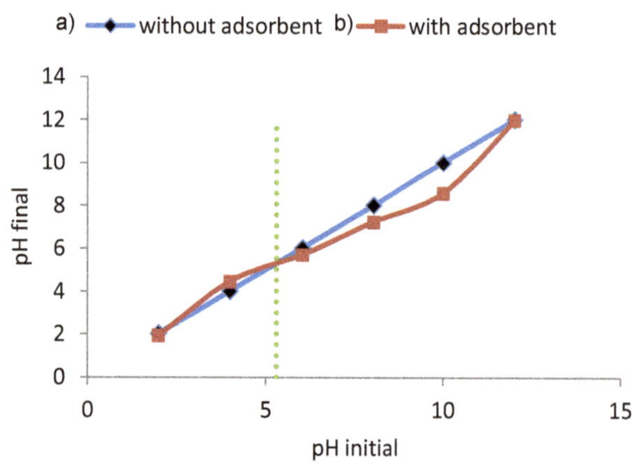

Figure 1. The plot of pHf vs pHi; a) without adsorbent, b) with adsorbent.

Discussion

By increasing the contact time, the efficiency of Co(II) removal increased and the removal of Co(II) increased from 54% to 59% when the time increased from 30 to 180 minutes. Figure 2 shows that the rate of adsorption was high at the initial time (<60 minutes) and decreased gradually while the adsorption rate had a constant trend, which is probably due to the large number of empty sites available on the adsorbents at the initial stage. Over time,

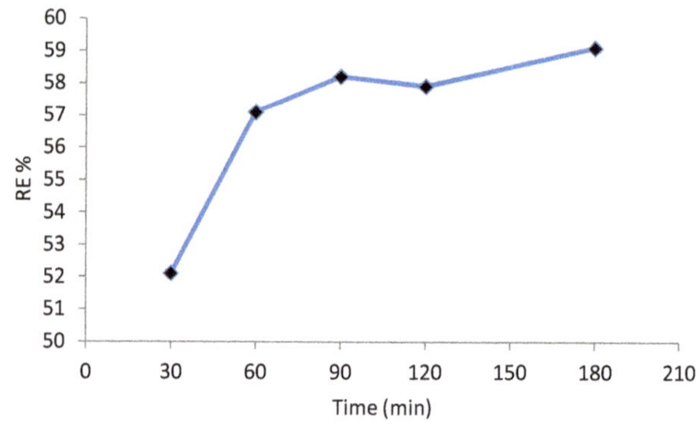

Figure 2. Effect of contact time on Co(II) removal at C_0 = 5 mg/l, pH = 9, m = 0.1 g.

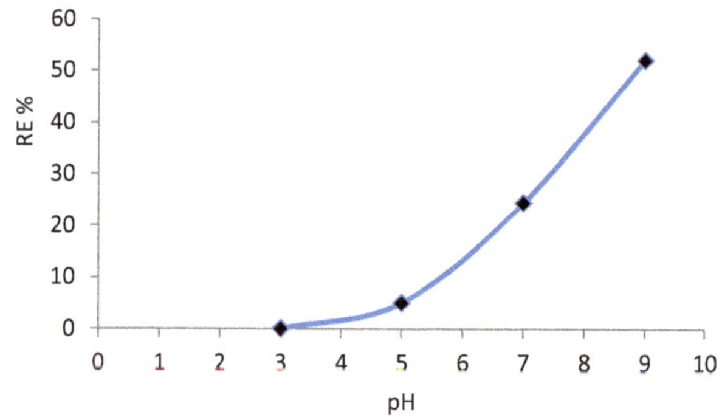

Figure 3. Effect of pH on Co(II) removal at C_0 = 5 mg/l, m = 0.1 g.

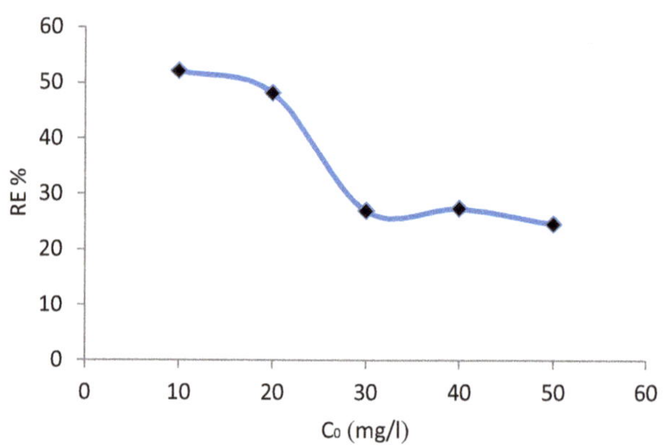

Figure 4. Effect of initial concentration on Co(II) removal at pH = 7, m = 0.1 g.

Table 1. Parameter values using pseudo-first order and pseudo-second order model for adsorption of Co(II) onto protein granules of chicken's feather

	Pseudo-first order			Pseudo-second order		
Linear equation	$t/q_t=1/q_e.t + 1/K_2q_e^2$			$\ln(q_e\text{-}q_t)= \ln q_e\text{-}K_1t$		
Metal ion	q_e	K_1	R^2	q_e	K_2	R^2
Co(II)	3.05	5.6×10^{-2}	0.998	1.11	5.6×10^{-2}	0.947

the accumulation of Co(II) ions on empty sites resulted to less number of active sites and increased the number of adsorbed particles from the solution, which led to a decrease in the adsorption rate of Co(II) from the solution. At the initial stage of the process, the concentration gradient increased between adsorbed material in the solution and surface adsorbent. This led to the increase of Co(II) adsorption in the initial stages. At times higher 60 minutes, the concentration decrease led to a decrease in the adsorption rate. These findings are consistent with those of Olgun and Atar (19) on copper and cobalt removal from aqueous solution by B wastes.

According to the results of kinetic investigations, the correlation coefficients in these models were more than 0.90 and the experiment data showed a good fit for the two models. Nevertheless, the maximum correlation coefficient was detected in the pseudo-first order model. So, the adsorption process followed the pseudo-first order model and this shows that the rate of adsorption depend on one component in the reaction. On the other hand, the adsorption rate was related to the concentration of Co(II) in the solution.

The results showed that variations of solution pH had a direct relationship with the removal efficiency. The maximum removal efficiency of Co(II) was obtained at pH = 9. According to Figure 1, the zero point charge (pH_{zpc}) of adsorbent was specified to be 5.3, then, at pH lower than pH_{zpc}, the surface charges became positive. It increased electrostatic repulsion force between Co(II) ions and adsorbent. So, the removal efficiency of Co(II) decreased at lower pH. At higher values of pH, greater than pH_{zpc}, due to OH^- attraction by the adsorbent, the adsorbent had negative charge which produced a powerful electrostatic force with positive ions of Co(II) and adsorbent. Therefore, Co(II) removal increased at higher values of pH. These findings are consistent with those of Abbas et al (3) on kinetic studies of cobalt absorption on apricot stone activated carbon and those of Yu et al (20) on the effect of pH, ion strength and fulvic acid on sorption and desorption of cobalt from aquatic solution by bentonite.

The rate of absorption was determined at five initial Co(II) concentration levels of 10, 20, 30, 40 and 50 mg/L. As can be seen in Figure 4, the greatest value of Co(II) removal was in the concentration of 10 mg/L. Increasing the initial concentration of Co(II) from 10 to 50 mg/L decreased the removal efficiency from 52% to 26%. At lower concentrations of Co(II), adsorption took place at empty sites on the adsorbent surface, due to the short length of diffusion path and presence of enough active sites, while in higher concentrations of Co(II), these empty sites became saturated and greater adsorption required penetration in deeper areas of adsorbent passing from a relatively long path which decreases the efficiency of adsorption. With increase in electrostatic repulsion forces between charges of Co(II) ions, the adsorption and removal efficiency decreased. In higher concentrations, some Co(II) ions were adsorbed on the surfaces of granules of feather; therefore, these ions made the great spaces around granules of adsorbent to become filled and increased the spatial prevention for other ions. These findings are consistent with those of Ahmadpour et al (21) on the removal of cobalt from aquatic solutions by almond green hull and those of Al-Shahrani (22) on the purification of waste water contaminated with cobalt using active bentonite.

Conclusion

Porous protein granules, the name for the residual chicken's feather remaining after slaughtering, was extracted and left as a waste product in an industrial poultry slaughterhouse. It was converted to an adsorbent and used for the removal of aqueous Co(II) from wastewater. The adsorption of Co(II) ions increased with an increase in solution pH and decreased with the initial concentration of Co(II) ions.

The results showed that the protein granules derived from a slaughterhouse can be effectively employed for decreasing the Co(II) concentration in wastewater. The results show that the removal efficiency of the porous granules derived from chicken's feather for Co(II) is comparable or more than the other available adsorbents. Also, it is a waste generated in the slaughter industry and can be easily available. Therefore, the cost of the porous granules prepared from chicken's feather would be cheaper than the commercially available adsorbents.

Acknowledgments

This project was financially supported by the Faculty of Public Health, Shahid Sadoughi University of Medical Sciences. The authors are grateful to the head of the Environmental Chemistry Laboratory, for his help.

Competing interests

The authors declare that they have no competing interests.

Authors' contributions

The study was directed by MF who performed all the experiments. MHSN is the first author and guided on the experimental methods and drafted the manuscript. All authors have read and approved the final manuscript.

References

1. Ehrampoush MH, Miri M, Salmani MH, Mahvi AH. Cadmium removal from aqueous solution by green synthesis iron oxide nanoparticles with tangerine peel extract. J Environ Health Sci Eng 2015; 13: 84.
2. Tchounwou PB, Yedjou CG, Patlolla AK, Sutton DJ. Heavy metal toxicity and the environment. EXS 2012; 101: 133-64.
3. Abbas M, Kaddour S, Trari M. Kinetic and equilibrium studies of cobalt adsorption on apricot stone activated carbon. J Ind Eng Chem 2014; 20(3): 745-51.
4. Prasad AS. Essential and Toxic Element: Trace Elements in Human Health and Disease. Elsevier; 2013.
5. Rivera-Utrilla J, Ferro-Garcia MA. Study of cobalt adsorption from aqueous solution on activated carbons from almond shells. Carbon 1987; 25(5): 645-52.
6. EPA. Environmental, Water Treatability Database, Toxicological Profile for Cobalt. U.S. Public Health Service and U.S. Environmental Protection Agency; 2004. http://www3.epa.gov/ttn/atw/hlthef/cobalt.html
7. Al-Shannag M, Al-Qodah Z, Bani-Melhem K, Qtaishat MR, Alkasrawi M. Heavy metal ions removal from metal plating wastewater using electrocoagulation: Kinetic study and process performance. Chem Eng J 2015; 260: 749-56.
8. Sargın İ, Arslan G. Chitosan/sporopollenin microcapsules: Preparation, characterisation and application in heavy metal removal. Int J Biol Macromol 2015; 75: 230-8.
9. Bhatnagara A, Sillanpaa M. Utilization of agro-industrial and municipal waste materials as potential adsorbents for water treatment—a review. Chem Eng J 2010; 157(2-3): 277–96.
10. Hashemian S, Saffari H, Ragabion S. Adsorption of cobalt(II) from aqueous solutions by Fe_3O_4/bentonite nanocomposite. Water Air Soil Pollut 2015; 226(1): 1-10.
11. Mittal A. Removal of the dye, Amaranth from waste water using hen feathers as potential adsorbent. Electron J Environ Agric Food Chem 2006; 5(2): 1296-305.
12. Gassner G. Conversion of avian feather-waste stream to useful products. Google Patents; 2000. http://www.google.com/patents/US6027608
13. Mittal A, Thakur V, Gajbe V. Adsorptive removal of toxic azo dye Amido Black 10B by hen feather. Environ Sci Pollut Res Int 2013; 20(1): 260-9.
14. Sun P, Liu ZT, Liu ZW. Particles from bird feather: a novel application of an ionic liquid and waste resource. J Hazard Mater 2009; 170(2-3): 786-90.
15. Zhou LT, Yang G, Yang XX, Cao ZJ, Zhou MH. Preparation of regenerated keratin sponge from waste feathers by a simple method and its potential use for oil adsorption. Environ Sci Pollut Res Int 2014; 21(8): 5730-6.
16. Eslahi N, Hemmatinejad N, Dadashian F. From feather waste to valuable nanoparticles. Part Sci Tech 2013; 32(3): 242-50.
17. Pandima Devi M, Muthukumaran C. Adsorption kinetics of acid brown dye from aqueous solution using biowaste. Asian J Sci Tech 2013; 4(8): 1-5.
18. Yang C, Guan L, Zhao Y, Yan Y. Sorption of Cu2+ and Zn2+ by Natural Biomaterial: Duck Feather. Appl Biochem Biotech 2007; 142(2): 168-78.
19. Olgun A, Atar N. Removal of copper and cobalt from aqueous solution onto waste containing boron impurity. Chem Eng J 2011; 167(1): 140-7.
20. Yu SM, Ren AP, Chen CL, Chen YX, Wang X. Effect of pH, ionic strength and fulvic acid on the sorption and desorption of cobalt to bentonite. Appl Radiat Isot 2006; 64(4): 455-61.
21. Ahmadpour A, Tahmasbi M, Bastami TR, Besharati JA. Rapid removal of cobalt ion from aqueous solutions by almond green hull. J Hazard Mater 2009; 166(2–3): 925-30.
22. Al-Shahrani SS. Treatment of wastewater contaminated with cobalt using Saudi activated bentonite. Alexandria Eng J 2014; 53(1): 205-11.

5

Antibiotics and heavy metals resistance patterns of *Enterococcus faecalis* and *faecium bacteria* isolated from the human and the livestock sources

Yaser Sharifi[1], Azadeh Abedzadeh[2], Atieh Salighe[2], Naser Kalhor[3], Mohammad Khodadad Motlagh[4], Ali Javadi[5*]

[1]MSc, Young Researchers and Elite Club, Qom Branch, Islamic Azad University, Qom, Iran
[2]MSc, Department of Microbiology, Qom Branch, Islamic Azad University, Qom, Iran
[3]MSc, Academic Center for Education Culture and Research, Qom Branch, Qom, Iran
[4]PhD Student of Bacteriology, Faculty of Veterinary Medicine, Shiraz University, Shiraz, Iran
[5]PhD Student, Students' Scientific Research Center, Tehran University of Medical Sciences, Tehran, Iran

Abstract

Background: *Enterococci* have emerged as a major cause of nosocomial infections and within this group, *Enterococcus faecalis* and *Enterococcus faecium* cause the majority of human and livestock *enterococcal* infections. In this article, we tried to determine antibiotics and metals resistance patterns of *E. faecalis* and *E. faecium* strains.

Methods: One hundred sixty different strains of *E. faecalis* and *E. faecium* were collected from livestock sewage and the human fecal waste during 15 months. Then bacterial antibiotics sensitivity tests were carried out using the Agar disc diffusion method.

Results: Generally, 100% of *E. faecalis* strains separated from human and livestock sources (i.e. sheep) showed penicillin (P)/ kanamycin (K)/ nitrofurantoin (N)/ loracarbef (L)/ Ciprofloxacin (Cc)/ ampicillin (AN)/ nalidixic acid (NA)/ sulfamethoxazole (S) antibiotics resistance patterns. In addition, 55% of isolated *E. faecium* showed P/S/AN/NA antibiotics resistance patterns. Each strain showed a resistance to at least two aminoglycoside antibiotics. However, *E. faecalis* strains from human and the livestock sources showed 94% and 100% of resistance to nitrofurantoin, respectively. The effects of different metal concentrations was evaluated in both strains. The agar dilution method was applied in this stage. Hg at 0.05 mmol/L of minimum inhibitory concentration (MIC) showed toxicity to both the human and livestock *Enterococcus* strains. Cadmium at 1 mmol/L and 0.5 mmol/L concentrations had the most toxicity to *E. faecalis* and *E. faecium* strains, respectively. Obviously, toxicity to bacteria is less than other metals. As a result, Zn/Ni/Cu/Co resistance pattern is suggested for both strains. Finally, antibiotics and heavy metals resistance patterns were monitored simultaneously.

Conclusion: Almost all *E. faecalis* strains isolated from humans and livestock showed antibiotics and heavy metals resistance patterns of P/K/L/Cc/S/AN/NA/Zn/Cu/Co simultaneously. Moreover, 55% of *E. faecium* strains showed similar antibiotics and heavy metals resistance patterns of P/S/AN/NA/Zn/Ni/Cu/Co.

Keywords: *E. faecalis*, Heavy metals, Antibiotics, Resistance

***Correspondence to:**
Ali Javadi
Email: alijavadi1388@gmail.com

Introduction

Enterococci species is gram positive, catalase enzyme negative, oxidase enzyme negative, non-spore forming and facultative anaerobic cocci with strains found in humans. The strains can be separated from animal, environmental human and livestock sources (1). Enterococci are a part of the conventional microorganism of humans and animals. They have been long recognized as necessary human pathogens and are getting progressively thus. The genus *Enterococcus* includes 17 species, though only a handful can cause clinical infections in humans. They have emerged as a serious cause of nosocomial infections and at intervals, this cluster, *Enterococcus faecalis* and *Enterococcus faecium* cause the bulk of human and therefore livestock enterococcal infections (2). These infections are also systematic together with tract, abdominal, wound infections, bacteremia and carditis. Since *E. faecalis* and *E. faecium* strains are capable of living under varied environmental conditions (such as temperature extremes and also the presence of digestive fluid salts) and since they acquire resistance to multiple antibiotics, these microorganism have become a significant cause of unhealthiness (3). *E. faecalis* and *E. faecium* are the most prevalent species cultured from humans, accounting for more than 90% of clinical isolates. Other enterococcal species known to cause human infection include *Enterococcus avium*, *Enterococcus gallinarum*, *Enterococcus casseliflavus*, *Enterococcus durans*, *Enterococcus raffinosus*, and *Enterococcus mundtii*.

E. faecium is responsible for most vancomycin-resistant

enterococci (VRE) infections. This can mean serious health problems, which include the lack of available antibiotics therapy for VRE infections since most VRE strains are resistant to multiple antibiotics besides vancomycin (e.g. aminoglycosides and ampicillin (4). Enterococci vary in both intrinsic and acquired resistance and are resistant to antibiotics thus making them vital medical building pathogens. In and of itself, *enterococci* are resistant to beta-lactam antibiotics as they contain penicillin-binding proteins (PBPs); thus, they are ready to synthesize parts of the cell-wall (5). In addition, it is known that contamination of surface waters has important effects on life dynamics and the ecology of their surroundings (6).

Recently many studies have been carried out on environmental *Enterococcus*. Iranian researchers in 2014 evaluated antibiotic and heavy metal resistance patterns in *enterococcal* species cultured from waters of 9 recreational areas in Iran (7). Researchers in 2011 identified that *E. faecalis* bacteria are resistant to heavy metals and antibiotics in surface waters of the Mololoa River in Tepic, Nayarit, Mexico. The results of their study showed that prevalence increased resistance to metals and antibiotics in *Enterococcus* spp. (8). In this study, antibiotics and heavy metals resistance patterns of *E. faecalis* and faecium bacteria isolated from the human and livestock sources are evaluated.

Methods

The measurement of bacteria strains sensitivity to antibiotics

In this study, the sensitivity test of two strains, *E. faecalis* and *E. faecium* to antibiotics isolated from human and livestock samples were carried out. The antibiogram disks were provided by MAST Company. To determine bacterial sensitivity to antibiotics, Kirby-Bauer laboratory diagnostic method was used (9). Antibiogram disks were used to determine sensitivity to *E. faecalis* and *E. faecium* by Disk diffusion method (Figures 1 and 2).

Determination of maximum growth time (exponential growth) of the isolated strains

This work was carried out based on spectrophotometry. First the wavelength of the device was adjusted according to the chart prepared for Muller Hinton medium (Blank). In this case, it is equal to 400 nm. Secondly, decrease some of culture medium containing bacteria once every 30 minutes (2 ml bacterial suspensions with 0.5 McFarland turbidity in 40 cc Muller Hinton broth medium) while the container is shaken by an incubator shaker at 37°C and 150 rpm speed (Figures 3, 4 and 5) (10).

Determination of toxicity to heavy metals

Bacteria resistance to heavy metals was determined by means of Agar-dilution method in Muller Hinton Agar plates at different concentrations (mmol/liter) of metals as follows: 10, 9, 8, 7, 6, 5, 4, 3, 2, 1, 0, 0.5, 0.2, 0.1, 0.05, 0.01, and 0.005. The plates were dried at 37°C for 30 minutes, inoculated by 10 μ of bacteria. Metal salts used in this study include zinc sulfate, copper sulfate, cobalt chloride,

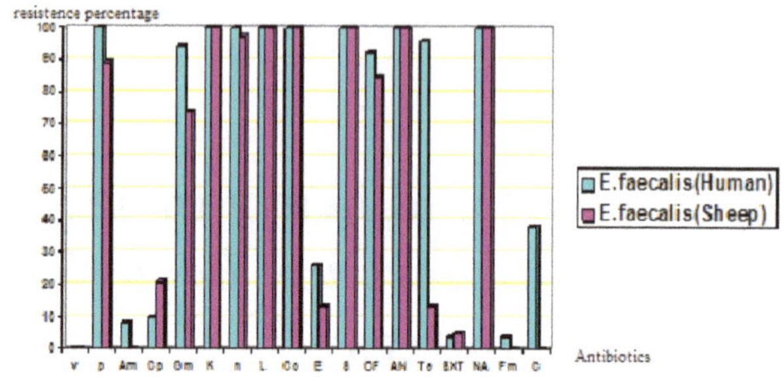

Figure 1. Comparing the percentages of antibiotics resistance of the human and livestock *E. faecalis.*

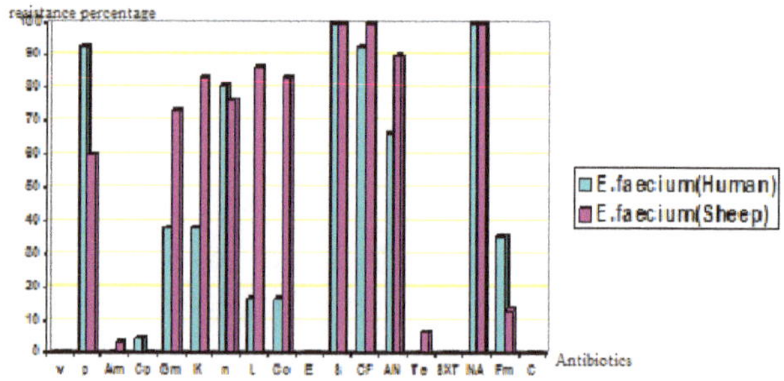

Figure 2. Comparing the percentages of antibiotics resistance of the human and livestock *E. faecium.*

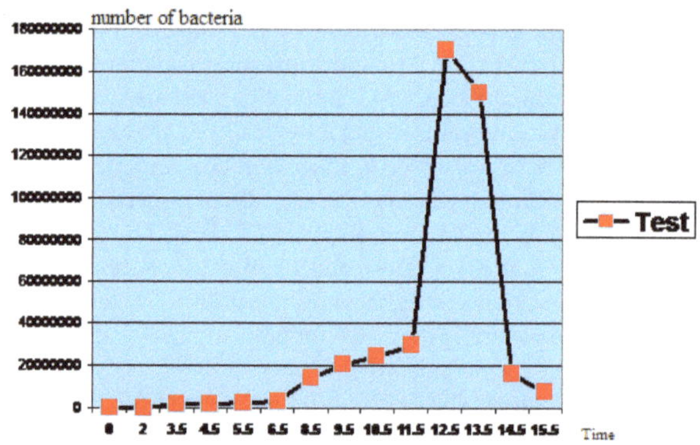

Figure 3. The growth curve of *E. faecalis* strains in the blank medium.

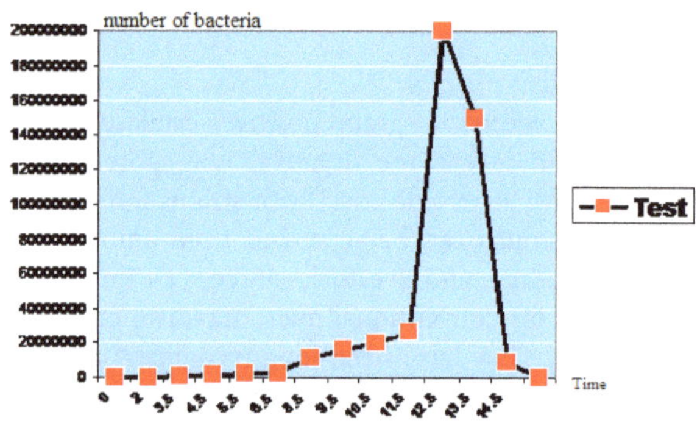

Figure 4. The growth curve of *E. faecium* strains in the blank medium

cadmium chloride, nickel sulfate and mercuric chloride, which were provided by Merck Company (Figure 6) (11). After preparing the Muller Hinton Agar plates containing different concentrations of metals, the suspension of certain bacterial strains in exponential growth phase were

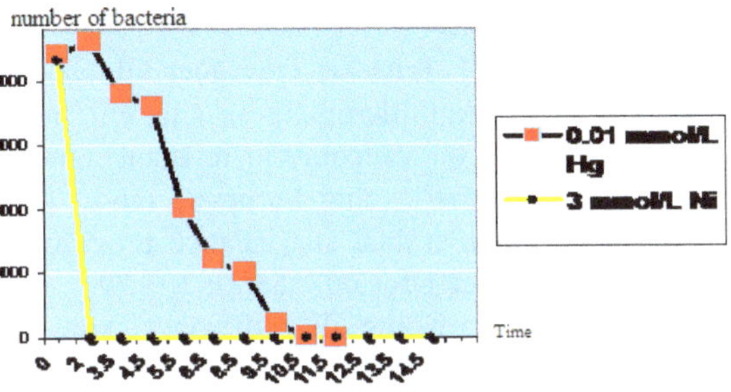

Figure 5. The growth curve of *E. faecium* strains in the medium with 0.01 mmol/L Hg and 3 mmol/L Ni.

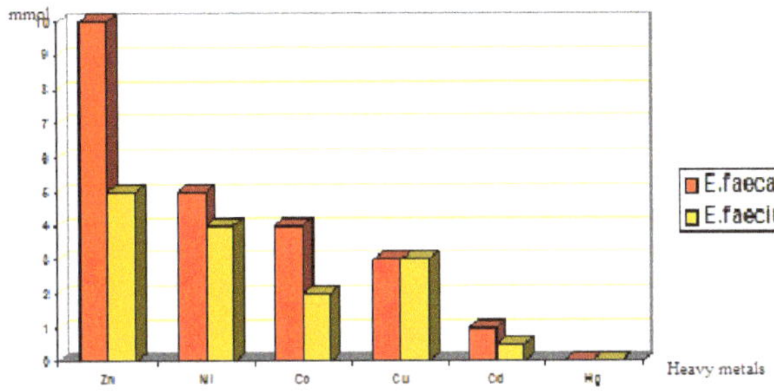

Figure 6. Comparing a metal MIC of the human strains and the livestock ones.

added in this stage and 10 UI drops numbered at certain intervals with regard to the number of strains and then they were placed on the medium containing metals (12). The plates were then incubated for 18 hours at 35°C. Finally, microbial biomass growth rate and minimum inhibitory concentration (MIC) of metals were controlled and recorded, respectively. In this case, the acceptable concentrations of metal ions were used to determine metals resistance.

Results

The determination of bacteria sensitivity related to antibiotics and the resistance pattern

The number and percentage of sensitive, semi-sensitive *E. faecalis* and *E. faecium* resistant spices are shown in Figures 1 and 2.

All species of *E. faecalis* and *E. faecium* that were isolated from the two sources were sensitive to vancomycin. This test plays a crucial role in the initial identification of *Enterococcus*. In this study, no resistant strains occurred. However, all strains of the two species have shown high resistance to penicillin. *E. faecalis* isolated from human and livestock sources showed 100% and 89.2% resistance. Moreover, *E. faecium* species isolated from the human and livestock sources showed 92.8% and 60% resistance to penicillin. On the other hand, the resistance of both species to ampicillin was equal to zero. This point is very important because *Enterococcus* strains sensitivity to ampicillin has been rarely reported.

All strains of both spices showed remarkable sensitivity to cotrimoxazole. Also high resistance to lincomycin and clindamycin in *Enterococcus* strains was observed.

However, *Enterococcus* strains of humans showed very low resistance. Resistance to nitrofurantoin, chloramphenicol, tetracycline, erythromycin, and ciprofloxacin was relatively low. Nevertheless, approximately 100% of human *E. faecalis* strains showed resistance to tetracycline.

The sensitivity to heavy metals - comparing the resistive patterns with the growth curves

Bacteria sensitivity to heavy metals and the growth of bacterial strains were determined in terms of Agar dilution method. The amounts of metal MIC of strains are shown in Figure 6 and curves.

The effectiveness of different concentrations of heavy metals on the growth of human and the livestock *Enterococcus* in terms of agar dilution method are shown in Figure 4.

As a result, the amounts of metal toxicity of both species are as follows:

E. faecalis: Hg> Cd> Cu> Co> Ni> Zn

E. faecium: Hg> Cd> Co> Cu> Ni> Zn

The relationship between antibiotic and metal resistance patterns:

As is observed *E. faecalis* and *E. faecium* strains showed resistance to multiple antibiotics. In addition, it recognized that these species are considerably resistant to four kinds of metals: cobalt, nickel, zinc, and copper.

All *E. faecalis* strains with P/K/N/L/Cc/S/AN/NA antibiotic and metal resistance patterns have Cu^{2+}/Mi^{2+}/Cu^{2+}/Co^{2+}/Zn^{+} patterns as well.

E. faecium strains metal resistance patterns were similar to *E. faecalis* strains. However, their antibiotics resistance patterns varied. human and the livestock sources showed 94% and 100% of resistance to Nitrofurantoin, respectively.

Nevertheless, *Enterococcus* strains from humans showed 40%, 64%, 70% resistance to GM, AN/NA, respectively.

The results show that all *E. faecalis* strains from human and/or livestock sources have simultaneously antibiotic and metal patterns thus: P/K/N/L/Cc/S/AN/NA/Zn^{2+}/Ni^{2+}/Cu^{2+}/C.

Almost 50 and 56% of *E. faecium* strains from human and livestock sources have simultaneously antibiotic and metal patterns of P/S/AN/NA/Zn/Ni/Cu/Co, respectively

Discussion

Determination of *Enterococcal* spp. from human and livestock sources may help to clarify their ecological characteristics (13). The research study on the coasts of the Caspian Sea, shows that the most predominant species are *E. faecalis* and *E. faecium* which is consistent with our study on animal and human sources (13). According to the study performed by de Oliveira and Pinhata, the most resistance to antibiotics was seen in *E. faecalis* and *E. faecium* species as the highest frequency of resistance observed against streptomycin and erythromycin in water samples and against erythromycin and tetracycline in sand samples (14). In the study of Issazadeh et al, all samples were found to be resistant to penicillin (7). In this study, each strain showed a resistance to at least two aminoglycoside antibiotics. However, *E. faecalis* strains from human and livestock sources showed 94% and 100% resistance to nitrofurantoin, respectively. Bacteria were isolated from the environment and livestock samples resistant to heavy metals. Kimiran-Erdem et al showed that 93% of isolates were resistant to Fe and Zn and 85% of them were resistant to Cr (15). In the study of Kermanshahi et al, the most resistance wasto Zn that the high MIC in 2 of the 3 industrial slopes were evaluated as 24 mMol thus belonging to *Coryneform* while *E. faecalis*, and the least amounts evaluated as 8 and 12 mMol belonged to *Citrobacter* and some Bacillus species, respectively (16). In this study, high resistance heavy metal in *E. faecalis* and *E. faecium* refer to Zn and low resistance heavy metal refer to Hg. Finally, the toxicity to mercury is higher than other metals.

Conclusion

According to the observed resistance to heavy metals and antibiotics and comparison with other studies, the bacteria can be spread in the environment and animal resources, therefore the control of this bacteria is very important in environmental health.

Acknowledgments

The authors would like to thank the research council of Qom Islamic Azad University.

Competing interests

The authors declare that they have no competing interests.

Authors' contributions

All authors were involved in study design, data collection, and article approval.

References

1. Low DE, Keller N, Barth A, Jones RN. Clinical prevalence, antimicrobial susceptibility, and geographic resistance patterns of enterococci: results from the SENTRY Antimicrobial Surveillance Program, 1997-1999. Clin Infect Dis 2001; 32 Suppl 2: S133-45.
2. de Perio MA, Yarnold PR, Warren J, Noskin GA. Risk factors and outcomes associated with non-Enterococcus faecalis, non-Enterococcus faecium enterococcal bacteremia. Infect Control Hosp Epidemiol 2006; 27(1): 28-33.
3. Lebreton F, Willems RJ, Gilmore MS. Enterococcus diversity, origins in nature, and gut colonization. In: Gilmore MS, Clewell DB, Ike Y, Shankar N. Enterococci: From Commensals to Leading Causes of Drug Resistant Infection. Boston: Massachusetts Eye and Ear Infirmary; 2014.
4. Willems R, Top J, Santen MV, Robinson DA, Coque TM, Baquero F, et al. Global spread of vancomycin-resistant Enterococcus faecium from distinct nosocomial genetic complex. Emerg Infect Dis 2005; 11(6): 821-28.
5. Deshpande LM, Fritsche TR, Moet GJ, Biedenbach DJ, Jones RN. Antimicrobial resistance and molecular epidemiology of vancomycin-resistant enterococci from North America and Europe: a report from the SENTRY antimicrobial surveillance program. Diagn Microbiol Infect Dis 2007; 58(2): 163-70.
6. Haferburg G, Kothe E. Microbes and metals: interactions in the environment. J Basic Microbiol 2007; 47(6): 453-67.
7. Issazadeh K, Razban S, Khoshkholgh M, Pahlaviani M. Isolation and identification of Enterococcus species and determination of their susceptibility patterns against antibiotics and heavy metals in coastal waters of Iran. Int J Adv Biol Biom Res 2014; 2(6):2026-2030.
8. Mondragón VA, Llamas-Pérez DF, González-Guzmán GE, Márquez-González AR, Padilla-Noriega R, Durán-Avelar Mde J, et al. Identification of Enterococcus faecalis bacteria resistant to heavy metals and antibiotics in surface waters of the Mololoa River in Tepic, Nayarit, Mexico. Environ Monit Assess 2011; 183(1-4): 329-40.
9. Patterson JE, Sweeney AH, Simms M, Carley N, Mangi R, Sabetta J, et al. An analysis of 110 serious enterococcal infections. Epidemiology, antibiotic susceptibility, and outcome. Medicine (Baltimore) 1995; 74(4): 191-200.
10. Byappanahalli MN, Shively DA, Nevers MB, Sadowsky MJ, Whitman RL. Growth and survival of Escherichia coli and enterococci populations in the macro-alga Cladophora (Chlorophyta). FEMS Microbiol Ecol 2003; 46(2): 203-11.
11. Maiga A, Diallo D, Bye R, Paulsen BS. Determination of some toxic and essential metal ions in medicinal and edible plants from Mali. J Agric Food Chem 2005; 53(6): 2316-21.
12. Wiegand I, Hilpert K, Hancock RE. Agar and broth dilution methods to determine the minimal inhibitory concentration (MIC) of antimicrobial substances. Nat Protoc 2008; 3(2): 163-75.
13. Arvanitidou M, Katsouyannopoulos V, Tsakris A. Antibiotic resistance patterns of Enterococci isolated from coastal bathing waters. J Med Microbiol 2001; 50(11): 1001-5.
14. de Oliveira AJ, Pinhata JM. Antimicrobial resistance and species composition of Enterococcus spp. isolated from waters and sands of marine recreational beaches in southeastern Brazil. Water Res 2008; 42(8-9): 2242-50.
15. Kimiran-Erdem A, Arslan EO, Sanli Yurudu NO, Zeybek Z, Dogruoz N, Cotuk A. Isolation and identification of enterococci from seawater samples: assessment of their resistance to antibiotics and heavy metals. Environ Monit Assess 2007; 125(1-3): 219-28.
16. Kermanshahi R. K, Ghazifard A, Tavakoli A. Identification of bacteria resistant to heavy metals in the soils of Isfahan Province. Iranian Journal Of Science And Technology Transaction A- Science 2007; 31(A1): 7-16. [in Persian]

6

Efficiency and kinetic modeling of removal of nutrients and organic matter from a full-scale constructed wetland in Qasr-e-Shirin, Iran

Abdolmajid Gholizadeh[1*], Mitra Gholami[2], Reza Davoudi[3], Ayoob Rastegar[4], Mohammad Miri[5]

[1]PhD Student, Department of Environmental Health Engineering, School of Public Health, Shahid Sadoughi University of Medical Sciences, Yazd, Iran
[2]Professor, Department of Environmental Health Engineering, School of Public Health, Iran University of Medical Sciences, Tehran, Iran
[3]MSc Of Environmental Health Engineering, Department of Environmental Health Engineering, School of Public Health, Iran University of Medical Sciences, Tehran, Iran
[4]PhD Student, Department of Environmental Health Engineering School of Public Health, Sabzevar University of Medical Sciences, Sabzevar, Iran.
[5]PhD Student, School of Public Health, Shahid Sadoughi University of Medical Sciences, Yazd, Iran

Abstract
Background: This study assessed the removal of organic material and nutrients from full-scale subsurface flow (SSF) constructed wetlands (CWs) followed by anaerobic stabilization ponds under environmental conditions.
Methods: The effluents were distributed evenly in 12 reed beds. Samples were taken twice monthly for a total of 6 months from several points in the wetland. Biochemical oxygen demand (BOD), chemical oxygen demand (COD), total suspended solids (TSS), and nutrient removal from the system and the longitudinal effect of the reed beds for removal of pollutions were determined. A full-scale model of flow, BOD, and nutrients in SSF in the CWs is presented.
Results: The flow rate and concentrations of parameters indicated that removal of organic matter and nutrients in the cold months decreased rather than in the hot months, as expected. The removal efficiency for BOD, COD, and TSS and the strongest biological interactions showed no uniform trends. The beds showed the highest removal rates in the first few meters of bed. The hybrid Monod-Plug flow regime and the Stover-Kincannon models showed the best fit for the kinetics of the processes. U_{max} in the Stover-Kincannon model was 3.64 mg/l.d for nitrogen and 0.24 mg/l.d for phosphorus. These values are very low, which indicates lower consumption and inefficiency of the system for removing nitrogen and phosphorus.
Conclusion: It can be concluded that the SSF in CWs are able to treat average wastewater as effectively as common mechanical systems at lower cost.
Keywords: Kinetic modeling, Nutrient, Organic matter, Wetland, Qasr-e-Shirin

***Correspondence to:**
Abdolmajid Gholizadeh
Email: gholizadeh_eng@yahoo.com

Introduction

Relatively simple wastewater treatment technologies can be designed to provide low cost sanitation and environmental protection while providing additional benefits from the reuse of water. Constructed wetlands (CWs) are essentially new developments for the utilization of wastewater and sludge. These wetlands provide cost-effective waste treatment options (1,2). For regulatory purposes under the Clean Water Act, the term "wetlands" denotes those areas inundated or saturated by surface or ground water, either permanently or seasonally. Wetlands vary widely according to regional and local differences in soils, topography, climate, hydrology, water chemistry, vegetation, and other factors such as disturbance by humans. These areas can be distinguished from other water bodies or land by their water levels and the types of vegetation and animal communities in the soil and at the surface (3). Different types of pollutant (nitrogen, phosphorus, organics, solids, metals, and coliforms) can be removed by CWs through a complex inter-connected system of plants, media, bulk water, and biomass population (4). They already function to filter water and trap sediment before it enters a body of water. As an interface between catchments and surface water, they play an important role in the control of water quantity and quality of surface water systems in general and in the reduction of diffuse pollution in catchments in particular (3,5).

CWs are an ideal environment to support the growth of organisms that break down pollutants in wastewater by biodegradation. The presence of bacteria is extremely important because they often act as catalysts for pollutant removal in subsurface flow (SSF) wetland bioreactors (6). In CWs, the wastewater is purified by the physical, chemical, and biological triple synergy of the natural ecosystem (7,8). CWs have the advantage of producing higher quality effluent without the input of fossil energy, thereby reducing operational costs (6,9,10). They are often categorized as surface flow (SF) CWs or SSF CWs. SSF CWs can be further subdivided into horizontal flow (HF) and vertical flow (VF) systems, depending on the direction of the flow of water through the porous media (sand or gravel) (8,11). The combination of a VF and HF system is known as a hybrid wetland and can be employed for the treatment of wastewater. This combination often optimizes nitrogen and organics removal with the presence of aerobic, anaerobic, and anoxic phases (12). SSF treatment wetlands are used worldwide to remove pollutants from wastewater because they are mechanically simple and require low operation and maintenance (O&M) in comparison with conventional wastewater treatment technologies (13). These systems are currently in limited use as unusual constructed reed beds in many parts of the world. Further and more detailed study of the method is needed.

Overviews of current developments on numerical modeling of SSF CWs based on modeling and model development have been presented by various authors (14-16); however, there is little information available about the use of wetlands as public remediation-sites for wastewater treatment purposes (17). This usage is not fully understood to date because of the lack of appropriate models. The most widely-employed modeling equations (Kickuth equation and KeC model) give only an exponential profile of inlet and outlet pollutant concentrations without considering the full range of pollutant variability in CWs (10,18).

Almost all available design guidelines are based on empirical rules of thumb, such as those using specific surface area requirements or simple first-order decay models. The main objective of numerical modeling is to obtain a better understanding of the processes governing the biological and chemical transformation and degradation occurring in CWs. Once reliable numerical models are developed and validated against experimental data, they can be used to evaluate and improve existing design criteria (8,19,20). Kinetic modeling and optimization of system parameters can be used to optimize the design, implementation, and O&M of such systems. The goal is to provide the highest efficiency filtration based on prior knowledge of local conditions at a reasonable cost. This will contribute to the development of technology for wastewater treatment using artificial reed beds. Langergraber (21) reported that few numerical models are able to describe the treatment processes in SSF CWs.

The present study investigated the removal efficiency of organic matter and nutrients and kinetically modeled the processes at work in a CW near the city of Qasr-e-Shirin in Kermanshah province in Iran. A wastewater treatment system consisting of an anaerobic stabilization pond combining with a SSF CW was selected for examination. In this system, wastewater passing from the screening and grit chamber enters two anaerobic stabilization ponds. The effluent from the anaerobic ponds is eventually distributed uniformly into 12 reed beds having similar physical and hydraulic forms. The effluent from the reed beds drains into a single collecting channel and flows into a chlorination pool. The pattern of removal of organic matter, nitrogen, phosphorus, and suspended solids on effluent quality are the main parameters.

Methods

Wetland characteristics

This empirical full-scale study was conducted over the course of 6 months that included both hot months (July, August, and September) and cold months (December, January, and February) at a wastewater treatment plant in Qasr-e-Shirin, Iran. This treatment plant having 14 hectares has been in operation since 2008. The projected plan for 2021 is to meet the requirements for a population of 30 thousand people.

Of the 12 subsurface constructed beds, 11 contain reeds (*Phragmites*) and one contains cattails (*Typha*). All beds have similar physical and hydraulic design conditions. The length of the beds is 25 m and the width is 125 m. Each bed contains 16 distribution tubes placed 7.5 m apart, providing a length to width flow ratio of 1:3. One reed bed was randomly selected for investigation (bed 11; Figure 1). All outlets were connected to a manhole from which the output was collected. The effluent of all reed beds flows into a single collecting channel and then into a chlorination pool. Substrate draining was facilitated by a 0.8% slope of the floor of the beds. The depth of gravels at beginning and end of the beds was 75 and 95 cm, respectively. Table 1 and 2 provide detailed characteristics of the ponds and reed beds.

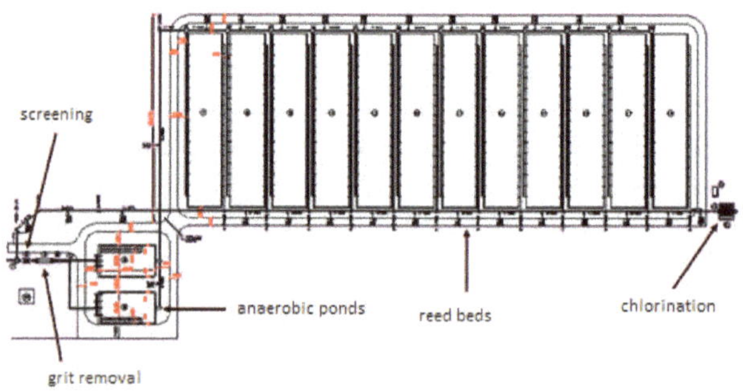

Figure 1. Unit arrangement at Qasr-e Shirin wastewater treatment plant.

Table 1. Characteristics of Qasr-e Shirin wetland beds

Parameter	Characteristic
System type	Subsurface horizontal flow
Type of plant	*Phragmites australis, Typhae*
Input flow to each bed	180 m^3/d
BOD_5 loading	450 kg/d
Total surface area	3,125 m^2
Length of each bed	25
Width of each bed	7.5
Length of all sand beds	25 m
Width of all sand beds	125 m
Average depth of sand bed	0.85 m
Bed porosity	35%
Hydraulic retention time	5.2 d
No. of input tubs in each bed	16
No. of output tubs in each bed	16
Floor type	Compacted clay
Distance between adjacent straws	5 m
Sand diameter	8-10 mm

Table 2. Characteristics of Qasr-e-Shirin anaerobic ponds

Parameter	Characteristic
Volume of each pond	4400 m^3
Area of each pond	880 m^2
Useful depth	5 m
Hydraulic retention time	4 d
Interval of sludge discharging	3 y
Floor type	Compacted clay: height 20 cm

Wetland efficiency

The temperature, pH, total biochemical on demand ($TBOD_5$), total chemicals on demand (TCOD), total suspended solids (TSS), total nitrogen (TN), and total phosphorus (TP) were determined using 24-h composite sampling 3 times/d (2-h samples) at 4 boreholes in the reed bed. The 3 samples were mixed (Figure 2) to provide a mean value. Sampling was done twice/mo over the 6 months study duration. A total of 252 samples were collected and analyzed. The reed bed effluent flow rate was also measured manually and ultrasonically using the existing Parshall flume.

The effect of bed length on pollutant removal was determined by selecting a bed of 25 m in width (bed 11) and dividing it into five parts at intervals of 5 m. Seven sampling sites were designated in consideration of the location of the 4 polyethylene sampling boreholes embedded in the bed and the entrance (anaerobic pond effluent) and exit points in the reed bed and entrance point to the treatment plant.

Data gathering

All experiments were performed at the Water and Wastewater Laboratory at the Western Higher Educational

Figure 2. Surveyed wetland and sampling points.

Complex of the Department of Civil Engineering and Sewage Research. Testing for BOD_5, COD, TSS, and TP were performed using the Standard Methods for the Examination of Water and Wastewater (22). TN tests were done using ready-to-use vials (Hatch) according to manufacturer recommendations (Table 3).

Kinetic modeling

Kinetic equations are applied to explain the transport behavior of the adsorbate molecules per time unit. Four hybrid models were used for kinetic modeling of organic matter, nitrogen and phosphorous removal as follows (24):

Combined first-order kinetic equation and flow regime Plug flow model

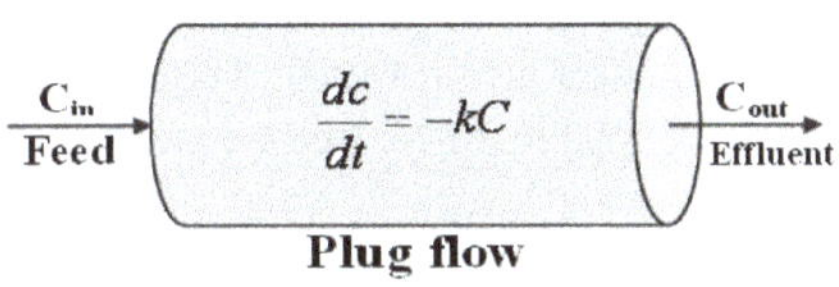

This creates the first design equation (Kickuth), which is the easiest and most commonly-used equation for design of wastewater stabilization ponds (WSPs):

$$A_h = \frac{Q(\ln C_{in} - C_{out})}{K_1} \tag{1}$$

where A_h is pond area (m^2), Q is flow rate (m^3/d), C_{in} is input BOD_5 concentration (mg/l), C_{out} is output BOD_5 concentration (mg/l), and k_1 is the velocity constant (m/d).

Combined first-order kinetic equation and CSTR model

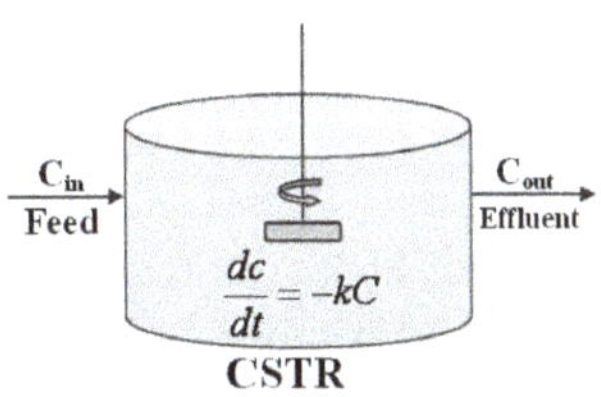

This flow pattern results in the simplified equation shown below. Areal rate constant k_1 (m/d) is used to correlate the influent and effluent N and organics in the SSF CWs (10):

$$k_1 = \frac{q(C_{in} - C_{out})}{C_{out}} \tag{2}$$

where q is the hydraulic loading rate (m/d).

Table 3. Methods used to measure parameters

Test	Method	Headline
COD	Reflux titration and Hatch vials for control	C 5220
BOD_5	5-d method and using OXITOP (WTW) and Hatch vials for control	B5210
TSS	Drying at 103-105°C	D 2540
TP	Total phosphorus method	P 4500
TN	Hatch vials and Nessler's reagent method	Ref. (23)

Combined Monod kinetic equation and Plug flow regime

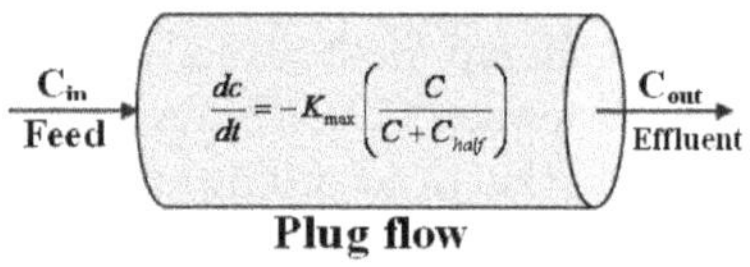

This equation is:

$$\frac{dc}{dt} = -k_{max}\frac{C_{in}}{C + C_{half}} \quad (3)$$

$$A_h \frac{Q(C_{in} - C_{out} + 60\ln(C_{in}/C_{out}))}{K_3} \quad (4)$$

where k_{max} is the maximum BOD_5 removed in pond regardless of temperature (g/m³.d), C_{half} is the wastewater BOD_5 when the organic matter is removed and equals half of k_{max} (60 mg/l; a common value for Monod's equation 3).

Combined Monod kinetic equation and continuously stirred reactors (CSTR) model

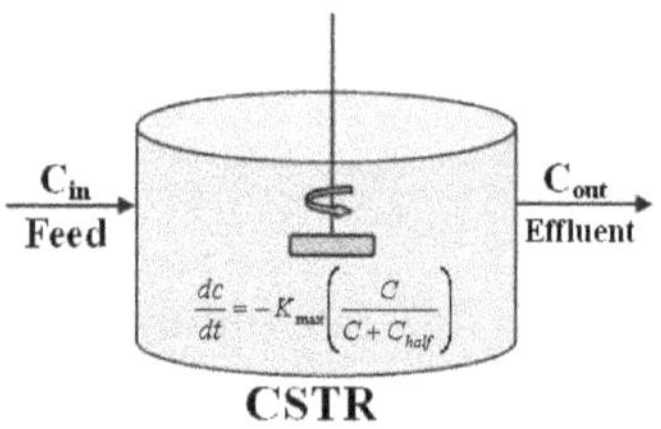

This is described by equation 5 below, which links inlet and outlet pollutant concentrations

$$\frac{C_{in} - C_{out}}{\tau} = K_{max}\frac{C_{out}}{C_{half} + C_{out}} \quad (5)$$

Equation 5 is expressed in terms of hydraulic retention time (unit/d) and k_{max} represents maximum pollutant mass removal per m³ of wetland volume per d. The relationship between hydraulic retention time, area (A, m²), depth (h, m), and porosity (e) of packed media and inlet discharge (Q, m³/d) can be expressed as:

$$\tau = \frac{Ahe}{Q} \quad (6)$$

Combining k_{max} (g/m³/d) with h and e results in areal maximum pollutant removal (k_2: g/m²/d) as expressed in Eq. 7 to correlate inlet and outlet pollutant concentrations as.

$$k_2 = \frac{q(C_{in} - C_{out})(C_{half} + C_{out})}{C_{out}} \quad (7)$$

Equation 7 can be used to predict N and organic degradation in SSF CWs. In the first step of NH_4-N transformation (NH_4-N to NO_2-N) during nitrification, the half saturation constant (C_{half} for Nitrosomonas degradation) was 0.05 mg/l (25). This value was used as C_{half} in equation (7) for nitrification. For denitrification of NO_3-N, the nitrate half saturation constant in the Monod kinetics was used at 0.14 mg/l (26). For heterotrophic BOD_5 removal, the recommended degradation half saturation constant is 60 mg/l for wastewater treatment (27). For COD, the COD half saturation constant used was 20 mg/l as suggested by Saeed et al (10) for sewage treatment.

Stover–Kincannon kinetic model

The Stover–Kincannon model was first employed to predict attached growth in wastewater treatment and was then modified and used to describe and predict the performance of the biological reactor. The linearized state of this model is (28):

$$\frac{1}{\frac{ds}{dt}} = \frac{V}{Q(S_o - S)} = \frac{K_B}{U_{max}}\frac{V}{QS_o} + \frac{1}{U_{max}} \quad (8)$$

where U_{max} is the maximum removal rate, K_B is the saturation rate, Q is the flow rate (m³/d), and V is the the volume of the reactor (m³).

Results

Performance evaluation of treatment plant

The data shows that the average flow rate and pollutant concentration in the cold months were lower than in the hot months, as expected. This is generally the reverse in urban areas without considering precipitation point. The current study indicates higher organics removal in SSF wetlands than in SF wetlands. During the study period, TSS and TP in the reed beds effluent were below standard levels (TSS = 12 mg/l; TP = 3.8 mg/l). The concentration of TN in the effluent was 17 mg/l (maximum 19.3; minimum 15.4 mg/l) which approaches effluent standards. The input and output of the parameters in each unit of the treatment system are shown in Table 4.

The amount of organic matter, even in the worst state (coldest month; January) was less than or close to standard levels. The proportion of BOD5, COD and TP eliminated from the anaerobic ponds and reed beds was almost equal, but in the hot months, the anaerobic ponds recorded higher removal efficiency than the reed beds (Figure 3). TN removal in the reed beds overall was greater than in the anaerobic ponds (Figure 4). Note that the total area used for the anaerobic ponds was 1760 m²; the area occupied by the plant beds was 37500 m² (21 times larger than the anaerobic ponds), which shows the importance of the anaerobic ponds and their contribution to refining.

Table 4. Parameters during hot, cold, and total periods of study for all treatment plant units

Location parameter	Input to treatment plant (mg/l) (anaerobic pond)			Output of anaerobic pond (mg/l) (input of reed bed)			Output of reed bed (mg/l) (treatment plant)		
	Min.	Max.	Mean	Min.	Max.	Mean	Min.	Max.	Mean
Total study period									
BOD_5	193	265	224	86	155	123	13	45	27
COD	348	422	375	153	244	211	29	71	51
TSS	177	222	199	42	106	72	6	17	12
TN	23	39.2	33	23.5	37.2	31	14.2	19.8	17
TP	5	7.5	6.4	4.2	6.4	5.1	3.3	4.4	3.8
Hot months									
BOD_5	193	235	220	86	128	109	13	30	19
COD	348	389	367	153	244	190	29	52	38
TSS	177	215	200	52	88	70.8	7	15	11
TN	31	39.2	34.38	29	35.9	31.4	14.2	17.8	16.5
TP	5	7.5	6.48	4.2	6.4	4.9	3.3	4.1	3.6
Cold months									
BOD_5	212	265	228	123	155	137	24	45	35
COD	357	422	383	218	243	232	54	71	64
TSS	179	222	198	51	96	73.2	6	17	13
TN	23	39.1	31.62	23.5	37.2	30.6	15.2	19.8	17.5
TP	5.7	7.1	6.32	4.4	6.4	5.3	3.5	4.4	4

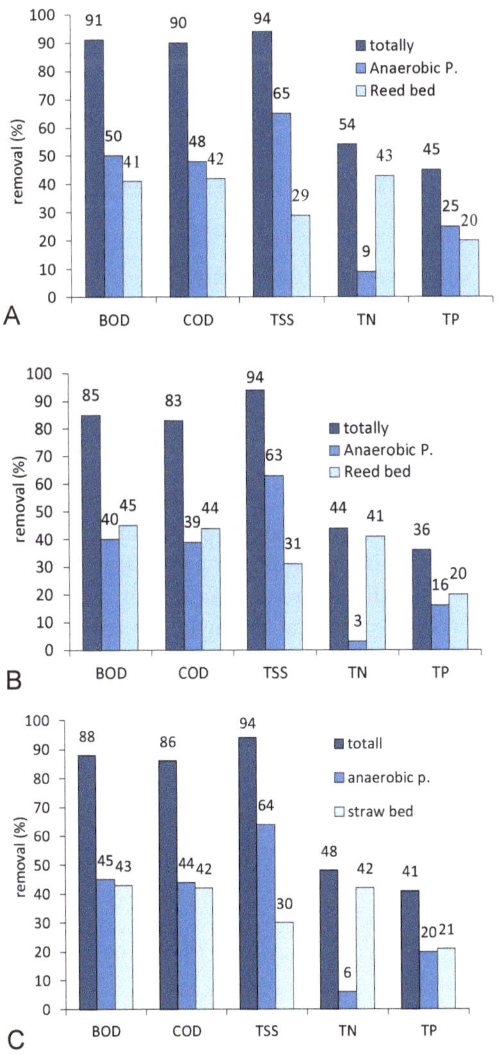

Figure 3. Average parameter removal in anaerobic ponds and reed beds in (A) hot months; (B) cold months; and (C) overall.

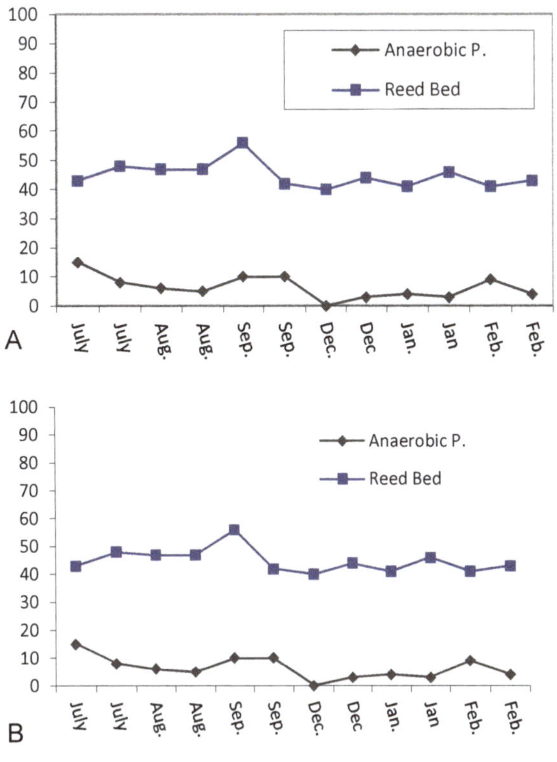

Figure 4. (A) Average TN removal; and (B) average TP removal in reed beds and anaerobic ponds for total study period.

Longitudinal effect on organic matter and nutrient removal

The BOD5, COD, TSS, TN, and TP removal showed no uniform trend in the reed bed (Figure 5). The highest removal efficiency was observed in the first few meters of the bed. The greatest biological interaction occurred in the first meters (27%, 30%, 47%, 13%, and 10%, respectively), but removal of the final 5 m of the bed was only

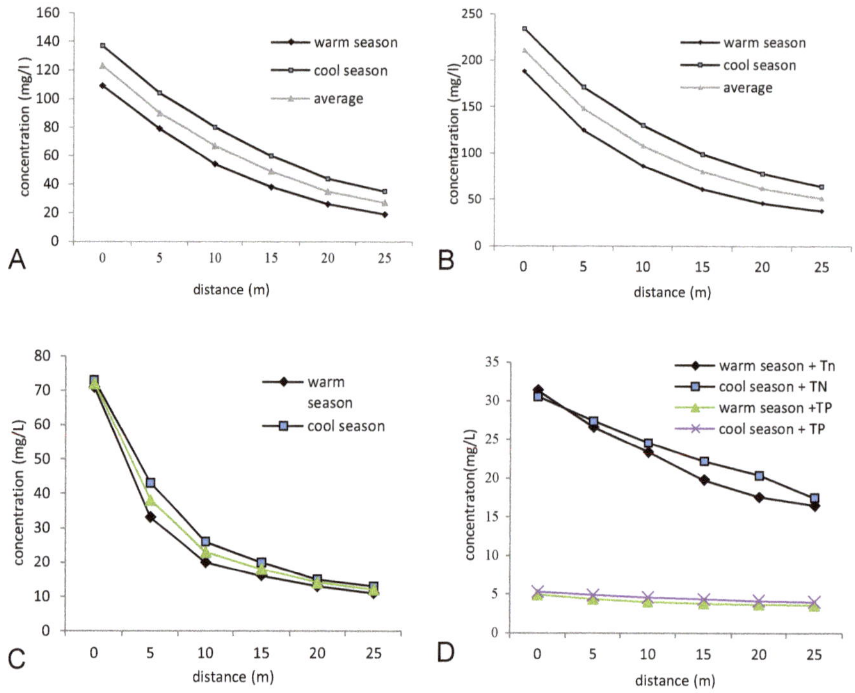

Figure 5. Average concentration of: (A) BOD5; (B) COD; (C) TSS; and (D) TP and TN in the hot and cold months along the length of the bed.

7%,5%, 3%, 6% and 2% for BOD5, COD and TSS, TN and TP, respectively. A greater amount of suspended BOD5 was removed than soluble BOD5 because of the porosity of the bed.

Evaluation of process kinetics

Organic matter

The hybrid Monod kinetics and Plug flow regime models showed good fit with the field data. To further investigate and achieve greater agreement, other valid models such as the Stover-Kincannon and pseudo-first-order kinetic models were examined (Table 5). The correlation coefficient (R^2) for the Stover-Kincannon was 0.95. The Stover-Kincannon model showed better fitness than the other model; R^2 was low for the other model, which indicates poor correlation with the field data.

The U_{max} (feed consumption rate) for BOD5 in the subsurface artificial straw bed at Qasr-e-Shirin was very low (50 mg/l.d). This low value indicates that the system (bed volume) was too large for this loading rate and that treatment plant has the capacity to receive more wastewater flow.

Nutrients

The Stover-Kincannon and first-order models were investigated for removal of N and P and the Stover-Kincannon model was determined to be the most applicable for removal of nutrients (Table 6). U_{max} for N was 3.64 mg/l.d and P was 0.24 mg/l.d, which is very low.

Discussion

CWs are typical of natural and environmentally-friendly systems using rooted water-tolerant plants and gravel or soil media to treat wastewater. As a green treatment technology, CWs have the unique advantage of producing higher effluent quality without the input of fossil energy, thereby reducing operation costs (6). The disadvantage of

Table 5. Kinetic parameters for removal of BOD5 from selected wetland

Equation	Parameter			
	R^2	k_1		Regression equation
Combined first-order and Plug flow model	0.484	0.079		Y = 0.079 x
Combined first-order and CSTR model	0.267	0.199		Y = 0.199 x
Combined Monod and Plug flow model	0.723	9.53		Y = 9.53 x
Combined Monod and CSTR model	0.393	16.76		Y = 16.76 x
Stover-Kincannon kinetic model	R^2	U_{max}	K_B	
	0.95	50	42.7	Y = 0.854 x -0.020

Table 6. Performance of kinetic models for TN and TP removal from wetland

Stover-Kincannon				
Coefficients	R^2	U_{max}	K_B	Regression eq.
N removal	0.859	3.64	13.45	Y = 3.696 × -0.275
P removal	0.754	0.24	1.87	Y = 7.791 × -4.251
First order				
Coefficients	R^2	k_1		Regression eq.
N removal	-0.15	0.138		Y = 0.138 ×
P removal	0.104	0.57		Y = 0.057 ×

HF wetlands include high area demand, clogging, sulfur transformation that can affect nitrification sensitivity, and loss of TP removal performance. Careful calculation of hydraulics are necessary for optimal oxygen supply and low ammonium oxidation.

The average flow rate and pollutant concentration in the cold months was lower than in the warm months. Because the boundary city of Qasr-e-Shirin is a border crossing point to the Holy Shrines in Iraq, the number of pilgrims traveling in autumn and winter is higher than in the hot summer months. This affects the quality and quantity of treatment plant influent.

The treatment plant has sufficient capacity, especially in summer, to treat the pollutants loaded into the system. The removal efficiency of COD in the hot months was somewhat lower than environmental standards, but did not meet the standard levels in the cold months. In the summer, wetland plants (macrophytes) typically grow in water or soil media subjected to oxygen deficiency (6). The removal of organics was higher in the SSF wetlands compared than in the SF wetlands, but the system exhibited poor TN removal, which agrees with the results of the previous studies (6,29). Several surveys conducted on mechanical systems and lagoons have reported removal efficiencies of 90% to 95% for BOD_5 and 90% to 95% for TSS, which agrees with the results of the present study. TN and TP removal were reported to be 10% to 20% and 15% to 25%, respectively, but these deficiencies have been better addressed in the present treatment system (27,30,31).

CWs provide an ideal environment that supports the growth of organisms that break down pollutants in wastewater by biodegradation. Bacteria, fungi, and algae are common organisms in wetlands (6). The physical components of wetland macrophytes (aerial tissue, plant tissue in water, and roots) contribute to wetland performance optimization to some extent (32).

Seswoya and Zainal (33) found that the SSF CWs performance for removal of high strength effluent in domestic wastewater is potentially good. The pollutant removal percentage depends upon the area of the CW. A large area is needed for higher removal. Chang et al (34) employed 2 pilot-scale integrated vertical-flow CWs with the plant species *Typha Orientalis* and *Arundo Donax* var. versicolor, and *Canna Indica* and *Pontederia Cordata*. The mean removal efficiencies, respectively, were 59.9% versus 62.8% for COD, 15.0% versus 12.8% for TN, and 52.0% versus 51.1% for TP. The mean mass removal rate (g m^{-2} d^{-1}) was 44.3 versus 46.4 for COD, 1.27 versus 1.08 for TN, and 0.393 versus 0.386 for TP, respectively. It was noted that nitrification was the limiting step for TN removal.

The organic matter content, even in the worst state (coldest month; January) was below or close to the standard level. This can be explained by the high performances in anaerobic ponds, as the first unit, which increased the treatment plant efficiency and reduced pollution-loading into the plant bed. The proportion of anaerobic ponds and reed beds for elimination of BOD5, COD, and TP were almost equal, but, in the hot months, the anaerobic ponds showed higher removal efficiency than the reed beds.

Saeed and Sun (35,36) and Tee et al (37) asserted that the substantial organic and TN removal observed in VF and HF systems is caused by: (*a*) atmospheric oxygen diffusion through porous media; and (*b*) leaching of C from the employed organic media to the bulk water, stimulating denitrification.

The performance disparity of CWs for TN and organics removal could be attributed to: (*a*) the presence of excessive organic compounds in wastewater that inhibits nitrification (38) because faster heterotrophic organic degradation depletes dissolved oxygen (DO) availability (10); and (*b*) lack of biodegradable organics often hinders classic denitrification metabolism from dependency on organic carbon in wetland systems (39). The current target of wetland performance optimization (40) was heavily dependent on overcoming the conflicting dependency, between TN and organics removal.

In SSF wetland systems, the transformation and removal of TN are accomplished by both classic and newly-discovered routes. The classic pathways include biological (ammonification, nitrification, denitrification, plant uptake, biomass assimilation, dissimilatory nitrate reduction), and physicochemical routes (ammonia volatilization, and adsorption). The newly-discovered nitrogen removal routes are solely dependent on microbiological metabolism such as partial nitrification-denitrification, anammox, and Canon (Figure 6). The presence of macrophytes is essential for wetlands to improve TN removal performance (6,41,42)

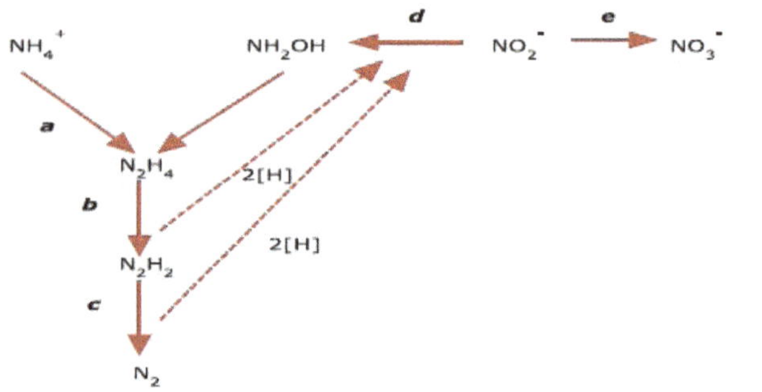

Figure 7. N removal kinetics using anammox. Step a: NH4 þ is oxidized to hydrazine by hydroxylamine; Steps b, c, d: Hydrazine is oxidized to N_2 and the reducing equivalents reduce nitrite to hydroxylamine; Step e: NO_3 forms as the by-product of anammox kinetic metabolism.

Longitudinal effect on organic matter and nutrient removal

BOD5, COD, TSS, TN and TP removal showed no uniform trends in the reed bed. The highest removal efficiency occurred in the first few meters of bed for which the highest level of biological interaction occurred. A greater amount of suspended BOD5 was removed than soluble BOD5 because of the porosity of the bed. The subsurface wetland systems acted similarly to a sand filter (29).

Process kinetics

Organic matter

The R^2 for Stover-Kincannon was 0.95, meaning that it showed a better fit than the other models. Since this model assesses entrance-loading against removed load, it is well-adapted for real-scale systems, such as adsorption, suspended growth, and attached growth and was first designed for attached-growth systems (43).

The value of U_{max} for a system of this size for approximately 6 days retention time was very low. This system was inefficient for removal of organic matter. This could be the result of improper design and the very broad surfaces compared to the small amounts of feed, which decreased the amount of biomass in the system (21,44).

Nutrients

The Stover-Kincannon and first-order models were investigated for removal of TN and TP. The Stover-Kincannon model was most applicable for removal of nutrients (Table 6). The U_{max} for TN was 3.64 and for TP was 0.24 mg/l.d. These values are very low, indicating decreased consumption and inefficiency of the system for removing TN and TP.

Most recent modeling efforts evaluate the performance of CWs for the treatment of wastewater. One approach is to model a CW as an infinite number of CSTRs. This assumes a wetland to possess many through-flow channels flanked by side regions of limited flow.

Conclusion

Anaerobic ponds integrated with subsurface CWs operated satisfactorily when compared with expensive, complex mechanical conventional systems that consume energy and are inefficient during cold months. The system was also more efficient for removal of TSS, TN and TP than the classic systems. Elimination of BOD5, COD and TSS in SSF CWs was significant. During the study period, TSS and TP in the reed bed effluent was below standard levels and the concentration of TN in the effluent was close to effluent standards.

The reed bed showed the highest removal efficiency in the first few meters because there was greater biological interaction in this area. The hybrid Monod-Plug flow regime and the Stover-Kincannon models showed the best fit with the kinetics of the process. It can be concluded that the SSF CWs have the ability to treat average wastewater about as effectively at lower cost than many common mechanical systems.

Acknowledgements

This study was funded and supported by Tehran University of Medical Sciences (TUMS). The authors appreciate the efforts of all those who were involved in this study.

Authors' contributions

All authors were involved in all stages of article preparation. We confirm that the article has not been published previously and is not under consideration for publication elsewhere. On behalf of all co-authors, the corresponding author bears full responsibility for this submission.

Competing interests

We affirm that the article is the original work of the authors and we have no conflict of interest to declare.

References

1. Crites RW, Middlebrooks EJ, Reed SC. Natural Wastewater treatment Systems. CRC Press; 2010.
2. Farzadkia M, Ehrampush M, Abouee Mehrizi E, Sadeghi S, Talebi P, Salehi A, et al. Investigating the efficiency and kinetic coefficients of nutrient removal in the subsurface artificial wetland of Yazd wastewater treatment plant. Environ Health Eng Manag J 2015; 2(1): 23-30.
3. Zhu JQ, Yu H, He QY. Development of wetland agriculture and its prospect. Third International Conference on Intelligent System Design and Engineering Applications (ISDEA); 2013. p. 596-601.
4. Fountoulakis MS, Terzakis S, Chatzinotas A, Brix H, Kalogerakis N, Manios T. Pilot-scale comparison of constructed wetlands operated under high hydraulic loading rates and attached biofilm reactors for domestic wastewater treatment. Sci Total Environ 2009; 407(8): 2996-3003.
5. Saeed T, Afrin R, Muyeed AA, Sun G. Treatment of tannery wastewater in a pilot-scale hybrid constructed

wetland system in Bangladesh. Chemosphere 2012; 88(9): 1065-73.
6. Saeed T, Sun G. A review on nitrogen and organics removal mechanisms in subsurface flow constructed wetlands: Dependency on environmental parameters, operating conditions and supporting media. J Environ Manage 2012; 112: 429-48.
7. Zhang T, Xu D, He F, Zhang Y, Wu Z. Application of constructed wetland for water pollution control in China during 1990–2010. Ecol Eng 2012; 47: 189-97.
8. Langergraber G, Giraldi D, Mena J, Meyer D, Peña M, Toscano A, et al. Recent developments in numerical modelling of subsurface flow constructed wetlands. Sci Total Environ 2009; 407(13): 3931-43.
9. Reyes-Contreras C, Hijosa-Valsero M, Sidrach-Cardona R, Bayona JM, Bécares E. Temporal evolution in PPCP removal from urban wastewater by constructed wetlands of different configuration: a medium-term study. Chemosphere 2012; 88(2): 161-7.
10. Saeed T, Sun G. Kinetic modelling of nitrogen and organics removal in vertical and horizontal flow wetlands. Water Res 2011; 45(10): 3137-52.
11. Haberl R, Grego S, Langergraber G, Kadlec RH, Cicalini AR, Dias SM, et al. Constructed wetlands for the treatment of organic pollutants. J Soils Sediments 2003; 3(2): 109-24.
12. Kadlec RH, Wallace SD. Treatment Wetlands. 2nd ed. USA: CRC Press, Taylor and Francis; 2009.
13. Knowles P, Dotro G, Nivala J, García J. Clogging in subsurface-flow treatment wetlands: Occurrence and contributing factors. Ecol Eng 2011; 37(2): 99-112.
14. Edwards JD. Industrial wastewater treatment, a guidebook. Boca Raton, FL: Lewis Publishers; 1995.
15. Nadavala SK, Swayampakula K, Boddu VM, Abburi K. Biosorption of phenol and o-chlorophenol from aqueous solutions on to chitosan-calcium alginate blended beads. J hazard Mater 2009; 162(1): 482-9.
16. Rubin E, Rodriguez P, Herrero R, Sastre de Vicente ME. Biosorption of phenolic compounds by the brown alga Sargassum muticum. J Chem Technol Biotechnol 2006; 81(7): 1093-9.
17. Rhee JS, Iamchaturapatr J. Carbon capture and sequestration by a treatment wetland. Ecol Eng 2009; 35(3): 393-401.
18. Azadi N, Falahzadeh R, Sadeghi S. Dairy wastewater treatment plant in removal of organic pollution: a case study in Sanandaj, Iran. Environ Health Eng Manag J 2015; 2(2): 73-7.
19. Rousseau DP. Performance of constructed treatment wetlands: model-based evaluation and impact of operation and maintenance [thesis]. Toegepaste Biologische Wetenschappen: Milieutechnologie; 2005. http://biomath.ugent.be/publications/download/rousseaudiederik_sum.pdf
20. Kadlec RH, Wallace S. Treatment Wetlands. CRC; 2008.
21. Langergraber G. Modeling of processes in subsurface flow constructed wetlands: a review. Vadose Zone J 2008; 7(2): 830-42.
22. Eaton AD, Franson MA. Standard methods for the examination of water & wastewater. American Public Health Association; 2005.
23. Hu L, Hu W, Dengb J, Li Q, Gaoa F, Zhua J, et al. Nutrient removal in wetlands with different macrophyte structures in eastern Lake Taihu, China. Ecol Eng 2010; 36(12): 1725-32.
24. Khusravi R, Khodadadi M, Gholizadeh A, Mehrizi EA, Shahriary T, Shahnia A. BOD5 removal kinetics and wastewater flow pattern of stabilization pond system in Birjand. Eur J Exp Biol 2013; 3(2): 430-6.
25. Verstraete W, Van Vaerenbergh E. Aerobic activated sludge. Biotechnology 1986; 8: 43-112.
26. Wiesmann U. Biological nitrogen removal from wastewater. Adv Biochem Eng Biotechnol 1994; 51: 113-54.
27. Tchobanoglous G, Burton FL, Stensel HD. Wastewater Engineering Treatment and Reuse. 4th ed. Boston: McGraw-Hill; 2003.
28. Ni SQ, Sung S, Yue QY, Gao BY. Substrate removal evaluation of granular Anammox process in a pilot-scale upflow anaerobic sludge blanket reactor. Ecol Eng 2012; 38(1): 30-6.
29. Vymazal J. The use constructed wetlands with horizontal sub-surface flow for various types of wastewater. Ecol Eng 2009; 35(1): 1-17.
30. Kumar NS, Min K. Phenolic compounds biosorption onto Schizophyllum commune fungus: FTIR analysis, kinetics and adsorption isotherms modeling. Chem Eng J 2011; 168(2): 562-71.
31. Hammer MJ. Water and Wastewater Technology. New Jersey: Prentice Hall Inc; 2008.
32. Vymazal J. Plants used in constructed wetlands with horizontal subsurface flow: a review. Hydrobiologia 2011; 674(1): 133-56.
33. Seswoya R, Zainal MY. Subsurface-flow constructed wetland: proposed design area for high strength effluent domestic wastewater. International Conference on Science and Technology Application in Industry & Education; 2010.
34. Chang JJ, Wu SQ, Dai YR, Liang W, Wu ZB. Treatment performance of integrated vertical-flow constructed wetland plots for domestic wastewater. Ecol Eng 2012; 44: 152-9.
35. Saeed T, Sun G. A comparative study on the removal of nutrients and organic matter in wetland reactors employing organic media. Chem Eng J 2011; 171(2): 439-47.
36. Saeed T, Sun G. Enhanced denitrification and organics removal in hybrid wetland columns: comparative experiments. Bioresour Technol 2011; 102(2): 967-74.
37. Tee HC, Lim PE, Seng CE, Nawi M. Newly developed baffled subsurfaceflow constructed wetland for the enhancement of nitrogen removal. Bioresour Technol 2012; 104: 235-42.
38. Lansing SL, Martin JF. Use of an ecological treatment

system (ETS) for removal of nutrients from dairy wastewater. Ecol Eng 2006; 28(3): 235-45.
39. Lavrova S, Koumanova B. Influence of recirculation in a lab-scale vertical flow constructed wetland on the treatment efficiency of landfill leachate. Bioresour Technol 2010; 101(6): 1756-61.
40. Stefanakis AI, Komilis DP, Tsihrintzis VA. Stability and maturity of thickened wastewater sludge treated in pilot-scale sludge treatment wetlands. Water Res 2011; 45(19): 6441-52.
41. Herouvim E, Akratos CS, Tekerlekopoulou A, Vayenas DV. Treatment of olive mill wastewater in pilot-scale vertical flow constructed wetlands. Ecol Eng 2011; 37(6): 931-9.
42. Leverenz HL, Haunschild K, Hopes G, Tchobanoglous G, Darby JL. Anoxic treatment wetlands for denitrification. Ecol Eng 2010; 36(11): 1544-51.
43. Karapinar Kapdan I, Aslan S. Application of the Stover–Kincannon kinetic model to nitrogen removal by Chlorella vulgaris in a continuously operated immobilized photobioreactor system. J Chem Technol Biotechnol 2008; 83(7): 998-1005.
44. Verschueren K. Handbook of Environmental Data on Organic Chemicals. John Wiley & Sons; 2009.

Dairy wastewater treatment plant in removal of organic pollution

Nammam Ali Azadi[1,2], Reza Ali Falahzadeh[3], Shahram Sadeghi[4*]

[1]Kurdistan Research Center for Social Determinants of Health, Medical School, Kurdistan University of Medical Sciences, Sanandaj, Iran
[2]Assistant Professor, Department of Epidemiology and Biostatistics, Medical School,Kurdistan University of Medical Sciences, Sanandaj, Iran
[3]PhD Student, Department of Environmental Health Engineering, School of Public Health, Shahid Sadoughi University of Medical Sciences, Yazd, Iran
[4]Student Research Committee, Kurdistan University of Medical Sciences, Sanandaj, Iran

Abstract
Background: Wastewater produced by a dairy in Sanandaj is a major source of environmental pollution threatening the city. The dairy uses activated sludge treatment to remove organic pollution from the wastewater. The present study evaluated the performance of this process and its compliance with national requirements for chemical oxygen demand (COD), biochemical oxygen demand (BOD), and total suspended solids (TSS) remaining in the plant effluent.
Methods: A total of 48 samples were obtained from the dairy inflow and outflow. The COD, BOD, and TSS were measured for each sample. The statistical sign test was used to assess the standards.
Results: The results showed that the average BOD, COD and TSS in the input wastewater was 292.25, 422.92, and 198.33 mg/l, respectively. The ratio of BOD/COD was 0.69, which indicates the capacity of biological treatment was high. The BOD decreased to 64.22 mg/l (92% removal), COD to 33.74 mg/l (92% removal), and TSS to 43.11 mg/l (94% removal) in the effluent, indicating significant removal of water contaminants. The statistical sign test showed that TSS ($P<0.0001$) and BOD ($P=0.031$) were incompliance with national standards, but COD exceeded standard threshold ($P=0.076$).
Conclusion: Activated sludge treatment showed a good performance for TSS removal, but was not reliable for removal of BOD and COD pollutants.
Keywords: Activated sludge system, Dairy wastewater, Sanandaj, Wastewater treatment, BOD, COD, TSS

***Correspondence to:**
Shahram Sadeghi
Email: Shahram.snna@yahoo.com

Introduction

Urbanization has increased industrial activity and the discharge of industrial wastewater into surface water. Water pollution decreases the quality of water for agricultural purposes and limits potable water supplies (1,2). Wastewater is water that has been contaminated and cannot be used for consumption. Sewage is waste transported by water from residential, office, commercial and industrial areas. Depending on the conditions, it is possible for groundwater, surface water and flood to be mixed (3). If untreated wastewater accumulates, partially decomposed organic materials can produce large quantities of malodorous gases (4).

Industrial wastewater treatment is one step necessary for elimination of environmental impacts (5). Physical, chemical and biological processes can be used for wastewater treatment. Biological processes are used extensively for treatment of wastewater with high organic loads. The oldest and most common biological process for sewage treatment is activated sludge (6,7). Activated sludge system is used in slaughterhouses and the dairy industry because it offers simple technology and high efficiency for removal of organic materials (8).

Milk and dairy industry wastewater usually contains ammonium (from amino acids) and phosphate (from casein) and is an eutrophication factor in the environment. Wastewater treatment facilities are expected to control these resources at in accordance with national standards. The parameters used to evaluate the performance of treatment plants are biochemical oxygen demand (BOD), chemical oxygen demand (COD) and total suspended solids (TSS) in the effluent (9,10).

Several studies have examined the performance of activated sludge system and have arrived at different conclusions about the effectiveness of this system. Zazouli et al (11) evaluated the performance of an activated sludge wastewater treatment system in Golestan province. The efficiency of this system for removal of BOD, COD and TSS from effluent reported to be 97%. They stated that effluent quality of 3 contaminants was in accordance with standards at most times of year; however, they did not report the standards and they did not apply statistical tests to support their claim.

Faizi et al (12) examined the performance of activated sludge for treatment of wastewater from the Sanandaj slaughterhouse and showed very weak performance of the system. They found that the mean level of all contaminants in the effluent was unexpectedly far above national standards. They reported that the mean level of BOD was 517.5 mg/l (487.5 mg/l greater than BOD standard of 30 mg/l) and of COD was 671.71 mg/l (227.35 mg/l greater than COD standard of 60 mg/l). The main source of industrial wastewater in the city of Sanandaj is a dairy that employs an activated sludge system. The present study investigated the performance of this system and its compliance with national standards.

Methods

In this study, 48 wastewater samples from treatment system inflow (24 samples) and outflow (24 samples) were collected. Figure 1 shows the Sanandaj dairy that uses activated sludge in its wastewater treatment plant. This plant comprises a screen for the influent, a primary settling chamber, an aeration basin, a secondary settling chamber, and a disinfection method that uses chlorine as a disinfectant. The sampling plan was collection of 2 samples per week (one from influent and another from effluent) over 6 months.

As an indicator of wastewater pollution, the COD, BOD, and TSS were measured in all samples. The open reflux method was applied to COD measurement and aqualytic BOD meter (model AL606, Germany) was used for measurement of BOD. TSS was measured by the gravimetric method. Potassium dichromate, mercury sulfate, silver sulfate, sulfuric acid, ferrous ammonium sulfate and Freon manufactured by Merck (Germany) used for COD measurement. Experimentation was conducted in accordance with standard methods (13). The national standards for each pollutant were obtained from the United States Environmental Protection Agency (EPA) (14). R statistical software was used to analyze the data (15).

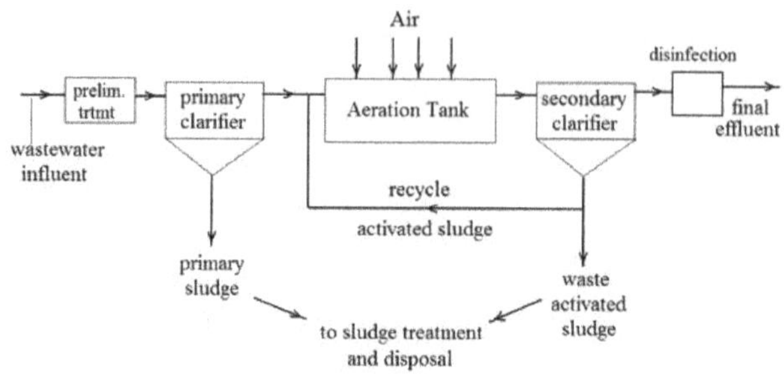

Figure 1. Flow diagram of wastewater treatment plant at Sanandaj dairy.

Results

Table 1 shows the preliminary results of analysis. The mean and standard deviation for COD for treatment inflow was estimated to be 422.92 and 19.89 mg/l, respectively (411.92 ±19.89 mg/l). The COD was the highest in September and at a minimum in October. The mean COD for treatment outflow decreased to just 33.74 mg/l, which indicates 92% removal from the effluent.

Comparison of the inflow and outflow rates for BOD and TSS showed 92% and 94% removal of contaminants respectively. Figure 2 shows the overall level of contaminants in the effluent. The horizontal axis represents the treatment inflow and the vertical axis represents treatment outflow. TSS showed less variation for both inflow and outflow and higher variation was recorded for BOD, especially for treatment outflow. The COD in the effluent is shown to be above the national standard for some periods of the year.

Figure 3 shows the level of wastewater contaminant for each month of the study period. The boxplot measures

Table 1. Mean BOD, COD and TSS from Sanandaj dairy treatment effluent

	TSS[a]		COD[a]		BOD[a]	
	Outflow	Inflow	Outflow	Inflow	Outflow	Inflow
July	8.50	192.12	72.00	224.32	11.90	237.60
August	15.0	190.00	24.34	435.00	35.00	264.30
September	10.0	198.88	30.00	450.00	14.00	300.00
October	12.98	198.00	42.25	430.22	16.00	280.00
November	14.80	200.00	40.56	390.00	28.96	321.11
December	7.50	211.00	61.10	420.00	30.00	350.53
Mean	11.46	198.33	33.74	422.92	22.64	292.25
SD	3.24	7.37	20.17	19.89	94.9	40.44

Abbreviations: COD, chemical oxygen demand; BOD, biochemical oxygen demand; TSS, total suspended solids.
[a] Each the average is the result of 4 measurements at 4 weeks of the month.

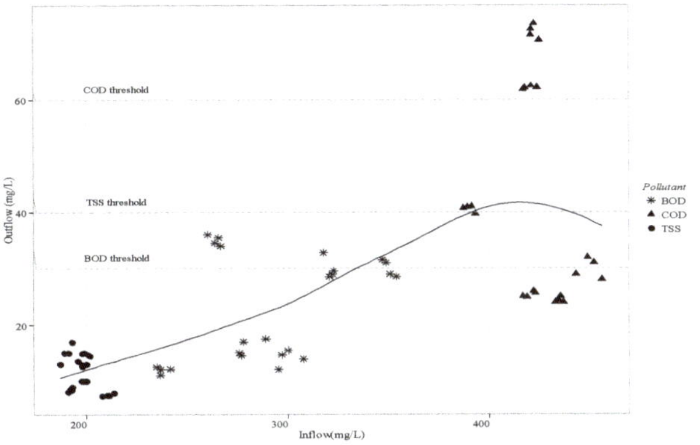

Figure 2. BOD, COD and TSS in Sanandaj in dairy wastewater. The dashed lines denote standard thresholds for each contaminant.

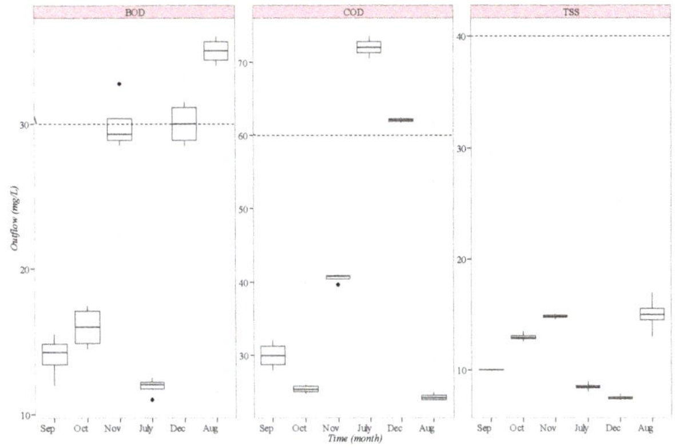

Figure 3. BOD, COD and TSS for Sanandaj dairy effluent per month. The dashed line denotes national standard.

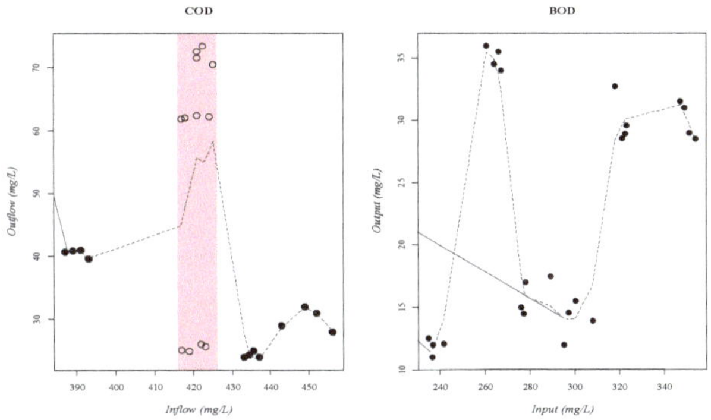

Figure 4. BOD and COD levels in Sanandaj dairy effluent. The dashed line denotes contaminant in a spline smoothing curve. The grey band in the left panel shows plant inflow.

the median and uncertainty around the median (length of box), which helps understand treatment functionality. Figure 3 shows the variation within and between samples during the study period. For example, in December, almost half of the samples were above threshold; in November, only 1 sample did not fall within the national standard range (denoted by a solid point). This sample was much different from others in this month and represented an unusual observation.

Figure 3 shows that TSS in the effluent was consistently low, which was not the case with the other pollutants. It appears that the activated sludge failed to maintain the BOD and COD at standard levels during some months. The level of COD and BOD in the wastewater treatment's outflow are displayed versus treatment's inflow in Figure 4. This Figure demonstrates how well the treatments does remove contaminations from the waste. The dotted lines in the graphs show the general trend of contaminant removal is non-linear. Note the behavior of treatment in removing COD at 420 mg/l inflow (highlighted area) where it is totally unpredictable; while the level of COD was bellow threshold in October (four bottom points in the bin) it was above the nominal allowed level on July and December (8 top points in the bin). This shows that the performance of the plant is unrelated to entrance inflow. Table 2 reports the results of the sign test for whether the treatments effluent met the national norms for each contaminant. Sign test was used since the assumption of normality was disrupted. National norms are given in the Table under the alternative hypothesis. Table 2 shows that the levels of BOD and TSS in the effluent met the norms but the level of COD did not.

Table 2. Comparison of non-parametric test for BOD, COD, and TSS for Sanandaj dairy treatment effluent with national norms

TSS[a]	COD[a]	BOD[a]
H_1:[b] Average is less than 40	H_1: Average is less than 60	H_1: Average is less than 30
P<0.0001	P=0.076	P= 0.031

Abbreviations: COD, chemical oxygen demand; BOD, biochemical oxygen demand; TSS, total suspended solids.

[a] Effluent from treatment plant released to surface water versus national standards.

[b] H_1 is alternative hypothesis.

Discussion

EPA guidelines recommend that COD concentration in treatment plant effluent should be least than 60 mg/l. COD removal efficiency using activated sludge was lowest in July and December and complied with the national standards during August, September, October and November. Input hydraulic shock is one possible reason for the poor performance of the plant in July and December. Vasseghian et al (16) showed that use of a UV reactor can increase COD removal efficiency.

The average output BOD was 22.64 mg/l, which is lower than the 30 mg/l permissible limit for discharge into surface water; however, the average did not fall into the standard range in August. The average BOD was 30 mg/l in December, which suggests that the level of activated sludge was not desirable. Mohapatra et al (17) showed that other treatments, such as advanced oxidation, can be effective for consistent compound removal.

Figure 3 shows the impressively low concentration of TSS in the system effluent for all months of study. These results comply with those from a study by Zazouli et al (11). The thinness of the box in represents the very low rate of output and shows consistent TSS removal. For example, samples taken during different weeks in July ranged from 8.04 to a maximum of 9 mg/l. Some change was also seen in the effluent for COD, but this was outside standards for 2 months of the year. In other words, in 33% of cases, activated sludge was ineffective for COD removal.

Figure 3 showed that the results for activated sludge were random in nature for BOD removal. The BOD level in the effluent was in accordance with national standards in July and August, but did not meet the standard in September and was again within national standards in November and December (slightly higher or lower). The output BOD in 29% of cases was outside national standards (30 mg/l). It can be inferred from Figure 3 that there was a lack of uniform treatment for the removal of all contaminants in each month. The emission concentration of BOD fell within standard range in July. COD did not fall within the standard range in August, but BOD was outside of the standard range. Only September showed all pollutants within national range.

Hassani et al (18) showed that a fixed activated sludge system can increase the performance of persistent material compounds such as heavy metals and fat. Figure 4 shows a lack of a predictable relationship between input and output pollutants. The output concentration of COD was nonstandard for time at 390 mg/l, but was within standard range at 435 mg/l for input at 420 mg/l (highlighted by gray bar). At the same concentrations, the purification system had a dual function. In some cases, the output fell within the standard range (COD output <60 mg/l); at other times, the COD was significantly higher than standard levels. Similar oscillations were also observed for BOD removal; as input BOD increased, the output of the system fell within standard range at times and outside standard range at others.

Table 2 indicates that the performance of the Sanandaj treatment plant was within standard range for removal of TSS and BOD, but was nonstandard for COD. Note that, although the sign test determined that output BOD was within standard range, this should be interpreted with caution. That the results are within standard range. If one unit of standard value decreased and is assumed to be 29 mg/l, the result of the signal test for output BOD does not comply with the standard range ($P > 0.05$). Table 2 showed that the treatment performance for BOD removal fell at the border of standard range in many cases or at a slight distance outside the standard. Porwal et al (19) concluded that the use of some microbial isolates from activated sludge can be effective for treatment of this kind of wastewater. Brennan (20) concluded that the use of an activated sludge wastewater treatment system with a high-rate anaerobic side stream reactor increased efficiency of removal of all biodegradable contaminants.

Conclusion

The levels of BOD, COD and TSS pollutants in the effluent of dairy wastewater at a treatment plant in Sanandaj were measured to study the performance of an activated sludge system. The results showed that this system was successful for removal of TSS and adequate for removal of the other 2 pollutants. Statistical analysis detected that the BOD concentration in the effluent was accordance with national standards, but caution should be used when interpreting these results because half of the samples were recorded at threshold level or slightly outside and 29% of samples fell outside the standard range for effluent BOD for release to surface water (not statistically significant). The performance of activated sludge for COD removal was outside the standard in 33% of cases, which is statistically significant. There was no special relationship between input and output of COD. The following steps are suggested to improve treatment efficiency:

1. Build an equalizer pond in the corner of the main equalizer pond to increase wastewater retention time and improve neutralization and chock control.
2. Require the dairy to perform primary treatment of sewage using storage tanks, dilution the sewage, or use of septic tanks as pretreatment and sand filtration as tertiary treatment.
3. Return the output sludge to the aeration pond.

Acknowledgements

The authors would like to express their gratitude toward the Kurdistan University of Medical Sciences for funding this research.

Competing interests

The authors declare that they have no competing interests.

Author contributions

SS conceived and designed the study. RAF and NA performed the literature search and wrote the manuscript. All authors participated in the data acquisition, analysis and interpretation. All authors critically reviewed, refined and approved the manuscript.

References

1. Sartaj M, Fathollahi F, Philizade Y. Investigation of pollutant sources and purification potential in Anzali talab. First National Conference in Civil Engineering; Tehran; 2005. [In Persian].
2. Khushwaha JP, Srivastava VC, Mall ID. An overview of various technologies for treatment of dairy wastewaters. Crit Rev Food Sci Nutr 2011; 51(5): 442-52.

3. Metcalf & Eddy. Wastewater Engineering: Treatment and Reuse. 4th ed. New Delhi: McGraw Hill; 2003.
4. Pescod MB. Wastewater treatment and use in agriculture. Irrigation and Drainage. Rome: FAO; 1992.
5. Alleman JE. The history of fixed-film wastewater treatment systems. http://web.deu.edu.tr/atiksu/ana52/biofilm4.pdf.
6. Novotny V, Imhoff KR, Olthof M, Krenkel PA. Karl Imhoff's Handbook of Urban Drainage and Wastewater Disposal. John Wiley & Sons; 1989.
7. Avcioglu E, Orhon D, Sözen S. A new method for the assessment of hetrothrophic endogenous respiration rate under aerobic and anoxic condition. Water Sci Tech 1998; 38(8-9): 95-103.
8. Dignac MF Urbain V, Rybacki D, Bruchet A, Snidaro D, Scribe P. Chemical description of extracellular polymers: implication on activated sludge floc structure. Water Sci Tech 1998; 38(8-9): 45-53.
9. Melidis P, Vaiopoulou E, Aivasidis A. Development and implementation of microbial sensors for efficient process control in wastewater treatment plants. Bioprocess Biosyst Eng 2008; 31(3): 277-82.
10. Cirja M, Ivashechkin P, Schäffer A, Corvini PF. Factors affecting the removal of organic micropollutants from wastewater in conventional treatment plants (CTP) and membrane bioreactors (MBR). Rev Environ Sci Biotechnol 2008; 7(1): 61-78.
11. Zazouli MA, Ghahramani E, Ghorbanian AlahAbad M, Nikouie A, Hashemi M. Survey of activated sludge process performance in treatment of Agghala industrial town wastewater in Golestan province in 2007. Iran J Health Environ 2010; 3(1): 59-66. [In Persain].
12. Faizi N, Ghafori S, Daihim M, Ghahramani E. Survey of activated sludge process performance in treatment of butchery in Sanandaj. 3th National Conference on Environmental Health; 2012. [In Persian].
13. American Public Health Association. Standard methods for the examination of water and wastewater. 20th ed. Washington, DC: American Public Health Association, 1998.
14. Gholami M, Sabzali A, Dehghani Fard E, Mirzaei R, Motalebi D. Comparative study of SMBR and extended aeration activated siudge processes in the treatment of strength wastewaters. Iran J Health Environ 2011: 4(3); 255-66. [In Persian].
15. Team RC. R: A Language and Environment for Statistical Computing. Vienna, Austria: R Foundation for Statistical Computing; 2013.
16. Vasseghian Y, Ahmadi M, Dolati F, Heydari A. Modeling and experimental prediction of wastewater treatment efficiency in oil refineries using activated sludge process. Journal of Chemical and Petroleum Engineering 2014; 48(1): 69-79.
17. Mohapatra DP, Brar SK, Tyagi RD, Picard P, Surampalli RY. Analysis and advanced oxidation treatment of a persistent pharmaceutical compound in wastewater and wastewater sludge-carbamazepine. Sci Total Environ 2014: 470-471: 58-75.
18. Hassani AH, Hossenzadeh I, Torabifar B. Investigation of using fixed activated sludge system for removing heavy metals (Cr,Ni and Ph) from industrial wastewater. J Environ Stud 2010; 36(53): 22-4.
19. Porwal HJ, Mane AV, Velhal SG. Biodegradation of dairy effluent by using microbial isolates obtained from activated sludge. Water Resources and Industry 2015; 9: 1-15.
20. Brennan A. Investigating pilot scale performance of an activated sludge wastewater treatment system with a high rate anaerobic side stream reactor. Environmental & Water Resources Engineering Masters Projects; 2012. http://scholarworks.umass.edu/cee_ewre/56

8

The role of soil layers in preventing ground water pollution with 17ß-estradiol hormone (E_2)

A'zam Golzari[1], Mohammad Mehdi Taghizadeh[2*]

[1]MSc Student in Environmental Engineering, Department of Environmental Engineering, Estahban Branch, Islamic Azad University, Estahban, Iran
[2]Assistant Professor, Department of Environmental Engineering, Estahban Branch, Islamic Azad University, Estahban, Iran

Abstract
Background: Estrogens include estoril (E_3), estradiol and estrone (E_1). These chemicals are produced in human and animal bodies as well as in synthetic chemicals (drugs). Estrogens can enter water sources in different ways. When these chemicals enter the human body through water and wastewater, they have the ability to mimic or disrupt the normal estrogen activities in humans and animals. Estrogens in wastewater are able to pass soil layers and contaminate groundwater. Therefore, in this study, the removal of the hormone 17ß-estradiol (E_2) as a representative of estrogens in three types of soils was studied. The selection was chosen in respect to the importance of entering the hormone into groundwater through the soil.
Methods: This study was an experimental study in which the removal of the hormone E_2 from different depths of three types of soils was experimented. The soils were consisted of two different textures, the silty sandy clay and the silty sand with gravel. The hormone E_2 was diluted and injected into the drilled holes. Soils were characterized in the soil mechanics laboratory. Hormone extraction from the soils was performed using a centrifuge and analyzed with the Elecsys device. The results were analyzed using the IBM SPSS version 22 software.
Results: The results showed that the removal rates of hormone E_2 in the three types of soils were higher than 99.5%, and the removal rate in the silty sand was more than the others. In all three soil samples, the removal rates in the first layer were high. The average injected hormone in the soil decreased from 3500 to 3112 ng/l. The results showed that the adhesion and plasticity of the soil had also affected the removal rates.
Conclusion: Results showed that the soil plays a significant role in the removal of E_2 hormone and this hormone was reduced or eliminated in the first layers of the soils. Thus, the risk of groundwater contamination is low.
Keywords: Estrogen, 17ß-estradiol hormone, Silty sandy clay, Silty sand with gravel

***Correspondence to:**
Mohammad Mehdi Taghizadeh
Email: mmt@yahoo.com

Introduction

Hormones are a class of chemicals which are produced in human and animal bodies which regulate some main activities. These materials exist in the body at very low levels and any decrease or increase can have many adverse effects on the body's natural status, which may result in the development of some important diseases in human. In fact, estrogen is not the name of a particular hormone and it refers to a class of hormones which includes estoril (E_3), estradiol (E_2) and estrone ($E_{1)}$. Estrogens are mainly made in the ovaries and are known as gynaecic hormones. Men produce a small amount of this hormone in their testicles. Estrogens are responsible for the growth and extension of the female reproductive system, creation of gynaecic attributes, increase of ossification activities, distribution of body fats in gynaecic form, skin softness, elegance and play a very important role during pregnancy and parturition. They enhance the beauty of women and draw attention to them (1).

Besides the human body, estrogens can be found in plants, synthetic chemicals, some pesticides like D.D.T, some medicinal combinations, detergents, plastics, ornamental materials, industrial materials and heavy metals like mercury, lead and cadmium (1,2). Estrogens entered the human body through water or wastewater have the capability to mimic or disturb the activity of natural existed estrogens in the body which have very dangerous consequences on human health such as testicle disorder syndrome, endometriosis, reduction of sperm number and very unfavorable complications on regeneration (infertility, cancers and metamorphosis) (2).

The appearance of gynandrous in fish and other aquatic

animals, the creation of unfavorable effects on the fertility of fish and mollusks are among the adverse effects of estrogens on aquatic animals (2,3). Besides the above unfavorable problems, another concern about estrogens is difficulty in specifying a threshold for their effects on more vulnerable populations (2).

Filtration methods such as activated sludge can be applied for removal of estrogens from wastewater resulted from refineries. Some filtration methods have removal efficiencies higher than 85% (2,4).

Studies have shown that treatment with chlorine and ozone are able to remove 80% to 95% and 95% to 99% of estrogen from water respectively (5).

Studies revealed that soil properties such as soil type and particle size have had effective roles on removal rates of impurities (6-9).The lateral flows of groundwater do not have any effects on the lateral relocation of any materials, but reinforce downward relocation (10). Soil depth is among parameters which affect material removal from soil (11). Parameters like volatility, contact surface with air, oxygen rate, soil porosity and moisture, primary density of impurity and atmospheric conditions, contact time and surface to depth ratio are among the factors affecting removal rates (12-15). The biologic removal of estrogens from soil is usually accomplished through the activity of bacterial or microbial consortiums and in some cases fungi and algae. Bacterial degradation is performed under aerobic and anaerobic conditions by iron, nitrate and sulfate bacteria (4). Biofilm layers also have the ability to remove E_2 from wastewater (16). It has been shown that soil augmentation with Bacillus sp. PS11 can improve the removal efficiency by 20% (7). Organic materials (humic) present in the soil can hinder the mobility of estrogenic hormones and cause reduction of their presence and toxicity (17-19). The oxidation of soil organic material can also be effective on soil removal ability (20). The types of pollutants present in effluents or wastewaters are effective on estrogen relocation in soil (21). The soil senectitude phenomenon is another mechanism of removing estrogens from the soil. During senescing, some processes such as sorption to soil particles, being stockpiled in the spot and penetration to regions of soil which are unavailable (micropores) and stranding in soil organic material occur (14).

E_2 degradation can be similar to first-order reactions which lie in the range of 0.006-0.6 h^{-1} with relative factors and exist in testing materials with an average of 0.026 h^{-1}. The calculated half-life of E_2 has an average of 47.6 h in soil samples and 39.8 h in samples from soil colloids (17). In the environmental studies of estrogenic analyses of soil samples, there are urgent needs for accurate, rapid, comprehensive and sensitive distinction methods, since these pollutants are present in very low amounts and can interfere with a few analyses. The extraction of estrogens from soil samples is a difficult task because these combinations have a low solubility in water and are nearly hydrophobic combinations. Estrogens strongly stick to soil and their separation rarely occurs (14). Due to these reasons, few assessment studies have investigated the level of estrogens in soil and their removing roles. All studies have been accomplished using high-performance liquid chromatography in their analysis. This study was performed with the aim of investigating the rate of 17β-estradiol (E_2) hormone removal from soil, using the human estradiol hormone recognition device, Elecsys.

Methods

This research was an experimental study in an open-air location. Hormone E_2 was used as a representative of estrogens. It was selected because of availability, low price and the existence of a measurement device for this hormone in medical recognition laboratories. Estradiol was diluted with diethyl ether solvent and reached a density of 3500 ng/L (14). This value is considered equal to the utmost density of estradiol extracted from surface runoffs (2). 10 cc of diluted hormone was poured into falcon tubes. Then three regions with different soils were selected and in each region, one part was chosen randomly and zoned. Thereafter, four holes were drilled in each region for more reliability and accuracy of results and statistical analysis. Four parts in each region were selected randomly, and some holes with diameter of 10 cm and depth of 60 cm were drilled (22). The type and characterization of each region's soil were determined in a soil mechanics laboratory. Soil saturation was performed by Lofran's method (23). After saturation and before hormone injection, a 10 g sample of soil from drilled holes was collected for determining the presence or lack of estradiol in soil samples and diluted estradiol hormone was injected into the holes. Samples were taken after 48 hours from the bottom of each hole in 5 cm distances from the bottom, i.e. depths of 65, 70, 75, 80, 85 and 90 cm (depths 0, 5, 10, 15, 25, and 30 cm of bottom wells) up to 10 g from each depth for hormone analysis (14,17). Collected samples were completely dried in environmental temperature and in the laboratory, each 10 g soil sample was added to a pipe containing 10 mg diethyl ether. Thereafter, samples were spun for 5 minutes and centrifuged at 150 rpm for 20 minutes, then the organic layer was collected and passed through a filter paper (14). The hormone extracted from the soil was analyzed using a measurement device of estradiol hormone of Elecsys mark which is an extremely sensitive device for human blood estradiol measurement. To test for device sensitivity, a diluted sample of 3500 ng/L was analyzed by the device which showed 2499 ng/L, the accuracy of this measurement system is about ±5 ng/L. For more insurance from analysis results, some samples were randomly analyzed twice and the device showed the same results. All obtained results were analyzed by IBM SPSS version 22 software.

Results

The results of soil mechanics experiments showed that experimented soils included a type of silty sandy clay (CL-ML1) with Soil liquidity limit (LL) = 24 and Soil plasticity limit (PI) = 4, a type of silty sand with the gravel (SM) with LL= 24 and PL = 3 and another type of silty sandy clay (CL-ML2) with LL= 27 and PI= 6. Table 1 shows the results of sample analysis after statistical analysis.

Table 1 shows the average and standard deviation (SD) values, along with minimum and maximum observed values and confidence interval of 95% for E_2 removal in three types of experimented soils. As noted in the amounts of Table 1, silty sand with gravel (SM) with an average of 3365.25 (ng/L), by comparing to other types, had higher estradiol removal rates and the rates obtained in this type of soil showed a less deviation from the data average. Figures 1 and 2 show the results for intuitional perception.

In order to compare the average E_2 hormone absorption level in three studied soils with the assumption of its distribution normality in each group and performance of Levene test for investigating equality and homogeneity of variances, the unilateral variance analysis test was used which is explained in Table 2. Based on the results of Table 2, no significant difference was observed between average E_2 hormone absorption level in three types of studied soils (F [2, 18] =1,936, $P>0.05$).

Table 3 shows the average and SD values, along with minimum and maximum observed values and CI of 95% for E_2 removal in various depths of the soil.

In order to compare the average E_2 hormone absorption levels in various depths of the soils and performance of long test for investigating equality and homogeneity of variances, the unilateral variance analysis test was used and its result is explained in Table 4. Based on the results of Table 4, there is a significant difference between average E_2 hormone absorption level in various depths of soils (F [6,14] = 6, 290, $P<0.05$).

The Dunnett test was performed in two by two comparisons of various depths with the final tested depth. The results of these tests are presented in Table 5 and groups which have significant difference with each other are marked with star mark. Based on the results obtained from Dunnett test, a significant statistical difference is observed between E_2 hormone removal rate at the well bottom and at 5 cm with 30 cm depth ($P<0.05$, but there was no significant difference in comparing other groups with the final studied depth ($P>0.05$).

Figure 3 has also been provided for comparing the effect of soil depth on estradiol removal rate and indicates an increase of removal rate with depth increase in three types of experimented soils.

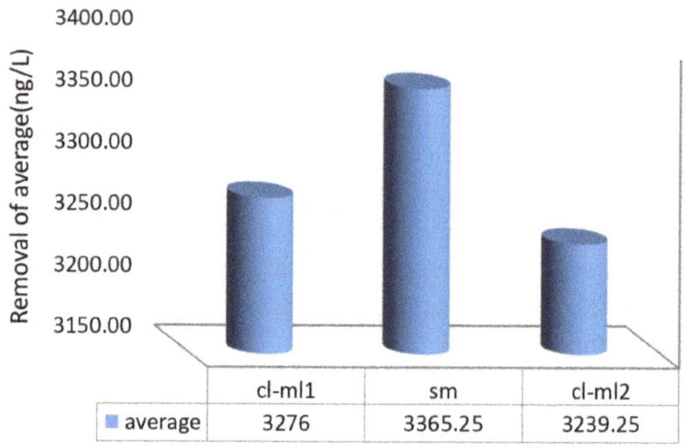

Figure 1. Estradiol removal average in three soil samples.

Discussion

This research investigated the effect of depth, type and properties of soil on estrogen removal rate. The obtained results showed that the soil has a very effective role in the removal of E_2 hormone and increase in penetration depth causes an increase in removal. The physical and chemical properties of soil such as particle size, plasticity limit and adhesion rate are among other effective parameters studied in this research. It was also shown that the removal average increases with increase in soil size; hence, soil samples with higher percentages of gravel and silt had higher penetration rates. Adhesion and plasticity limit in two types of soils with similar physical properties are among the factors affecting penetration rate. In this study, two types of silty sandy clay, the sample with higher adhesion and plasticity limit showed higher penetration rate. Ebrahimi et al (11) in their study investigated hydrocarbon pollution emission, they concluded with increase in penetration depth, the pollution rate rapidly decreases and this observation was also clearly observed in the present study.

Gitipoor et al (24) in their study concluded that clay soils due to the special properties like high surface to volume ratio and high swell and penetration quality have always

Table 1. Values of descriptive indexes for estradiol removal in three types of soil (ng/L)

Type of soil	N	Mean	SD	SE	95% CI for mean		Minimum	Maximum
					Lower bound	Upper bound		
CL-ML1	7	3276.0000	127.83787	48.31817	3157.7697	3394.2303	3059.00	3451.00
SM	7	3365.2500	80.63937	30.47882	3290.6710	3439.8290	3291.75	3451.00
CL-ML2	7	3239.2500	150.64915	56.94003	3099.9228	3378.5772	2985.50	3414.25
Total	21	3293.5000	128.84564	28.11642	3234.8502	3352.1498	2985.50	3451.00

Table 2. Variance analysis test result in comparing the average estradiol removal in three types of soil (ng/L)

Source change	Sum of Squares	df	Mean Square	F	P value
Between Groups	58781.625	2	29390.813	1.936	0.173
Within Groups	273242.375	18	15180.132		
Total	332024.000	20			

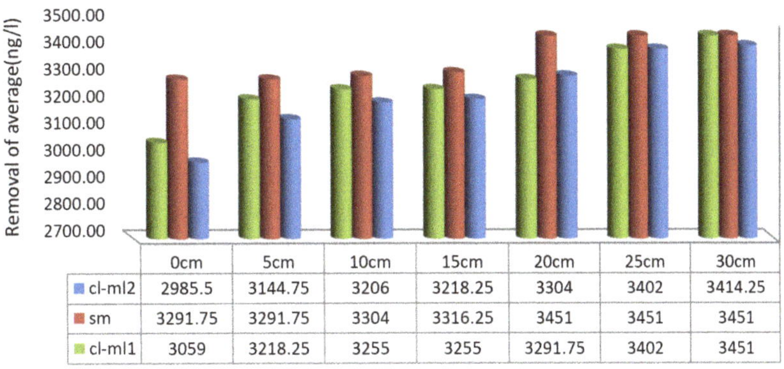

Figure 2. Comparison of Estradiol removal average in three soil samples.

Table 3. Levels of descriptive indexes for Estradiol removal in various depths of soil (ng/L)

Depth (cm)	N	Mean	SD	SE	95% CI for mean		Minimum	Maximum
					Lower bound	Upper bound		
0	3	3112.0833	159.87697	92.30501	2714.9269	3509.2397	2985.50	3291.75
5	3	3218.2500	73.50000	42.43524	3035.6659	3400.8341	3144.75	3291.75
10	3	3255.0000	49.00000	28.29016	3133.2773	3376.7227	3206.00	3304.00
15	3	3263.1667	49.50779	28.58333	3140.1825	3386.1508	3218.25	3316.25
20	3	3348.9167	88.61868	51.16402	3128.7757	3569.0577	3291.75	3451.00
25	3	3418.3333	28.29016	16.33333	3348.0567	3488.6100	3402.00	3451.00
30	3	3438.7500	21.21762	12.25000	3386.0425	3491.4575	3414.25	3451.00
Total	21	3293.5000	128.84564	28.11642	3234.8502	3352.1498	2985.50	3451.00

Table 4. The result of various analysis test in comparison of average estradiol removal in three types of soil (ng/L)

Source change	Sum of squares	df	Mean square	F	P value
Between Groups	242186.583	6	40364.431	6.290	0.002
Within Groups	89837.417	14	6416.958		
Total	332024.000	20			

Table 5. The results of Dunnett test for two by two comparisons of various depths with the final tested depth

(I) Depth (cm)	(J) Depth (cm)	Mean Difference (I-J)	SE	*P* value	95% CI	
					Lower bound	Upper bound
0	30	-326.67*	65.40	.001	-517.15	-136.17
5	30	-220.50*	65.40	.021	-410.99	-30.00
10	30	-183.75	65.40	.060	-374.24	6.74
15	30	-175.58	65.40	.076	-366.07	14.90
20	30	-89.83	65.40	.577	-280.32	100.65
25	30	-20.41	65.40	.999	-210.90	170.07

been under consideration by environmental researchers for organic pollutant penetration. Among clay soils, bentonite types have special significance due to particular links between cells. Also, in the present study the rates of hormone removal in silty sandy clay soils were high.

In the study of Mahmoodi et al (25), the rate of heavy metals removal from soils with smaller size particles was higher.

Cai et al (9) investigated the role of clay coarseness and softness on the rate of material removal. In this study, the role of soil cohesion on removal rate was investigated.

In study of Zadebafghi et al (15), they investigated parameters affecting the efficiency increase of cyanide natural decay and introduced parameters like volatility, contact surface with materials, soil porosity and moisture, primary concentration and atmospheric conditions among factors affecting cyanide removal. They also concluded that for a distinctive soil volume polluted with cyanide, with increase of depth to surface ration, the amount of cyanide remaining in the soil decreases faster and higher primary density causes an increase of removal efficiency. In another study, Essandoh et al (6) reached the same conclusion. Fan et al (19) showed that E_2 through abiotic chemical processes was converted to a polar unknown combination. Through physical and chemical processes, this hormone was oxidized to E_3. Other results showed that soil organic materials can inhibit the mobility of estrogenic hormones present in soil and as a result reduce their avail-

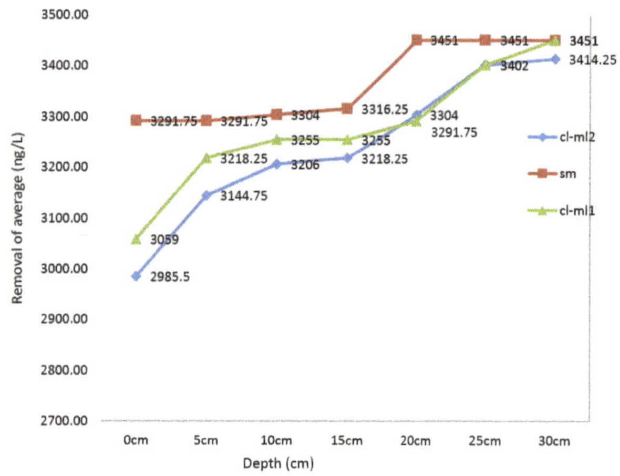

Figure 3. The effect of soil depth on Estradiol removal rate.

ability and toxicity. Based on the above research, we can conclude that estradiol used in this research was removed from the soil as a result of the chemical and physical reactions which occurred in the soil.

Cajthaml et al (4) studied the process of estrogens microbial removal in soil and showed that the process of removing these matters from soil, wastewater and septic is usually accomplished through the activity of bacteria and microbe consortiums and in some cases through fungi and seaweeds. Degradation is performed by bacteria under aerobic and anaerobic conditions and by iron, nitrate and sulfate bacteria and also through adsorption. In the performed study, the destruction of estradiol is likely to be achieved under aerobic and anaerobic conditions. Stanford et al (21) concluded that E_1 and E_2 present in septic effluent entering into the soil can be slaked with removal of surfactants containing phenyl polyethoxylate and total organic carbon. In this study, estradiol was diluted with aquapura and no other effective matter existed.

Chun et al (14) compared three methods of extraction of E_2 in sand, bentonite and organic-rich silt loam and concluded that bentonite and silt loam had the highest efficiency of E_2 extraction with diethyl ether method.

Chun et al (14) found that the extraction of estrogens from soil samples is a difficult task, since these combinations have little solubility and are nearly hydrophobic combinations. Estrogens strongly stick to soil and their separation rarely occurs.

Their final hypothesis was that diethyl ether due to having a polarity like estrogens, probably increases extraction efficiency. In the mentioned research, soil senectitude phenomenon was introduced as one of the probable reasons for E_2 not being extractable from the soil. During this phenomenon, inter soil passive processes included sorption to soil particles, being stockpiled in the spot and penetration to soil regions that are not available (microprose) as well as stranding in soil organics (14).

Based on this research, diethyl ether was used for hormone extraction from soil and the obtained result showed that the amount of extracting hormone from experiments soils is very low. Prater et al (17) declared the average half-life of E_2 removal as 47.6 h in soil samples and 39.8 h in samples of soil colloids.

Conclusion

Based on the results of the present study, soil plays an effective role in removing the E_2 hormone and as a result, estrogens which have entered into the soil are reduced or removed in the primary layers of the soil. Therefore, the probability of polluting groundwater in regions where wastewater containing estrogens had passed through is very low.

Acknowledgments

Special thanks to Mr. Abdi for statistical analysis and professors of the Faculty of Pharmacy and Soil Mechanics Laboratory staffs of Kerman province who supported this study.

Competing interests

The authors have no competing interests.

Author's contributions

MMT conceived and designed the study while AZ performed the literature search and wrote the manuscript. Both authors participated in data acquisitions analysis and interpretation. Both authors critically reviewed, refined and approved the manuscript.

References

1. Ryan KJ. Biochemistry of aromatase: significance to female reproductive physiology. Cancer Res 1982; 42(8 Suppl): 3342s-3344s.
2. Wise A, O'Brien K, Woodruff T. Are oral contraceptives a significant contributor to the Estrogenicity of drinking water? Environ Sci Technol 2011; 45(1): 51-60.
3. Wagner M, Oehlmann J. Endocrine disruptors in bottled mineral water total estrogenic burden and migration from plastic bottles. Environ Sci Pollut Res Int 2009; 16(3): 278-86.
4. Cajthaml T, Kresinová Z, Svobodová K, Sigler K, Rezanka T. Microbial transformation of synthetic estrogen 17alpha-ethinylestradiol. Environ Pollut 2009; 157(12): 3325-35.
5. Pereira RO, Postigo C, de Alda ML, Daniel LA, Barceló D. Removal of estrogens through water disinfection processes and formation of by-produts. Chemosphere 2011; 82(6): 789-99.
6. Essandoh HM, Tizaoui C, Mohamed MH. Removal of estrone (E1), 17β-estradiol (E2) and 17α-ethinylestradiol (EE2) during soil aquifer treatment of a model wastewater. Sep Sci Techno 2012; 47(6): 777-87.
7. Djokic L, Narancic T, Biocanin M, Saljnikov E, Casey E, Vasiljevic B, et al. Phenol removal from four different natural soil types by Bacillus sp. PS11. Appl Soil Ecol 2013; 70: 1-8.
8. Shirani H, Kord M, Sayyad GA, Naghavi H. Simulating

bromide transport in disturbed soil columns using Hydrus-1d model. Quarterly Watershed Management Researches (Pajouhesh-va-Sazandegi) 2011; 24(3): 20-31. [in Persian].
9. Cai P, Huang Q, Jiang D, Rong X, Liang W. Microcalometric studies on the adsorption of DNA by soil colloidal particles. Colloids Surf B Biointerfaces 2006; 49(1): 49-54.
10. Kamon M, Endo K, Kawabata J, Inui T, Katsumi T. Two-dimensional DNAPL migration affected by groundwater flow in unconfined aquifer. J Hazard Mater 2004; 110(1–3): 1-12.
11. Ebrahimi S, Shayegan J, Malakuti MJ, Baybordi M, Ghodousi J, Akbari A. Study of Hydrocarbon contamination distribution in around of sarkhun refinery soil. Water and Soil Conservation 2009; 16(4): 101-24. [In Persian].
12. Prater JR, Horton R, Thompson ML. Reduction of estrone to 17 β-estradiol in the presence of swine manure colloids. Chemosphere 2015; 119: 642-5.
13. Czajka CP, Londry KL. Anaerobic biotransformation of estrogens. Sci Total Environ 2006; 367(2–3): 932-41.
14. Chun S, Lee J, Geyer R, White DC. Comparison of three extraction methods for 17beta-estradiol in sand, bentonite, and organic-rich silt loam. J Environ Sci Health B 2005; 40(5): 731-40.
15. Zadebafghi A, Taebi A, Afyuni M. Investigation of the Effective Parameters on the Cyanide Fate in Soi. Journal of Water and Wastewater 2007; 58(17): 44-53. [In Persian].
16. Iasur-Kruh L, Hadar Y, Minz D. Isolation and bioaugmentation of an estradiol-degrading bacterium and its integration into a mature biofilm. Appl Environ Microbiol 2011; 77(11): 3734-40.
17. Prater JR. Impact of environmental colloidson the transport of 17β-estradiol in intact soil cores [Thesis]. Iowa: Iowa State University; 2012.
18. Soares AA, Albergaria JT, Domingues VF, Alvim-Ferraz Mda C, Delerue-Matos C. Remediation of soils combining soil vapor extraction and bioremediation: benzene. Chemosphere 2010; 80(8): 823-8.
19. Fan Z, Casey FX, Hakk H, Larsen GL. Persistence and fate of 17beta-estradiol and testosterone in agricultural soils. Chemosphere 2007; 67(5): 886-95.
20. Romero A, Santos A, Cordero T, Rodríguez-Mirasol J, Rosas JM, Vicente F. Soil remediation by Fenton-like process: Phenol removal and soil organic matter modification. Chem Eng J 2011; 170(1): 36-43.
21. Stanford BD, Amoozegar A, Weinberg HS. The Impact of co-contaminants and septic system effluent quality on the transport of estrogen sand nonylphenols through soil. Water Res 2010; 44(5): 1598-606.
22. Monzavi MT. Wastewater Treatment. 6th ed. Tehran: Tehran University publications; 1996. p. 70-1. [In Persian].
23. Ghanbari A. Site Investigation in Geotechnical Engineering. 3rd ed. Tehran: Pajuheshi-e-Noavaran-e-Sharif publications; 2009. p. 155-62.
24. Gitipoor S, Abolfazlzade M , Hosseinpoor MA. Stady of hydrocarbon phylic specification of modified and ordinary bentonite in conformity with wariation of space between layer in confront with water, benzene, MTBE and BTEX componds. Journal of Environmental Sciences and Technology 2010; 4(47): 11-20. [In Persian].
25. Mahmoudi S Mohammadi J Naderi M. Statistical and local distribution some heavy elements in size classes of surface soil particles of south Esfahan. Journal of Water and Soil Conservation 2013; 20(2): 1-2. [In Persian].

9

Optimizing functions of coagulants in treatment of wastewater from metalworking fluids: Prediction by RSM method

Hamzeh Ali Jamali[1], Maryam Moradnia[2,3*]

[1]Department of Environmental Health Engineering School of Health, Qazvin University of Medical Sciences, Qazvin, Iran
[2]Department of Environmental Health Engineering, School of Health Isfahan University of Medical Sciences, Isfahan, Iran
[3]Health Network of Divandareh, Kurdistan University of Medical Sciences, Sanandaj, Iran

Abstract
Background: Coagulation is a treatment procedure for metalworking fluids (MWFs). This study aimed to optimize coagulation using four coagulants and compare the results.
Methods: In this research, the outputs of a coagulation procedure in chemical oxygen demand (COD) removal, turbidity, and the release of oil were investigated using four coagulants, ferric chloride ($FeCl_3$), ferric sulfate ($Fe_2(SO_4)_3$, calcium chloride ($CaCl_2$), and aluminum sulfate ($Al_2(SO4)_3$. To optimize the process, a central composite design (CCD) and response surface methodology (RSM) were used.
Results: Under optimal conditions for using calcium chloride (coagulant concentration 4.2 g/L and pH 3.71), COD and turbidity removals and oil released were respectively 93%, 96.9%, and 31.8 mL. COD and turbidity removals and oil released using aluminum sulfate as a coagulant under optimal conditions (5.36 g/L, pH 4.64) were 75.7%, 89.9%, and 28.9 mL, respectively. With respect to ferric sulfate under optimal conditions (6.25 g/L, pH 3.4), COD and turbidity removals were 60% and 84.1%, respectively and oil released was 16.1 ml. COD and turbidity removals and the greatest amount of oil released by ferric chloride under optimal conditions (3.16 g/L, pH 3.5) were 54.4%, 84.7%, and 15.9 ml, respectively.
Conclusion: Compared with other coagulants in the treatment of MWFs, calcium chloride achieved the highest removal efficiency in eliminating pollution indicators.
Keywords: Wastewater treatment, Metalworking fluids, Response surface methodology, Optimization

***Correspondence to:**
Maryam Moradnia
Email:
maryam.moradnia2000@gmail.com

Introduction

Industrial metalworking fluids (MWFs) have several applications in metalworking processes such as machining and lathe operations. While cooling down the manufactured parts and machining tools, these fluids lubricate and remove segregated work piece materials (1). More than 300 substances are employed in the production of different types of MWFs. The microbial degradation of water-miscible MWFs creates a wide range of problems, such as color changes in emulsion, lower quality workpieces, and unsatisfactory functioning of machining equipment. Likewise, high degrees of microbial loading may lead to health problems for the workers. MWFs lose their properties due to prolonged use and must be replaced with fresh MWFs (2). As a result, a large volume of wastewater containing chemical oxygen demand (COD) with high turbidity is generated; this discharge causes environmental risks and needs to be treated to meet discharge-related standards (3). Some treatment methods, such as chemical coagulation (4), and biological processes (5) were used to treat this kind of wastewater. Due to high concentrations of oil and antimicrobial agents, this type of wastewater cannot be thoroughly treated by means of biological processes; however, chemical coagulation is an appropriate procedure for treatment. The proper application of this process depends on careful scrutiny in selecting the amount of coagulant and the pH. Classical methods have some limitations, like being highly time-consuming because of the experiments and the high cost of consumables that can be removed by statistical experimental designs. Furthermore, in this method, it is impossible to examine the interactions among variables (6,7). Response surface methodology (RSM) is a proper approach that also omits these limitations.

This study was conducted in 2014 at the School of Public Health of Qazvin University of Medical Sciences. It aimed to optimize the treatment of MWFs wastewater with four coagulants, ferric chloride, ferric sulfate, calcium chloride and aluminum chloride (aluminum sulfate), and compare their effectiveness in treatment. The main focus of this study was on optimization and modeling of the response and predictions. The optimization of parameters, modeling of response, and predictions were done by RSM.

Methods

In this study, four coagulant solutions (Merck, Germany), ferric chloride ($FeCl_3$), ferric sulfate ($Fe_2(SO_4)_3$), calcium chloride ($CaCl_2$), and alum ($Al_2(SO_4)_3$, were prepared and used during the study. Coagulation and flocculation processes were conducted by means of a conventional jar-test (made in Iran) apparatus with impellers equipped with six 1.5 cm × 3.5 cm rectangular blades. Jar testing steps include rapid and slow mix with mixing speeds of 90 and 30 rpm, respectively. Times were set at 1 and 20 minutes for each step, respectively, while 30 minutes was considered as settling time. Central composite design (CCD) and RSM were applied to evaluate the relations between three responses (COD removal efficiency, turbidity, and amount of oil released) and two variables (amount of coagulant and pH). Design Expert software (Stat-Ease Inc., Minneapolis, USA) was utilized for designing, statistical modeling, and optimization of the variables influencing the elimination of responses.

Before designing the research, the range of coagulant amounts to use and the pH were obtained through some preliminary experiments. The coagulant amounts ranged from 0.1 g/L to the limit for appropriate reduction efficiency rates for COD and turbidity and increasing the amount of oil released. To find an effective range of pH, a wide range from 2 to 12 was examined.

The ranges of pH and coagulant amounts were pH=2-6 and 3-6 g/L for $FeCl_3$, pH=3-7 and 3-7 g/L for Al_2SO_4, pH=2.5-5.5 and 4.5-6 g/L for Fe_2SO_4, and pH=2-5 and 2.5-5.5 g/L for $CaCl_2$. The maximum, median, and minimum values (-1, -0.5, 0, +0.5, +1) of each variable (including coagulant amounts (A) and pH (B)) were specified based on CCD (Table 1).

In this study, the experimental design for each coagulant was performed based upon the full factorial design as 13 ($2^k + 2k$), where k is the number of factors (=2) and 5 additional experiments were performed as the reproducibility of center point. Totally 52 experiments were carried out for four coagulants.

The second-order model equation used to describe the system behavior can be stated as the following:

$$Y = \beta_0 + \sum_{i=1}^{K} \beta i \cdot Xi + \sum_{i=1}^{K} \beta ii \cdot Xi^2 + \sum_{i i \le j}^{K} \sum_{j}^{K} \beta ij . Xi . Xj + \cdots + e \quad (1)$$

where Y, i, j, β, e, and k are the response variable, linear constant, quadratic constant, regression constant, random error, and number of parameters, respectively.

Analysis of variance (ANOVA) was performed for model selection. Significant model terms were identified at 95% when the modified model was obtained through the elimination of non-significant factors.

For graphic analysis of the data and to determine the interactive effects between the independent variables of the process and responses, ANOVA was performed. The quality of the model fitting was determined by R^2 and adjusted R^2 determination coefficient, and the Fischer test (F-test) was used for statistical significance; $P \le 0.05$ was determined as the significance level.

Following the analysis of experimental data, optimal conditions were determined graphically. Two-dimensional plots for the four coagulants ($FeCl_3$, $Fe_2(SO_4)_3$, $CaCl_2$ and $Al_2(SO_4)_3$) were plotted based on the effects of the two factors (coagulant amount and pH) in five levels. Furthermore, numerical optimization was performed for the four coagulates, selecting "in range" for both factors (coagulant dose and pH) and putting all responses in maximized mode. With numerous responses, the optimum conditions where all variables concurrently met the desirable removal criteria could be plotted graphically by overlaying the response surface contours in an overlay diagram. Graphical optimizations for various parameters were determined by overlaying the contour plots. To

Table 1. Independent variables and experimental levels for response surface study

Independent variable	Coded levels				
	-1	-0.5	0	+0.5	+1
			$FeCl_3$		
A: Dose (g/L)	2.00	3.00	4.00	5.00	6.00
B: pH	2.00	3.00	4.00	5.00	6.00
			$CaCl_2$		
A: Dose (g/L)	2.50	3.25	4.00	4.75	5.50
B: pH	2.00	2.75	3.50	4.25	5.00
			$Fe_2(SO_4)_3$		
A: Dose (g/L)	4.50	5.25	6.00	6.75	7.50
B: pH	2.50	3.25	4.00	4.75	5.50
			$Al_2(SO_4)_3$		
A: Dose (g/L)	4.50	5.25	6.00	6.75	7.50
B: pH	2.50	3.25	4.00	4.75	5.50

confirm the agreement of the results obtained from the model and the real conditions using each coagulant, two extra experiments were performed utilizing the optimal circumstances.

Results

Based on the experimental results, CCD values for the four coagulants are presented in Table 2. According to the obtained results, the highest COD removal efficiency rates of $FeCl_3$, $CaCl_2$, Alum, and $Fe_2(SO_4)_3$ were 56%, 96.1%, 81%, and 67%, respectively. The highest percentages of turbidity removal by the four mentioned coagulants were 98%, 100%, 94%, and 97.1%, respectively. The amounts of oil released were 17.2 mL, 36 mL, 30 mL, and 15 mL, respectively.

The results obtained from ANOVA for the removal of COD and turbidity, the amount of oil released, and the modified models are presented in Table 3.

Using $CaCl_2$ under optimal conditions (coagulant concentration = 4.2 g/L; pH = 3.3), COD and turbidity removal rates and the amount of oil released were 93%, 96.9%, and 31.8 ml, respectively. COD and turbidity removal rates and the amount of oil released by aluminum sulfate as coagulant under optimal conditions (coagulant concentration = 5.36 g/L; pH = 4.64) were 75.7%, 89.9%, and 28.9 ml, respectively. With respect to $Fe_2(SO_4)_3$ under optimal conditions (6.25 g/L, pH 3.4), COD and turbidity removal rates and the amount of oil released were 60%, 84.1%, and 16.1 ml, respectively. COD and turbidity removal rates and the highest amount of oil released by $FeCl_3$ under optimal conditions (3.16 g/L and pH 3.5) were 54.4%, 84.7%, and 15.9 ml, respectively. The desirability of the models for each coagulant was found to be 93.6%, 87.6%, 91.2%, and 85%, respectively.

The contour plots for the effective factors in reducing COD and turbidity and increasing the amount of oil released by the four coagulants are plotted in Figures 1 to 3.

By overlaying the four contour plots, the optimum removal efficiencies for COD and turbidity, and the amount of oil released were obtained and are shown in Figure 4.

Discussion

The lack of fit (LOF) defines the variation of data around the fitted model. If the model does not fit the data well, this test will be significant (8). In this study, the lack of fit test was not significant for any of the responses, indicating a good fit of the data to the model.

The coefficient of determination (R^2) expresses the quality of the quadratic polynomial model fit. This parameter shows the ratio of the sum of squares of regression (SSR)

Table 2. CCD values for the four coagulants

Run	Parameter		Responses			Run	Parameter		Responses		
	pH	Dose (g/L)	COD (%)	Turbidity (%)	Oil (mL)		pH	Dose (g/L)	COD (%)	Turbidity (%)	Oil (mL)
$FeCl_3$						$Al_2(SO_4)_3$					
1	4.00	3.00	51.00	84.40	14.40	1	5.50	5.00	62.50	94.00	24.00
2	4.00	4.00	50.00	90.00	14.00	2	4.00	3.00	20.00	30.00	5.00
3	2.00	6.00	28.00	52.50	7.00	3	5.00	6.00	65.30	80.00	28.00
4	4.00	4.00	48.00	94.00	12.00	4	5.00	5.00	78.30	92.00	29.00
5	2.00	2.00	56.00	50.00	17.20	5	5.00	5.00	78.00	89.00	29.00
6	6.00	2.00	16.00	56.00	3.00	6	4.00	7.00	58.80	58.00	25.00
7	4.00	4.00	49.30	89.00	13.50	7	6.00	7.00	20.00	49.00	3.00
8	4.00	4.00	50.00	88.00	15.00	8	4.50	5.00	68.50	88.00	30.00
9	4.00	4.00	54.50	98.00	16.00	9	5.00	4.00	66.50	83.00	21.00
10	5.00	4.00	35.50	85.00	9.80	10	6.00	3.00	25.40	51.50	5.00
11	3.00	4.00	50.50	79.00	15.00	11	5.00	5.00	64.10	91.00	17.00
12	6.00	6.00	14.00	66.10	3.00	12	5.00	5.00	81.00	90.00	27.00
13	4.00	5.00	40.50	86.00	13.00	13	5.00	5.00	80.60	93.00	27.00
$CaCl_2$						$Fe_2(SO_4)_3$					
1	3.50	4.00	90.50	94.00	30.00	1	4.00	6.00	67.00	97.10	15.00
2	3.50	4.00	89.00	94.00	29.00	2	2.50	4.50	24.50	38.00	5.40
3	3.50	4.75	88.00	93.00	27.00	3	3.25	6.00	54.00	80.00	12.10
4	4.25	4.00	90.00	96.00	30.00	4	4.00	6.00	65.00	97.00	14.00
5	2.75	4.00	94.00	96.00	28.00	5	4.00	6.00	61.00	97.00	12.50
6	3.50	4.00	96.10	100	36.00	6	5.50	4.50	19.00	30.00	4.50
7	3.50	4.00	92.60	98.00	32.00	7	4.00	5.25	49.00	81.00	11.40
8	3.50	3.25	89.00	90.00	29.00	8	4.00	6.00	56.00	95.00	13.00
9	5.00	5.50	67.00	80.00	13.00	9	5.50	6.75	53.00	85.50	13.50
10	5.00	2.50	46.40	62.00	1.00	10	2.50	7.50	25.00	35.00	5.20
11	2.00	5.50	60.50	80.00	14.00	11	5.50	7.50	51.50	61.00	10.20
12	3.50	4.00	91.00	94.00	31.00	12	4.75	6.00	59.00	86.00	12.40
13	2.00	2.50	65.20	75.00	6.00	13	4.00	6.00	63.00	97.10	14.00

Table 3. ANOVA results for removal of COD and turbidity and the amount of oil released using the four coagulants

Response	Final equation in term of code factor	P	PLOF	R^2	Adj.R^2	AP	CV	SD
	$FeCl_3$							
COD	4896/ 83/7– A – 1367/ B + 65/ AB- 1586/B^2	<0.0001	0.30	0.98	0.96	22.56	6.71	2.81
Turbidity	9019/ + 502/ B – 2356/B^2	<0.0001	0.60	0.97	0.94	15.37	4.94	3.87
Oil released	1401/ – 243/A – 462/B - 255/AB – 594/ B^2	<0.0001	0.90	0.96	0.94	17.68	10.15	1.19
	$CaCl_2$							
COD	9288/ + 3 42/A -318/ B + 6933/ AB -2346/ A^2	<0.0001	0.42	0.98	0.97	24.00	3.44	2.80
Turbidity	982/ + 544/ A 89/2- B + 325/ AB- 1996/A^2	<0.0001	0.78	0.97	0.95	21.00	2.72	2.41
Oil released	3151/ + 422/ A	<0.0001	0.40	0.96	0.93	15.20	12.10	2.84
	$Fe_2(SO4)_3$							
COD	6042/ + 778/ A +522/ B + 8 AB 39/26–A^2	0.006	0.43	0.96	0.93	14.56	8.70	4.30
Turbidity	36.72 + A -4.67 B 8.5 + AB 25.74 $A^2$26.74 – B^2	<0.0001	0.11	0.98	0.97	23.90	5.18	3.88
Oil released	13.22 1.45 + A 0.94 +B +1.47 AB 4.02 – B^2	0.0044	0.81	0.97	0.94	16.51	7.45	0.82
	$Al_2(SO_2)_3$							
COD	7424/- 809/B – 1105/ AB	0.0002	0.60	0.94	0.91	12.43	11.10	6.56
Turbidity	9132/ + 533/ A + 344/ B – 762/ AB – 4107/ A^2	<0.0001	0.06	0.99	0.98	32.57	3.65	2.78
Oil released	2665/ + 478/ A – 556/ B – 55/ AB	0.0018	0.91	0.91	0.84	9.21	19.40	4.00

to total sum of squares (SST) (9,10). The high value of R^2 shows the satisfactory agreement of the experimental data with the quadratic model. In all cases of this study, the values of R^2 were found to be higher than 0.9 for all responses; thus, it can be said that the quadratic model has a good fitness for the data.

In other words, AP compares the mean prediction error with the range of values predicted at design points. Ratios above 4 show the adequacy of the models (11). In this research, the AP values for all responses were greater than 4, indicating the high predictive ability of this model in the obtained results.

The coefficient of variation is a measure of the residual variation of the data relative to the size of the mean. It is the standard deviation divided by the dependent mean and expressed as a percentage. The coefficient of variation (CV) is very low, indicating that the experimental values have a very high degree of precision and a good deal of reliability. Typically, a model will be considered as reproducible when its CV value is within the acceptable range of 10% (12). In this study, the values of CV for all responses except the amount of oil released and the removal of COD by alum were not greater than 10%, which demonstrates high precision in the results and the reproducibility of the model.

Figure 4A illustrates an overlay scheme of the optimum condition for $CaCl_2$. According to this diagram, the optimal area for the three responses, COD, turbidity, and the amount of oil released, was a concentration range of 3-4.5 g/L and a pH range of 2.75-4. In this area, the

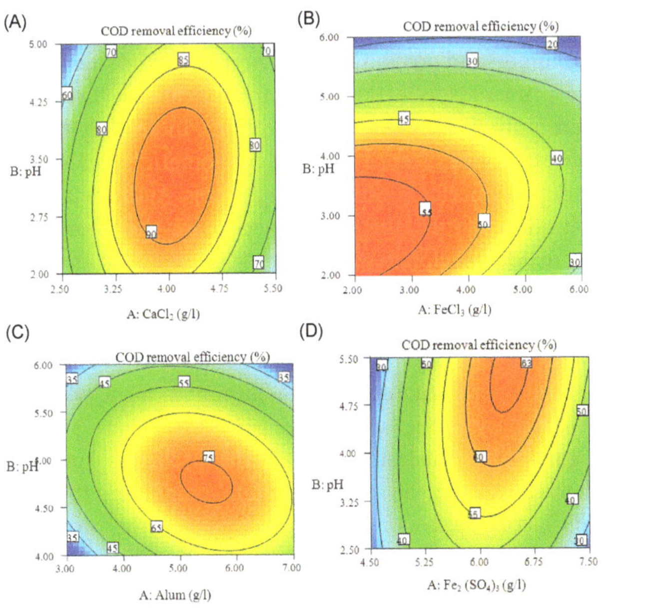

Figure 1. The contour plots for the removal of COD using (A) $CaCl_2$, (B) $FeCl_3$, (C) alum, and (D) $Fe_2(SO_4)_3$.

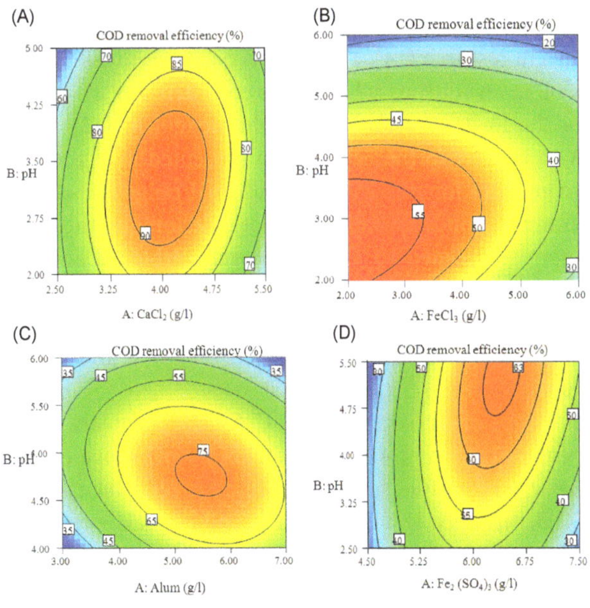

Figure 2. The contour plots for the removal of turbidity using (A) $CaCl_2$, (B) $FeCl_3$, (C) alum, and (D) $Fe_2(SO_4)_3$.

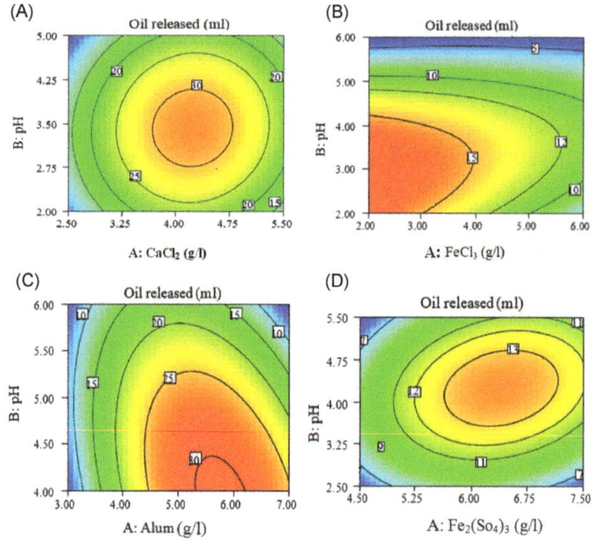

Figure 3. The contour plots for the amount of oil released using (A) $CaCl_2$, (B) $FeCl_3$, (C) alum, and (D) $Fe_2(SO_4)_3$.

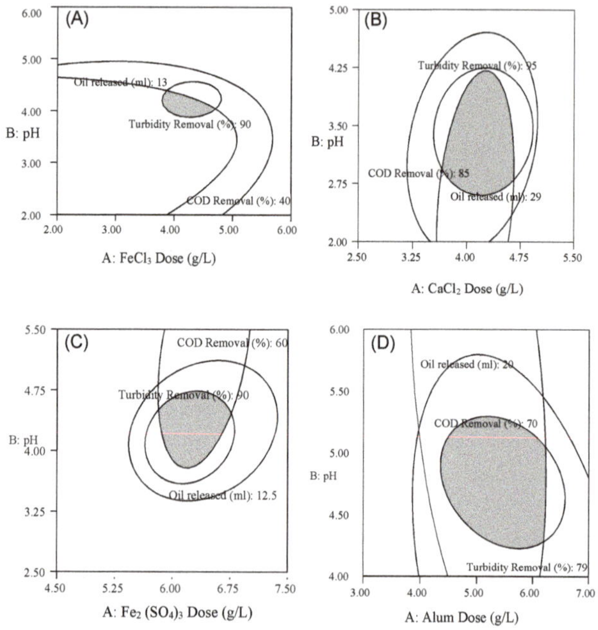

Figure 4. The overlaying plots for the removal of COD and turbidity, and the amount of oil released using (A) $CaCl_2$, (B) $FeCl_3$, (C) alum, and (D) $Fe_2(SO_4)_3$.

minimum removal values for COD and turbidity and the amount of oil released were 85%, 95%, and 30 mL, respectively. The results of a research on the application of aluminum chloride and $CaCl_2$ in a combined treatment method incorporating MWF emulsion destabilization and vacuum filtration illustrated that aluminum chloride left a minimal effect on the destabilization of MWF emulsion, while a 0.05 mol/L concentration of $CaCl_2$ increased the efficiency of this combined system in removing COD up to 90% (13). This confirms the desirability of using aluminum chloride in the treatment of MWFs. Similarly, the results from the application of $CaCl_2$ along with a cationic coagulant as an emulsion breaker in the treatment of industrial wastewater containing oily materials demonstrated that in a concentration of 1 g/L of this chemical compound, the COD removal efficiency was 90%, which was largely consistent with the results of the current research (14). The lack of complete emulsification of oil in an aqueous solution was considered as the reason for the lower consumption of this coagulant (15).

With respect to the $FeCl_3$ graph presented in Figure 4B, a concentration of 3.25-5 g/L and a pH range of 3-4 were optimum conditions of the process factors for the three responses related to COD removal, turbidity, and the amount of oil released. In order to achieve this condition, the minimum removal of COD and turbidity and the amount of oil released were selected 40%, 85%, and 13 mL, respectively.

As can be observed in Figure 4C, for the coagulant alum, a concentration of 4.5-5.5 g/L within a pH range of 4.5-5.4 was optimum, and the minimum removal rates of COD, turbidity, and the amount of oil released were 70%, 85%, and 20 mL, respectively. The treatment of industrial MWFs using electrocoagulation with aluminum electrode indicated that increasing pH will raise the removal rates of turbidity and COD. In this case, the removal efficiencies for COD and turbidity were 92% and 99%, respectively, indicating the good effectiveness of aluminum in removing these two contaminants from industrial wastewater. Canizares et al reported that pH has a strong influence on the COD removal from oil-water emulsion by alum destabilization. They reported that the best results are only attained in a pH range of 5-9; that is somewhat consistent with the current results (16). Aluminum species present in an aqueous solution depend on the aluminum concentration and pH. In this point of attitude, it is more important to take into account that the aqueous chemistry of aluminum is especially complex, as it involves the formation of polymeric hydroxo-aluminum ions and monomeri and precipitates of aluminum hydroxide. Thus, the primary species are monomeric, cationic hydroxo-aluminum species in acidic conditions. Increasing the pH causes the coexistence of these monomeric species with rising polymeric cations and precipitates. At the neutrality condition of pH, the main species are the aluminum hydroxide precipitates that can be completely charged; however, increases in pH produce the reversal charge, leading to negatively-charged precipitates. Additional increases in the pH cause the dissolution of precipitates to monomeric anionic hydroxo-aluminum form, which is the primary species at alkaline pHs (17).

Figure 4D illustrates an overlay scheme for the characterization of the optimum point for the $Fe_2(SO_4)_3$ coagulant. For the three responses, the optimum conditions were a concentration of 5.25-6.75 g/L and a pH range of 3.7-4.7. In order to achieve this condition, the minimum removal rates of COD and turbidity and the amount of oil released selected were 60%, 90%, and 13.5

Table 5. Verification of experimental results at optimum conditions

Responses		Cacl$_2$	Fecl$_3$	Fe$_2$(SO$_4$)$_3$	Al$_2$(SO$_4$)$_3$
Experimental results	COD removal efficiency (%)	97.00	50.00	62.00	74.10
	Turbidity removal efficiency (%)	98.00	81.00	86.00	85.00
	Oil released (mL)	34.20	14.30	25.00	27.00
Model response	COD removal efficiency (%)	93.00	54.40	60.00	75.70
	Turbidity removal efficiency (%)	96.90	84.70	84.10	89.90
	Oil released (mL)	31.80	15.90	28.90	28.90
Error and standard deviation	COD removal efficiency (%)	4±1.50	-4.4±1.7	2±1.10	-1.6±0.25
	Turbidity removal efficiency (%)	1.1±0.50	-3.7±1.1	1.9±0.30	-4.9±0.60
	Oil released (mL)	2.4±1.65	-1.6±0.40	-3.9±0.55	-1.9±0.50

mL, respectively. Tir and Moulai-Mostefa used alum and $Fe_2 (SO_4)_3$ for waste oil/water emulsion treatment. Their results indicated reductions in turbidity and COD by the mentioned coagulants. The effectiveness of wastewater treatment is better with alum than with the addition of $Fe_2 (SO_4)$3; this result is in agreement with the current results. These coagulant agents, which usually help to increase separation effectiveness, cause a physicochemical effect which is used to destabilize emulsion and increase droplet size. In fact, an increase in metal sulfate concentration leads to the acceleration of flocculation and, consequently, an increase in the size of flocs (18). Zhang et al compared the performances of $CaCl_2$, sodium chloride, $FeCl_3$, polymeric ferric sulfate, ferrous sulfate, and polyaluminum chloride as emulsion breakers in the treatment of MWFs. The results demonstrated that with an equal consumption of each coagulant (3 g/L), the COD removal efficiencies of each coagulant were 85%, 58%, 49%, 40%, 46%, and 46%, respectively (19). The attachment of adsorbing macromolecules to more droplets by means of bridging flocculation is the main mechanism of destabilization. This mechanism inducts electrically-charged species as reagents, and these species can join through attractive electrical forces with oppositely charged active sites which are present on the droplet surface. Furthermore, to bridge flocculation, some researchers have indicated the oil removal from emulsified effluents through adsorption properties by increasing metal hydroxides in the coagulation-flocculation method (1-3). The efficiencies of oil removal were reported as 65%, 85%, 60%, 53%, 50%, and 50%, respectively. These results were in agreement with those of the current study and indicated the good performance of $CaCl_2$ compared to other coagulants. The application of coagulants such as 8 g/L of polyferric sulfate in pH 6, 10 g/L of $FeCl_3$ in pH 8, and 12 g/L of $Fe_2 (SO_4)_3$ in pH 7.5 for the treatment of leachate, which is considered a type of industrial wastewater, suggested that the efficiencies of these coagulants in the removal of turbidity and COD were respectively 56.38% and 89.79% for polyferric sulfate, 68%, 65%, and 98.85% for $FeCl_3$, and 55.87% and 94.13% for $Fe_2 (SO_4)_3$, indicating the high rate of consumption of common coagulants in the treatment of wastewater (20).

The results of the present study showed that the percentage of turbidity removal depended on the initial concentration of turbidity in MWFs. For $CaCl_2$, the optimal concentration was 6%, while for alum, oil was better separated from emulsion in lower concentrations (2%-4%). Since the practical concentration of MWFs in workshops is commonly 6% and $CaCl_2$ leaves less environmental impacts compared to alum, $CaCl_2$ is considered a higher priority in the destabilization of oil emulsions. Optimal value for consumption of $CaCl_2$ in the current work was 3.6 g/L, and turbidity was reduced to an acceptable value (20 NTU). COD reduction was relatively low in comparison with the results of this research, which may due to the type of MWF wastewater being studied. To confirm the results obtained from the experimental data, three additional experiments were performed for each coagulant and the results were compared. Verification of the experimental results at optimum conditions for the four coagulants are shown in Table 4.

The results indicate good agreement between the predictive data obtained from the models and the experimentally measured data (independent data).

Conclusion

Physical-chemical methods are recommended for the pretreatment of MWFs; among them, coagulation-flocculation is one of the plain and usual methods. Although many coagulants are available for the treatment of water and wastewater, selecting the most effective coagulant for a specific wastewater still depends greatly on the results of laboratory jar testing. Calcium chloride as an uncommon coagulant for the treatment of wastewater was investigated in this study for MWFs wastewater treatment. Alum, ferric chloride, and ferric sulphate as conventional coagulants were applied to the same wastewater for performance comparison. The results showed calcium chloride had a higher removal efficiency in the reduction of organic load and suspended solids of MWF compared with the other three coagulants used for such industrial wastewater treatment. Furthermore, the results showed good agreement between experimental data and the predicted equation.

Acknowledgments

This study was funded by Qazvin University of Medical

Sciences, Department of Environmental Health Engineering (IR.QUMS.REC.28.44.5997 code of ethics). The authors would like to express their gratitude to the authorities of Qazvin University of Medical Sciences for providing the experimental facilities in the laboratory of the Environmental Health Engineering Department.

Competing interests

The authors declare that they have no competing interests.

Authors' contributions

All authors contributed equally to data collection and article approval.

References

1. Byers JP. Metalworking Fluids. 3rd ed. Boca Raton: CRC Press; 2016.
2. Grossi M, Ricco B. A portable electronic system for in-situ measurements of oil concentration in MetalWorking fluids. Sens Actuators A Phys 2016; 243: 7-14. doi: 10.1016/j.sna.2016.03.006.
3. Jagadevan S, Graham NJ, Thompson IP. Treatment of waste metalworking fluid by a hybrid ozone-biological process. J Hazard Mater 2013; 244-245: 394-402. doi: 10.1016/j.jhazmat.2012.10.071.
4. Demirbas E, Kobya M. Operating cost and treatment of metalworking fluid wastewater by chemical coagulation and electrocoagulation processes. Process Saf Environ Prot 2017; 105: 79-90. doi: 10.1016/j.psep.2016.10.013.
5. Thill PG, Ager DK, Vojnovic B, Tesh SJ, Scott TB, Thompson IP. Hybrid biological, electron beam and zero-valent nano iron treatment of recalcitrant metalworking fluids. Water Res 2016; 93: 214-21. doi: 10.1016/j.watres.2016.02.028.
6. Khuri AI, Mukhopadhyay S. Response surface methodology. Wiley Interdiscip Rev Comput Stat 2010; 2(2): 128-49. doi: 10.1002/wics.73.
7. Moradnia M, Panahi fard M, Dindarlo K, Ali Jamali H. Optimizing potassium ferrate for textile wastewater treatment by RSM. Environ Health Eng Manag J 2016; 3(3): 137-42. doi: 10.15171/ehem.2016.12.
8. Montgomery DC. Design and Analysis of Experiments. 8th ed. John Wiley Sons; 2008.
9. Emamjomeh MM, Jamali HA, Moradnia M. Optimization of Nitrate Removal Efficiency and Energy Consumption Using a Batch Monopolar Electrocoagulation: Prediction by RSM Method. J Environ Eng 2017; 143(7): 401-12. doi: 10.1061/(ASCE)EE.1943-7870.0001210.
10. Hasan HA, Abdullah SRS, Kamarudin SK, Kofli NT. Response surface methodology for optimization of simultaneous COD, NH4+-N and Mn2+ removal from drinking water by biological aerated filter. Desalination 2011; 275(1): 50-61. doi: 10.1016/j.desal.2011.02.028.
11. Arslan-Alaton I, Akin A, Olmez-Hanci T. An optimization and modeling approach for H2O2/UV-C oxidation of a commercial non-ionic textile surfactant using central composite design. J Chem Technol Biotechnol 2010; 85(4): 493-501. doi: 10.1002/jctb.2315.
12. Yu Y, Zhou X, Wu S, Wei T, Yu L. High-yield production of the human lysozyme by Pichia pastoris SMD1168 using response surface methodology and high-cell-density fermentation. Electron J Biotechnol 2014; 17(6): 311-6. doi: 10.1016/j.ejbt.2014.09.006.
13. Gutierrez G, Lobo A, Benito JM, Coca J, Pazos C. Treatment of a waste oil-in-water emulsion from a copper-rolling process by ultrafiltration and vacuum evaporation. J Hazard Mater 2011; 185(2): 1569-74. doi: 10.1016/j.jhazmat.2010.10.088.
14. Moscoso F, Deive FJ, Villar P, Pena R, Herrero L, Longo MA, et al. Assessment of a process to degrade metal working fluids using Pseudomonas stutzeri CECT 930 and indigenous microbial consortia. Chemosphere 2012; 86(4): 420-6. doi: 10.1016/j.chemosphere.2011.10.012.
15. Tong K, Zhang Y, Chu PK. Evaluation of calcium chloride for synergistic demulsification of super heavy oil wastewater. Colloids Surf A Physicochem Eng Asp 2013; 419: 46-52. doi: 10.1016/j.colsurfa.2012.11.047.
16. Canizares P, Martinez F, Jimenez C, Saez C, Rodrigo MA. Coagulation and electrocoagulation of oil-in-water emulsions. J Hazard Mater 2008; 151(1): 44-51. doi: 10.1016/j.jhazmat.2007.05.043.
17. Canizares P, Garcia-Gomez J, Martinez F, Rodrigo MA. Evaluation of a simple batch distillation process for treating wastes from metalworking industries. J Chem Technol Biotechnol 2004; 79(5): 533-9. doi: 10.1002/jctb.1021.
18. Tir M, Moulai-Mostefa N. Optimization of oil removal from oily wastewater by electrocoagulation using response surface method. J Hazard Mater 2008; 158(1): 107-15. doi: 10.1016/j.jhazmat.2008.01.051.
19. Zhang W, Xiao P, Wang D. Central treatment of different emulsion wastewaters by an integrated process of physicochemically enhanced ultrafiltration and anaerobic-aerobic biofilm reactor. Bioresour Technol 2014; 159: 150-6. doi: 10.1016/j.biortech.2014.02.067.
20. Liu X, Li XM, Yang Q, Yue X, Shen TT, Zheng W, et al. Landfill leachate pretreatment by coagulation–flocculation process using iron-based coagulants: optimization by response surface methodology. Chem Eng J 2012; 200-202: 39-51. doi: 10.1016/j.cej.2012.06.012.

Qualitative analysis of Orzooiyeh plain groundwater resources using GIS techniques

Mohsen Pourkhosravani*

Assistant Professor, Department of Geography Shahid Bahonar University of Kerman, Kerman, Iran

Abstract

Background: Unsustainable development of human societies, especially in arid and semi-arid areas, is one of the most important environmental hazards that require preservation of groundwater resources, and permanent study of qualitative and quantitative changes through sampling. Accordingly, this research attempts to assess and analyze the spatial variation of quantitative and qualitative indicators of Orzooiyeh groundwater resources in the Kerman province by using the geographic information system (GIS).

Methods: This study attempts to survey the spatial analysis of these indexes using GIS techniques besides the evaluation of the groundwater resources quality in the study area. For this purpose, data quality indicators and statistics such as electrical conductivity, pH, sulphate, residual total dissolved solids (TDS), sodium, calcium; magnesium and chlorine of 28 selected wells sampled by the Kerman regional water organization were used.

Results: A comparison of the present research results with standard of Industrial Research of Iran and also the World Health Organization (WHO) shows that, among the measured indices, the electrical conductivity and TDS in the chosen samples are higher than the national standard of Iran and of the WHO but other indices are more favourable.

Conclusion: Results showed that the electrical conductivity index of 64.3% of the samples have an optimal level, 71.4% have the limit of Iran national standard and only 3.6% of them have the WHO standard. The TDS index, too, did not reach national standards in any of the samples and in 82.1% of the samples this index was on the national standard limit. As per this index, only 32.1% of the samples were in the WHO standards.

Keywords: Spatial analysis, Water resources, Quality indices, Orzooiyeh.

***Correspondence to:**
Mohsen pourkhosravani
Email: pourkhosravani@uk.ac.ir

Introduction

According to environmental conditions in arid and semi-arid areas, groundwater is among the most important sources of drinking water and agriculture in these areas. Groundwater is an important part of the ecosystem renewable water resources that has suffered qualitative and quantitative changes because of improper management of extraction that has directly or indirectly caused the destruction of other resources. Increase population and the overdraw of groundwater in recent years pose a serious risk to the quantity and quality of limited water resources. This has made planning to protect water resources so much more important. The main purpose of qualitative surveys is the preservation of public and consumer health. For this, making adequate water available, and adapting it to the physical, chemical, biological and radiological codification standards are necessary (1). It should be noted that changes in groundwater quality is a function of the physical and chemical indicators that are also strongly influenced by the geological structure of the area and human activities (2). In general, today, the unstable development of human societies is one of the most important environmental hazards that make imperative the preservation of groundwater resources and a permanent study of its qualitative and quantitative changes through sampling (3). Because groundwater's quality has been measured in a scale of time and space, and its properties cannot be assumed in fixed time and space, the preparation salinity and mineral maps can be an important step towards the proper utilization of the water resource (4). Various researchers have done extensive investigations in relation to the quantity and quality of groundwater in different regions according to the importance of groundwater resources and sustainable development of arid and semi-arid areas. A study of temporal and spatial changes in the Mehriz Baharan plain shows that the groundwater in 48% of the study area has a strict level and in 52% have moderate to low levels (5).

Moreover, assessment and mapping of the qualitative and quantitative destruction of ground water resources in the Firuzabad plain show that 93% of the study area faces the extreme danger of the destruction of groundwater resources (6). Too assessment of groundwater quality and its suitability for drinking and agricultural uses in the Oshnavieh area, showed fluoride and nitrate are within the permissible limits for human consumption and crops as per the international standards (7). Another study evaluated the qualitative and quantitative changes of groundwater resources in the Amol and Babol plains, showing a slight decrease in the groundwater level and an improvement in the water quality (8). Analytical methods of assessing underground water quality are mainly based on classical statistics and graphical methods involving a limited number of samples and lack of spatial correlation of samples with qualitative parameters. A change from classical to systematic attitudes nowadays has led to a growing interest in the use of geostatistical methods when doing spatial analysis of quantity and quality in groundwater resources. Geostatistics is a calculation process in which the value of a quantity at a given point is based on the weight assigned to clear information about adjacent parts. Since the quantity and quality of groundwater has temporal and spatial variations, the geostatistical techniques are important for taking into account the spatial correlation of data and expressing them in the form mathematical models (9). Generally, among geostatistical methods, the kriging method using the low-variance coefficient is more important in the evaluation of qualitative and quantitative changes of groundwater resources (10). Accordingly, this research attempts to assess and analyze the spatial variation of quantitative and qualitative indicators of Orzooiyeh groundwater resources in the Kerman province by using the geographic information system (GIS).

Methods

Orzooiyeh is a tectonic plain situated to the south-west of Kerman between 21' 56° to 35' 56° eastern longitude and between 27' 28° to 45' 28° northern latitude (Figure 1). This plain probably emerged in the late Pliocene and early Pleistocene and due to compression phases. The value of the longitudinal axis has a north-west to south-east trend and is structurally in Sanandaj-Sirjan zone and Zagros. Hydrologically, the Orzooiyeh plain is the part of the entire basin that flows into the Persian Gulf and the average height of the basin's mountainous part is 1770 m, and 1100 m is the average height of the sliding Dolatabad. This study is based on descriptive and analytical methods and field study. In order to spatially analyse quality of the Orzooiyeh plane groundwater resources, annual indices of qualitative data of electrical conductivity, pH, SO_4, residual total dissolved solids (TDS), sodium, calcium, magnesium, and chlorine related to 28 selected wells were gathered from the Kerman regional water organization samples in 2011 by doing a complete chemical analysis. Results of this study were achieved by a comparison of the standards of the World Health Organization (WHO) (11) and the Institute of Standards and Industrial Research of Iran (ISIRI) (12). In the next step, the zoning of groundwater quality was conducted by the kriging method on the basis of various indicators and using Arc GIS software. In geostatistics we can estimate the quantity value in another point with known coordinates by having the quantity value in known coordinate. This is possible provided the coordinate of an unknown value placed in the domain that has a spatial structure (13). Estimation of variable with distribution and spatial structure and evaluation of error with this estimation is called kriging (14). This estimator has the following characteristics:

1. The estimated values for the unknown points are the linear combination of adjacent samples. So:

$$\mu K = \sum_{i=1}^{n} \lambda_i X_i$$

Where $\mu\kappa$ is the kriging estimation and λ_i is the allocated weights vector to samples. In other words, an estimated value is had by assigning weights λ_i to known amounts measured in adjacent points (x_i).

2. It is unbiased, meaning, its mathematical expectation is equal to the actual average (wμ) samples.

$E = (\mu\kappa - \mu w) = 0$

3. The mean square error has the least possible value. That means:

$E[(\mu\kappa - \mu w)^2] = amin$

In this method, after measuring the amount of variable for several samples with regard to their situation, the amount

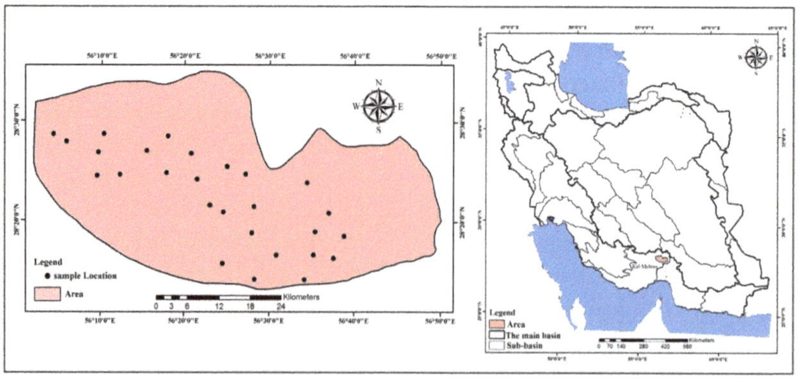

Figure 1. Location of the study area and sampling wells.

of variable of non-sampling points is also estimated by considering their position and the distance from common points (15). One of the most important characteristics of the kriging method is that, for each estimate, the relative error can be calculated and the number of points in the adjacent is usually about 10 points. So we can estimate and calculate the confidence interval for each estimated amount. Since in classical methods these operations are ineffective, the zoning maps of each indicator were prepared with the help of Arc GIS software.

Results

Table 1 shows the descriptive statistical characteristics of quality indicators measured from the study area of underground sources indicators by comparing these indices using standard WHO standards and Industrial Research Organization of Iran.

The results show the maximum amount of electrical conductivity near the centre of the area relates to well No. 8a and the lowest is for well No. 28. The amount of dissolved solids in the sample wells is also similar to electrical conductivity, with a maximum of 3991 mg/L in well No. 26, and a minimum of 320 mg/L in well No. 28. According to the table results, the highest level of جH in water resources studied is related to wells No. 20, 22 and 25 at the rate of 7.8 and the lowest is for wells No. 9 and 28 with جH 7. On the other hand, the highest amount of sulphate is in Well No. 8 (20.2 mg/L) and the lowest rate is for well number 28 (0.8 mg/L). The maximum amount of chlorine relates to well No. 8 (17.8 mg/L) and the lowest for well number 28 (0.5 mg/L). The results show, the maximum amount of sodium relates to Well No. 1 with 67.2 mg/L and the lowest amount to well No. 28 with 15.1 mg/L. The highest amount of magnesium is for well No. 8 with the amount of 11.6 mg/L and the least is for well No. 1 with a rate of 0.7 mg/L. On the other hand, the highest amount of calcium is for well No. 8 with the amount of 14.4 mg/L and the least is for well No. 1 and 2 with a rate of 1.5 mg/L.

Generally, the table indicates that the values of indicators pH, magnesium, calcium, sulphate, chloride and sodium in groundwater resources in the study area were lower than

Table 1. Statistical data of groundwater quality indicators and compared to the standard values

No.	Sample well		Depth (m)	ph	mic-mho	mEq/L					Na%
					Ec	Mg	Ca	TDS	SO4	Cl	
1	Sarghaz Dolatabad		120.0	7.3	656	0.7	1.5	426	1.1	1.8	67.2
2	Dolatabad Esa Safi nataj		110.0	7.3	818	2.5	1.5	532	2.2	2.8	50
3	Dolatabad Mahmod dolati		120.0	7.4	660	1.4	1.6	429	1.9	1.8	53.8
4	Ghaderabad Abdollah Shafiei		110.0	7.5	850	2.2	3.6	533	3.3	2.4	34.8
5	Aliabad Shamshirbar		130.0	7.3	1088	3.3	3.1	707	3.6	4.2	42.9
6	Rahmatabad		120.0	7.3	2250	8.2	4.8	1463	13.8	7	48
7	Vakilabad Honarestan		120.0	7.3	1295	4.5	2.5	842	3.9	5.5	45.7
8	Kachalabd Keshavarzan		130.0	7.4	4160	11.6	14.4	2704	20.2	17.8	38.1
9	Damaneh Tiraft		120.0	7	2870	8.2	8.8	1866	14.4	14	45.9
10	Nosratabd		110.0	7.2	1100	5.4	2.1	715	4.8	3.5	36.4
11	Vakilabad Mostazafan		100.0	7.5	1399	4.6	3.4	909	4.3	6.8	44.1
12	Aliabd Sepahani		120.0	7.5	2160	7.9	3.1	1404	12.1	7.2	52.2
13	Soltanabd masoudi		100.0	7.4	1495	4.6	3.8	972	4.3	7.9	44.7
14	Allahabad		120.0	7.4	1728	5.7	3.9	1123	4.9	9.3	46.4
15	Takhte khaje		120.0	7.3	1928	8	6	1253	9.3	8.4	32.4
16	Aliabad Shamshirbar		110.0	7.5	1023	2.9	1.9	665	4.4	2.5	56.8
17	Chilabad		130.0	7.5	2520	9.8	3.2	1638	14.4	9	52.6
18	Dolatabad Didar bahrami		100.0	7.1	935	2	2.4	608	3.9	3	53.2
19	Hoseinabad Zabihollahe Bahari		100.0	7.7	995	3.5	2	647	4	2.5	45
20	Ghadeabad Mansouri		110.0	7.8	910	2.3	1.9	592	3.7	2.4	56.3
21	Allahabad Dasht berm		110.0	7.7	1385	7	2	900	8.5	3.2	41.2
22	Dolatabd Ahmad Ghanji		100.0	7.8	677	1.7	1.7	440	2.2	1.8	48.5
23	Ghaleno Alireza Saberi		120.0	7.5	2160	5.8	5.8	1404	4.2	15	47.3
24	Vakilabad Ghasem Maket		120.0	7.4	2770	7.8	8.6	1801	10.3	14	39.9
25	Gholnaabad Keramat Almasi		110.0	7.8	836	2.9	1.6	543	3.8	2.1	47.1
26	Ghaemabad Manochehr Tavakoli		130.0	7.7	614	3	2.4	3991	3.9	2	44.3
27	Tale Yatim Vakilabad		120.0	7.7	751	1.6	3.4	488	2.5	1.5	33.3
28	Hosienabad Kalian		120.0	7	492	1.5	3	320	0.8	0.5	15.1
Average			115.4	7	1447	4.4	3.7	1022	5.8	5.4	44.4
Minimom			100	7	492	0.7	1.5	320	0.8	0.5	15.1
Maximum			130	7.8	4160	11.6	14.4	3991	20.2	17.8	67.2
Iran national standard		Optimal level		6.5-8.5	1500	30	75	100	250	250	200
		Standard limit		6.5-9	2000	150	250	1500	400	400	200
World Health Organization				6.5-8.5	600	50	75	600	200	200	200

those set by the WHO and the Iran Industrial Research Organization standards, and groundwater resources have a very good grade for these indices. However, according to the results in this table, the electrical conductivity in 64.3% of samples has an optimal level, and 71.4% has the Iran national standard, and only 3.6% of these resources of electrical conductivity have the WHO standard. TDS have not been optimal in the light of national standards in any of the samples, and 82.1% of these samples have standard limit. The value of this index in 32.1% of samples is of standard value. In general, according to the results shown in the table, average indices of electrical conductivity and TDS in the groundwater resources of the study area conform to the national standard of Iran, and average value of these indices is higher than the optimal level of the national standard and also of the WHO.

Discussion

As mentioned, the zoning of the qualitative indices of groundwater resources in the study area was done using the kriging method by the Arc GIS software. Figure 2 shows the zoning of study area based on electrical conductivity. According to this figure, the amount of electrical conductivity of the area bordering the centre has increased.

Table 2 also shows classes and degrees of risk in electrical conductivity.

It is obvious from a comparing of the mean of electrical conductivity with class and degree of the risk shown in this table that groundwater resources in the study area have a middle class, based on these indicators. But according to the zoning plan, this water resources index is of the central part of the region that is in the severe-hazard class.

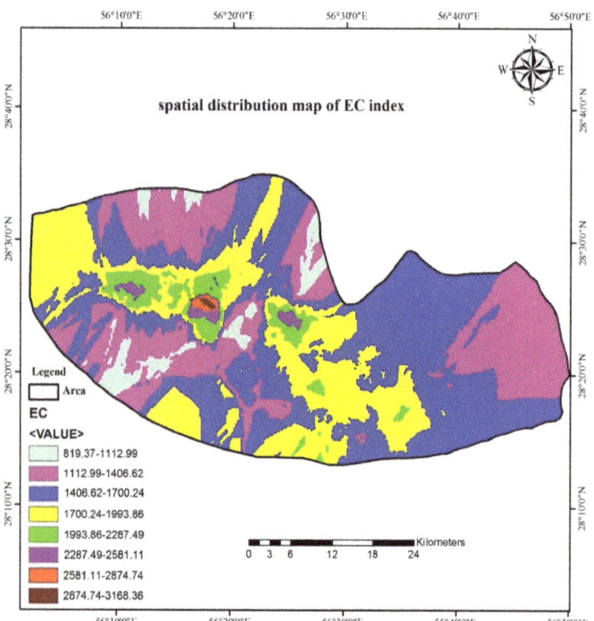

Figure 2. Zoning of electrical conductivity.

Table 2. Classes and degrees of risk in electrical conductivity

Degree of risk	Class of risk	Domain (μSiemens/cm)
1	No risk	0-250
2	Low risk	250-750
3	Mean	750-2250
4	Severe risk	2250-5000
5	Severe hazard risk	≥5000

In American salinity laboratory, water with electrical conductivity less than 953 μSiemens/cm is appropriate and can be used to irrigate salt-sensitive plants. Water with 753 to 953 μSiemens/cm salinity requires proper drainage management. Water with 753 to 9953 μSiemens/cm salinity and water with electrical conductivity of more than 9.953 μSiemens/cm is not suitable for irrigation unless tolerant plants are planted and leaching is made possible with proper drainage (16).

Figure 3 shows zoning of TDS in the study area. The TDS in the water except suspended sediments, colloidal and dissolved gases include mineral salts with a lesser concentration of organic substances such as carbonate, bicarbonate, chloride, sulphate, nitrate, sodium, potassium, calcium and magnesium (17). The higher the amounts of these substances in water, the greater will be the water's electrical conductivity and salinity. Moreover, according to the zoning map of dissolved solids in the groundwater resource of the area, the more we move from the border area towards the centre, the amounts of soluble substances increases.

In general, electrical conductivity represents the cationic and anionic salts dissolved in water, which can accelerate electricity flow in water. The more the electrical conductivity of water sample, the greater is the amount of salts to be is found in it, and can be an indication of poor quality. The electrical conductivity of water depends on soluble salts. Hence, high electrical conductivity of water indicates high levels of soluble salts and ions in it.

Figure 4 shows that the amount of pH increases from the border to the centre and somewhat to the east. As can be

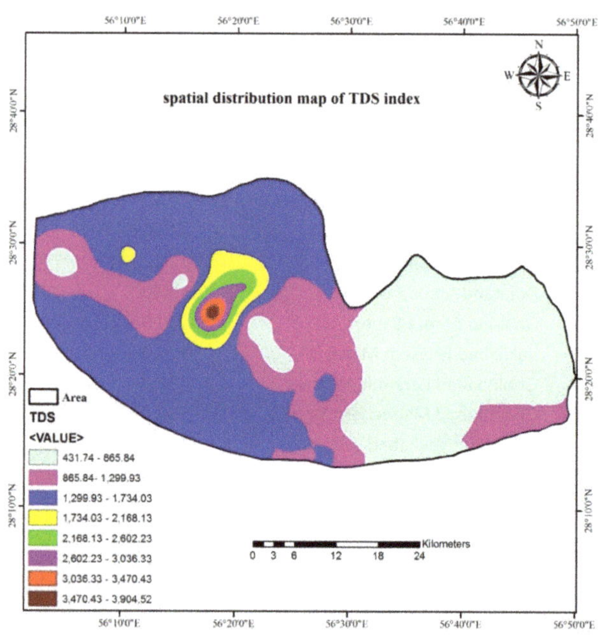

Figure 3. Zoning of dissolved solids.

seen, the water resource of the area in general deals with good ph. It should be noted that, although pH does not directly affect consumer heath, it is one of the factors affecting water quality and the guideline value is not based on health effects (18).

According to the sulphate zoning plan in the study area (Figure 5) the maximum amount of sulphate in the groundwater resources is found at the centre of the region. Probably the amount of sulphate concentration has increased between layers of calcareous rocks of gypsum compounds and sulphated compounds with the interaction of these rocks with groundwater and their dissolution in water.

Usually, a large amount of clay and gypsum increases sulphate concentration (19). However, the results show the level of sulphate in the groundwater under study is much lower than the WHO standards and the limit set by the ISIRI.

According to the zoning map of chlorine in the study area (Figure 6), a higher indication of chlorine is seen in the northern and central regions and a decline in its presence in groundwater towards the east. The results show the presence of fine-grained sediments in this part of the region, which are among factors increasing the presence of chloride in water resources.

According to Figure 7, a higher sodium index pertains to the southern and south-eastern parts of the study area, with its concentration decreasing as we move to the north-

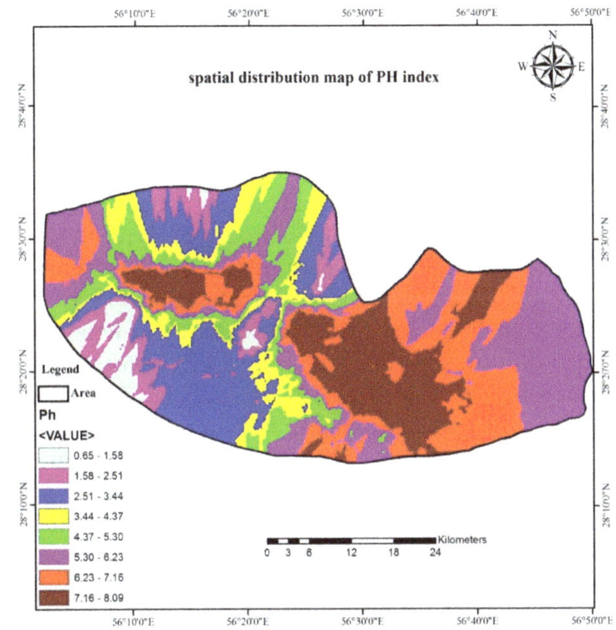

Figure 4. Zoning of pH.

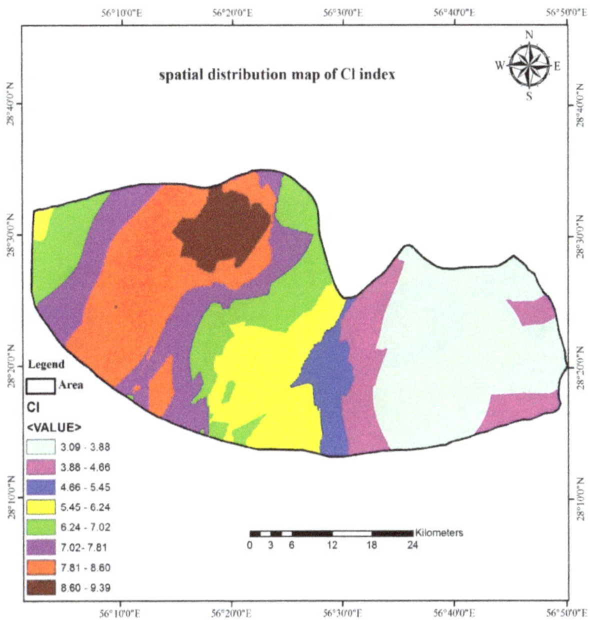

Figure 6. Zoning of chlorine.

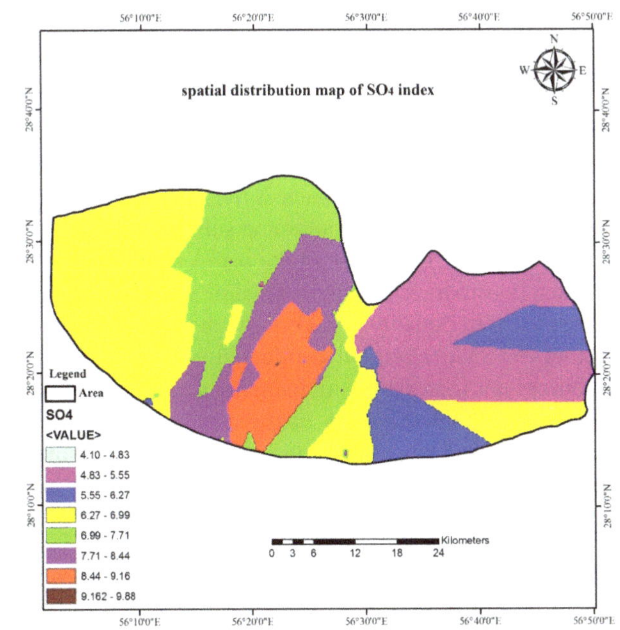

Figure 5. Zoning of sulphate.

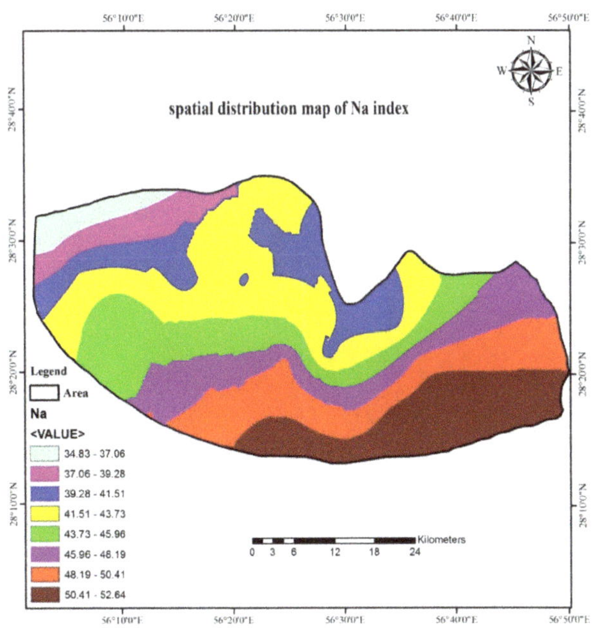

Figure 7. Zoning of sodium.

west region.

Generally, uncontrolled exploitation of groundwater resources is increasing the amount of magnesium. With a fall in groundwater levels, waters with salinity turn bitter, a taste contributed by the magnesium factor.

Moreover, as shown in Figure 8, there is a higher concentration of magnesium in the vicinity of the centre and north of it, with presence declining from this part to the border region, and especially in the eastern area.

According to the spatial distribution of calcium (Figure 9), the maximum amount of calcium is found in the western part of the study area, while its concentration keeps diminishing as we move towards the east. Chile and limestone formations in the western part amount increases

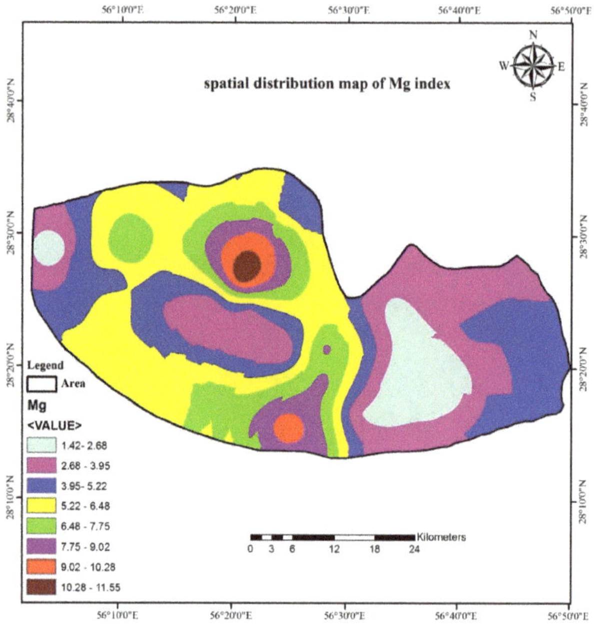

Figure 8. Zoning of magnesium.

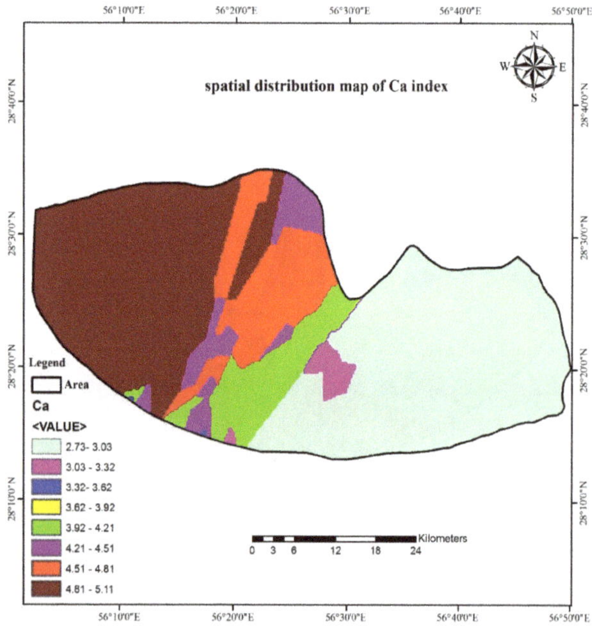

Figure 9. Zoning of calcium.

Table 3. As standard elements in groundwater resources

Scale	SO4	Ca	TDS	mg	Cl
Good	<3.02	<5	<500	<3	<4.92
Appropriate	3.02- 5.83	5-10	500-1000	3-5	4.92-9.85
Inappropriate	5.83- 12.08	10-20	1000-2000	5-10	9.85-19.71
Bad	12.08- 23.95	20-40	2000-4000	10-20	19.71-39.43

the presence of this element in the underground water resources. However, value of this element in the region's water resources is reduced with the expansion of alluvial formations towards the centre and the eastern region.

Table 3 shows the standard elements in groundwater resources.

A comparison of the results of this table and the results of this study shows that the average amount of these elements in groundwater resources in the study area is within acceptable limits. Only the average amount of TDS is inadequate.

Conclusion

Quantification of water quality is an integral part of scientifically based water resources management (20). Because water quality monitoring not only improves water quality but also involves economic value in the process of producing clean water and is an important factor in reducing production and water treatment costs. Some of these salts in water are essential for human health, while amounts higher than the standard limit pose a health hazard. Hence, clean water guarantees public health, and the first step in understanding water is a survey of water quality parameters (21). Accordingly, this research evaluated the qualitative indices and zoning of these indices in the study area of groundwater resources. It should be noted that the concentration of ions in groundwater is usually dependent on the spatial position of each point and the geographical coordinates of such points. In the study area, the flow of groundwater is from the north and east to the south and west of the plains. Hence, the more it moves from the north and the north-west to the south and south-east and especially the basin centre, composed of fine-grained, the quality of underground water keeps deteriorating. In general, the existence of two salt domes such as the Zagros Thrust and spring spas in the basin, containing sulphur compounds, and, in addition, the nature of the plains inside the basin serving as evaporation plain with saline soil and high salt causes the water quality to improve with basic flow. These factors have led electrical conductivity of groundwater resources to increase from 69 μSiemens at the basin entrance to 2000 μSiemens in the centre. According to what was said above, results show that between measured indices the two indices of electrical conductivity and total dissolved salts increase from the standards of Iran national standard organizations and the WHO. Also the pH of water resources increase and quality resources find neutral and slightly alkaline desire from the centre of the area toward East. Thus the water resources quality in the study area drops from the north and north-west as one moves towards the centre, south and south-east basin due to topographic and geological characteristics of the area.

Acknowledgments

Special thanks to Kerman regional water organization that offered the qualitative data for the accomplishment of this research.

Competing interests

None to be declared.

Author' contribution

MP is the single author of the paper.

References

1. Rahmani Z, Gholami M, Khoshnevis Zadeh A, Rezayee Kalantari R. Investigation of Buin Zahra drinking water resources quality by using of GWQI. J Alborz Univ Med 2013; 3(2): 147-55. [In Persian].
2. Pourakbar M, Mosaferi M, Shaker Khatibi M, Moradi A. Groundwater quality assessment from a hydrogeochemical viewpoint a case study of Sarab county. J Water Wastewater 2015; 26(3): 116-26. [In Persian].
3. Zreabyaneh H. Spatial analysis of groundwater resources quality parameters of Hamadan–Bahar plain. Geography and Environmental Hazards 2014; 2(8): 65-86. [In Persian].
4. Heydari E, Ekrami M, Nasabpour S. Study temporal and spatial groundwater quality using geostatistical (case study: Yazd–Ardakan plain). Desert Ecosystems Engineering Journal 2015; 4(6): 63-70. [In Persian].
5. Sahebjalal E, Dehghani F, Tabatabaizadeh M. Investigating spatio-temporal variations of groundwater quality using Kriging method. Journal of Science and Technology of Agriculture and Natural Resources 2013; 17(65): 51-61. [In Persian].
6. Masoudi M, Barzegar S. Assessment and mapping of qualitatative and quantitative severity degradation of groundwater resourses using the modified IMDPA desertification model and GIS. A Case Study: Firuz-abad plain of Fars province. Journal of Irrigation & Water Engineering 2015; 5(20): 86-95. [In Persian].
7. Aghazadeh N, Asghari Mogaddam A. Assessment of groundwater quality and its suitability for drinking and agricultural uses in the Oshnavieh area, Northwest of Iran. Journal of Environmental Protection 2010; 1(1): 30-40. doi: 10.4236/jep.2010.11005.
8. Afzali A, Shahedi K. Investigation on Trend of Groundwater Quantity-Quality Variation in Amol-Babol Plain. Journal of Watershed Management Research 2014; 5(10): 144-56. [In Persian].
9. Sun Y, Kang S, Li F, Zhang L. Comparison of interpolation methods for depth to groundwater and its temporal and spatial variations in the Minqin oasis of northwest China. Environ Model Softw 2009; 24(10): 1163-70. doi: 10.1016/j.envsoft.2009.03.009.
10. Theodossiou N, Latinopoulos P. Evaluation and optimization of groundwater observation networks using the kriging methodology. Environ Model Softw 2006; 22(7): 991-1000. doi: 10.1016/j.envsoft.2005.05.001.
11. World Health Organization (WHO). Guidelines for Drinking-Water Quality. 3rd ed. Geneva: WHO; 2008.
12. Institute of Standard and Industrial Research of Iran. Drinking Water: Physical and Chemical Characteristics. 5th ed. Iran: Institute of Standard and Industrial Research of Iran; 2009. [In Persian].
13. Asakereh H. Kriging Application in Climatic Element Interpolation A Case Study: Iran Precipitation in 1996. Journal of Geography and Development 2008; 6(12): 25-42. [In Persian].
14. Davis JC. Statistics and Data Analysis in Geology. New York: John Wiley & Sons; 1973. p. 550.
15. Salehi H, Zaynivand H. Survey of the quality of underground water for drinking and agriculture, and select the most appropriate spatial interpolation method for special interpolation (Case study: West of the city of Marivan). Ecohydrology Magazine 2014; 3: 153-66. [In Persian].
16. Moqimi H, Zare Eghbalabadi M. Qualitative zoning of heart aquifer (Yazd province) for agricultural purposes using GIS. The First National Conference on the Use of Advanced Spatial Analysis Models (remote sensing and GIS) in Spatial Planning; 2014 March 5-6; Islamic Azad University of Yazd. [In Persian].
17. Sedaghat M. Land and Water Resources (Groundwater). Tehran: Payam Noor University Press; 2008. p. 110-121. [In Persian].
18. WHO. The guidelines: a framework for safe drinking-water. In: WHO, ed. Guidelines for Drinking Water Quality. 3rd ed. Geneva: WHO; 2006. p. 6-8.
19. Hudak PF, Sanmanee S. Spatial patterns of nitrate, chloride, sulfate, and fluoride concentrations in the Woodbine Aquifer of north-central Texas. Environ Monit Assess 2003; 82(3): 311-20. doi: 10.1023/A:1021946402095.
20. Parparov A, Gal G, Hamilton D, Kasprzak P, Ostapenia A. Water quality assessment, trophic classification and water resources management.Journal of Water Resource and Protection 2010; 2(10): 907-15. doi: 10.4236/jwarp.2010.210108.
21. Shariat Panahi M. Principle of quality and treatment of water and wastewater. 5th ed. Tehran: Tehran University Press; 1998. p. 114-29. [In Persian].

Adsorption of diazinon from aqueous solutions onto an activated carbon sample produced in Iran

Zeynab Akbarlou[1], Vali Alipour[2*], Mohsen Heidari[2], Kavoos Dindarloo[3]

[1]MSc of Environmental Health Engineering, Student Research Committee, Department of Environmental Health Engineering School of Health, Hormozgan University of Medical Sciences, Bandar Abbas, Iran
[2]Assistant Professor of Environmental Health Engineering, Department of Environmental Health Engineering, School of Health, Hormozgan University of Medical Sciences, Bandar Abbas, Iran
[3]Associate Professor of Environmental Health Engineering, Department of Environmental Health Engineering, School of Health, Hormozgan University of Medical Sciences, Bandar Abbas, Iran

Abstract
Background: Considering the severe health and environmental hazards caused by the entry of diazinon toxin into water resources, its removal is very important. Given the high costs of imported adsorbents, this research attempted to evaluate, for the first time, the efficiency of Iranian activated carbon in removing diazinon from aqueous solutions.
Methods: In this batch experimental study, the effects of contact time (5-90 minutes), adsorbent concentration (0.5-30 g/L), initial concentration of diazinon (5-50 mg/L), and pH (3-10) on the adsorption of diazinon onto the activated carbon were evaluated. Concentrations of diazinon were measured using a high pressure liquid chromatography (HPLC) instrument. The specific surface area of the adsorbent was determined by BET and BJH methods. The experimental adsorption data was analyzed using Langmuir and Freundlich isotherm models. Pseudo first-order and pseudo second-order kinetics models were employed to determine kinetics. Moreover, data was analyzed using SPSS version 19, and Pearson correlation and analysis of variance (ANOVA) tests were performed at a significance level of 0.05.
Results: The optimum contact time, sorbent dose, and pH were 30 minutes, 10 g/L, and 5, respectively. The adsorbent could adsorb diazinon with a removal efficiency of 92.5% under optimum conditions (initial concentration: 20 mg/L). The experimental data was better described by the pseudo-second order kinetic and Langmuir isotherm. A maximum adsorption capacity of 71.4 mg/g was calculated by the Langmuir isotherm model.
Conclusion: With respect to the high adsorption capacity of Iranian activated carbon, this sorbent can be considered an efficient adsorbent for the removal of diazinon from aqueous solutions.
Keywords: Diazinon, Adsorption, Activated carbon, High pressure liquid chromatography

***Correspondence to:**
Vali Alipour
Email: v_alip@yahoo.com

Introduction

Insecticides and pesticides are among the most serious toxins found in various water resources (1). The widespread use of pesticides to control pests and develop agriculture has resulted in the pollution of surface water and groundwater, because these pesticides enter water resources through various routes including wastewater disposal, direct washing of pesticides, surface runoff, agricultural drainage water, erosion, aerial application, and others. This pollution has become a serious environmental problem in recent years (2-4). The entrance of these pollutants into water resources, especially sources of drinking water, can have adverse effects on human and environmental health, because they are highly resistant to environmental factors, very soluble in water, and toxic to living organisms (5,6). The extent of water pollution by toxins depends on the distance between the water resource and the sprayed area, the amount of precipitation, soil permeability, soil type, and other factors (7).

The degree to which people are adversely affected by the toxins depends on the quality of the chemicals, the length of time they are used, the duration of human exposure to them, the concentration of the pesticide, and the degree of its toxicity (8,9). Results of some research have shown that pesticide application causes human health complications, including abortion, mental retardation, defective fetuses, and potential carcinogenesis and mutation (10,11). Moreover, the results of some studies suggest that environmental factors, especially the use of pesticides, are reasons for the increased prevalence of diabetes (12,13).

To date, international organizations have classified more than 56 pesticides as carcinogenic agents for laboratory animals. According to reports from these organizations, the use of some pesticides has been banned and the use of others has been limited in some countries (10).

Since the use of organochlorine insecticides has been banned, organophosphate pesticides are increasingly being used all over the world to control pests (1,8,9). However, organophosphate pesticides are relatively volatile and enter the human body through contract with the skin and through the respiratory system (13). They suppress cholinesterase, overstimulate neurons, and cause neurological side effects (9,11,13,14).

Diazinon is one organo-phosphorus insecticide used to control a broad spectrum of pests in agriculture and at home (15). The World Health Organization (WHO) has classified it as a class II pesticide. Its toxicity for aquatic organisms is 350 ng/L, and its lethal dose for humans is about 90-444 mg/kg body weight (15-18). According to German standards, its maximum acceptable residue level is 1 ppm (15).

The need to remove contaminant pesticides from water resources and the high costs that will result in case of failure to take proper action about this has instigated the use of various methods and advanced oxidation processes including Fenton-like, Fenton, photo Fenton, UV/H_2O_2, UV/TiO_2, UV/O_3, iron nanoparticles, usual oxidation, coagulation and flocculation, precipitation, membrane filtration, microfiltration and reverse osmosis, nanofiltration, biological degradation, and adsorption processes. Advanced oxidation and membrane methods are not affordable, require large amounts of coagulants thus producing large amounts of sludge, and has low efficiency in the removal of bio-degradable matter. Adsorption methods are most commonly employed because of their low costs and easy operation (4,19-27).

One method for the removal of pesticides is the use of adsorbents, the most important of which is activated carbon (27-30). Because of its high porosity, activated carbon provides a large surface for the adsorbable material. The adsorption process is either physical or chemical. Covalent bonds and van der Waals forces are involved in physical and chemical adsorption, respectively (31).

Considering that diazinon is used extensively in Iran for various health-related and agricultural purposes, and given its persistence in the environment, worries remain concerning the adverse effects of this pollutant on human health and the environment. Moreover, the high cost of importing activated carbon has prevented its removal from vast areas that have been exposed to this pesticide. Therefore, in line with the policies of resistive economy and national self-sufficiency, the use of domestically produced products has received top priority. This research purposed to study the efficiency of domestically produced activated carbon in removing diazinon from aqueous solutions as a function of contact time, adsorbent dosage, pH, and initial concentrations of diazinon parameters.

Materials and methods

This batch study was carried out on a laboratory scale. Diazinon with 99% purity was purchased from the German Company Sigma Aldrich, and activated carbon produced in Iran was obtained from domestic companies. Stock diazinon solution at a concentration of 1000 mg/L was prepared by dissolving 0.1 g of diazinon in 100 mL of pure methanol (high pressure liquid chromatography - HPLC grade). All experiments were carried out in batch reactor (Erlenmeyer flasks). Contact time (5, 15, 30, 45, 60, and 90 minutes), dose of activated carbon (0.5, 1, 3, 5, 10, 20, and 30 g/L), initial diazinon concentration (5, 10, 20, 30, and 50 mg/L), and pH (3, 5, 7, 9, and 10) were the studied variables. The optimum value for each variable was determined by altering it while maintaining other variables constant. After adding activated carbon to the samples on a shaker at the speed of 200 rpm, their volumes were raised to 100 mL. The samples were centrifuged at 3000 rpm for 10 minutes, passed through a 0.45 μm pore size syringe filter, and injected into an HPLC instrument (model Shimadzu) to determine their concentrations. The chromatography conditions for diazinon analysis were isocratic reversed-phase in a C-18 column with a water/methane mobile phase using a 70/30 mixture, UV detector at wavelength of 220 nm, and flow rate of 1 mL/min (15). Specific surface area, pore volume, and average pore diameter were determined using BET and BJH methods. SPSS version 19 was used to calculate Pearson correlation coefficient and analysis of variance (ANOVA) at a significance level of 0.05 to analyze the data, and Excel 2013 was used to draw the diagrams.

Results

Characteristics activated carbon

Results of BET and BJH experiments performed to determine the characteristics of the activated carbon produced in Iran are presented in Table 1.

Based on the two afore-mentioned methods, the mean of the studied PAC specific area was 136.5 m^2/g; the BJH method was used to measure the micro pores (the main pores in adsorption), and a high portion of micropores specific area to total specific area (69%) was noted. It was also noted that the studied PAC had a good quality in this area.

Effects of contact time

Figure 1 shows the effects of contact time on the removal of diazinon by activated carbon produced in Iran at the

Table 1. Characteristics of the studied activated carbon

Specific area (BET method)	Specific surface area (BJH method)	Total pore volume	Average pore size
161.53 m^2/g	111.53 m^2/g	0.2-16 cm^3/g	4.9919 nm

constant concentration of 20 mg/L and neutral pH. As shown in the diagram, the percentage of diazinon removed improved with increases in contact time up to 30 minutes; the maximum removal rate of 48.7% was obtained at 60 minutes contact time, while at a contact time of 30 minutes, the removal rate was 48.2%. Clearly during the last 30 minutes, no adsorption occurred, and following that, no tangible difference in removal percentage was noticed; in other words, adsorption had reached equilibrium. Therefore, the optimum contact time was considered 30 minutes. The diagram indicates the effect of contact time on the quantity of diazinon adsorbed onto the activated carbon.

Effects of the adsorbent dose

The effects of adsorbent dose on removal percentage are shown in Figure 2. The diagram suggests that percentage removal improved with increases in adsorbent dose. Increases were remarkable up to the dose of 10 g/L, beyond which no conspicuous increases in removal percentage were noticed (i.e. the extra percentage increases in removal efficiency were not cost-effective). Therefore, the optimum dose in this study was considered 10 g/L. Figure 2 shows the effects of adsorbent dose on adsorption capacity. Results indicated that adsorption capacity declines with increases in adsorbent dose because of the rise in removal percentage.

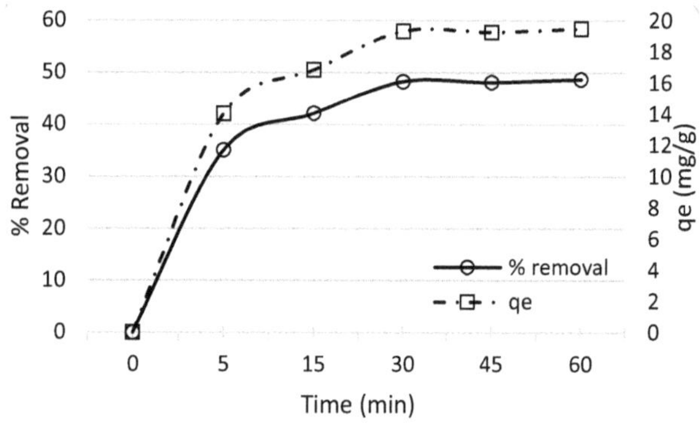

Figure 1. The effects of contact time on adsorption capacity of the adsorbent and on diazinon removal percentage (dose = 0.5 g/L; pH = 7; C_0 = 20 mg/L)

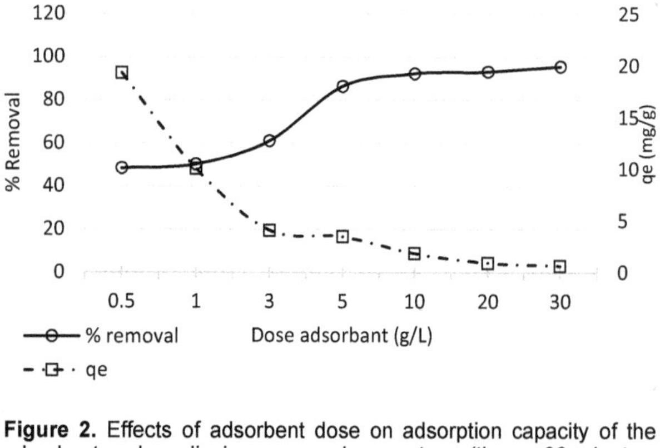

Figure 2. Effects of adsorbent dose on adsorption capacity of the adsorbent and on diazinon removal percentage (time = 30 minutes; C_0 = 20 mg/L; pH = 7).

Effects of pH

Figure 3 shows the effects of pH on removal percentage. The results indicated that percentages of removal were suitable at all pH values except for neutral pH (pH=7). Furthermore, the effects of pH on adsorption capacity were satisfactory at all pH values except for neutral pH.

Effects of initial concentration of diazinon

The effects of initial concentration on removal percentage are shown in Figure 4. As shown in the diagram, removal percentage decreased and adsorption capacity improved with increases in initial diazinon concentration. At the initial concentration of 5 mg/L, adsorbent dose of 10 g/L, and pH of 5, adsorption efficiency was 100%. The minimum adsorption capacity (about 84.47%) was observed at the initial diazinon concentration of 50 mg/L, which was acceptable considering the high initial concentration and the most critical removal efficiency.

Adsorption isotherm models

Langmuir and Freundlich isotherm models were used to study adsorption isotherm models. The linear form of Langmuir (Eq. 1) and Freundlich (Eq. 2) models are as follows:

$$\frac{1}{q_e} = \frac{1}{a.b}\left(\frac{1}{C_e}\right) + \frac{1}{b} \quad (1)$$

$$\log q_e = \log k_f + 1/n \log C_e \quad (2)$$

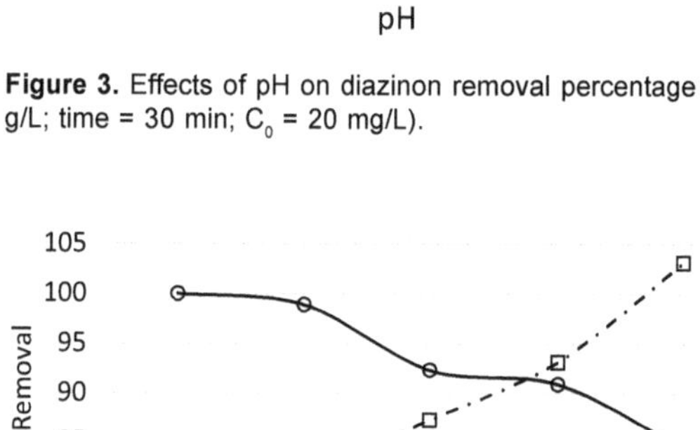

Figure 3. Effects of pH on diazinon removal percentage (dose = 10 g/L; time = 30 min; C_0 = 20 mg/L).

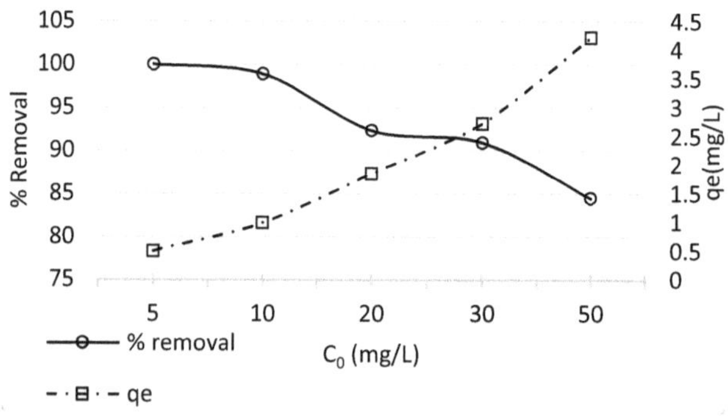

Figure 4. Effects of initial diazinon concentration on adsorption capacity of the adsorbent and on diazinon removal percentage (time = 30 min; dose = 10 g/L; pH = 5)

where q_e is adsorption capacity at equilibrium (mg/g), C_e is equilibrium concentration of adsorbate (mg/L), b is the maximum monolayer capacity (mg/g), a is constant absorption related to absorption energy (at a given temperature), and k_f and n are constants of Freundlich related to the capacity and absorption rate, respectively. The isotherm plots are shown in Figure 5. Furthermore, the coefficients of Langmuir and Freundlich adsorption models are presented in Table 2.

Kinetic studies of adsorption

Data obtained concerning adsorption was studied using pseudo-first order and pseudo-second order models to determine reaction kinetics.

Correlation coefficients for the pseudo-first order and pseudo-second order models were $R^2=0.3461$ and $R^2=0.997$, respectively. Results of the present research conform to those of the pseudo-second order model shown in Figure 6. Based on Table 3, the values of the calculated qe (qe_{cal}) and empirical qe (qe_{epx}) are very close to each other, which indicates it is more fitted to the pseudo-second order model. This model was selected as the model for describing the kinetic behavior of the activated carbon produced in Iran for removal of diazinon from aqueous solutions.

Discussion

Equilibrium time and effects of adsorbent dose, pH, and initial concentration and Langmuir and Freundlich isotherm models were studied in this research. Results of BET experiments showed that the activated carbon produced in Iran was suitable because of its large specific surface area. As shown in Figure 1, removal efficiency improved with increases in contact time to 30 minutes, because increased contact time provided greater opportunities for the adsorbent to come into contact with diazinon ions and adsorb them. Adsorption was rapid at first, but then gradual, possibly due to the presence of active unoccupied sites at the early stages of adsorption, the number of which declined with the passage of time; it may also be due to the reduced possibility of adsorption caused by the repulsive force between diazinon and activated carbon (1,30,31). Statistical studies on adsorbent dose showed that there was a significant relationship between adsorbent concentration and removal efficiency ($P=0.038$); the extent of adsorption improved, removal efficiency increased, and adsorption capacity decreased with increasing adsorbent mass. As with increases in the adsorbent's mass, a larger surface area of the adsorbent became available, and the number of active adsorption sites increased; therefore, a greater number of diazinon molecules were transferred from the solution to the surface of the adsorbent, thus increasing removal efficiency (1,28). Removal efficiency increased when higher doses of the adsorbent were used, but this increase resulted in reduced adsorption capacity. This phenomenon can be attributed to the utilization of the available surface in the unsaturated state, which leads to the undesirable use of available spaces. This point is economically important in designing large-scale facilities (32,33). As shown in Figure 2, the adsorbent dose of 10 g/L was more cost effective and resulted in satisfactory removal efficiency.

Although the optimum pH was 5, results of pH tests indicated a high removal efficiency at all pH values except for neutral pH; this finding is supported by the results of Moussavi et al, Gupta et al, and Salman et al (1,28,29). Since the pka of diazinon is 2.6 (1), then it is negatively charged at pH > 2.6. On the other hand, the pHzpc of the activated carbon was 7. Therefore, in pH values between 3-7, there was an electrostatic attraction between the pesticide species and the surface of the sorbent. At a pH of 7, there was no charge on the surface of the sorbent and

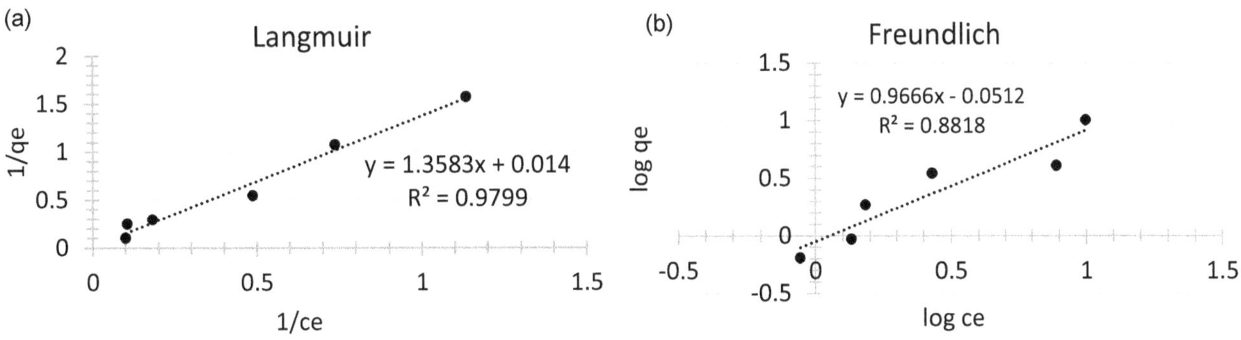

Figure 5. Adsorption isotherm diagrams: Langmuir and Freundlich models.

Table 2. Parameters of the Langmuir and Freundlich adsorption models

Langmuir model	Linear equation	b (mg/g)	a (L/mg)	R_L	R^2
	1/qe = 1.3583 (1/ce) + 0.014	71.42	0.0103	0.0007	0.9799
Freundlich model	Linear equation		N	K (L/mg)	R^2
	Log (qe) = 0.9666Log(ce)- 0.0512		1.034	0.888	0.8818

Table 3. Constants of the pseudo-first and pseudo-second order models for diazinon adsorption on activated carbon

Kinetic models	R^2	K_1	K_2	qe_{exp}	qe_{cal}
Pseudo-first order	0.3461	0.0291	-	19.30861	5.01734
Pseudo-second order	0.997	-	0.097	19.30861	18.2149

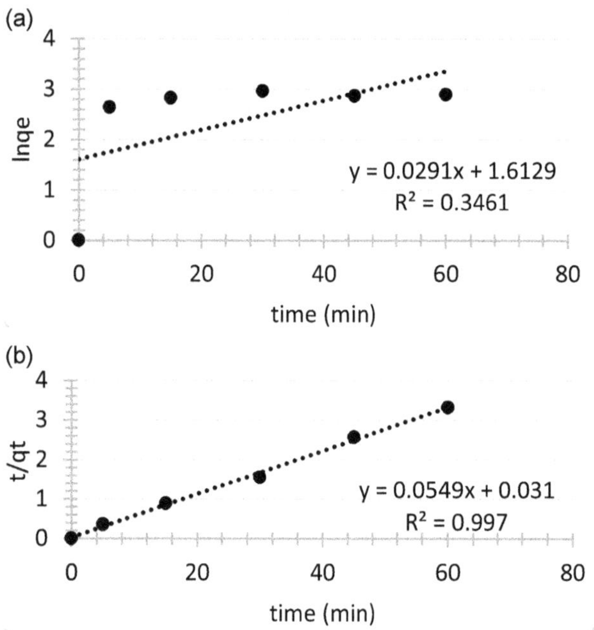

Figure 6. Diagram of adsorption kinetics: (a) pseudo-first order model; (b) pseudo-second order model.

adsorption was reduced. At higher pHs, the adsorption of diazinon was increased. Under basic conditions, other adsorption mechanisms such as hydrogen bonding may be at play instead of electrostatic attraction.

As shown in Figure 4, removal efficiency decreased with increases in initial diazinon concentration, because there were active adsorption sites at low concentrations of the pollutant. However, the number of these active sites declined and the sites were saturated with increases in the concentration of the pollutant, and desorption could take place, which led to reduced removal efficiency. Statistical studies also confirmed this. Pearson correlation coefficient was (r = - 0.83), while it must be between -1 and 1 to suggest correlation. The negative sign indicated an inverse correlation. Moreover, the $P = 0.003$ indicates there was a significant relationship between the initial concentration and removal efficiency. Maximum removal efficiency was obtained with an initial concentration of 5 mg/L. Adsorption capacity also improved with increases in initial concentration, probably because of increases in the number of molecules that competed with each other to be transferred to the active adsorption sites (34). Results of the present research are in agreement with those of other studies (26,29,35).

Using the correlation coefficient R^2 as the basis for selection of the obtained data, the present research was more in agreement with Langmuir isotherm, and R_L = 0.007 represented a desirable rate of adsorption (Table 2). According to the Langmuir isotherm, the maximum adsorption capacity of Iranian activated carbon was 71.42 mg/g. Langmuir model shows whether the adsorbent has one layer and expresses the uniformity of the adsorbent's surface. Fitness to Langmuir model suggests that the active adsorption sites were identically distributed on the surface of the activated carbon produced in Iran. In the study of Ioannidou et al, similar results were obtained that agreed with those of the present study (31). Moussavi et al reported a maximum adsorption capacity of 250 mg/g for diazinon adsorption onto a NH4Cl-assisted activated carbon (NAC) prepared from a waste biomass. It is noteworthy that no modification was done on the activated carbon used in the present study (1).

Results of the present study indicated the kinetics of diazinon adsorption on activated carbon produced in Iran was better described by the pseudo-second order model. Results of the present study conform to those found by Moussavi et al and Ayranci and Hoda (1,36).

Conclusion

This research attempted to study critical conditions. Therefore, a high concentration of the pollutant (20 mg/L) was used as the optimum concentration so that the efficiency of the activated carbon produced in Iran could be evaluated under critical conditions. Results showed that the efficiency of diazinon removal from aqueous solutions attained by the activated carbon produced in Iran was 92.53% under optimum conditions. Considering the large specific surface area of this activated carbon, it can be introduced as a very good adsorbent. The optimum conditions were a contact time of 30 minutes, initial diazinon concentration of 20 g/L, activated carbon dose of 10 g/L, and pH of 5. Results of the present study showed that the kinetic behavior of the adsorbent in a contact time of 30 minutes conformed to the pseudo-second order model. Moreover, study of the isotherm models indicated they conformed well to Langmuir model; i.e. the active adsorption sites were homogeneously distributed on the surface of the adsorbent.

Acknowledgements

The authors of this article would like to express their gratitude to the honorable Research and Technology Deputy of Hormozgan University of Medical Sciences for their financial support of this project.

Competing interests

The authors affirm that this article is their original work and have no conflicts of interest to declare.

Authors' contributions

All authors were involved in all stages of the article. On behalf of the co-authors, the corresponding author bears full responsibility for this submission.

References

1. Moussavi G, Hosseini H, Alahabadi A. The investigation of diazinon pesticide removal from contaminated water by adsorption onto NH 4 Cl-induced activated carbon. Chem Eng J 2013; 214: 172-9. doi: 10.1016/j.cej.2012.10.034.
2. Vymazal J, Březinová T. The use of constructed wetlands for removal of pesticides from agricultural runoff and drainage: a review. Environ Int 2015; 75: 11-20. doi: 10.1016/j.envint.2014.10.026.
3. El Bakouri H, Morillo J, Usero J, Ouassini A. Potential use of organic waste substances as an ecological technique to reduce pesticide ground water contamination. J Hydrol 2008; 353(3): 335-42. doi: 10.1016/j.jhydrol.2008.02.019.
4. Wang Q, Lemley AT. Oxidation of diazinon by anodic Fenton treatment. Water Res 2002; 36(13): 3237-44. doi: 10.1016/S0043-1354(02)00041-6
5. Khoadadi M. Determination of residual concentrations of phosphorus and carbamate pesticides and organiphosphate pesticides in drinking water sources in Hamedan. 12th National Conference of Environmental Health Iran; Nov 3-5, 2009. Tehran: Shahid Beheshti University of Medical Sciences and Health Services; 2009.
6. Ohno K, Minami T, Matsui Y, Magara Y. Effects of chlorine on organophosphorus pesticides adsorbed on activated carbon: desorption and oxon formation. Water Res 2008; 42(6-7): 1753-9. doi: 10.1016/j.watres.2007.10.040.
7. Jafari S. Investigation of pesticides effect on groundwater quality of shemiran villages [dissertation]. Tehran: Islamic Azad University, Science and Research; 2010. [In Persian].
8. Zhou Q, Sun X, Gao R, Hu J. Mechanism and kinetic properties for OH-initiated atmospheric degradation of the organophosphorus pesticide diazinon. Atmos Environ 2011; 45(18): 3141-8. doi: 10.1016/j.atmosenv.2011.02.064.
9. Čolović M, Krstić D, Petrović S, Leskovac A, Joksić G, Savić J, et al. Toxic effects of diazinon and its photodegradation products. Toxicol Lett 2010; 193(1): 9-18. doi: 10.1016/j.toxlet.2009.11.022.
10. Mostafalou S, Abdollahi M. Current concerns on genotoxicity of pesticides. Int J Pharmacol 2012; 8(6): 473-4. doi: 10.3923/ijp.2012.473.474.
11. Pirsaheb M, Dargahi A, Hazrati S, Fazlzadehdavil M. Removal of diazinon and 2, 4-dichlorophenoxyacetic acid (2, 4-D) from aqueous solutions by granular-activated carbon. Desalination and Water Treatment 2014; 52(22-24): 4350-5. doi: 10.1080/19443994.2013.801787.
12. Mostafalou S, Abdollahi M. The role of environmental pollution of pesticides in human diabetes: Int J Pharmacol 2012; 8(2): 139-140. doi: 10.3923/ijp.2012.139.140.
13. Wright DA, Welbourn P. Environmental Toxicology. Cambridge: Cambridge University Press 2002.
14. Androutsopoulos VP, Hernandez AF, Liesivuori J, Tsatsakis AM. A mechanistic overview of health associated effects of low levels of organochlorine and organophosphorous pesticides. Toxicology 2013; 307: 89-94. doi: 10.1016/j.tox.2012.09.011.
15. Rezaei Kalantary R, Dadban Shahamat Y, Farzadkia M, Esrafili A, Asgharnia H. Photocatalytic degradation and mineralization of diazinon in aqueous solution using nano-TiO2 (Degussa, P25): kinetic and statistical analysis. Desalination and Water Treatment 2015; 55(2): 555-63. doi: 10.1080/19443994.2014.928795.
16. Li PC, Swanson EJ, Gobas F. Diazinon and its degradation products in agricultural water courses in British Columbia, Canada. Bull Environ Contam Toxicol 2002: 69(1): 59-65. doi: 10.1007/s00128-002-0010-0.
17. Farzadkia M, Esrafili A, Baghapour MA, Shahamat Dadban Y, Okhovat N. Degradation of metronidazole in aqueous solution by nano-ZnO/UV photocatalytic process. Desalination and Water Treatment 2014; 52(25-27): 4947-52. doi: 10.1080/19443994.2013.810322.
18. Shemer H, Linden KG. Degradation and by-product formation of diazinon in water during UV and UV/H 2 O 2 treatment. J Hazard Mater 2006; 136(3): 553-9. doi: 10.1016/j.jhazmat.2005.12.028.
19. Humbert H, Gallard H, Suty H, Croué JP. Natural organic matter (NOM) and pesticides removal using a combination of ion exchange resin and powdered activated carbon (PAC). Water Res 2008; 42(6-7): 1635-43. doi: 10.1016/j.watres.2007.10.012.
20. Khodadadi M, Samadi MT, Rahmani AR. Comparison between the efficiency of advanced oxidation process and coagulation for removal organophosphorus and carbamat pesticides. Iran J Health Environ 2011; 4(3): 277-88.
21. Navarro S, Fenoll J, Vela N, Ruiz E, Navarro García G. Removal of ten pesticides from leaching water at pilot plant scale by photo-Fenton treatment. Chem Eng J 2011; 167(1): 42-9. doi: 10.1016/j.cej.2010.11.105.
22. Rodriguez-Mozaz S, Ricart M, Köck-Schulmeyer M, Guasch H, Bonnineau C, Proia L, et al. Pharmaceuticals and pesticides in reclaimed water: efficiency assessment of a microfiltration–reverse osmosis (MF–RO) pilot plant. J Hazard Mater 2015; 282: 165-173. doi: 10.1016/j.jhazmat.2014.09.015.
23. Samadi MT, Khodadadi M, Rahmani AR, Allahresani A, Saghi MH. Comparison of the efficiency of simultaneous application of UV/O_3 for the removal of organophosphorus and carbamat pesticides in aqueous solutions. Water and Wastewater 2010; 21(1): 69-75. [In Persian].
24. Sanches S, Crespo MTB, Pereira VJ. Drinking water treatment of priority pesticides using low pressure UV photolysis and advanced oxidation processes. Water Res 2010; 44(6): 1809-18. doi: 10.1016/j.watres.2009.12.001.
25. Sanches S, Penetra A, Rodrigues A, Ferreira E, Cardoso VV, Benoliel MJ, et al. Nanofiltration of hormones and pesticides in different real drinking water sources. Sep Purif Technol 2012; 94: 44-53. doi: 10.1016/j.seppur.2012.04.003
26. Ameri A, Srafili A, Farzadkia M, Kermani M, Hajizadeh A. Fenton and Fenton performance review processes for the removal of diazinon in aqueous solutions. 16th National Conference on Environmental Health; Oct 1-3, 2013. Tabriz: Tabtiz University of Medical Sciences; 2013.
27. Foo KY, Hameed BH. Detoxification of pesticide waste via activated carbon adsorption process. J Hazard Mater 2010; 175(1-3): 1-11. doi: 10.1016/j.jhazmat.2009.10.014.
28. Gupta V, Gupta B, Rastogi A, Agarwal Sh, Nayak A. Pesticides removal from waste water by activated carbon prepared from waste rubber tire. Water Res 2011; 45(13): 4047-55. doi: 10.1016/j.watres.2011.05.016.
29. Salman JM, Njoku VO, Hameed BH. Adsorption of pesticides from aqueous solution onto banana stalk activated carbon. Chem Eng J 2011; 174(1): 41-8. doi:10.1016/j.cej.2011.08.026.
30. Srivastava B, Jhelum V, Basu DD, Patanjali PK. Adsorbents for pesticide uptake from contaminated water: a review. J Sci Ind Res 2009; 68(10): 839-50.
31. Ioannidou OA, Zabaniotou AA, Stavropoulos GG,

Islam MA, Albanis TA. Preparation of activated carbons from agricultural residues for pesticide adsorption. Chemosphere 2010; 80(11): 1328-36. doi: 10.1016/j.chemosphere.2010.06.044.

32. Ong ST, Lee CK, Zainal Z. Removal of basic and reactive dyes using ethylenediamine modified rice hull. Bioresour Technol 2007; 98(15): 2792-9. doi: 10.1016/j.biortech.2006.05.011.
33. Sulak MT, Demirbas E, Kobya M. Removal of Astrazon Yellow 7GL from aqueous solutions by adsorption onto wheat bran. Bioresour Technol 2007; 98(13): 2590-8. doi: 10.1016/j.biortech.2006.09.010.
34. Kumar Meena A, Rajagopal C, Mishra GK. Removal of heavy metal ions from aqueous solutions using chemically (Na2S) treated granular activated carbon as an adsorbent. J Sci Ind Res 2010; 69(6): 449-453.
35. Akbarzadeh S, Kolahi S. Removal of pesticides parathion, diazinon and carbaryl from drinking water samples to help iron nanoparticles. 16th National Conference on Environmental Health; Oct 1-3, 2013.Tabriz: Tabtiz University of Medical Sciences; 2013.
36. Ayranci E, Hoda N. Adsorption kinetics and isotherms of pesticides onto activated carbon-cloth. Chemosphere 2005; 60(11): 1600-7. doi: 10.1016/j.chemosphere.2005.02.040.

12

Catch per unit efforts and impacts of gears on fish abundance in an oxbow lake ecosystem in Eastern India

Dipankar Ghosh[1], **Jayanta Kumar Biswas**[2*]

[1]PhD of Ecological Studies, Department of Ecological Studies, University of Kalyani, Kalyani, West Bengal, India
[2]Associate Professor of Ecological Studies, Department of Ecological Studies and Coordinator of International Centre for Ecological Engineering, University of Kalyani, Kalyani, West Bengal, India

Abstract
Background: Oxbow lakes are abundant in indigenous fishes, but they are subject to unsustainable fishing practices, potential overexploitation, and indiscriminate use of fine-meshed fishing gear. To quantify the catch per unit effort (CPUE) and impact of fishing gears on fish abundance, a survey was carried out in an oxbow lake in eastern India.
Methods: The gear-wise CPUE for fish caught in per unit hour of operation was calculated by dividing the total sampling gear catch in biomass, which is the observed value of fish caught by a particular gear, by the total sampling effort hours. A value of $P < 0.05$ was accepted as statistical significance.
Results: Average annual values of the CPUE of triangular push nets, gill nets, long lines, seine nets, drag nets, stationary dip nets, cone-framed cast nets, and line and hook were calculated as 328.34, 4.12, 36.71, 572.92, 3928.57, 237.78, 235.80, and 0.44 grams of fish per hour of operation, respectively. All the 8 different gears exhibited lower CPUE during monsoon and post-monsoon seasons than in the pre-monsoon season. The line and hook was dominant (>71%). Cone-framed cast net hauled the maximum catch in biomass (31.51%), and gill nets contributed the maximum number of fish (64.92%). The lower CPUE values of line and hook, gill net, cone-framed cast net and long lines identified them as the most harmful among all gears.
Conclusion: Indiscriminate use of gear, particularly line and hook, gill nets, cone-framed cast nets, and long lines, demands regulations and preventions concerning such gear to obtain higher fish abundance.
Keywords: Fish abundance, Fishing gears, CPUE, Oxbow lake

***Correspondence to:**
Jayanta Kumar Biswas
Email: biswajoy2000@yahoo.com

Introduction

The oxbow lakes were formed by the Ganga River. It supports rich biodiversity and offers a livelihood and nutritional security, yet its conservation is less studied in India. Fish abundance in oxbow lakes is severely impacted by human activities like over-exploitation and the indiscriminate use of fine-meshed fishing gear, which always forces the development of systematic and appropriate planning for biodiversity conservation and management strategies (1-3). Among the different types of fishing gear studied (4,5), those like gill nets and seine nets exhibit highly destructive effects (6-9) on fish abundance. The catch per unit effort (CPUE) is a measure of stock density, physical and financial productivity, and an indicator of the efficiency of a fishing operation. Varied CPUE means from different aquatic systems across the world have been reported as being 1.5-14.0 kg/d (6); 2.04 to 48.99 kg/unit/d (9); 0.58 kg/ha/hr and 0.08 kg/ha/hr/purse seine (10); 20-310 g/h/gear (11); 0.95-15.25 kg/unit/d (12); 2.6-62.4 kg/fisher/d (13). The average daily catch per fisher has been reported as 1.43 kg with different fishing efforts (no of gear/day): push nets (6.75), gill nets (3), cast nets (8), lift nets (5.25), current nets (3.75), and hook (67) (11). Fishing durations (h/d) of gill nets (6-12), push nets (0.5-2), drag nets (2-4), lift nets (2-6), and hooks (2-5) have also been reported with the highest density of fishers and catches in the monsoon season, and the lowest was reported in the pre-monsoon season (5,14). However, not enough information is available on the estimation and impact of CPUE of different fishing gear on the patterns of native fish species abundance in the oxbow lake of the Nadia district, India in particular. As there is a need to develop systematic and appropriate planning for fish conservation and management strategies for oxbow lakes, the current study aimed to quantify the CPUE and impact of 8 different fishing gears on fish abundance in a tropical oxbow lake ecosystem of the Ganga river basin in eastern India.

Materials and Methods

Study area

The Chhariganga oxbow lake (abandoned, fractioned, and derived from the river Ganga), located in the Nakashipara development block of Nadia district, West Bengal, India, was randomly selected. This lake is situated at 23.5800°N, 88.3500°E, about 90 km away from the Kalyani University campus in Nadia and nearly 40 km away from the Tropic of Cancer towards the north. It is a fresh-water, semi-closed oxbow lake and receives water from the Ganga river during the monsoon season through a narrow channel at the northeast corner of a loop in the river. This lake spreads over an area of 58.28 ha and has an annual average depth of 2.6 m. It also stores rain water. The catchment area of the oxbow lake is nearly 600 ha (Figure 1). Climate changes create three distinct annual seasons in this region: the monsoon or rainy season generally runs from July to October; the jute retting period lies normally during August and September; the post-monsoon or winter season runs from November to February; and the pre-monsoon or dry season runs from March to June. Occasional flooding of the surrounding banks occurs during the monsoon season. The oxbow lake is subjected to all forms of human activities, including jute retting during the monsoon season, agriculture, and fishing. It is the only source of irrigation water for the neighboring agricultural communities.

Fish sampling and analysis

Sample fishing was carried out on several random occasions using the expertise of local fisher folk and different types of gear (triangular push nets, gill nets, long lines, a seine net, a drag net, a stationary dip net, and a cone-framed cast net all with varying mesh sizes, and a line and hook) (1-3), which allowed the sampling of a range of fish sizes and minimized the bias due to specific gears (sample size 24). Each gear was operated for a period of 4 to 24 hours at different sites of the oxbow lake, bringing the total mean efforts per day to 65, 44, 77, and 95 with gear densities of 2565, 5161, 2957, and 10683, and total sampling gear efforts to 3648, 5200, 3411, and 12259 during the pre-monsoon, monsoon, and post-monsoon seasons, and the full year, respectively. Fishes were sorted by their numbers and weighed.

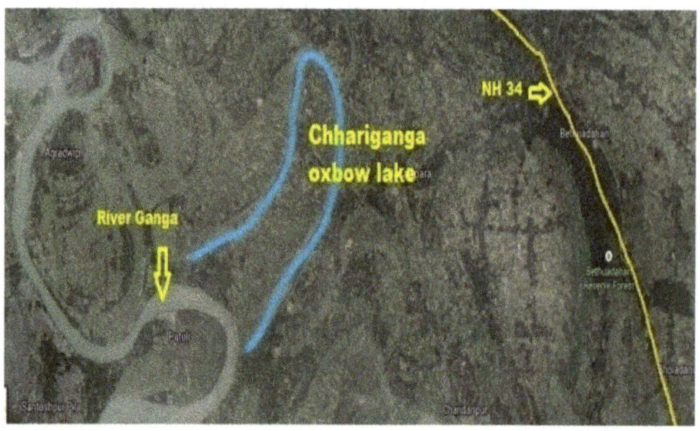

Figure 1. Map showing study area (Modified after being downloaded from google.com on 12-05-2016).

Studies of different fishing gear used in sample fishing

Eight different gears, namely triangular push nets, gill nets, long lines, seine nets, drag nets, stationary dip nets, cone-framed cast nets, and line and hook, which had different dimensions and specifications and varying mesh sizes were considered for fishing in the present study (Table 1). Neither the stationary dip net during the pre-monsoon season (like the triangular push nets or seine nets in Komor, a fish aggregating device (FAD) nor the cone-framed cast net during the monsoon season was seen in operation at the oxbow lake. The drag net was seen to be operated only during the monsoon season. Sample fishing gear densities were based on their appearances observed in the oxbow lake for the three seasons; accordingly, seasonal sampling fishing was randomly performed. Line and hook gear was most dense among all forms, whereas the seine net in Komor was used only once during the year. Eight different fishing gears known to contribute major fish catches in the Chhariganga oxbow lake were studied in detail (Table 1).

Calculation of catch per unit effort

In the current study, gear-wise CPUE for fish caught per unit hour of operation was calculated (1,3) by dividing the total sampling gear catch in biomass (TSGCB) or number (TSGCN), which is the observed value of fish caught by a particular gear, by total sampling effort hours (TSEH). TSEH is calculated as the product of average sampling effort hour of operation of a particular gear per day (SEHPD) and total numbers of such gear used, i.e. sampling gear density (SGD).

$$\text{CPUE (g/h or n/h)} = (\text{TSGCB or TSGCN}) \div \text{TSEH}$$

Where total sampling effort hours (TSEH)=SEHPD (sampling effort hours/day/gear) x SGD (sampling gear density), g = gram, n = number, h = hour.

Calculation of catch per gear effort

Similarly, the gear-wise catch per gear effort (CPGE) for fish caught per unit effort (or attempt or operation) was calculated [1; 3] by dividing total sampling gear catch (TSGCB or TSGCN) by the total sampling gear effort (TSGE). TSGE was calculated as the product of putting an average sampling effort per day (SEPD) of a particular gear and total numbers of such gear used, i.e. SGD.

$$\text{CPGE (g/e or n/e)} = (\text{TSGCB or TSGCN}) \div \text{TSGE}$$

where, total sampling gear effort (TSGE) = SEPD (sampling effort/day/gear) x SGD (sampling gear density), g=gram, n=number, e=operation effort.

The overall catch per unit or gear effort (CPUE or CPGE), a measure of relative abundance (n or g/h and n or g/e), was calculated by dividing the total catch in number (n) or wet biomass (g) from gear (s) by total hours (h) or efforts (e) of operations of gear(s) used during those three seasons and the year. Average mean fish density (n/m^3) and standing biomass (g/m^3) was calculated by dividing the total number (n) or wet biomass (g) of fish encountered in an area (m^3) operated by gear (s) for each season.

Statistical analyses determined the mean, standard deviation (SD), and degree of relationships with the help of MS Excel software. The results were then presented in textual, tabular, and graphical form. The level of $P<0.05$ was accepted as statistical significance.

Results

Season-wise and gear-wise CPGE and CPUE are furnished in Table 2. The annual average values of CPGE were calculated as 45.46, 41.19, 403.84, 13750.03, 4583.33, 163.98, 71.77, and 4.36 grams of fish per effort; CPGE (5.45, 4.58, 1.67, 1444.50, 174.17, 30.83, 0.08 and 0.04 fish per effort); CPUE (328.34, 4.12, 36.71, 572.92, 3928.57, 237.78, 235.80, and 0.44 grams of fish per hour of operation), and CPUE (39.33, 0.46, 0.15, 60.19, 149.29, 44.70, 0.27, and 0.004 fish per hour of operation), respectively, for triangular push nets, gill nets, long lines, seine net, drag net, stationary dip net, cone-framed cast net, and line and hook. The lowest CPGE and CPUE values were observed with the line and hook (*Nal Borshi*) during all seasons. The CPUE of triangular push nets during the post-monsoon season was reduced by 14%, to 301.26 g/h from 350 g/h during the pre-monsoon season. Similarly, the pre-monsoon CPUE (g/h) of gill nets was decreased by 1st/4th during monsoon season and by half during the post-monsoon season. The monsoon CPUE (g/h) of the net was also reduced by 37%. In the present study, the pre-monsoon CPUE (g/h) of long lines was decreased by 3rd/4th during monsoon season and by 79% during the post-monsoon season when the monsoon CPUE (g/h) of the gear was also reduced by 1st/5th. The CPUE of the stationary dip nets of the monsoon season was sharply decreased by 2nd/3rd during the post-monsoon season. The pre-monsoon CPUE of the cone-framed cast net was also sharply reduced by 53% during the post-monsoon season. The CPUE of the gear line and hook was drastically reduced by more than 50% during the post-monsoon season of the year in the present study.

Seasonal variations in relative gear dominance and relative catch composition

Maximum fishing duration was observed for seine nets (24 h/d), and the minimum was observed for push nets (4-5 h/d). Drag nets (7 h/d), stationary lift nets (18-22 h/d),

Table 1. Gears used in sample fishing

Common name	Vernacular name	Dimensions and attributes	Mesh size (mm)	EGA (sqm)	AOP (sqm) PRM	MON	POM	Year
Triangular push net	Thela Jaal	(1 sqm = 0.5× 1 m × 2 m), operated mainly under water hyacinth covered area	5.0-15.0	1	16620	-	22710	19665
Gill nets	Fansh Jaal	(45 m × 0.75 m), made of monofilament, 10-12 no net each for 70-110 fishers	≥ 22.0	33.75	183380	455050	277290	305240
Long lines	Daun	With 80-100 baits, mainly carnivores caught, operates in coverage area (200 sqm = 200 m × 1 m)	-	200	183380	455050	277290	305240
Seine net	Komor	Fish caught from a type of microsanctuary or FAD created by tree branches	3.0-10.0	333	200000	-	300000	250000
Drag net	Ber Jaal	(50 m × 6 m)	10.0-12.0	300	-	500000	-	500000
Stationary dip net	Bashaal Jaal, Dhenki Jaal	(72 sqm = 0.5 × 12 m × 12 m)	≥ 5.0	72	-	455050	277290	366170
Cone framed cast net	Chaabi Jaal	(7 sqm = 3.14 × 1.5 m × 1.5 m)	≥10.0	7	200000	-	300000	250000
Line and hook	Nal Borshi	Area (1200 sqm = 1000 m × 1.2 m) operates in a line of with 200-250 baits, mainly carnivores caught	-	5.33	183380	455050	277290	305240
Total	8				966760	2320200	1731870	2301555

Abbreviations: EGA, effective gear area; PRM, pre-monsoon; MON, monsoon; POM, post-monsoon; AOP, area of operation; (-), no operation.

Table 2. Fishing gears employed and their CPUE and CPGE

Season	Gear	Triangular push nets	Gill nets	Long lines	Seine net	Drag net	Stationary dip net	Cone framed cast net	Line and hook	Total
PRM		1	24	5	2	-	-	33	2500	2565
MON	SGD	-	24	10	-	1	1	-	5125	5161
POM		1	24	21	2	-	1	33	2875	2957
YR		2	72	36	4	1	2	66	10500	10683
PRM		28	1	1	1	-	-	33	1	65
MON	SEPD	-	1	1	-	6	35	-	1	44
POM		37	1	1	1	-	23	13	1	77
YR		33	1	1	1	6	29	23	1	95
PRM		5	10	11	24	-	-	7	10	67
MON	SEHPD	-	10	11	-	7	22	-	10	60
POM		4	10	11	24	-	18	7	10	84
Year		4.5	10	11	24	7	20	7	10	94
PRM		28	24	5	2	-	-	1089	2500	3648
MON	TSGE	-	24	10	-	6	35	-	5125	5200
POM		37	24	21	2	-	23	429	2875	3411
YR		65	72	36	4	6	58	1518	10500	12259
PRM		5	240	55	48	-	-	231	25000	25579
MON	TSEH	-	240	110	-	7	22	-	51250	51629
POM		4	240	231	48	-	18	231	28750	29522
YR		9	720	396	96	7	40	462	105000	106730
PRM		1750	1324	6100	27500	-	-	74250	13090	124014
MON	TSGCB	-	1005	3100	-	27500	7500	-	25625	64730
POM		1205	637	5338	27500	-	2011	34690	7033	78414
Year		2955	2965	14538	55000	27500	9511	108940	45748	267158
PRM		124	141	20	1876	-	-	79	108	2348
MON	TSGCN	-	126	21	-	1045	1304	-	230	2726
POM		230	63	19	3902	-	484	45	56	4799
Year		354	330	60	5778	1045	1788	124	394	9873
PRM		62.50	55.16	1220.00	13750.04	-	-	68.18	5.24	15161
MON		-	41.86	310.00	-	4583.33	214.29	-	5.00	5154
POM	CPGE (g/e)	32.57	26.54	254.20	13750.02	-	87.43	80.86	2.45	14234
Season Average		47.53	41.19	594.74	13750.03	4583.33	150.86	74.52	4.23	19246
Year average		45.46	41.19	403.84	13750.03	4583.33	163.98	71.77	4.36	19064
PRM		4.43	5.88	4.00	938.00	-	-	0.07	0.04	952
MON		-	5.25	2.10	-	174.17	37.26	-	0.04	219
POM	CPGE (n/e)	6.22	2.63	0.90	1951.00	-	21.04	0.10	0.02	1982
Season Average		5.32	4.58	2.33	1444.50	174.17	29.15	0.09	0.04	1660
Year average		5.45	4.58	1.67	1444.50	174.17	30.83	0.08	0.04	1661
PRM		350.00	5.52	110.91	572.92	-	-	321.43	0.52	1361
MON		-	4.19	28.18	-	3928.57	340.91	-	0.50	4302
POM	CPUE (g/h)	301.26	2.65	23.11	572.92	-	111.72	150.17	0.24	1162
Season Average		325.63	4.12	54.07	572.92	3928.57	226.32	235.80	0.42	5348
Year average		328.34	4.12	36.71	572.92	3928.57	237.78	235.80	0.44	5345
PRM		24.80	0.59	0.36	39.08	-	-	0.34	0.004	65
MON		-	0.53	0.19	-	149.29	59.27	-	0.004	209
POM	CPUE (n/h)	57.50	0.26	0.08	81.29	-	26.89	0.19	0.002	166
Season Average		41.15	0.46	0.21	60.19	149.29	43.08	0.27	0.004	295
Year average		39.33	0.46	0.15	60.19	149.29	44.70	0.27	0.004	294

Abbreviations: SGD, sampling gear density; SEPD, sampling effort per day per gear; TSGE, total sampling gear effort; SEHPD, sampling efforts hrs per day per gear; TSEH, total sampling effort hours; TSGCB, total sampling gear catch in biomass; TSGCN, total sampling gear catch in number; CPGE, catch per gear effort (g/e or n/e), TSGC/TSGE, or TSGCN/TSGE; CPUE, catch per unit effort (g/h or n/h)=TSGCB/TSEH or TSGCN/TSEH; g, gram; n, number; e, effort; h, hour; PRM, pre-monsoon; MON, monsoon; POM, post-monsoon, (-), no operation.

and line and hook (10 h/d) had the highest and lowest gear operator densities during the post-monsoon and monsoon seasons, respectively. The seine net followed by the gill net were seen to be used the most and the drag net was used the least by gear operators in the present study. The stationary lift net and long lines (Daun Borshi) made fourth and fifth position, respectively, in the oxbow lake of the present study. Seasonal variations in relative gear dominance (RGD) revealed the gear line and hook made its dominant presence (>71%) in the lake throughout the year in terms of numbers (Table 3). The stationary dip net showed maximum relative catch composition both in biomass (RCB) and numbers (RCN) throughout the year except during the pre-monsoon season when the cone-framed cast net hauled in the maximum catch in biomass (31.51%) and the gill nets contributed the maximum number of fish (64.92%). The seine net operation appeared to be the least dense among all gears throughout the year except during the monsoon season when the drag net had the least operation. Annual variations in relative fish abundance both in terms of catch biomass and numbers in relation to dominance and CPUE of different fishing gears are furnished in Figure 2.

Discussion

Gear density was inversely related with gear catch throughout the year for the biomass and for the numbers with the exception of the pre-monsoon season. Seasonal variations in correlation coefficients of gear densities and gear catches during all three seasons of the year are given in Table 4. Gear densities among all seasons were strongly correlated, whereas gear catches in biomass of the three seasons were weakly correlated. Gear catches in numbers were also strongly correlated in those three seasons, like its density, with the exception of the post-monsoon season. All the gears in the present study exhibited lower CPUE during the monsoon and post-monsoon seasons compared to the pre-monsoon season. This might be due to over-exploitation by the fishers and swelled water spread area due to influx from the Ganga River. Fishers in the present study generally were observed to use multiple gears enhancing the density of the gear operators. That was extended to increase their earnings through catching without increasing the number of fishers in the oxbow lake. The present findings regarding different fishing gears are in full or partial agreement with the results of other similar studies (4-6,8,9,11,14).

The mean CPUE of gears employed in the Chhariganga oxbow lake under study varied due to the gear dimensions, density and pressure, luring ingredients, the catchers' preferences of fishing places and, of course, their experience and expertise. The CPUE showed fluctuating trends during the study period due to the influx of fish along with flood waters entering the oxbow lake during the monsoon season and subsequently the heavy fishing pressure and jute retting (1-3) during the monsoon and post-monsoon seasons. However, the fish availability and catch decreased in the post-monsoon season and increased again in the pre-monsoon season. Except the line and hook, gill net, cone-framed cast net, and long lines, other gears had a moderately higher annual CPUE for both fish biomass and numbers in the present study, indicating the comparatively more harmful impacts of those gears on fish abundance. This result concurs partially with the findings of other studies (1,15).

The CPUEs of all the fishing gears of the studied oxbow lake ecosystem were much lower than the values obtained in the study on floodplain *Beels* in Bangladesh (15). This result may be attributable to the variations in their geographical and anthropological parameters. The findings of the current study regarding the CPUE of different fishing gears are also in full or partial agreement with the results of other similar studies (6,9-13).

The oxbow lake in the present study is a moderately to poorly productive body of water the fish abundance of which is decreasing (2,3). Fishing nets were found to be the most harmful among all studied gears (1). The indiscriminate use of fishing gear with varied mesh sizes

Table 3. Seasonal variations in relative gear dominance and relative catch composition (%)

Gear catch	PRM			MON			POM			Year total		
	RGD	RCB	RCN	RGD	RCB	RCN	RGD	RCB	RCN	RGD	RCB	RCN
Triangular push nets	0.60	20.79	30.61	-	-	-	0.73	26.74	33.84	0.53	12.66	15.13
Gill nets	26.72	29.33	64.92	18.23	19.12	18.89	21.74	17.45	11.45	21.25	21.99	23.07
Long lines	0.40	9.66	0.66	0.86	6.71	0.36	0.55	4.23	0.10	0.65	7.09	0.32
Seine net	0.28	1.30	1.84	-	-	-	0.26	1.82	1.71	0.22	0.82	0.81
Drag net	-	-	-	0.02	7.22	2.16	-	-	-	0.02	3.29	1.17
Stationary dip net	-	-	-	0.33	56.85	77.87	0.55	32.91	52.51	0.46	33.38	58.72
Cone framed cast net	0.94	31.51	0.70	-	-	-	0.85	11.28	0.10	0.71	12.56	0.13
Line and hook	71.06	7.41	1.27	80.56	10.09	0.71	75.30	5.57	0.29	76.15	8.22	0.66
Total	100	100	100	100	100	100	100	100	100	100	100	100

Abbreviations: RGD, relative gear dominance; RCB, relative catch biomass; RCN, relative catch numbers; PRM, pre-monsoon; MON, monsoon; POM, post-monsoon.

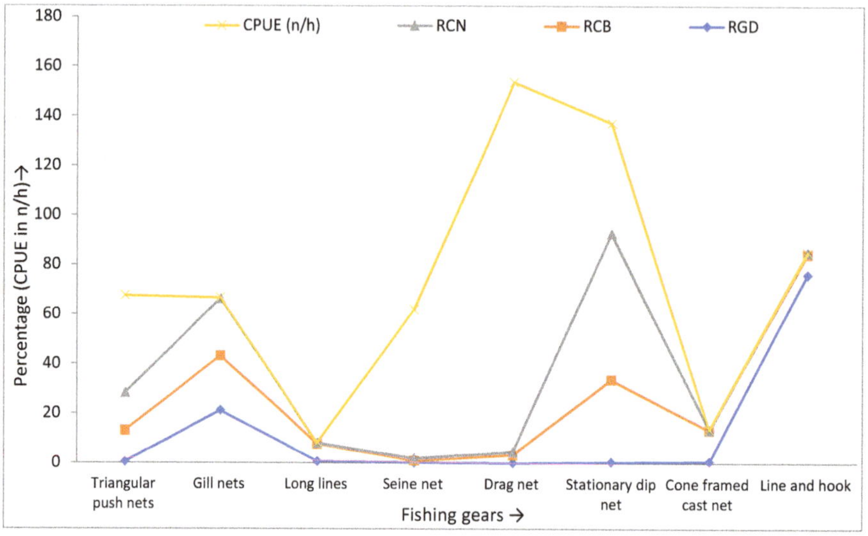

Figure 2. Annual variations in fish abundance with dominance and CPUE of different fishing gears.
Abbreviations: RGD, relative gear dominance; RCB, relative catch biomass; RCN, relative catch numbers; CPUE, catch per unit effort (n/h); n, number; h, hour.

Table 4. Seasonal variations in correlation coefficients of gear density and catch

	Gear Density			Gear Catch (Biomass)			Gear Catch (numbers)		
	MON	POM	YR	MON	POM	YR	MON	POM	YR
PRM	0.986	0.996	0.995	0.94	0.66	0.86	1.00	0.56	0.99
MON		0.998	0.998		0.97	0.93		1.00	0.99
POM			1.000			0.85			0.90

should be regulated and prevented so as to arrest the decreasing trend in fish abundance in the present oxbow lake. The findings from the present study will reveal the way for the planning and management of sustainable fisheries and the conservation of these natural resources at a national level.

Conclusion

Indiscriminate use of fishing gears, particularly the line and hook, gill net, cone-framed cast net, and long lines, demand regulations and preventions to provide for a higher fish abundance in the present oxbow lake in years to come.

Authors' contributions

All authors contributed equally and participated in the collection, analysis, and interpretation of the data. All authors critically reviewed, refined, and approved the

Competing interests

The authors declare that they have no competing interests.

Acknowledgements

Authors acknowledge the research facilities and/or other help and cooperation extended by Department of Ecological Studies, University of Kalyani, Department of Fisheries, Government of West Bengal and Kutirpara Fishermen Cooperative Society associated with Chhariganga Oxbow lake.

References

1. Ghosh D, Biswas JK. Erosion of fish diversity: ranking degree of dangers of unsustainable fishing gears in a Tropical Oxbow lake in eastern India. Int J Chem Biol Sci 2017;3(12):1-17.
2. Ghosh D, Biswas JK. Fish productivity: Assessing sustainability in a tropical Oxbow lake of Nadia district, West Bengal, India. Archives of Agriculture and Environmental Science 2017;2(1):6-20.
3. Ghosh D, Biswas JK. Fish Fauna faces anthropogenic double trouble: erosion of fish diversity in tropical Oxbow Lake of the Ganga River Basin in Eastern India. Journal of Biodiversity and Endangered Species 2017;5(2):188. doi: 10.4172/2332-2543.1000188.
4. Chakravartty P, Sharma S. Different types of Fishing Gears used by the Fishermen in Nalbari district of Assam. International Journal of Social Science & Interdisciplinary Research 2013;2(3):177-91.
5. Zahan N. Management of Beel Fishery: A Special Reference to Chapaigachi Beel of Kushtia [dissertation]. Mymensingh: Bangladesh Bangladesh Agricultural University; 2013.

6. Joadder AR. An Ecological Study on the Beel Joshi (Rajshahi District), Northern Bangladesh. Journal of Fisheries International 2009;4(2):23-9.
7. Rahman S, Mazid MA, Kamal M, Hossain MA, Hossain MS. Study on fishing gears, species selectivity toward gears and catch composition of BSKB *Beel*, Khulna, Bangladesh. Bangladesh Journal of Fisheries Resources 1999;3(1):25-32.
8. Saberin IS, Reza MS, Ayon NJ, Kamal M. Estimation of size selectivity of fish species caught by different gears in the old Brahmaputra River. Journal of the Bangladesh Agricultural University 2013;11(2):359-64. doi: 10.3329/jbau.v11i2.19940.
9. Sayeed A, Hashem S, Salam MA, Hossain MA, Wahab MA. Efficiency of fishing gears and their effects on fish biodiversity and production in the Chalan Beel of Bangladesh. European Scientific Journal 2014;10(30):294-309.
10. Sayduzzaman AK, Das M, Hasan MR. Catch assessment of indigenous and exotic carp species of Nasti baor. Bangladesh Journal of Fisheries Research 1999;3(2):113-22.
11. Ahmed N, Rahaman MM, Rahaman MM. Fish catch assessment Maljhee Kangsa Flood plain in Bangladesh. Pak J Biol Sci 2005;8(3):396-400. doi: 10.3923/pjbs.2005.396.400.
12. Ahmed MS. Assessment of fishing practices on the exploitation of the Titas floodplain in Brahmanbaria, Bangladesh. Turkish Journal of Fisheries and Aquatic Sciences 2008;8:329-34.
13. Novaes JL, Carvalho ED. Analysis of artisanal fisheries in two reservoirs of the upper Paraná River basin (Southeastern Brazil). Neotrop Ichthyol 2013;11(2):403-12. doi: 10.1590/S1679-62252013005000002 .
14. Siddiq A, Miah I, Ahmed ZF, Asadujjaman M. Present status of fish, fishers and fisheries of Dogger Beel in Hajigonj Upazila, Chandpur, Bangladesh. Journal of Aquatic Science 2013;1(2):39-45. doi: 10.12691/jas-1-2-3.
15. Rahman M, Sayeed A, Rasul G, Mondal N, Majumdar BC, Azad Shah A. Impact of fishing gear on fish biodiversity of *Hakaluki haor* in Bangladesh. International Journal of Fisheries and Aquatic Studies 2016;4(6):257-62.

13

Removal of blue cat 41 dye from aqueous solutions with ZnO nanoparticles in combination with US and US-H_2O_2 advanced oxidation processes

Sohrab Golmohammadi[1], Mohammad Ahmadpour[2], Aliakbar Mohammadi[3], Azim Alinejad[4], Nezam Mirzaei[5], Mansour Ghaderpoori[6*], Afshin Ghaderpoori[7]

[1]MSc of Water and Wastewater, Environmental Engineering (Trends Water and Wastewater), Kurdistan Rural Water & Wastewater CO (Kamyaran), Sanandaj, Iran

[2]Instructor, Department of Environmental Health Engineering, School of Public Health, Maragheh University of Medical Science, Maragheh, Iran

[3]Instructor, Department of Environmental Health Engineering, School of Public Health, Neyshabur University of Medical Sciences, Neyshabur, Iran

[4]Ph.D of Environmental Health Engineering, Social Determinants in Health Promotion Research Center, Hormozgan University of Medical Sciences, Bandar Abbas, Iran

[5]Assistant Professor of Environmental Health Engineering, Environmental Health Research Center, Kurdistan University of Medical Sciences, Sanandaj, Iran

[6]Ph.D of Environmental Health Engineering, Students Research Committee, Shahid Beheshti University of Medical Sciences, Tehran, Iran

[7]Bachelor of Environment Health Engineering, Department of Environmental Health Engineering, Shahid Beheshti University of Medical Sciences, Tehran, Iran

Abstract

Background: The purpose of the present study was to assess the efficiency of ultrasound/hydrogen peroxide processes and ultrasound/hydrogen peroxide/ZnO nanoparticles in the removal of blue cat 41 dye from aqueous solutions.

Methods: ZnO nanoparticles were prepared using the hydrothermal method. Variables including pH, concentration of ZnO nanoparticles, initial dye concentration, concentration of hydrogen peroxide, and contact time were investigated.

Results: H_2O_2 alone is not effective in dye removal. In conditions of H_2O_2=20 mg/L and US=30 kHz, removal efficiency rates of 6.5%, 23.5%, 30%, 51.8%, and 55%, respectively, were obtained. The maximum removal efficiency rate was obtained at the nanoparticle concentration of 3 g/l. Also, removal efficiency was reduced when the initial dye concentration was increased.

Conclusion: The combination of nanoparticles and US and H_2O_2 is very effective in removing blue cat 41 dye. As a result, photo catalytic processes can be considered to effectively remove environmental pollutants.

Keywords: Advanced oxidation process, Ultrasonic waves, ZnO nanoparticles, Hydrothermal, Hydrogen peroxide, Dye

***Correspondence to:**
Mansour Ghaderpoori
Email: mghaderpoori@gmail.com

Introduction

Dye is an environmental pollutant that enters the environment through different sources such as textile, paper and pulp, printing, leather, cosmetics, and food industries (1,2). Problems in the environment such as toxicity reduce light penetration in water and the photosynthetic activity of aquatic organisms (3). Coloring processes like conventional treatments, coagulation, and adsorption are not very effective (4). According to a previous study, advanced oxidation processes (AOPs) like UV, UV-H_2O_2, Fenton, TiO_2, ZnO-UV, and ZnO-H_2O_2-UV are very effective in removing environmental pollutants (4,5). In this research, the strong oxidizing agent hydrogen peroxide and the catalyst ZnO were used in the presence of ultrasonic waves as the radiation source (H_2O_2-US and US). Results indicate this could be a promising process for the removal of toxic and harmful organic compounds from water and wastewater. Also, considering the application of ZnO nanoparticles as a catalyst in the presence of US and H_2O_2/US for dye removal and the high corresponding threshold of ZnO (425 nm), the process of (ZnO/H_2O_2/US) for dye removal is an environmentally-friendly method (6). The main advantage of this process is its inherent destructive importance. Moreover, the ZnO nano-photo-catalyst is non-toxic and has a high relative chemical stability. Among the other advantages of using this reactor in the treatment of waters contaminated with dye compounds is its high capacity for absorption of the solar spectrum, low cost, and high efficiency in removing organic molecules from both acidic and basic environments (7). Thus far, this process has been used to remove different contaminants from such environments. In 2008, Wang et al and in 2004, Yu et al used this process to remove acid red B, rhodamine B, and rhodamine dyes (8,9). Many different physical and chemical methods such as activated carbon adsorption processes (10,11), absorbing resins (12), and the hydrogen peroxide oxidation process have been used to remove dyes; these processes, however, have some inherent limitations such as high costs, inefficacy, complexity, the formation of dangerous residual products, and high energy requirements. Therefore, it is necessary to find an effective method for the treatment of dye. Among the new oxidation methods known as AOPs, chemical oxidation using ultrasound in the presence of hydrogen peroxide is considered a promising technique. Ultrasound is described as the generator of very active OH°, HOO°, and H° radicals (13). These radicals perform oxidation and revival quickly. In the last decade, researchers have reported the successful applications of ultrasonic process/hydrogen peroxide for dye removal from wastewater (14,15). The effect of ultrasonic radiation on removing organic contaminants from organic solutions can be explained physically and chemically. In physical terms, the ultrasonic waves clean solid particles, reduce particle size, and transfer mass (16). In chemical terms, cavitation occurs when ultrasonic waves are radiated on the solution. A severe collision of cavitation bubbles leads to the emergence of hot spots that have extremely high temperatures, and these hot spots can generate hydroxyl radicals (17). The present study purposed to assess the efficiency of ultrasound/hydrogen peroxide processes and ultrasound/hydrogen peroxide/ZnO nanoparticles in the removal of blue cat 41 dye from aqueous solutions.

Methods

Advanced oxidation was performed using Elema (made in Germany) with a frequency of 37 kHz and in the presence of ZnO nanoparticles and hydrogen peroxide to remove the dye from liquid environments. Experiments were basic-practical work conducted on a laboratory scale. ZnO nanoparticles were prepared by the hydrothermal method. For testing, the stock solution for removing blue cat 41 (1000 mg/L) was first prepared by dissolving 1 g dye in deionized water and refrigerating it to avoid variations in concentration. Different concentrations of the stock solution were prepared and the standard method was used to determine the most appropriate wavelength for measuring the concentrations of the studied dye. Finally, the wavelength which achieved maximum absorption for the desired dye was determined using the main solution to be most appropriate. Different concentrations of the desired dye solution were prepared, and their absorption rates were at the optimum wavelength were determined. Then, by drawing the standard curve, an unknown dye concentration was obtained in the testing steps. The variables pH, concentration of ZnO nanoparticles, initial dye concentration, hydrogen peroxide concentration, and contact time were also assessed. In the first step of the US process, the optimum pH was obtained such that under constant conditions, the dye concentration of 50 mg/L, contact time of 60 minutes, nanoparticle concentration of 1 g/L, and removal efficiency rates at pH 4, 9, and 7 were assessed, and optimum pH was achieved. In the next step, the optimum nanoparticle concentration was obtained; under constant conditions, the obtained pH, dye concentration of 50 mg/L, contact time of 60 min, and nanoparticle concentrations of 0.5, 1, 2.5, and 3 g/L were used, and the percentage of removal was calculated. Afterwards, the effect of contact time was investigated; samples with the optimum pH, optimum nanoparticle concentration, contact times (10, 30, 60, 90, and 120 minutes), and a dye concentration of 50 mg/L were prepared and the removal percentage of each was calculated. To investigate the effect of initial blue cat 41 dye concentration, samples with an optimum pH, optimum nanoparticle concentration, optimum contact time, and different dye concentrations (20, 50, 100, 150, and 200 mg/L) were then prepared, and the removal percentage of each was determined. In the final step, all the above variables were re-assessed in the presence of hydrogen peroxide, and its optimum concentration was determined. A total of 250 samples were tested. All experiments were carried out according to standard methods (13). To evaluate and analyze the results, MS Excel 2013 software and statistical tests of analysis of variance (ANOVA) and t-test were used in SPSS 20 software.

Synthesis method of ZnO nanoparticles

In a Teflon liner, a 2M concentration of ZnO was prepared in 10 cc of normal NaOH and 1 cc of surfactant (N-butyl amine) was added to it. Then, the liner was placed in the steel autoclave (Memmert model, UNE 400 type, Germany) and placed in an oven set at 120°C for 12 hours. The contents of the liner were moved to a beaker, washed five times using deionized water, and dried at laboratory temperature(18,19). ZnO nanoparticles were prepared and could be used for analysis. To approve the formed ZnO nanoparticles, x-ray diffraction (XRD) spectrometer was used. A scanning electron microscope (SEM) was also applied to determine the pore diameter and shape of the ZnO nanoparticles.

Results

Figures 1 and 2 show the structural characteristics and the XRD pattern of as-synthesis ZnO particles, respectively. The results of placing H_2O_2 solutions in contact with con-

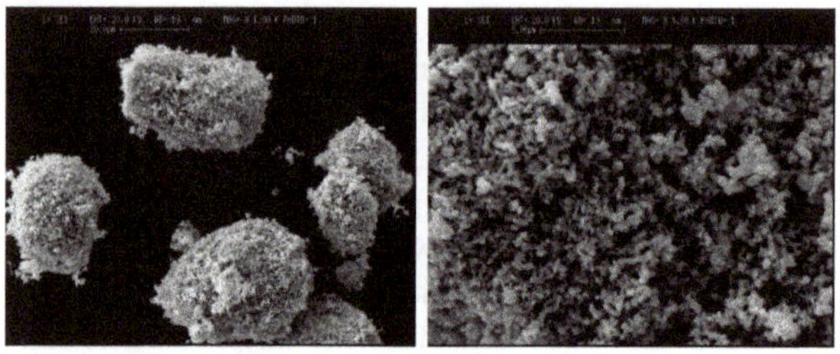

Figure 1. SEM image for ZnO nanoparticles.

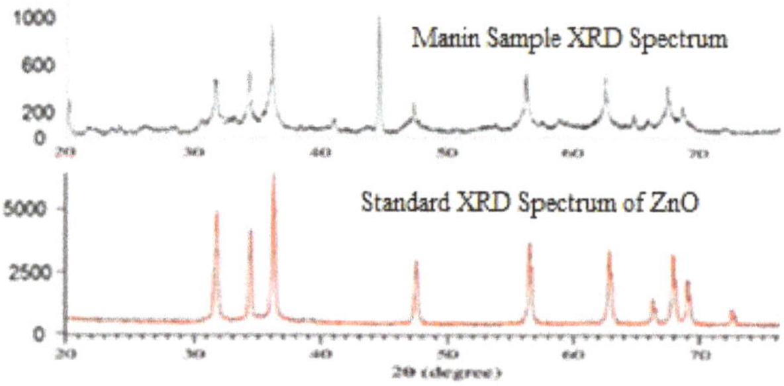

Figure 2. XRD spectrum of the synthesized ZnO nanoparticles and standard spectrum of ZnO nanoparticles

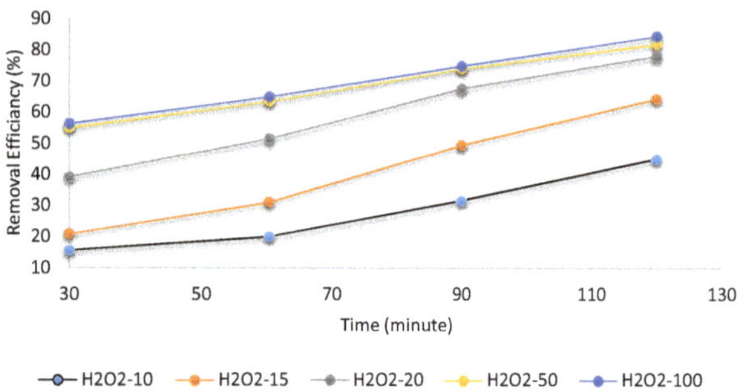

Figure 3. Efficiency variations of ultrasonic process in the removal of blue cat 41 dye with hydrogen peroxide concentration changes (initial dye concentration: 50 mg/L; ultrasonic intensity: 30 kHz and pH = 4)

centrations of 50 mg/L blue cat 41 for 60 minutes and through an ultrasonic process are presented in Figure 3. Figure 4 shows the variations in efficiency of the ultrasonic process in removing blue cat 41 dye with changes in the ZnO nanoparticle concentration. Figure 5 shows the effects of variations in the initial concentration of bleaching on the efficiency of the US- H_2O_2-ZnO nanoparticle process in removing blue cat 41 dye. Figure 6 presents the variations in efficiency of the ultrasonic and hydrogen peroxide process with pH changes. The results of kinetic studies of blue cat 41 degradation with ZnO-US and ZnO-US-H_2O_2 processes are shown in Figure 7.

Discussion

Characteristics of nanoparticles synthesized by the hydrothermal method

The average particle size of synthesized nanoparticles was 88 nm. An electron microscope was used to identify the shape and size of ZnO particles. Figure 1 shows that particle size was 80 nm; hence, the particle size distribution was appropriate (20,21). XRD analysis was used to approve the crystal structure and purity of ZnO nanoparticles. The results showed that copper had 99.6% purity and 0.4% impurity. The comparison of the XRD spectrum and ZnO standard spectrum, presented in Figure 2, demonstrated their good correspondence.

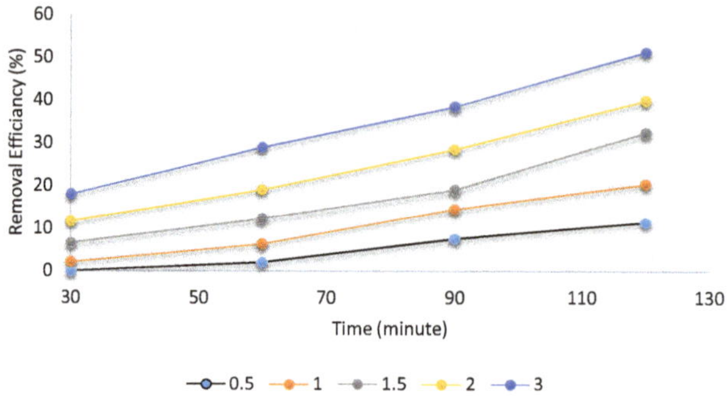

Figure 4. Efficiency variations of ultrasonic process in removal of blue cat 41 dye with changes in ZnO nanoparticle concentrations (initial dye concentration: 50 mg/L, ultrasonic intensity: 30 kHz and pH = 4)

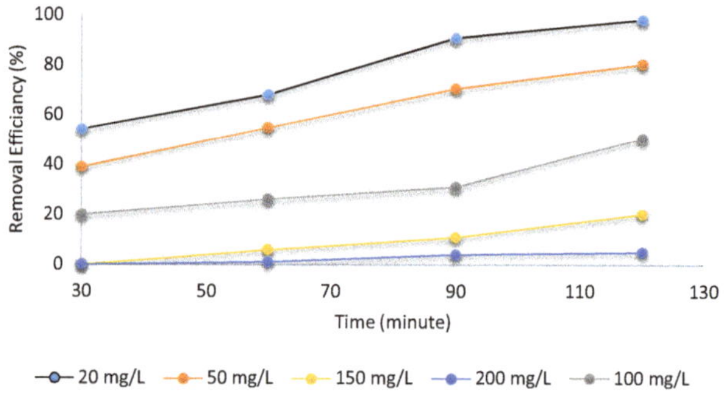

Figure 5. Effect of variations in initial concentration of bleaching on the efficiency of US-H_2O_2-ZnO nanoparticle process in removal of blue cat 41.

Investigating the effect of hydrogen peroxide alone and in combination with ultrasonic process for removing blue cat 41

The effects of hydrogen peroxide on the removal of blue cat 41 dye were determined by preparing a concentration of 50 mg/L of dye, exposing it to hydrogen peroxide at a concentration of 50 mmol/l, and sampling it at test times (10, 30, 60, 90, and 120 minutes). Results showed that, during these times, the initial dye concentration was unchanged, and hydrogen peroxide alone did not affect dye removal; these results are in agreement with those of other studies (22). In order to obtain the best concentration of hydrogen peroxide (optimum concentration), solution concentrations of 10, 15, 20, 50, and 100 mmol/l were prepared and placed in contact with 50 mg/L concentrations of blue cat 41 for 60 minutes. Then the ultrasonic process was performed, and the results are presented in Figure 3. For the prepared hydrogen peroxide concentrations, removal efficiency rates were 20%, 31%, 51.5%, 63.5%, and 65%, respectively. The maximum removal efficiency was obtained at a concentration of 100 mmol/l. The increased dye removal efficiency obtained by adding hydrogen peroxide could be described as follows: By increasing the hydrogen peroxide concentration in the solution, the hydroxyl radicals (OH^o) that play the main role in decomposition of dye and other contaminants are always increased; the collision probability of these very active radicals with the target contaminants is also increased, and decomposition and removal are conducted (23-25). In fact, some of the mentioned studies have reported that when the concentration of hydroxyl radicals exceeds a particular concentration, it has no noticeable effect on removal, which is attributed to the re-combination of hydroxyl radicals. In other words, when the concentration increases beyond a certain level, the hydrogen peroxide itself removes the hydroxyl radicals (through their re-combination) (22,26,27). Ghaderpoori et al determined that the maximum detergent removal efficiency was obtained at a concentration of 40 mg/L, and increased concentrations had no great effect on the increase in removal rate (5). However, this phenomenon was not observed in the present study.

Effects of ZnO nanoparticles on removal of blue cat 41 in ultrasound (US) and hydrogen peroxide (US- H_2O_2) processes

To determine the best dose of ZnO nanoparticles in the ultrasonic-hydrogen peroxide process, the concentrations of 0.5, 1, 2.5, and 3 g/L nanoparticles were investigated. A solution of 50 mg/L blue cat 41 and hydrogen peroxide with a concentration of 20 mmol/l was prepared and nanoparticles were exposed to contact, as demonstrated in Figure 4. As can be seen, in the ultrasonic process for the above-mentioned nanoparticle concentrations, the removal efficiency rates were 2%, 6.4%, 12.3%, 19.2%, and 29% for 60 minutes; these rates increased with increased doses of ZnO nanoparticles (direct relationship). The effect of using ZnO nanoparticles in the combination process of ultrasound and hydrogen peroxide was investigated by preparing a solution of 50 mg/L blue cat 41 and hydrogen peroxide with a concentration of 20 mmol/l at pH 4; various concentrations of nanoparticles were then added and assessed. The results are presented in Figure 4. As can be seen, for the concentrations of 0.5, 1, 2.5, and 3 g/L of ZnO nanoparticles in the mentioned conditions (hydrogen peroxide concentration of 20 mg/L and ultrasonic intensity of 30 kHz), removal efficiency rates were 6.5%, 23.5%, 30%, 51.8%, and 55%, respectively, and maximum removal efficiency in ZnO nanoparticles was obtained at the concentration of 3 g/L. Ghaderpoori et al reported that increased concentrations of ZnO nanoparticles from 2 g caused the penetration of ultraviolet radiation into the solution to decrease because of increased opacity; as a result, the concentration of hydroxyl radicals was also decreased and, eventually, the efficiency of the process was reduced (5). However, in some studies, the positive effect of nanoparticle concentration (as a catalyst) has been mentioned. Masombaigi et al reported that improved catalyst performance at higher concentrations was related to more active places on the catalyst surface and possibly the stronger ultraviolet radiation effect on it. In this study, as mentioned earlier, removal efficiency was increased with increased concentrations of nanoparticles (28). By comparing the results in two US-ZnO and US-H_2O_2-ZnO processes (Figures 4 and 5), it can be found that the second process (US- H_2O_2-ZnO) was much more effective than the ultrasonic/ZnO nanoparticle process in the removal of blue cat 41; removal was increased by 69%, 73%, 54%, 63%, and 47% for the used ZnO nanoparticle concentrations, respectively, at 60 minutes. Also, it was observed by performing *t* tests that the difference between the two processes was significant (P value < 0.014), i.e. the combination process of ultrasonic-peroxide hydrogen-ZnO nanoparticles was more efficient in removing blue cat 41 than the US-ZnO process.

Effect of initial blue cat 41 concentration on the combination process of ultrasound and hydrogen peroxide

The effect of increasing bleaching concentration on the efficiency of the ultrasonic process in combination with ZnO nanoparticles and hydrogen peroxide was investigated. Concentrations of 20, 50, 100, 150, and 200 mg/L were prepared and tested in the above process. Results are shown in Figure 5. As can be seen, the mentioned efficiency rates for the dyes at the mentioned concentrations at 60 minutes was 68%, 55%, 26%, 5%, and 1%, respectively; by increasing initial dye concentration, removal efficiency was decreased (increasing dye concentration had an inverse relationship with dye removal). Thus, at the concentration of 200 mg/L, 99% of the initial dye exited the system without being removed, because by increasing the dye concentration, competition between the constituting interfaces became possible by decomposition and dye molecules, and this disturbance was increased in the large amounts of the generated interfaces at high initial dye concentrations (29,30). Also at high dye concentrations, the catalyst's active places were covered with negatively-charged dye molecules. The absorption of ultraviolet radiation by dye molecules also played an important role in reducing the bleaching rate. These factors decrease the generation of hydroxyl radicals in the catalyst surface (31,32). Therefore, dye removal efficiency in this process is higher for diluted solutions.

pH variations in removal of blue cat 41 in the combination process of ultrasound and hydrogen peroxide

Because of the amphoteric behavior of most semiconductor oxides, pH is an important parameter in these reactions. Solution pH affects the speed of the reactions that occur on the surface of semiconductor particles by influencing the properties of the surface charge (33). Dye removal relative to pH variations in acidic, neutral, and basic conditions (pH of 4, 7, and 9) was investigated. A solution of blue cat 41 with a 50 mg/L concentration and hydrogen peroxide with a concentration of 20 mmol/l was prepared, and pH variations on the process efficiency were assessed. The results are shown in Figure 6. As can be observed, the process efficiency rates relative to pH variations at point 4 for the periods of 30, 60, 90, and 120 minutes were equal to 39%, 51.8%, 67.5%, and 78%, respectively. Among the studied pH values, the best removal efficiency was obtained under acidic conditions; removal efficiency rates at 60 minutes at this pH were 17.95% and 36.29% greater than those of pH 7 and 9, respectively. Li et al used pH values of 1 to 10 to decompose red, and the results showed that the ultrasonic-hydrogen peroxide process had maximum decomposition power in acidic conditions. The removal efficiency rates of red 24 at pH=1 and pH=10 were 98.09% and 33.9%, respectively (17). ANOVA test results revealed that these differences were significant, and process efficiency was significantly higher at pH=4 than at other pH values (*P* value < 0.001). The variations in process efficiency with pH changes in the ultrasonic process were also evaluated. The results showed that removal rates at pH 4, 7, and 8 were 18.2%, 16.1%, and 6.8%, respectively. In this process, the maximum removal amount was also obtained at pH = 4, and ANOVA test results demonstrated that removal efficiencies at these pH values were different and statistically significant (*P* value < 0.001). pH = 4 was effective for the removal of blue cat 41.

Kinetic studies of blue cat 41 degradation

Figure 7 shows the results of kinetic studies of blue cat 41 degradation with ZnO-US and ZnO-US- H_2O_2 processes. The kinetic model of the pseudo first order reaction was used (34).The kinetics constants of blue cat 41 are shown in Table 1. Study results showed that US, US- H_2O_2, and ZnO-H_2O_2-US follow pseudo first order kinetics. The pseudo first order reaction had the highest coefficient of determination among the studied processes (35). As illustrated in Table 1, the K1 constant increased in US, US-

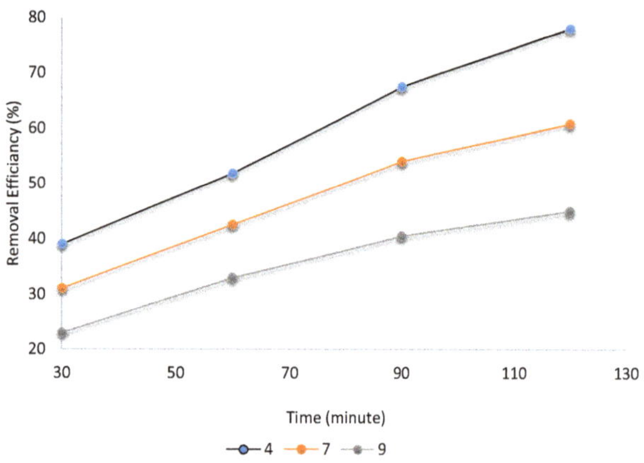

Figure 6. Efficiency variations of ultrasound and hydrogen peroxide process with pH changes.

Table 1. Kinetics constants of the photo-catalytic degradation of blue cat 41 from aqueous solutions. (Initial con. = 20 mg/L, pH = 4)

Kinetics models	R^2	K (min^{-1})
US	0.9751	0.0077
US-H_2O_2	0.9492	0.0096
US-H_2O_2-ZnO	0.9875	0.0134

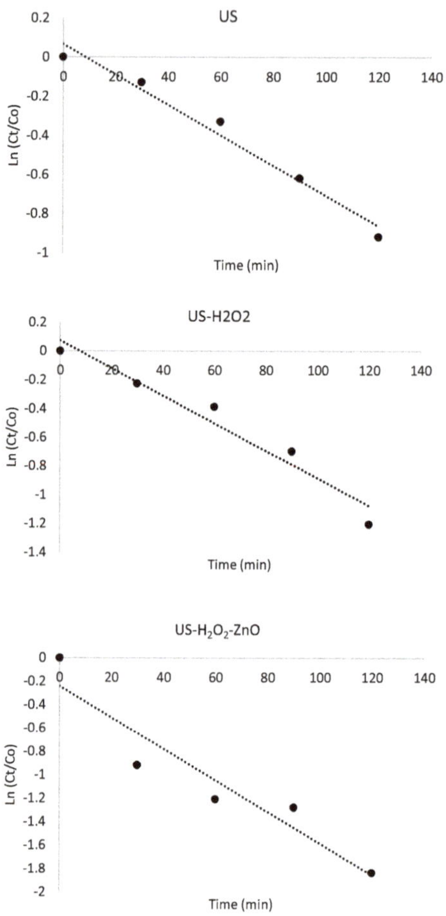

Figure 7. Degradation kinetics for removal of blue cat 41 from aqueous solutions.

H_2O_2, and US-H_2O_2-ZnO processes, respectively. With an increase in the K1 constant, the degradation rate of blue cat 41 also increased (36). Due to the synergistic effect of combination processes like AOPs, the rate of degradation of a pollutant can be increased. The synergistic effect of AOPs has been assessed in various studies (37-39).

Conclusion

In studying the combination process of ultrasound/hydrogen peroxide/ZnO nanoparticles, the following results were obtained:

- Hydrogen peroxide alone has no effect on dye removal.
- Process efficiency in dye removal is decreased by increasing initial dye concentration.
- The efficiency of the above-mentioned process in dye removal is higher at acid pH (≅4).
- Removal efficiency is increased by increasing retention time.

Competing interests

The authors of this article declare that they have no competing interests.

original work. It has not been published, nor is it under review in another journal, and it is not being submitted for publication elsewhere.

Authors' contributions

All authors contributed equally and were involved in the study design, data collection, and article approval.

Acknowledgments

The authors would like to express our sincere thanks to Kurdistan University of Medical Sciences for the use of laboratory facilities.

Reference

1. Massoudinejad MR, Ghaderpoori Azari M, Rezazadeh Azari M. The removal of COD and color from textile industry by chlorine hypochlorite. Int J Adv Sci Technol 2015; 76: 35-42.
2. Abechi SE, Gimba CE, Uzairu A, Kagbu JA, Ocholi OJ. Equilibrium adsorption studies of methylene blue onto palm kernel shell-based activated carbon. International Refereed Journal of Engineering and Science 2013; 2(5): 38-45.
3. Fungaro DA, Grosche LC, Pinheiro AS, Izidoro JC, Borrely SI. Adsorption of methylene blue from aqueous solution on zeolitic material and the improvement as toxicity removal to living organisms. Orbital 2010; 2(3): 235-47.
4. Fahmi MR, Abidin CZ, Rahmat NR. Characteristic of colour and COD removal of azo dye by advanced oxidation process and biological treatment. International Conference on Biotechnology and Environment Management; Singapore; vol. 18; 2011. p. 13-8.
5. Ghaderpoori M, Dehghani MH. Investigating the removal of linear alkyl benzene sulfonate from aqueous solution by ultraviolet irradiation and hydrogen peroxide process. Desalination and Water Treatment 2016; 57(32): 15208-12.
6. Daneshvar N, Rasoulifard MH, Khataee AR, Hosseinzadeh F. Removal of C.I. Acid Orange 7 from aqueous solution by UV irradiation in the presence of ZnO nanopowder. J Hazard Mater 2007; 143(1-2): 95-101.
7. Edvard JC. Investigation of color removal by chemical oxidation for three reactive textile dyes and spent textile dye wastewater. Virginia: Department of Civil and Environmental Engineening Blacksburg, MSESE; 2000.
8. Wang J, Jiang Z, Zhang Z, Xie Y, Wang X, Xing Z, et al. Sonocatalytic degradation of acid red B and Rhodamine B catalyzed by nanosized ZnO powder under ultrasonic irradiation. Ultrason Sonochem 2008; 15(5): 768-74.
9. Yu D, Cai R, Liu Z. Studies on the photodegradation of Rhodamine dyes on nanometer-sized zinc oxide. Spectrochim Acta A Mol Biomol Spectrosc 2004; 60(7): 1617-24.
10. Ahmedna M, Marshall WE, Rao RM. Surface properties of granular activated carbons from agricultural by-products and their effects on raw sugar decolorization. Bioresour Techno 2000; 71(2): 103-12 .
11. Simaratanamongkol A & Thiravetyan P. Decolorization of melanoidin by activated carbon obtained from bagasse bottom ash. J Food Eng 2010; 96(1): 14-7.
12. Achaerandio I, Güell C, López F. Continuous vinegar decolorization with exchange resins. J Food Eng 2002; 51(4): 311-7.
13. Adewuyi YG. Sonochemistry: environmental science and engineering applications. Ind Eng Chem Res 2001; 40(22): 4681-715.
14. Zhang ZM, Zheng HL. Optimization for decolorization of azo dye acid green 20 by ultrasound and H_2O_2 using response surface methodolog. J Hazard Mater 2009; 172(2-3): 1388-93.
15. Xie JH, Shena MY, Nie SP, Li C, Xie MY. Decolorization of polysaccharides solution from Cyclocarya paliurus (Batal.) Iljinskaja using ultrasound/H_2O_2 process. Carbohydr Polym 2011; 84(1): 255-261.
16. Yano J, Matsuura JI, Ohura H, Yamasaki S. Complete mineralization of propyzamide in aqueous solution containing TiO_2 particles and H_2O_2 by the simultaneous irradiation of light and ultrasonic waves. Ultrason Sonochem 2005; 12(3): 197-203.
17. Li M, Li JT, Sun HW. Decolorizing of azo dye Reactive red 24 aqueous solution using exfoliated graphite and H2O2 under ultrasound irradiation. Ultrason Sonochem 2008; 15(5): 717-23.
18. Ibupoto ZH, Khun K, Eriksson M, AlSalhi M, Atif M, Ansari A, et al. Hydrothermal growth of vertically aligned ZnO nanorods using a biocomposite seed layer of ZnO nanoparticles. Materials 2013; 6(8): 3584-97.
19. Ni YH, Wei XW, Hong JM, Ye Y. Hydrothermal preparation and optical properties of ZnO nanorods. Materials Science and Engineering B 2005; 121(1-2): 42-7.
20. Kołodziejczak-Radzimska A, Jesionowski T. Zinc oxide—from synthesis to application: a review. Materials 2014; 7(4): 2833-81.
21. Kołodziejczak-Radzimska A, Markiewicz E, Jesionowski T. Structural characterization of ZnO particles obtained by the emulsion precipitation method. J Nanomater 2012; 2012: 656353.
22. García-Morales MA, Roa-Morales G, Barrera-Díaz C, Martínez Miranda V, Balderas Hernández P, Pavón Silva TB. Integrated advanced oxidation process (ozonation) and electrocoagulation treatments for dye removal in denim effluents. Int J Electrochem Sci 2013; 8: 8752-63.
23. Iordache Nechita MT, Aelenei N, Rosca I, Apostolescu G, Peptanariu M. Sonochemical enhancement of cyanide ion degradation from wastewater in the presenceof hydrogen peroxide. Pol J Environ Stud 2003; 12(6): 735-7.
24. Sarla M, Pandit M, Tyagi DK, Kapoor JC. Oxidation of cyanide in aqueous solution by chemical and photochemical processes. J Hazard Mater 2004; 116(1-2): 49-56
25. Shokoohi R, Mahvi A, Bonyadi Z, Samarghandi MR, Karimi M. The use of sonochemical technology for cyanide removal from aqueous solutions in the presence of hydrogen peroxide. Water and Wastewater 2010; 22(3): 32-7. [In Persian].
26. Movahedian MH, Rezaee R. Investigating the efficiency of advanced photochemical oxidation (AOP) technology in degradation of direct azo dye by UV/H_2O_2 process. Water Wastewater 2006; 59(4): 75-83.
27. Daneshvar N, Rabbani M, Modirshahla N, Behnajady MA. Critical Effect of Hydrogen Peroxide Concentration in Photochemical Oxidative Degradation of C.I.Acid Red 27 (AR27). Chemosphere 2004; 56(10): 895-900.
28. Masombaigi H, Rezaee R, Nasiri A. Photocatalytic degradtion of methylene Blue using ZnO nano - particles. Iranian Journal of Health and Environment 2009; 2(3): 188-95.

29. Mahmoodi NM, Rayat Tari KH, Borhany S, Arami M, Nourmohamadian F. Decolorization of colored wastewater containing Azo acid dye using photo-fenton process:operational parameters and a comparative study. Journal of Color Science and Tecnologhy 2008; 2(1): 31-40. [In Persian].
30. Mahmoodi NM, Arami M, Limaee NY, Tabrizi NS. Kineties of heterogeneous photocatalytic degradation of reactive dyes in an immobilized TiO_2 photocatalytic reactor. J Colloid Interf Sci 2006; 295(1): 159-64.
31. Konstantinou K, AlbanisAT. TiO_2-assisted photocatalytic degradation of Azo dyes in aqueous solution: kinetic and mechanistic investigations: a review. Applied Catalysis B: Environmental 2004; 49(1): 1-14
32. Shu HY. Degradation of dyehouse effluent containing C.I. Dirat Blue 199 by processes of ozonation , UV/H_2O_2 and in sequence of ozonation with UV/H_2O_2. J Hazard Mater 2006; 133(1-3): 92-98.
33. Faraji H, Mohamadi AA, Soheil Arezomand HR, Mahvi AH. Kinetics and equilibrium studies of the removal of blue basic 41 and methylene blue from aqueous solution using rice stems. Iran J Chem Chem Eng 2015; 34(3): 33-42.
34. Massoudinejad MR, Ghaderpoori M, Shahsavani A, Amini MM. Adsorption of fluoride over a metal organic framework Uio-66functionalized with amine groups and optimization with responsesurface methodology. J Mol Liq 2016; 221: 279-286.
35. Eslami A, Bahrami H, Asadi A, Alinejad A. Enhanced sonochemical degradation of tetracycline by sulfate radicals. Water Sci Technol 2016; 73(6): 1293-300.
36. El-Zomrawy AA. Kinetic studies of photoelectrocatalytic degradation of Ponceau 6R dye with ammonium persulfate. Journal of Saudi Chemical Society 2013; 17(4): 397-402.
37. Sun W, Chen L, Zhang Y, Wang J. Synergistic effect of ozonation and ionizing radiation for PVA decomposition. J Environ Sci (China) 2015; 34: 63-7.
38. Jun H, Jianlong W, Rong C. Degradation of 4-chlorophenol in aqueous solution by γ-radiation and ozone oxidation. Science in China Series B 2006; 49 (2): 186-92.
39. Agustina TE, Ang HM, Vareek VK. A review of synergistic effect of photocatalysis and ozonation on wastewater treatment. Journal of Photochemistry and Photobiology C: Photochemistry Reviews 2005; 6(4): 264-73.

14

The effects of the natural coagulant *Moringa oleifera* and alum in wastewater treatment at the Bandar Abbas Oil Refinery

Mohsen Dehghani[1*], Mohammad Hossein Alizadeh[2]

[1]Assistant Professor of Environmental Engineering and Science, Department of the Environmental Engineering, Islamic Azad University Bandar Abbas Branch, Bandar Abbas, Iran
[2]M.Sc. in Water and Wastewater Engineering, Islamic Azad University Bandar Abbas Branch, Bandar Abbas, Iran

Abstract
Background: The refining process generates large volumes of wastewater containing a variety of chemical contaminants. The use of natural substitutes in treating wastewater which have fewer harmful effects is considered an effective step towards protecting the environment and sustaining the development of these industries. This study focused on the use of *Moringa oleifera* and alum at the Wastewater Unit at Bandar Abbas Refinery.
Methods: This study was performed in 2014 in a laboratory using jar apparatus. These experiments were conducted in batch system and effective parameters including pH, coagulant dose and contact time were investigated on the wastewater obtained from Bandar Abbas Oil Refinery.
Results: The jar test experiment showed that *M. oleifera* at 70 mg/L, optimum temperature, pH, and mixing speed could remove 38.60% of chemical oxygen demand (COD), 63.70% of turbidity, and 62.05% of total suspended solids (TSS). Also, alum at 40 mg/L removed COD, turbidity, and TSS by 51.72%, 92.16%, and 85.26% respectively from the refinery wastewater. Moreover, when *M. oleifera* and alum was used together with a 2:1 dosage ratio (alum at 80 mg/L and *M. oleifera* at 70 mg/L), they will remove COD, turbidity, and TSS by up to 50.41%, 86.14%, and 81.52% respectively.
Conclusion: The use of *M. oleifera* as a natural coagulant is important in treating refinery wastewater not only from an environmental but also an economic point of view.
Keywords: *Moringa oleifera*, Alum, Bandar Abbas Refinery, Industrial wastewater

***Correspondence to:**
Mohsen Dehghani
Email: Dehghani933@gmail.com

Introduction

Rapid population growth and industrial development in the countries of the world are degrading the environment through the uncontrolled growth of urbanization and industrialization, as well as contributing to the destruction of natural habitats. The growing population and the environmental deterioration pose the challenge of sustained development without environmental damage. Nowadays, chemical compounds in industrial production is increasingly used and many of these substances have found their way into nature with adverse impacts, which can gradually destroy the environment (1). For example, the vast development of hydrocarbon-processing industries and the extensive consumption of petroleum and gas derivatives in most big industries have increased the danger of oil pollution in the natural environment (2). One of the problems with chemical industries is the presence of effluents containing emulsifiable petroleum products which pollute the environment enormously (3).

Wastewater from the petroleum industry contains organic and inorganic compounds, dissolved and suspended solids (4). The methods of treating wastewater include activated carbon adsorption, oxidation, chemical coagulation/flocculation, electrochemical methods, membrane techniques, and biological treatment processes, which are frequently used to treat textile effluents (5). In the process of wastewater treatment, especially during the flocculation and coagulation processes, various chemical materials are used. The effectiveness of these chemicals as coagulants is well recognized; however, there are some disadvantages in connection with using these coagulants, such as the cost of the process, the operation, and the maintenance and production of large volumes of sludge (6). In this regard, the use of substitute materials with fewer harmful effects can be a step taken to protect the environment and human health (7). Compounds extracted from plant and animal tissues are among these substitute materials (8). The use of these natural compounds as coagulant aids in some cases also is cost-effective (9).

The utilization of natural herbal substances in water and wastewater treatment is not a new idea (10). Natural biopolymers having applications in the water and wastewater industries include starches (potato, wheat, corn, and rice starches), sodium alginate, *Moringa oleifera*, calcium alginate, chitosan, tannin, natural gums and xanthan (7).

Moringa oleifera habitats originate from around the Dead Sea and extend in scattered regions along the Red Sea coast to northern Somalia and around the Arabian Peninsula, up to the mouth of the Persian Gulf states, United Arab Emirates, Saudi Arabia, and Palestine (11). *M. oleifera* grows in arid, semi-arid, and humid areas and has been named a "miracle tree" owing to its several benefits (12). Javanshir first published *M. oleifera* in Iranian flora in 1993. Major distribution areas of the species include Mt Bashagerd in Nikshahr through to south Jazmorian (11). Sutherland et al have suggested that *M. oleifera* seeds do not constitute a serious hazard to human health (13). Their use in new technologies of wastewater treatment is, therefore, very promising. The advantages of using *M. oleifera* are: a positive cost-effectiveness, water treated without extreme pH, and a high level of biodegradability (14). These advantages are especially the case in rural areas.

The use of *M. oleifera* for wastewater treatment is perhaps one of its more interesting usages. Although there are many previous papers investigating its utilization as a natural adsorbent for special pollutant removal (12), the seeds of this tropical tree have a high amount of proteins that act like cationic polyelectrolytes once they are added to raw water (15).

Moringa oleifera seeds are also used as a primary coagulant in wastewater treatment due to the presence of a water-soluble cationic coagulant protein able to reduce the turbidity, chemical oxygen demand (COD), and Total dissolved solids (TDS) of the wastewater treated. *M. oleifera* has been shown to be one of the most effective uses as a primary coagulant for water and wastewater treatment (7,16-21).

The aim of this study was to compare the efficiency of *M. oleifera* seed and alum in total suspended solids (TSS), COD, and turbidity removal in the wastewater unit at Bandar Abbas Refinery as one method for treating wastewater in the petroleum industries in developing countries.

Methods

Sampling

In this study, the coagulation performance of *M. oleifera* and alum in treating wastewater at Bandar Abbas Refinery was investigated.

First, wastewater samples were taken daily from the outlet of the equalization pool near the inlet into the dissolved air flotation (DAF) pool of the refinery for the duration of one month. Table 1 presents the four parameters studied and their mean values. All measurement methods of the parameters were based on the standards for water and wastewater experiments (22).

Preparing *Moringa oleifera*

Moringa oleifera seeds were obtained from Busher province (Herbarium number 53054) and Moringa powder and the extract was prepared and kept at room temperature. To prepare the powder, the seeds were placed in an oven at 35°C for 5 hours to make sure they were dry. The seeds were dehusked by hand, and then turned into a soft powder with a particle size of about 600 μm using a moulinex grinder. One gram of the powder was added to one liter of doubled distilled water and mixed using a stirrer for 3 minutes at 120 rpm. The extract was then passed through 0.45-micron filter paper to obtain a homogeneous solution free of suspended materials. This solution was freshly prepared each time it was used in water treatment, and kept in a cool place with a maximum temperature of 20°C to prevent changes in its pH and viscosity (23).

Measurement methods and laboratory equipment

For the jar test, 1 L of the wastewater was poured in beakers and quickly mixed using a stirrer at a high speed (120 rpm) for 1 minute, and then mixed slowly at 40 rpm for 20 minutes, followed by the precipitation process which took 30 minutes (24). To measure turbidity, a turbidity meter and the standard solutions 1 NTU, 10 NTU, and 100 NTU were employed by using the ASTM method. TSS was measured based on the SM 2540D method by using a vacuum oven with an Erlenmeyer flask (Oven-Vacuum Erlen), and COD measurements were carried out based on the ASTM DI252-06 method by using vials manufactured by reputable companies and a COD reactor (25-27). All experiments were performed at the laboratory of Bandar Abbas Oil Refinery Company.

Methodology

The jar test was performed first with a constant concentration of wastewater and varying concentrations of *M. oleifera* (10, 20, 40, 60, 70, and 80 mg/L) to determine the optimum dosage.

The pH test was then carried out at the pH values of 5, 5.5, 6, 6.5, 7, 7.5, 8, 8.5 and 9 to find the optimum pH value. Considering the weather conditions in Bandar Abbas, the jar test was performed at the optimum values of *M. oleifera* and pH at 5 different temperatures (20, 25, 30, 35, and 40°C) to determine the optimum temperature for

Table 1. Qualitative characteristics of the wastewater obtained from the equalization pool at the refinery

Parameter	Textile effluent	
	Range	Average± SE
COD (mg/L)	390-440	435 ± 18.40
TSS (mg/L)	85-120	95 ± 8.08
Turbidity (NTU)	29-41	37 ± 3.57
pH	4.5-9	8.7 ± 0.67

Abbreviations: COD, chemical oxygen demand; TSS, total suspended solids; SE, standard error.

Table 2. The range of parameters examined in this study

pH	Temperature (°C)	Rapid mix (rpm)	Dosage of *Moringa oleifera* (mg/L)	Dosage of Alum (mg/L)
5, 5.5, 6, 6.5, 7, 7.5, 8, 8.5, 9	20, 25, 30, 35, 40	220, 260, 300	10, 20, 40, 60, 70, 80	40

wastewater treatment.

In the next stage, rapid mixing was performed at three different speeds (220, 260, and 300 rpm) for one minute and the results were compared to determine the effects of mixing speed on treatment efficiency (Table 2).

Considering the fact that, at present, the Bandar Abbas Oil Refinery Company uses alum at 40 mg/L for the inoculation in its wastewater treatment, the experiment was repeated at optimum values to combine speed, temperature, and pH with alum at 40 mg/L to determine treatment efficiency.

Finally, the experiment was repeated at the optimum values to combine speed, temperature, and pH using the optimum concentration of alum and *M. oleifera* with the ratios of 1:2, 1:1, and 2:1, respectively.

Results

In this study, the efficiency of coagulant materials *M. oleifera*, alum, and the simultaneous use of alum and *M. oleifera* in the removal of turbidity, TSS, and COD was examined. The results of the jar test at various *M. oleifera* concentrations (10, 20, 40, 60, 70, and 80 mg/L) are presented in Figure 1. The optimum concentration of *M. oleifera* is 70 mg/L, which reduced the COD of wastewater by 17.24% (from 435 to 360 mg/L), turbidity by 52.43% (from 37 to 19.4 NTU), and TSS by 46.31% (from 95 to 44 mg/L).

Figure 2 shows the effects of various pH values at the constant *M. oleifera* concentration of 70 mg/L. As seen in this figure, the optimum pH is 6 where COD, turbidity, and TSS decreased by 39.78%, 62.16%, and 61.05% respectively.

In the jar test with *M. oleifera* concentration at 70 mg/L and pH 6 at 5 different temperatures (20, 25, 30, 35, and 40°C), the optimum temperature for removal efficiency of 20°C was obtained (Figure 3). Considering the results of the jar test at the three levels of rapid mixing, the removal efficiency at the mixing speed of 300 rpm was the highest. This removed COD, turbidity, and TSS by 38.60%, 63.70%, and 62.05% respectively (Figure 4).

After obtaining the optimal condition of the natural coagulant *M. oleifera* (pH: 6, temperature: 20°C, and mixing speed: 300 rpm), in the next stage the experiment was performed with the chemical compound alum at 40 mg/L. As seen in Figure 5, alum removed COD, turbidity, and TSS from the refinery wastewater with a higher efficiency than *M. oleifera* (51.72%, 92.16%, and 85.26% respectively).

Finally, an experiment was performed under optimum conditions (constant pH: 6, constant temperature: 20°C, and mixing speed: 300 rpm) in which mixtures of the two coagulants alum and *M. oleifera* were used. The re-

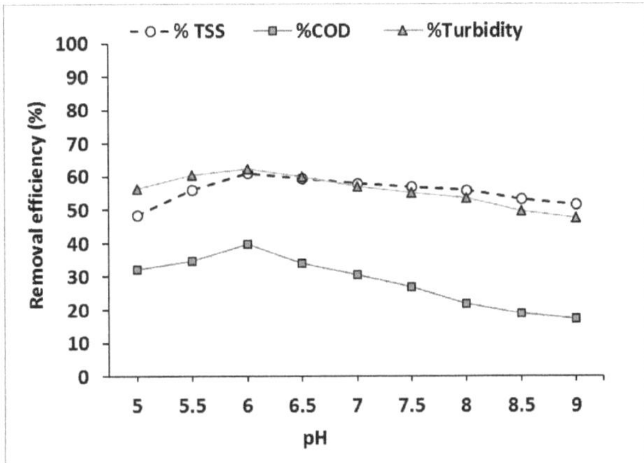

Figure 2. The effects of various pH values with a constant *Moringa oleifera* concentration of 70 mg/L on the efficiency of the removal of COD, turbidity, and TSS from petroleum refinery wastewater.

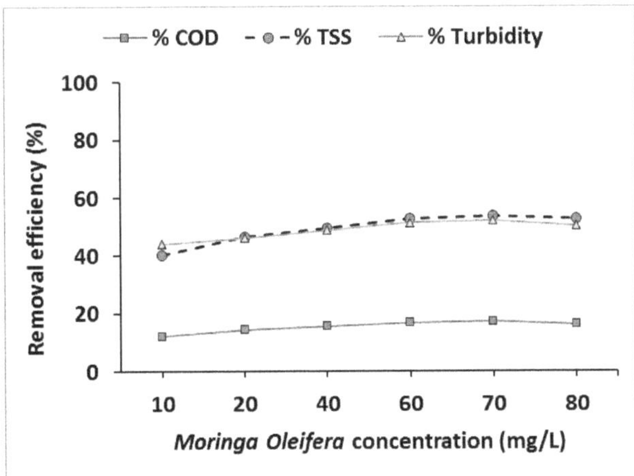

Figure 1. The effects of various *Moringa oleifera* concentrations on the efficiency of the removal of COD, turbidity, and TSS from petroleum refinery wastewater.

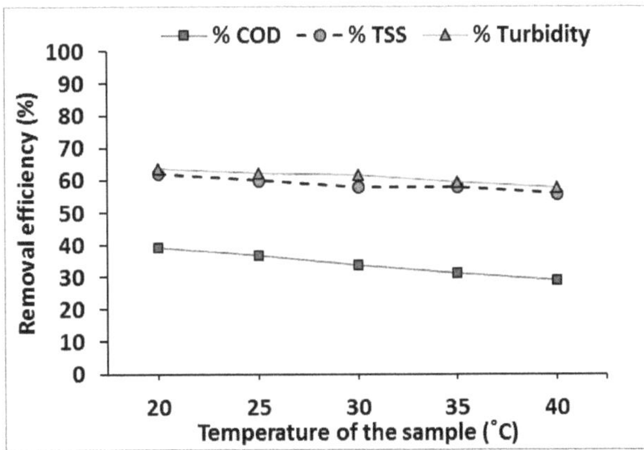

Figure 3. The effects of various temperatures with a constant *Moringa oleifera* concentration of 70 mg/L and pH value of 6 on the efficiency of the removal of COD, turbidity, and TSS from petroleum refinery wastewater.

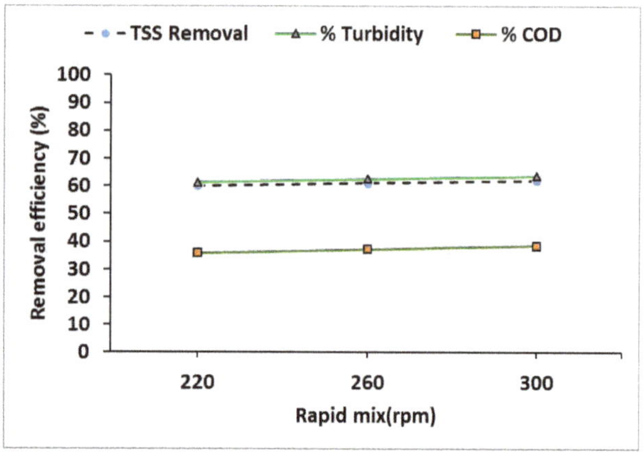

Figure 4. The effects of various mixing speeds with a constant *Moringa oleifera* concentration of 70 mg/L, constant pH value of 6, and constant temperature of 20°C on the efficiency of the removal of COD, turbidity, and TSS from petroleum refinery wastewater.

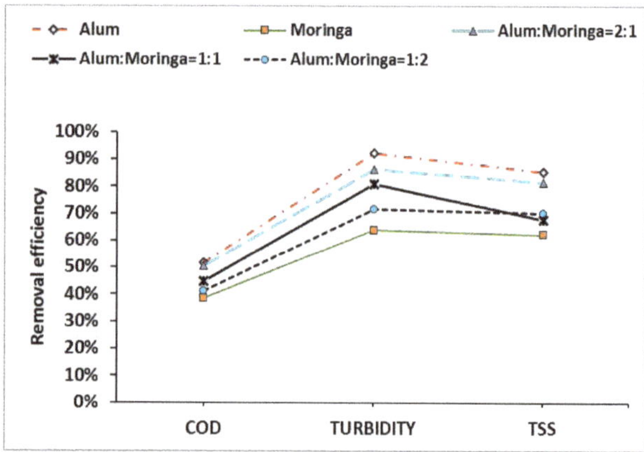

Figure 5. Efficiencies of alum and *Moringa oleifera* on the efficiency of the removal of COD, turbidity, and TSS from petroleum refinery wastewater.

sults showed that, if they are used with the optimal ratio of 2:1 (alum at 80 mg/L and *M. oleifera* at 70 mg/L), they will remove turbidity, COD, and TSS by up to 86.14%, 50.41%, and 81.52% respectively. Moreover, a mixture in the optimal ratio of 1:1 (alum at 40 mg/L and *M. oleifera* at 70 mg/L) removed the three parameters by 80.71% for turbidity, 44.70% for COD, and 67.50% for TSS, while a mixture in the ratio of 1:2 (40 mg/L alum and 140 mg/L *M. oleifera*) reduced the parameters by lower percentages (Figure 5).

Discussion

The results of this research indicated that *M. oleifera* has a good removal efficiency turbidity, COD, and TSS by 63.70%, 38.60%, and 62.05% respectively. Under optimum conditions (pH: 6, temperature: 20°C, and *M. oleifera* dosage: 70 mg/L) it can improve the quality of the refinery wastewater.

Figure 1 showed the effects of various *M. oleifera* dosage on the efficiency of the removal of COD, turbidity, and TSS. An analysis of variance (ANOVA) analysis was carried out to evaluate the significance of the difference between the mean of measured values. This statistical analysis did not confirm any significant difference for all parameters ($P=0.086$, $P=0.064$, and $P=0.069$ respectively for COD, TSS and turbidity).

The evaluation of pH effectivness in the removal of COD, turbidity, and TSS due to the addition of *M. oleifera* is presented in Figure 2 and shows that the maximum removal occured with pH in the range 5–6.5 (turbidity, COD and TSS reduces below 82%). The results were in line with some previous research studies (28). The basic amino acids present in the protein of *M. oleifera* would act as a proton acceptor and result in the release of hydroxyl functional group making the solution basic. This accounted for the basic pH values observed for *M. oleifera* treatments compared with alum treatments. These results also require high-tech equipment to ensure the preciseness of the alum dosage.

Both coagulants have high turbidity removal values (≥60%). The optimal dosage of *M. oleifera* was 70 mg/L, obtaining a reduction of turbidity of 63.70%. The mechanism of coagulation with the seeds of *M. oleifera* consists of absorption and neutralization of the colloidal positive charges that attract the negatively charged impurities in water (29). *M. oleifera* reduced the turbidity of abattoir wastewater from 218.4 NTU to 68 NTU representing a 68.86% reduction (30).

It was found that *M. oleifera* seed extract at 10–30 mg/L and pH of 6–8 removed more than 90% of turbidity (31). Also, another study revealed that the removal of turbidity was 50%-87.5% and 12%-62.5% for chitosan and coagulant protein of *M. oleifera* respectively (20), which was almost equal to the result obtained in the experiments performed on refinery wastewater (a 42% reduction in turbidity by alum). In addition, in other researches at optimum concentrations, alum, poly aluminum chloride (PAC), and *M. oleifera* removed turbidity by 99.5%, 99%, and 99.7% respectively (19). In these studies, it was found that natural biopolymer *M. oleifera* could be used to remove coagulum and turbidity. Also, similar results from highly contaminated waters were reported (31).

In the research that was performed at the optimum dose, pH value, and temperature, *M. oleifera* was used as a natural coagulant on synthetic samples containing turbidity of 40–200 NTU. It was found that the optimum *M. oleifera* dosage was 30–55 mg/L at pH value of 6.5, and temperatures below 15°C (21). It was reported that overall COD removal was 50% at both 50 and 100 mg/L of *M. oleifera* dosage. When 50 and 100mg/L seed dosages were applied in combination with 10 mg/L of alum, COD removal increased to 58 and 64% respectively (14).

These results are compatible with those found in the study carried out with samples taken from the wastewater unit of Bandar Abbas Refinery (where the optimum dosage, pH, and temperature were found to be 70 mg/L, 6, and 20°C respectively). Muthuraman and Sasikala showed that among 3 natural coagulants (*M. oleifera, Strychnos potatorum* and *Proteus vulgaris*), *M. oleifera* seed extracts had the highest performance in turbidity removal (99%) at pH 7 using 250 mg/L *M. oleifera* (7). Javid et al showed that only ferric chloride and aluminum sulphateamong the chemical coagulants (aluminum sulphate, ferric chloride, PAC and lime) were highly efficient in the removal of turbidity. Turbidity removal using aluminum sulfate within the pH range 4.5–5 was 99.33% (turbidity reduced from 166 to 1.1 NTU). In the process of ozonation to wastewater of chemical treatment with aluminum sulphatewithin the above range of time, COD was reduced from 22500 mg/L to 4800 mg/L (6). De Paula et al showed that with a combination of coagulant $Al_2(SO_4)_3$ and *M. oleifera* more than 90% of turbidity was removed, and a ratio of 20:80 (w/w) was obtained for *M. oleifera* and $Al_2(SO_4)_3$ and their results are in agreement with those of the present research (32). Abedini and Alipour indicated that under optimized absorption conditions (pH: 5, sorbent: 1.5 g, temperature: 45°C, Cd concentration: 2 ppm and exposure time: 180 minutes), the removal rate of Cd was 80% by using *M. oleifera* seed powder (33).

Muralimohan et al showed that the 50:50 blended coagulant *M. oleifera* and alum has a higher removal efficiency with respect to turbidity, TSS, TDS and COD; while the result of this study was showed in the ratio of 1:2 alum and *M. oleifera* the removal efficiency was optimal (34).

Although alum is considerably more efficient than *M. oleifera* in removing these parameters (COD, turbidity, and TSS reduced by 51.72%, 92.16%, and 85.26% respectively), the economic and environmental aspects of its use must be considered. Therefore, replacing alum (the material used in the wastewater treatment unit at the refinery in Bandar Abbas) with a natural biopolymer such as *M. oleifera*, or utilizing these two together, can be useful and effective. Moreover, the use of natural coagulants can substantially save money on the cost of chemical materials and reduce to a minimum the sludge produced and, hence, require less disposal. Some natural compounds have coagulant properties, and learning about these properties and understanding their exact mechanisms of action can lead to their correct application.

Conclusion

Considering the fact that *M. oleifera* coagulum can be locally produced, its use in water purification should be encouraged. This is likely to reduce the high cost of the current water treatment processes. It is recommended that a combination of alum and *M. oleifera* in different proportions be investigated to establish their effectiveness in treating wastewater. From the economic point of view, since *M. oleifera* trees grow naturally in the deprived region of Beshagard in Hormozgan province, the promotion of its cultivation can help the economy of the local communities, in addition to expanding the plant cover in the region. This valuable tree is found in about one million hectares of the lands in southern Iranian provinces (Sistan and Baluchestan, Hormozgan, and Fars provinces) and, with planning, one million hectares of grade 3 rangelands in these provinces can be annually turned into valuable land by planting *Moringa* trees. In this regard, industrial units such as refineries and petrochemical complexes are required to abide by regulations related to the protection of the environment, while the use of a natural biopolymer could substantially help to protect the environment as these biopolymers are biodegradable.

Suggestions

1. Natural biopolymers such as *M. oleifera* should be used, singly or in combination with another suitable coagulant, in other wastewater treatment units in the petroleum, gas, and petrochemical industries, and in urban wastewater treatment.
2. Other natural biopolymers, including chitosan, tannin, starches, etc., should be utilized as coagulants in the petroleum, gas, and petrochemical industries and to treat urban wastewater.
3. Research should be conducted on the optimum use of sludge resulting from the utilization of biopolymers.
4. Various biopolymers should be used as coagulants and coagulant aids in the Iranian water and wastewater treatment industries.

Acknowledgments

The authors appreciate the support provided by Islamic Azad University Bandar Abbas Branch and the Bandar Abbas Oil Refinery.

Competing interests

We affirm that this article is the original work of the authors and they have no conflict of interest to declare.

Authors' contribution

All authors were involved in all stages of the article. On behalf of the co-authors, the corresponding author bears full responsibility for this submission.

References

1. Rajkumar D, Palanivelu K. Electrochemical treatment of industrial wastewater. J Hazard Mater 2004; 113(1-3): 123-9. doi: 10.1016/j.jhazmat.2004.05.039.
2. Zouboulis AI, Avranas A. Treatment of oil-in-water emulsions by coagulation and dissolved-air flotation. Colloids Surf A Physicochem Eng Asp 2000; 172(1-3): 153-61. doi: 10.1016/S0927-7757(00)00561-6.
3. Rajakovic V, Aleksic G, Radetic M, Rajakovic L. Efficiency of oil removal from real wastewater with different sorbent materials. J Hazard Mater 2007; 143(1–2): 494-9. doi: 10.1016/j.jhazmat.2006.09.060.
4. Qasim SR, Motley EM, Zhu G. Water Works Engineering: Planning, Design, and Operation. New York, NY: Prentice Hall PTR; 2000.
5. Yin CY. Emerging usage of plant-based coagulants for water

and wastewater treatment. Process Biochemestry 2010; 45(9): 1437-44. doi:10.1016/j.procbio.2010.05.030.
6. Javid AH, Malakootian M, Mehdipour M. Investigation of ozone and coagulant material's (aluminum sulfate, ferric chloride, poly aluminum chloride and lime) efficiency in "Kerman Kork" industry wastewater treatment. Environmental Health Engineering and Management Journal 2015; 2(1): 1–6.
7. Muthuraman G, Sasikala S. Removal of turbidity from drinking water using natural coagulants. J Ind Eng Chem 2014; 20(4): 1727-31. doi: 10.1016/j.jiec.2013.08.023.
8. Fatombi JK, Lartiges B, Aminou T, Barres O, Caillet C. A natural coagulant protein from copra (*Cocos nucifera*): isolation, characterization, and potential for water purification. Sep Purif Technol 2013; 116: 35-40. doi: 10.1016/j.seppur.2013.05.015.
9. Prasad RK. Color removal from distillery spent wash through coagulation using *Moringa oleifera* seeds: use of optimum response surface methodology. J Hazard Mater 2009; 165 (1-3): 804-11. doi: 10.1016/j.jhazmat.2008.10.068.
10. Azadi NA, Falahzadeh RA, Sadeghi S. Dairy wastewater treatment plant in removal of organic pollution: a case study in Sanandaj, Iran. Environmental Health Engineering and Management Journal 2015; 2(2): 73-7.
11. Mirzaei Nadoushan H, Asadi Karam F. Moringa: A Miracle of Nature. Iran: Research Institute of Forests and Rangelands of Iran; 2010.
12. Ghebremichael K, Gunaratna K, Henrikson H, Burmer H, Dalhammar G. A simple purification and activity assay for the coagulant protein from *Moringa oleifera* seed. Water Res 2005; 39(11): 2338-44. doi: 10.1016/j.watres.2005.04.012.
13. Sutherland JP, Folkard GK, Grant WD. Natural coagulants for appropriate water treatment: a novel approach. Waterlines 1990; 8(4): 30-2.
14. Bhuptawat H, Folkard GK, Chaudhari S. Innovative physico-chemical treatment of wastewater incorporating Moringa oleifera seed coagulant. J Hazard Mater 2007; 142(1-2): 477-82.
15. Araújo CST, Melo EI, Alves VN, Coelho NM. *Moringa oleifera* Lam. seeds as a natural solid adsorbent for removal of Ag(I) in aqueous solutions. Journal of Brazilian Chemical Society 2010; 21(9): 1727-32. doi: 10.1590/S0103-50532010000900019.
16. Mohammad Beygi K, Vasheghani Farahani E. Comparison of different biopolymers performance in water treatment. Iranian Journal of Polymer Science and Technology 2006; 18(5): 285-90.
17. Liew AG, Noor MJ, Muyibi SA, Fungara AM, Muhammed TA, Iyuke SE. Surface water clarification using *M. oleifera* seeds. Int J Environ Stud 2006; 63 (2): 211-9. doi: 10.1080/00207230500117670.
18. Bhatia S, Othman Z, Ahmad AL. Coagulation–flocculation process for POME treatment using *Moringa oleifera* seeds extract: optimization studies. Chem Eng J 2007; 133(1-3): 205-12. doi: 10.1016/j.cej.2007.01.034.
19. Yazdani V. Evaluating of performance of powdered seeds of some species of *Moringa* tree (family *Moringaceae*) in comparison with alum and poly aluminum chloride (PAC) in urban wastewater treatment [dissertations]. Hamedan: Bu-Ali Sina University; 2009. [In Persian].
20. Mehdinejad MH, Bina B, Nikaeen M, Movahedian Attar H. Effectiveness of *Moringa Oleifera* Coagulant Protein and Chitosan as natural coagulant aids in removal of colloidal particles and bacteria from turbid waters. Journal of Gorgan University of Medical Sciences 2009; 11(3): 60-9. [In Persian].
21. Pritchard M, Craven T, Mkandawire T, Edmondson A, O'Neill JG. A study of the parameters affecting the effectiveness of *Moringa oleifera* in drinking water purification. Phys Chem Earth 2010; 35(13–14): 791-7. doi: 10.1016/j.pce.2010.07.020
22. Greenberg AE, Classer LS. Standard Method for the Examination of Water and Wastewater. USA: APHA, AWWA and WEF; 2002.
23. Katayon S, Noor MJ, Asma M, Ghani LA, Thamer AM, Azni I, et al. Effects of storage conditions of *Moringa oleifera* seeds on its performance in coagulation. Bioresour Technol 2006; 97(13): 1455-60. doi: 10.1016/j.biortech.2005.07.031.
24. USEPA. Enhanced Coagulation and Enhanced Precipitative Softening Guidance Manual. Washington, DC: EPA; 1999.
25. ASTM. Standard Test Method for Turbidity of Water (Withdrawn 2007). West Conshohocken, PA: ASTM International; 2000.
26. ASTM. Standard Test Methods for pH of Water. West Conshohocken, PA: ASTM International; 2012.
27. ASTM. Standard Test Methods for Chemical Oxygen Demand (Dichromate Oxygen Demand) of Water. West Conshohocken, PA: ASTM International; 2012.
28. Bhuptawat H, Folkard GK, Chaudhari S. Innovative physico-chemical treatment of wastewater incorporating Moringa oleifera seed coagulant. J Hazard Mater 2007; 142(1-2): 477-82. doi: 10.1016/j.jhazmat.2006.08.044.
29. Sotheeswaran S, Nand V, Matakite M, Kanayathu K. *Moringa oleifera* and other local seeds in water purification in developing countries. Res J Chem Environ 2011; 15(2): 135-8.
30. Lagasi JE, Agunwamba JC, Aho M. Comparative studies on the use of ordinary and de-oiled *Moringa oleifera* in the treatment of abattoir waste water. Int J Eng Sci 2014; 3(2): 1-7.
31. Bina B, Shasavani A, Asghare G, Hasanzade A. Comparison of water turbidity removal efficiencies of *Moringa oleifera* seed extract and poly-aluminum chloride. Water Waste Journal 2007; 18(1): 24-33. [In Persian].
32. De Paula HM, de Oliveira Ilha MS, Andrade LS. Concrete plant wastewater treatment process by coagulation combining aluminum sulfate and *Moringa oleifera* powder. J Clean Prod 2014; 76(1): 125-30. doi: 10.1016/j.jclepro.2014.04.031.
33. Abedini S, Alipour V. Cadmium removal from synthetic wastewater by using *Moringa oleifera* seed powder. Environmental Health Engineering and Management Journal 2015; 2(4): 157-63.
34. Muralimohan N, Palanisamy T, Vimaladevi MN. Experimental study on removal efficiency of blended coagulants in textile wastewater treatment. Int J Res Eng Technol 2014; 2(2): 15-20.

Efficiency in phenol removal from aqueous solutions of pomegranate peel ash as a natural adsorbent

Ali Asghar Najafpoor[1], Sina Dousti[2*], Ahmad Joneidi Jafari[3], Ahmad Hosseinzadeh[4]

[1]Associate Professor, Health Sciences Research Center, Department of Environmental Health Engineering, School of Health, Mashhad University of medical Sciences, Mashhad, Iran
[2]MSc in Environmental Health Engineering, Student Research Committee, School of Health, Mashhad University of medical Science, Mashhad, Iran
[3]Professor of Environmental Health Engineering, Department of Environmental Health Engineering, Center for Environmental Health Technology, School of Health, Iran University of Medical Sciences, Tehran, Iran
[4]MSc Student of Environmental Health Engineering, Student Research Committee, Department of Environmental Health Engineering, School of Health, Mashhad University of Medical Sciences, Mashhad, Iran

Abstract
Background: Phenol is an organic pollutant found in industrial effluents that is very toxic to humans and the environment. This study used pomegranate peel ash as a natural absorbent to remove phenol from aqueous solutions.
Methods: In this study, pomegranate peel ash in different doses was used as a new adsorbent for the removal of phenol. The effects of contact time, pH, adsorbent dose and initial phenol concentration were recorded. Then, the adsorption data was described with Langmuir and Freundlich adsorption isotherms; Excel software was used for data analysis.
Results: The highest percentage of phenol adsorption was observed at pH=7. The optimum amount of adsorbent was 0.6 g/l, and after 120 minutes, the process reached an equilibrium state. The adsorption of phenol decreased following an increase in the pH of the solution. It was also observed that contact time significantly affected the rate of phenol adsorption. The experimental data fit much better in the Freundlich (R^2=0.9056) model than in the Langmuir (R^2=0.8674) model.
Conclusion: Pomegranate peel ash has the potential to be utilized for the cost-effective removal of phenol from aqueous solutions.
Keywords: Phenol removal, Pomegranate peel ash, Aqueous solution

***Correspondence to:**
Sina Dousti
Email: sina_dosty@yahoo.com

Introduction

The presence of phenol and its derivatives in water and wastewater is a major concern because of their hazards to human health and the environment. Phenol enters the environment naturally from the decomposition of algae and plants; however, the presence of phenolic compounds in surface waters is caused mainly by industrial wastewaters (1). Phenolic compounds comprise one of the most common groups of pollutants in wastewaters produced by resins, plastics, explosives, steel, leather, etc. Organic contaminants that enter bodies of water in various concentrations are able to react with phenolic compounds and produce chlorinated phenolic compounds. Phenol is known to be a solvent and an antiseptic substance and is used as an additive to increase the efficiency of disinfection processes; such chlorinated substances can also be produced through processes (1,2). The Environmental Protection Agency (EPA) classifies these compounds as one the most important groups of priority pollutants because of their high toxicity, carcinogenic properties, bioaccumulation, and low biodegradation potential, and the adverse effects they have on human life (1-3). Health effects resulting from exposure to phenol are dependent upon absorption rate and exposure time. These effects range from mucosal and skin irritation, systemic toxicity, decreased blood pressure, and increased heart rate to coma (4). One study showed that the consumption of between 8 and 15 mg of phenol can lead to death in humans (5). Additionally, phenolic compounds in drinking water resources disinfected by chlorine compounds may create chlorophenol compounds. These substances may make some undesirable changes in the odor and taste of water, making it unsuitable for consumption (6). Moreover, phenol is considered to be a major carcinogenic agent. Even low concentrations of it may lead to significant health concerns. Phenol can penetrate cells and cause cytoplasmic coagulation; phenolic contaminants can damage sensitive cells (7-9). Therefore, the detection, identification, and determination of phenolic compounds in wastewaters are very important. The proportion of such pollutants in industrial wastewaters has to be considered before the wastewater is discharged into water flows. Based on the aforementioned reasons, the EPA has determined a maximum concentration of 1 mg/L for phenol in industrial effluents discharged into surface waters (10,11).

Various processes are used for the removal of phenolic compounds from aqueous solutions, including extraction, activated carbon adsorption, distillation with steam, electrochemical methods, irradiation, advanced oxidation, chemical oxidation, adsorption, and biological treatments. Most of these methods have disadvantages such as the high cost of treatment or the need for additional treatments. None of these approaches have been successful in completely removing phenol compounds from aqueous solutions; each suffers from one or more limitations, such as the high cost, the need for additional treatment, the production of dangerous byproducts, low efficiency, and limited applicability to various concentrations of pollutants (12).

Extensive research has been carried out to study the phenol removal potential of each method. Many biological and non-biological processes, such as polymerization, electrocoagulation, extraction, photo-decomposition, advanced oxidation process, and adsorption, have been employed by various researchers for phenol removal. The enzyme horseradish peroxide has been used for phenol polymerization. In optimal conditions, a phenol polymerization efficiency of about 90% was observed. During electro-coagulation, phenol was removed by the combined effects of sweep coagulation and adsorption. At a high current density and a solution pH of 7, the remarkable removal rate of 92% after 2 hours was achieved. Adsorption by various adsorbents is another attractive alternative. It is advantageous that a wide range of adsorbents are available. Many adsorbents are waste materials from agricultural and other industrial operations, such as distilleries. Phenol removal rates ranging from 88% to 95% have been observed using various adsorbents. Adsorption seems to be the most widely-studied operation for phenol removal (13).

The adsorption process, one of the most important methods, is widely used for the removal of organic and inorganic pollutants from aqueous solutions (14). Different adsorbents have been used for the removal of phenol from aqueous solutions; among them, commercial activated carbon has a great ability to absorb phenol from aquatic environments, although carbon production and its regeneration process has a high cost. Therefore, the economic aspects of absorption processes need to be considered. In recent years, researchers have attempted to replace commercial activated carbon with low-cost adsorbents (15).

The present study investigated the performance and efficiency of ash made from pomegranate peel for the removal of phenol from aqueous solutions.

Notably, this is the first time pomegranate peel ash as a natural adsorbent for phenol removal is being studied.

Methods

Preparation of absorbent

To produce the pomegranate peel ash, pomegranate skins were washed with distilled water to remove impurities and then dried at 105°C for 24 hours. The dried pomegranate peels were burned at 500°C for 2 hours to be turned into ash. Then the chopped dried skin was poured into a Chinese plant. Ash was produced after crushing operations by standard sieves ranging between 60- and 200-mesh sizes. The prepared adsorbent was poured into a closed container. No other chemical or physical operation was performed before adsorption on the adsorbent.

Preparation of phenol solution

All analytical grade chemicals used in the present study were provided by Merck. Phenol specific for laboratory analysis was purchased with a purity of 99.9% and a molecular mass of 94.11 g/mol. To prepare the stock solution (1000 mg/L), the required amount of phenol was dissolved in distilled water. Solution concentrations for the experiments were obtained by diluting the stock solution. Because of the volatile nature of phenol, all precautions were taken to prevent its evaporation from the samples during preparation and subsequent experiments. Normal HCL and NaOH 0.1 were applied for pH adjustment.

Adsorption experiments and analysis of samples

Phenol was measured with a spectrophotometer (Model T80 UV/VIS) at a wavelength of 500 nm. At first, the calibration curve was constructed by assaying standard solutions of phenol in concentrations of 5 to 100 mg/L; then the phenol concentrations of samples were calculated based on the obtained curve (16,17).

The effects of various parameters, including initial pH, adsorbent dose, contact time, and initial phenol concentration, on phenol removal were investigated. All tests were accomplished at room temperature (23±2°C). Experiments were performed in a closed flask of 250 mL capacity and on a magnetic shaker. In the experiments, a specific proportion of adsorbent was added to flasks containing 100 mL sample and stirred at 120 rpm. After the contact time elapsed, all samples were filtered through Whatman filter paper (No. 42), phenol concentrations in the filtrated solutions were measured, and the amount of phenol adsorbed (mg/g) was calculated using the following equation. Final results are reported as the average of the two measurements. The range of variables is presented in Table 1.

$$q_e = \frac{(C_0 - C_e)V}{M}$$

where, C_0 = is the initial concentration of phenol, C_e = is the phenol concentration after the test was performed and

Table 1. Range of parameters examined in this study

Contact time (min)	Temperature (°C)	The concentration of phenol (mg/L^{-1})	Dose of adsorbent (g/L^{-1})	pH
15, 30, 60, 90, 120, 150, 180, 210	23±2°	10, 20, 30, 40, 50, 80, 100	0.1, 0.2, 0.3, 0.4, 0.5, 0.6,0.7	2-12

a state of balance was achieved, V = is the volume of the solution (L), and M = is the amount of adsorbent used (g).

Adsorption isotherm

The equilibrium isotherms are very important in designing adsorption systems. To estimate the adsorption characteristics of an adsorbent, the isotherm adsorption of that adsorbent with a specific adsorbate is accomplished. The concentration variation method was used to calculate the adsorption characteristics of the adsorbent and the process. It is mainly carried out by selecting an appropriate concentration range of the adsorbate with a fixed mass of adsorbent. The ratio of the adsorbent mass and the volume of the dye solution were applied. For this purpose, two adsorption models (Langmuir and Freundlich) were used.

Langmuir Isotherm

The Langmuir equation is based on the assumption that the maximum adsorption occurred because of a saturated monolayer of adsorbate molecules on the adsorbent surface (18). The non-linear equation of the Langmuir isotherm is as follows:

Original equation: $q_e = \frac{q_m b C_e}{1 + bC_e}$

Linearized equation: $\frac{1}{q_e} = \frac{1}{bq_m C_e} + \frac{1}{q_m}$

Freundlich Isotherm

The Freundlich isotherm is based on multilayer adsorption on the heterogeneous surface of the adsorbent containing an unequal amount of energies. It is not limited to a monolayer adsorption, but is also applied for multilayer adsorption. The non-linear equation of the Freundlich isotherm is shown below (18), and the linearized and original equations of these models are as follows:

Original equation: $q_e = k_f C_e^{\frac{1}{n}}$

Linearized equation: $1nq = 1nk + \frac{1}{n} 1nC_e$

where q_e is the ratio of adsorbed adsorbate per specific adsorbent (mg/g), C_e is the residual adsorbate concentration (mg/L), n is the constant of the Freundlich equilibrium related to bond energies between the adsorbate and the adsorbent, k_f is the constant of the Freundlich equilibrium indicative of bond strength (mg/g), q_m is the proportion of the required adsorbate to form a monolayer (mg/g), and b is the constant of the Langmuir equilibrium indicative of the energy of adsorption (l/mg).

Analysis and statistical methods

Statistical analysis was done using SPSS software version 19 for windows. In addition, the coefficients of these two models along with the correlation coefficients (R^2) were calculated in Excel software. The results for these equations were fitted and their goodness of fit was evaluated.

Results

Effect of pH on adsorption efficiency

To determine the effects of initial pH on phenol adsorption by pomegranate peel ash and the optimal pH, phenol adsorption was surveyed as a function of hydrogen ion concentration. In each series of test samples, the initial pH was determined to be in the range of 2-12, absorbent at 0. 4 g per 100 mL was used, and the initial phenol concentration in all samples was 50 mg/L. Figure 1 shows the effect of initial pH on the performance by of pomegranate skin ash on phenol adsorption. As clearly indicated in the graph, the highest removal of phenol occurred in pH = 7.

Effect of adsorbent dose

As can be seen in Figure 2, there is a direct relationship between adsorbent dose and phenol removal efficiency. Phenol removal percentages increased from 69.8% to 97.06% at adsorbent doses of 0.1 and 0.7 g/L, respectively. However, the proportion of adsorption capacity was considered as the phenol adsorbed per gram of adsorbent re-

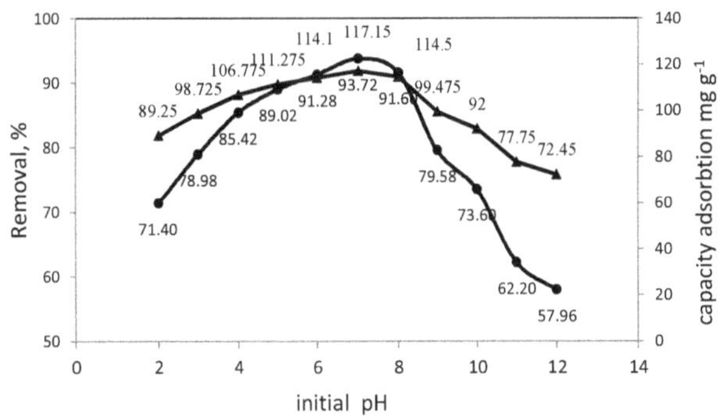

Figure 1. Effect of initial pH on phenol adsorption by pomegranate peel ash

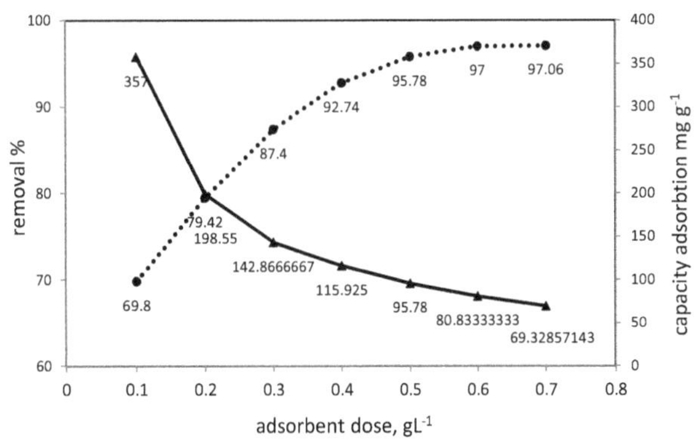

Figure 2. Effect of adsorbent dose on phenol adsorption by pomegranate peel ash.

duced. Based on the results, the adsorbent dose of 0.6 g/L was chosen as the optimal dose of adsorbent and was used in other parts of the study.

Effect of contact time

The effect of contact time on the adsorption process at the initial phenol concentration of 50 mg/L and 0.6 g/L of the adsorbent dose was conducted for times of 15 to 210 minutes. As Figure 3 shows, adsorption capacity and the percentage of phenol removal by adsorbent have a significant direct relationship with contact time in the early stages of adsorption. The highest elimination efficiency (95% to 96%) and adsorption capacity (79 to 97 mg/g) were obtained after 120 minutes. Therefore, an exposure time of 120 minutes was selected as the optimal equilibrium time.

Effect of initial phenol concentration:

As shown in Figure 4, the results show that when the initial phenol concentration increases from 10 to 100 mg/L, adsorption capacity of 16 mg/g increases from 10 to 148 mg/g, while the removal of phenol at the same time shows a reverse trend.

Table 2 shows the parameters of the isotherm and kinetic equations in phenol adsorption by pomegranate peel ash. As shown in Table 2, the phenol adsorbed into the pomegranate peel ash can be described by both Freundlich and Langmuir isotherms (Figures 5 and 6).

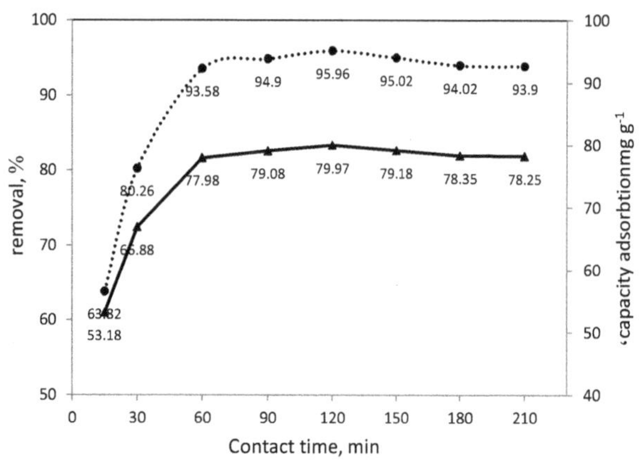

Figure 3. Effect of contact time on phenol adsorption by pomegranate peel ash.

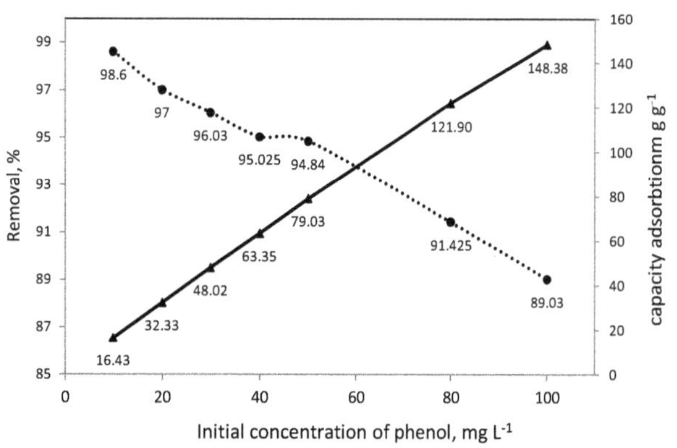

Figure 4. Effect of initial phenol concentration on phenol adsorption by pomegranate peel ash.

Table 2. Parameters regarding the adsorption isotherm models for phenol adsorption on pomegranate peel ash temperature of 23 ± 2°C

Langmuir isotherm			Freundlich isotherm		
q_m(mg g^{-1})	K_l(L kg^{-1})	R^2	K_f	n	R^2
344.86	0.143	0.8674	48.41	1.516	0.9056

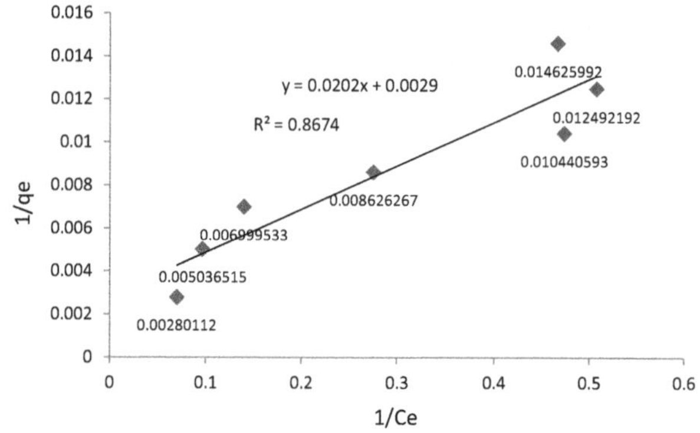

Figure 5. Langmuir isotherm for phenol adsorption by pomegranate peel ash (R^2 = 0.8674).

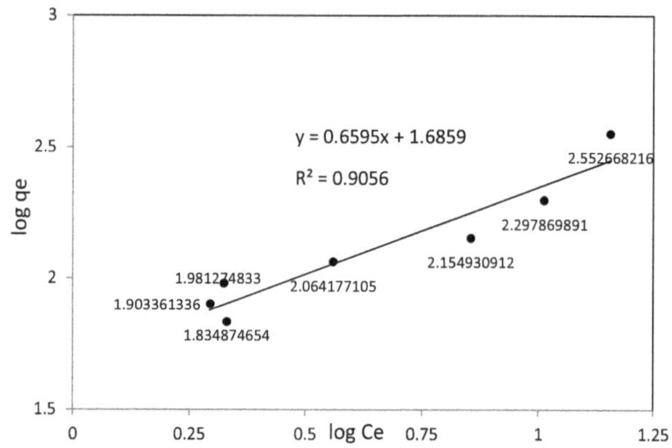

Figure 6. Freundlich isotherm for phenol adsorption by pomegranate peel ash (R^2 = 0.9056).

Discussion

The effect of pH on adsorption efficiency plays an important role in the process of adsorption and also has a key role in absorption capacity (19). In general, removal efficiency decreases have been reported in low and high pH values (20).

The direct relationship between pH and phenol removal efficiency may be caused by the presence of H^+ ions. Since phenol ions can be neutralized by the hydrogen ions, their adsorption is reduced on polar adsorbents. Moreover, the process efficiency is reduced when pH is increased above 7. This reduction in efficiency at higher pH values may be because of the fact that at a higher pH, phenol is in the form of salt and loses its negative charges. Thus, the adsorption of phenol will face serious difficulties. In addition, OH ions inhibit the absorption of phenol by the adsorbent (17). Rengaraj et al studied the removal of phenol by palm kernels, and their results agree with the results of the present study (20).

Effects of adsorbent dose

One of the most important parameters of the adsorption process is adsorbent dose. As shown in Figure 2, the percentages of phenol removal with increases in the amount of adsorbent experienced an upward trend. This can indicate that there are still active sites on the surface of the adsorbent that are unsaturated. Data from this study indicated clearly that the greater the rate of adsorbent was, the more efficient phenol removal would be. This could be because an increase in the proportion of adsorbent provides more active surface area for adsorption. Thus, when more absorbent is added, the remaining amount of pollutants will stay constant. The results showed that the phenol elimination efficiency encountered no changes at adsorbent doses higher than 0.6 g/L (21). Although phenol removal efficiency increased with increasing doses of adsorbent, the amount of phenol adsorbed per gram adsorbent decreased. Because of absorption by active sites, the concentration of the phenol solution reduced the saturated absorbent surface. Conversely, increasing the amount of adsorbent caused aggregation and adhesion to them (2).

Effects of contact time

Contact time is one of the most important parameters in determining the adsorption process equilibrium time. The results showed that at first, the removal of adsorbing phenolic compounds was high. However, the removal efficiency was gradually reduced because of the equilibrium time. This phenomenon can be attributed to the abundance of available spaces in the adsorbent at first. Over time, because of the repulsive force between the molecules of the absorber and the liquid phase, the remaining empty spaces of the adsorbent surface are hardly occupied. Similar findings have been reported by Calvete et al (22), Cengiz and Cavas (23), and SenthilKumar et al (24).

Effects of initial phenol concentration

The direct relationship between initial phenol concentration and phenol removal efficiency is interpreted as follows: the mass transfer driving force is increased with the rise of phenol concentration, and thus soluble phenolic molecules flow to the planchette liquid absorbent layer. Finally, the adsorbent surface increases (25,26). The reduction in phenol removal efficiency that occurs over this process may be owing to the decrease in active sites of the adsorbent surface. Similar results for phenol adsorption from aqueous solutions by clay and other absorbents have been recorded (27-30).

Conclusion

Adsorption is known as a vigorous and well-established procedure for treating industrial effluents. This study investigated the potential of pomegranate peel ash as an adsorbent in removing phenol from aqueous solutions. The effects of the four variables of initial pH, contact time, adsorbent dose, and phenol concentration as well as the adsorption isotherm were studied. The results indicated that the highest rate of phenol adsorption was observed at pH = 7, the optimum amount of adsorbent was 0.6 g/L, and the process reached the state of equilibrium after 120 minutes. Based on the results of the current study, pomegranate peel ash has a good potential for the removal of phenol from aqueous solutions.

Acknowledgments

This project was financially supported by the Faculty of Public Health, Mashhad University of Medical Sciences. The authors are grateful to the head of the Environmental Chemistry Laboratory for his help.

Competing interests

The authors declare that they have no competing interests.

Authors' contributions

AAN conceived the study and supervised development of work, helped in data interpretation and manuscript evaluation. SD designed and Performed analysis on all samples, interpreted data, wrote manuscript and acted as corresponding author. AJJ helped in data interpretation and manuscript evaluation. AH helped to evaluate and edit the manuscript.

References

1. Yapar S, Ozbudak V, Dias A, Lopes A. Effect of adsorbent concentration to the adsorption of phenol on hexadecyl trimethyl ammonium-bentonite. J Hazard Mater 2005; 121(1-3): 135-9.
2. Busca G, Berardinelli S, Resini C, Arrighi L. Technologies for the removal of phenol from fluid streams: a short review of recent developments. J Hazard Mater 2008; 160 (2-3): 265-88.
3. Ersoz A, Denizli A, Sener I, Atilir A, Diltemiz S, Say R. Removal of phenolic compounds with nitrophenol-imprinted polymer based on pi-pi and hydrogen-bonding interactions. Separation and Purification Technology 2004; 38(2): 173-9.
4. Karel V. Handbook of Environmental Data on Organic Chemicals, Vol 2. 4th ed. Canada: John Wiley and Sons Inc; 2000.
5. Raynold J, Martindale EF. The Extra Pharmacopocia Disinfectants and Antiseptics. 28th ed. Landan; 1982.
6. Rengaraj S, Moon SH, Sivabalan R, Arabindoo B, Murugesan VJ. Removal of phenol from aqueous solution and resin manufacturing industry wastewater using an agricultural waste: rubber seed coat. Hazard Mater 2002; 89(2-3): 185-96.
7. Bódalo A, Gómez E, Hidalgo AM, Gómez M, Murcia MD, López I. Nanofiltration membranes to reduce phenol concentration in wastewater. Desalination 2009; 245(1-3): 680-6.
8. Aksu Z, Yener J. A comparative adsorption/biosorption study of monochlorinated phenols onto various sorbent. Waste Manage 2001; 21(8): 695-702.
9. Viraraghavan T, De Maria Alfaro F. Adsorption of phenol

from wastewater by peat fly ash and bentonite. Hazard Mater 1998; 57(1-3): 59-70.
10. Balasubramanian A, Venkatesan S. Removal of phenolic compounds from aqueous solutions by emulsion liquid membrane containing Ionic Liquid [BMIM]$^+$[PF6]$^-$ in Tributyl phosphate. Desalination 2012; 289: 27-34.
11. Parliamentary Office Affairs. The environment laws and regulations. Tehran: EPA Publications; 2001.
12. Li HQ, Han HJ, Du MA, Wang W. Removal of phenols thiocyanate and ammonium from coal gasification wastewater using moving bed biofilm reactor. Bioresource Technol 2011; 102(7): 4667-73.
13. Kulkarni SJ, Kaware JP. Review on research for removal of phenol from wastewater. International Journal of Scientific and Research Publications 2013; 3(4): 1-5.
14. Nevskaia DM, Santianes A, Muñoz V, Guerrero-Ruíz A. Interaction of aqueous solutions of phenol with commercial activated carbons: an adsorption and kinetic study. Carbon 1999; 37(7): 1065-74.
15. Lin K, Pan J, Chen Y, Cheng R, Xu X. Study the adsorption of phenol from aqueous solution on hydroxyapatite nanopowders. Hazard Mater 2009; 161(1): 231-240.
16. Glesceria LA, Greenberg E, Eaton AD. Standard Methods for the Examination of Water and Wastewater. 31st ed. Washington DC: APHA; 2005.
17. Rasoulifard MH, Taheri Ghazvini N, Farhangnia E, Heydari A, Doost Mohammadi S. Removal of direct yellow 9 and reactive orange 122 from contaminated water using chitosan as a polymeric bioadsorbent by adsorption process. Color Science and Technology 2010; 4: 17-23. [In Persian].
18. Abedini S, Alipour V. Cadmium removal from synthetic wastewater by using Moringa oleifera seed powder. Environmental Health Engineering and Management Journal 2015; 2(4): 157-63.
19. Rengaraj S, Moon SH, Sivabalan R, Arabindoo B, Murugesan V. Agricultural solid waste for the removal of organics: adsorption of phenol from water and wastewater by palm seed coat activated carbon. Waste Manage 2002; 22(5): 543-8.
20. Rengaraj S, Sivabalan R, Arabindoo B, Murugesan V. Adsorption kinetics of o-cresol on activated carbonfrom palm seed coat. Indian Journal of Chemical Technology 2000; 7: 127-31.
21. Chakravarty P, Sarma NS, Sarma HP. Removal of lead(II) from aqueous solution using heartwood of Areca catechu powder. Desalination 2010; 256(1–3): 16-21.
22. Calvete T, Lima EC, Cardoso NF, Dias SLP, Pavan FA. Application of carbon adsorbents prepared from the Brazilian pine-fruit-shell for the removal of Procion Red MX 3B from aqueous solution—kinetic, equilibrium, and thermodynamic studies. Chem Eng J 2009; 155(3): 627-36.
23. Cengiz S, Cavas L. Removal of methylene blue by invasive marine seaweed: Caulerpa racemosa var cylindracea. Bioresour Technol 2008; 99(7): 2357-63.
24. SenthilKumar P, Ramalingam S, Sathyaselvabala V, Dinesh Kirupha S, Sivanesan S. Removal of copper(II) ions from aqueous solution by adsorption using cashew nut shell. Desalination 2011; 266(1-3): 63-71.
25. Caturla F, Martín-Martínez JM, Molina-Sabio M, Rodriguez-Reinoso F, Torregrosa R. Adsorption of substituted phenols on activated carbon. J Colloid Interface Sci 1988; 124(2): 528-34.
26. Imagawa A, Seto R, Nagaosa Y. Adsorption of chlorinated hydrocarbons from air and aqueous solutions by carbonized rice husk. Carbon 2000; 38(4): 628-630.
27. Nayak PS, Singh BK. Removal of phenol from aqueous solutions by sorption on low cost clay. Desalination 2007; 207(1–3): 71-9.
28. Bhatnagar A. Removal of bromophenols from water using industrial wastes as low cost adsorbents. J Hazar Mater 2007; 139(1): 93-102.
29. Ghaedi M, Tashkhourian J, Amiri A, Sadeghian B, Equilibrium, kinetic and thermodynamic study of removal of reactive orange 12 on platinum nanoparticle loaded on activated carbon as novel adsorbent. J Chem Eng 2011; 28(12): 2255-61.
30. Najafpoor AA, Vojoudi Z, Dehghani MH, Changani F, Alidadi H. Quality assessment of the Kashaf river in north east of Iran in 1996-2005. J Applied Sciences 2007; 7(2): 253-7.

16

Hydrogen sulfide and organic compounds removal in municipal wastewater using ferrate (VI) and ultraviolet radiation

Amirreza Talaiekhozani[1*], Zeinab Eskandari[2], Marzieh Bagheri[3], Mohammad Reza Talaei[4], Malihe Salari[5]

[1]Assistant Professor, Civil Engineering Department, Jami Institute of Technology, Isfahan, Iran
[2]MSc Student, Chemical Engineering Department, Jami Institute of Technology, Isfahan, Iran
[3]Assistant Professor, Chemical Engineering Department, Jami Institute of Technology, Isfahan, Iran
[4]Associate Professor, Chemical Engineering Department, Shiraz University, Shiraz, Iran
[5]MSc Student, Civil Engineering Department, Jami Institute of Technology, Isfahan, Iran

Abstract
Background: In this study, ferrate (VI) and ultraviolet (UV) radiation were employed to remove hydrogen sulfide from municipal wastewater resulting in a reduction in chemical oxygen demand (COD). Although ferrate (VI) and UV have been used individually for the removal of a few pollutants from urban and industrial wastewater, there exists no study to investigate the effectiveness of simultaneous utilization of both methods for the removal of hydrogen sulfide and reducing COD.
Methods: This study aims to compare the application of UV, ferrate (VI) and UV/ferrate (VI) for the removal of hydrogen sulfide and COD from municipal wastewater in batch mode. Moreover, the effect of many parameters such as ferrate (VI) concentration, temperature, hydraulic retention time (HRT) and pH on ferrate (VI) oxidation power, were investigated.
Results: The results of this study demonstrated that for pH less than 2, higher pollutant removal efficiency was obtained. COD removal efficiency could increase up to 68% by adding 1.68 mg/L of ferrate (VI), almost 100% of hydrogen sulfide was removed by the same concentration of ferrate (VI). Both hydrogen sulfide and COD removal efficiencies increased as temperature increased to 50°C; nevertheless, further increase in temperature had negative effect on the removal efficiency. The use of UV/ferrate (VI) increased the removal efficiency of both hydrogen sulfide and COD when compared with the use of UV and ferrate (VI) individually. UV method was not effective in the removal of hydrogen sulfide.
Conclusion: The research findings shed new light on wastewater treatment systems employing UV/ferrate (VI) to decrease both the hydrogen sulfide and COD of municipal wastewater. This new findings will assist in the inaccurate design and effective operation of such systems which can be employed to maintain or improve environmental quality.
Keywords: Ferrate (VI), Hydrogen sulfide removal, Organic compounds removal, Ultraviolet

***Correspondence to:**
Amirreza Talaiekhozani
Email: atalaie@jami.ac.ir

Introduction

Anaerobic reactors are among the most important wastewater treatment systems which are capable of removing recalcitrant organic compounds. Anaerobic reactors have many advantages like high pollutant removal efficiency, low energy consumption and less sludge production when compared to aerobic reactors. As a result, anaerobic reactors are common to be used for a diversity of industrial wastewater. Nevertheless, anaerobic systems suffer from odor production (1,2). Controlling odor production in wastewater collection and treatment systems is one of the most important research areas in the field of environmental engineering (3). Hydrogen sulfide is a malodorous compound emitted from wastewater collection and treatment systems (4). This gas is extremely toxic for humans at levels above 500 ppm (1). In addition to odor problem, hydrogen sulfide gas causes corrosion of facilities (1). Wastewater collection systems made of concrete are seriously damaged by microbial corrosion (5,6). Corrosion in wastewater collection systems occurs because different forms of sulfur (such as mineral and organic sulfur) are changed to hydrogen sulfide (1). The activities of sulfur-reducing bacteria, which exist in anaerobic conditions caused as a result of oxygen deficiency, produce hydrogen sulfide (7). Hydrogen sulfide evaporates from wastewater and accumulates in the upper space of wastewater collection pipes (8). Sufficient oxygen and moisture in the upper space of wastewater collection pipes generate a perfect condition for growing sulfur oxidizing bacteria, which absorb hydrogen sulfide and convert it to sulfuric acid (9). This is the reason behind the extensive corrosion problems in concrete pipes (1,7,8). Coagulation, oxidation and disinfection are three important processes in water and wastewater treatment technology (10). Coagulant compounds agglomerate non-settling colloidal particles in the form of large particles. These large particles can be eliminated by settling or filtration. Disinfection processes have been designed to remove microorganisms harmful to human health (11). Sometimes, disinfectants are used not only for the removal of microorganisms but also for odorous compounds (4). Several different oxidizers and disinfectants like chlorine, sodium hypochlorite, chlorine dioxide and ozone are used in water and wastewater treatment (12). Chlorine oxidizes wastewater organic compounds and produces harmful by-products called trihalomethanes. Although chlorine dioxide and ozone do not suffer from secondary pollutant production, they are expensive (13). As environmental regulations for releasing wastewater into the environment are being more strict annually, new efficient coagulants, oxidizers and disinfectants are needed to meet these stringent restrictions. Ferrate (VI) is an oxidizing compound capable of fulfilling both tasks of disinfection and degradation, while it does not produce any harmful secondary pollutant (6,14). Ferrate (VI) is the most powerful oxidant ever known (4). In acidic conditions, it is the most powerful oxidant ever known in water and wastewater treatment systems (12). Ferrate (VI) is converted into ferric(III) which is an coagulant reagent (15). So, ferrate (VI) can be utilized simultaneously, in a single unit for oxidation, disinfection and coagulation. There are different methods for the production of ferrate (VI) such as wet method, dry method and electrolysis method (16). Since ferrate (VI) solution is an unstable product, wet method is not a suitable method for the production of ferrate (VI) (17). Dry method has risk of explosion, so, it is not suitable for use in industrial usage. Electrochemical method is a simple method for producing ferrate (VI) (7). Talaiekhozani et al reported that in electrolysis method, the highest amount of ferrate (VI) production was observed when 14M sodium hydroxide solution was used (4).

This study aims to evaluate the effectiveness of ferrate (VI) and UV methods in the removal of hydrogen sulfide and chemical oxygen demand (COD) from municipal wastewater. Ferrate (VI) was produced employing electrolysis method. In this study, the effects of different parameters like pH, temperature, COD and hydrogen sulfide concentrations and hydraulic retention time (HRT) on the removal efficiency were investigated.

Materials and Methods

Production of Ferrate (VI)

Two rectangular iron with dimension of 60×24 mm and thickness of 0.63 mm were employed as anode and cathode electrodes, respectively. The DC voltage employed was within the range of 1–24 V. An electrolysis system was assembled to produce ferrate (VI) as demonstrated in Figure 1. In this study, 56 g of sodium hydroxide was dissolved in 100 mL of distilled water to prepare a 14M solution. Thereafter, the electrolysis container as illustrated Figure 1 was filled with 100 mL of 14M sodium hydroxide. Subsequently, electrodes were charged using DC current with voltage of 9 V and amperage of 1 A for 30 minutes. Given that ferrate (VI) is converted to ferric (III) as time goes by, the prepared ferrate (VI) solution must be used as soon as possible.

Determination of removal efficiency

As the first step of this study, 10 L of wastewater were obtained from the municipal wastewater collection system of Fooladshahr, Iran. Collected wastewater was carried in a cool box to maintain the temperature at 4°C and was transferred to the chemistry laboratory of Jami Institute of Technology, Isfahan, Iran as soon as possible. Thereafter, the collected wastewater's hydrogen sulfide and COD concentration were measured. Subsequently, 8 Erlenmeyer flasks, each containing 40 mL of collected wastewater, were prepared. Thereafter, certain amounts of the ferrate (VI) solution were added to each Erlenmeyer flask to prepare ferrate (VI) concentrations of 0.56, 0.7, 0.84, 0.98, 1.26, 1.4, 1.54 and 1.68 mg/L. Given that ferrate (VI) oxidizing ability is significantly increased for pH of less than 2, the pH of all solutions in the flasks was adjusted to 2 by the addition of a suitable amount of hydrochloric acid. After stirring for 30 minutes, the hydrogen sulfide and COD concentrations of all solutions in the Erlenmeyer flasks were determined. Finally, hydrogen sulfide and COD removal efficiencies were calculated using Eq. 1, as follows:

$$RE = \left(\frac{C_f - C_i}{C_f}\right) \times 100 \quad (1)$$

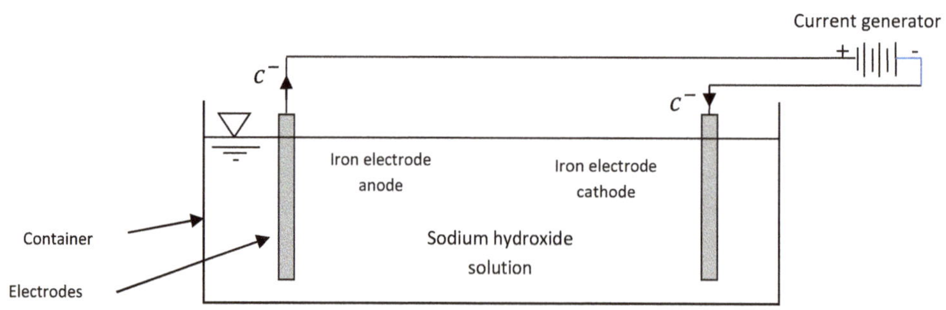

Figure 1. Schematic of the electrochemical cell for the synthesis of ferrate(VI).

Evaluation of hydraulic retention time effect

To the best of our knowledge, based on an extensive literature survey, there exists a shortage on the information about the efficiency of hydrogen sulfide and COD removal by ferrate (VI). So, in this section of the study, the effect of HRT on the removal efficiency of hydrogen sulfide and COD was investigated. For this purpose, 6 Erlenmeyer flasks, each containing 40 mL of wastewater, were prepared. A certain amount of ferrate (VI) was added to each Erlenmeyer flask to obtain a ferrate (VI) concentration of 0.5 mg/L. Subsequently, hydrogen sulfide and COD concentrations in the Erlenmeyer flasks were measured at HRTs of 10-60 minutes. Eq. 1 was utilized to calculate hydrogen sulfide and COD removal efficiency.

Evaluation of pH effect

Providing low-pH condition is vital in retaining ferrate (VI) oxidizing power as high as possible (4). In this section, the effect of pH on ferrate (VI) ability to remove hydrogen sulfide and COD was investigated. For this purpose, 6 Erlenmeyer flasks, containing 40 mL of the wastewater each, were prepared. Thereafter, a certain amount of ferrate (VI) was added to each Erlenmeyer flask to increase ferrate (VI) concentration to 0.5 mg/L. Subsequently, by adding suitable amounts of hydrochloric acid to each Erlenmeyer flask, the pH values of 1.4, 5.5, 7.2, 8.6, 10.9 and 13.2 were prepared. All Erlenmeyer flasks were stirred for 15 minutes at 23°C and then their hydrogen sulfide and COD concentrations were measured. Eq. 1 was used for calculating hydrogen sulfide and COD removal efficiency.

Evaluation of temperature effect

Temperature is an effective factor influencing the rate of chemical reaction. In this section, the effect of temperature on hydrogen sulfide and COD removal by ferrate (VI) was investigated. Temperature can be effective on the degradation rate of ferrate (VI) to ferric (III) and the oxidation rates of hydrogen sulfide and other organic compounds. So, this section was categorized into 2 sections, 1) evaluating the effect of temperature on the rate of degradation of ferrate (VI) and 2) evaluating the effect of temperature on the removal of hydrogen sulfide and COD.

In first set of experiments, 6 Erlenmeyer flasks, containing 40 mL of the wastewater each, were prepared. Thereafter, ferrate (VI) was added to each Erlenmeyer flask in the amount needed to increase its concentration to 0.5 mg/L. Erlenmeyer flasks were preserved at temperatures of 23.5, 45, 60, 70 and 80°C for 15 minutes. Finally, the concentrations of ferrate (VI) were determined in all Erlenmeyer flasks.

For the second set of experiments, 6 Erlenmeyer flasks, containing 40 mL of the wastewater each, were prepared. Thereafter, ferrate (VI) was added to each Erlenmeyer flask in the amount needed to increase its concentration to 0.5 mg/L. Erlenmeyer flasks were preserved at temperatures of 23.5, 45, 60, 70 and 80°C for 15 minutes. Given that the increase in temperature may cause evaporation of volatile organic compounds in wastewater, each Erlenmeyer flasks was considered as a control sample. Finally, the concentrations of hydrogen sulfide and COD were determined in all Erlenmeyer flasks. Eq. 1 was utilized to calculate hydrogen sulfide and COD removal efficiency.

Evaluation of UV/ferrate (VI) method effectiveness

In this section, 3 sets of experiments were performed. In first set of experiments, 6 Erlenmeyer flasks, containing 100 mL of the wastewater each, were prepared. Thereafter, a certain amount of ferrate (VI) was added to each Erlenmeyer flask to increase ferrate (VI) concentration to 0.5 mg/L. Subsequently, by adding a suitable amount of hydrochloric acid, the pH value of each Erlenmeyer was increased to 2. Thereafter, concentrations of hydrogen sulfide and COD in the Erlenmeyer flasks were measured at HRTs of 10–60 minutes.

In the second set of experiments, 6 containers holding 100 mL of the wastewater each were prepared. Thereafter, the containers were located under UV beam at 8 W power. The distance between the containers and the UV lamp was 15 mm. Subsequently, the concentrations of hydrogen sulfide and COD were measured at HRTs between 1 and 45 minutes.

In the third set of experiments, 6 containers, each containing 100 mL of the wastewater, were prepared. Thereafter, a certain amount of ferrate (VI) was added to each Erlenmeyer flask to increase ferrate (VI) concentration to 0.5 mg/L. Then, by adding a suitable amount of hydrochloric acid, the pH value of each Erlenmeyer flask was increased to 2. Subsequently, the containers were beamed by a UV lamp. Finally, the concentrations of hydrogen sulfide and COD at HRTs between 1 and 45 minutes were measured.

Analytical methods

The concentration of wastewater organic compounds was measured by the COD method, according to standard methods for examining water and wastewater (6). Iodometric titration was employed to measure hydrogen sulfide in wastewater (18). In iodometric titration method, sulfide was reacted with iodine in acidic condition. Thereafter, the amount of iodine remaining was measured by titration with sodium thiosulfate in the presence of starch as an indicator. The end of the titration was indicated by change in solution color from blue to light-straw color. A blank is treated exactly the same way as the samples. Thereafter, Eq. 2 was used for calculating hydrogen sulfide.

$$H2S\left(\frac{mg}{l}\right)=\frac{(A\times B)-(C\times D)\times 16000}{mL\,sample} \quad (2)$$

where, A is used iodine in mL, B is normality of iodine solution, C is used sodium thiosulfate in mL, and D is normality of sodium thiosulfate. The amount of ferrate (VI) was measured by the methods introduced by

Talaiekhozani et al (15).

pH value was determined using a AZ pH meter. All chemicals were obtained from Merck (Germany). Microscopic photos of the electrodes were taken by a Chinese metallurgical microscope Model VJ-2006B.

Results

Ferrate (VI) can be produced using three different methods, namely dry method, wet method and electrolysis method (4). Dry method for ferrate (VI) production is not a popular method owing to the risk of explosion. Although wet method is not dangerous, ferrate (VI) produced employing this method is highly unstable. The electrolysis method is simple, allowing *in situ* ferrate (VI) production. In this study, the effect of electrolysis process on electrode morphology was investigated using metallurgical microscope. Figure 2 illustrates the microscopic appearance of the electrodes before and after use showing high roughness on the electrode after the electrolysis process is completed. This roughness on the electrodes is created when iron is changed into ferrate (VI) and released into water.

Production of ferrate (VI) can generate a purple color in water. The colors of water before and after electrolysis are illustrated in Figure 3.

In this study, the effect of the presence of various metallic salts, like $FeCl_2$, NaOH, NaCl, $BaCl_2$ and $MgCl_2$, on the production of ferrate (VI) by electrolysis was investigated. The results are illustrated in Figure 4A.

Figure 4B illustrates the effect of temperature on the production of ferrate (VI). Figure 4C shows the effect of voltage on the production of ferrate (VI) and Figure 4D demonstrates the variation of hydrogen sulfide and COD removal efficiencies with concentrations of ferrate (VI).

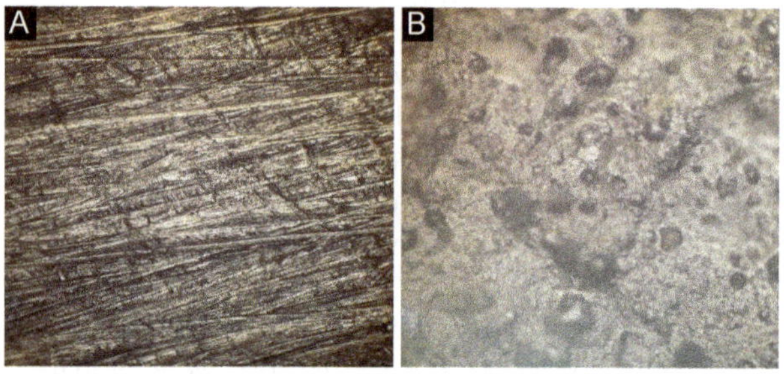

Figure 2. Microscopic figures of an iron electrode (A) before and (B) after electrolysis.

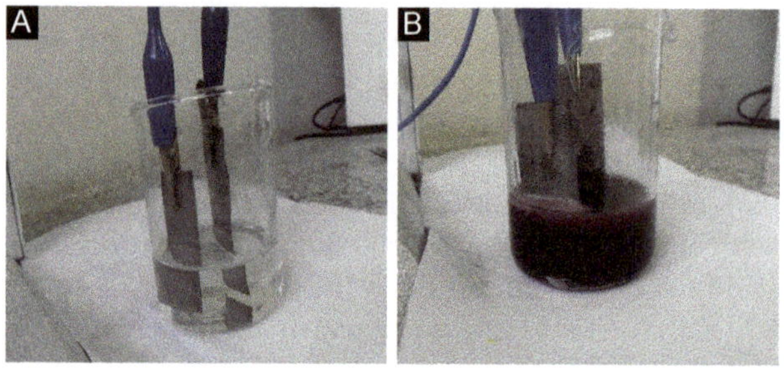

Figure 3. Change in the color of the electrolyte solution (A) before and (B) after electrolysis process.

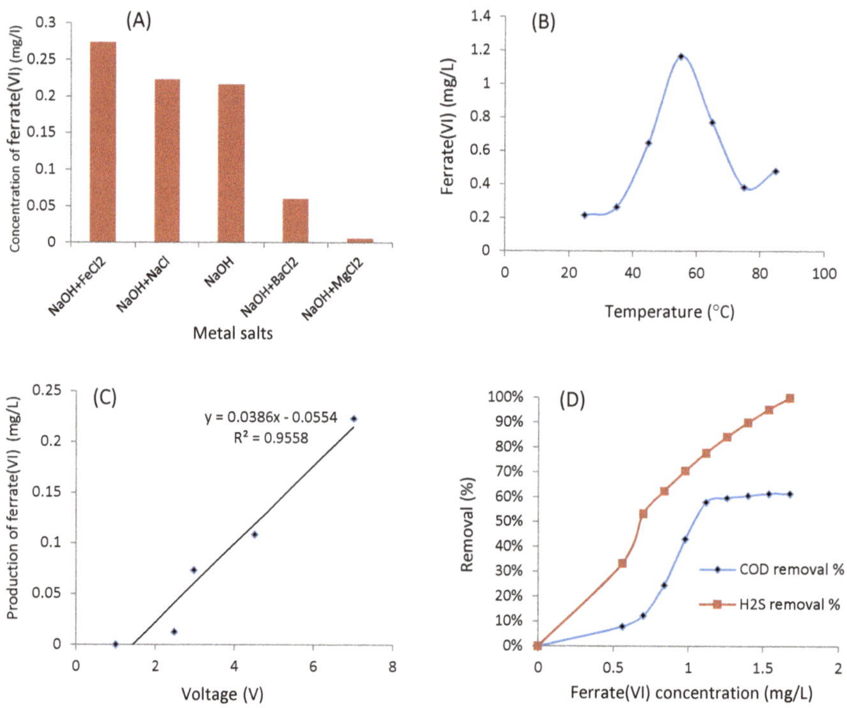

Figure 4. (A) Effect of the presence of metallic ions on ferrate(VI) production during 30 minutes of electrolysis; (B) Effect of temperature on ferrate(VI) production; (C) Effect of voltage on ferrate(VI) production and (D) Efficiency of hydrogen sulfide and COD removal with different concentrations of ferrate(VI).

Furthermore, the effects of pH, HRT and temperature on COD and hydrogen sulfide removal are shown in Figure 5A, 5B, 5C and 5D, respectively. Ferrate (VI) is degraded over time (Figure 6A and 6B) that the rate of its degradation under different temperatures can be seen in Figure 6A. Given that UV is a powerful oxidizing method, its effect on COD and hydrogen sulfide removal was investigated. The results of this investigation are illustrated in Figure 7. These results indicate that using combination of UV/ ferrate (VI) can improve the removal of both COD and Hydrogen sulfide.

Discussion

Ferrate (VI) production

In this study, the effect of the presence of a set of salt such as $FeCl_2$, NaCl, $BaCl_2$ and $MgCl_2$ as well as NaOH on the rate of ferrate (VI) production was investigated. This effect is important because in real-world wastewater treatment applications, there exist constituents other than Na and Fe ions in ferrate (VI) production mixture, especially in case of using industrial-grade NaOH. The results are illustrated in Figure 4A. The amount of ferrate (VI) production can be increased (20%) by the addition of $FeCl_2$ to the electrolyte basic solution. Such an increase in ferrate (VI) production supports the idea that Fe^{6+} production is a two-step process. In first step, Fe^{3+} is produced and then converted to Fe^{6+}. Both steps occur on the surface of cathode electrode. Thus, the addition of Fe^{3+} to the solution can enhance the production rate. This mentioned notion needs more detailed investigation to be confirmed. Another reason that can be attributed to such an increase in ferrate (VI) production rate is the rise in ionic strength due to the addition of an ionic compound. Increasing ionic strength can enhance electrical conductivity leading to an increase in electrical current and hence ferrate (VI) production rate (6). Nevertheless, this idea can be rejected by examining the results of ferrate (VI) production rate when ionic compounds other than $FeCl_3$ are used. For example, the addition of $MgCl_2$ and $BaCl_2$ deteriorated the production rate although the ionic strength of the solution was raised.

Temperature is an affecting parameter on the chemical reactions. In this study, a 14M solution of NaOH was electrolyzed for 15 minutes under various temperatures to investigate the effect of temperature on ferrate (VI) production. Figure 5B demonstrates that the increase in temperature in the range of 25 to 55°C led to an increase in ferrate (VI) production. Nevertheless, further increase in temperature led to a dramatic reduction of ferrate (VI) production rate. So, the optimum temperature for the production of ferrate (VI) employing electrolysis method is approximately 55°C.

Voltage is another effective parameter on the production of ferrate (VI) when utilizing the electrolysis method. The increase in voltage can increase the production rate of ferrate (VI) as demonstrated in Figure 4C. Increasing voltage results in increasing electrical current that passed through the circuit and hence ferrate (VI) production rate.

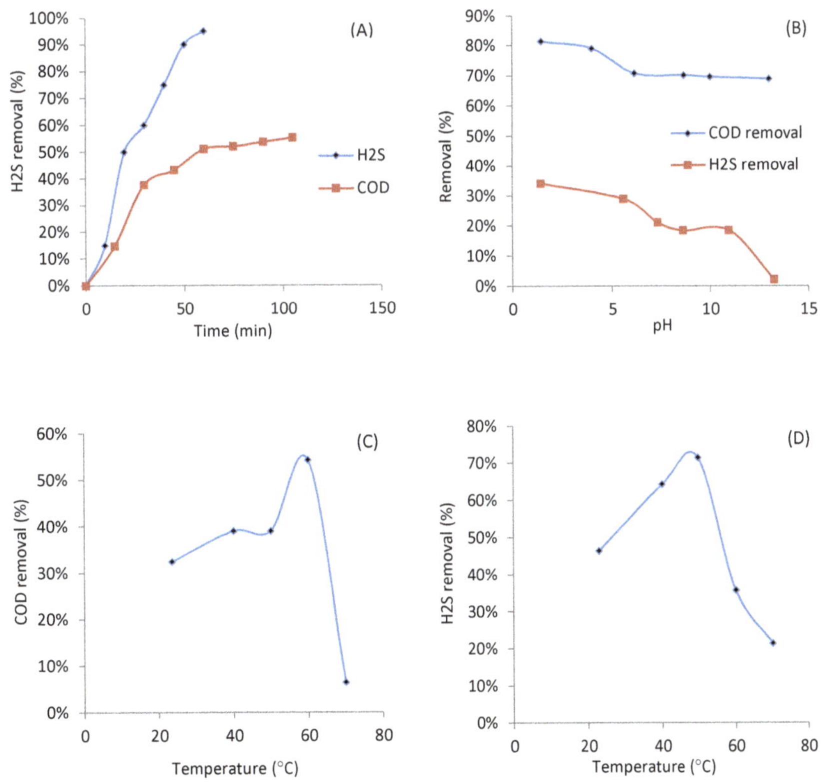

Figure 5. (A) Efficiency of hydrogen sulfide and COD removal over time; (B) Efficiency of hydrogen sulfide and COD removal at different pH values; (C) The variation of COD removal efficiency by using ferrate(VI) under different temperatures and (D) Variation of hydrogen sulfide removal efficiency by using ferrate(VI) under different temperatures.

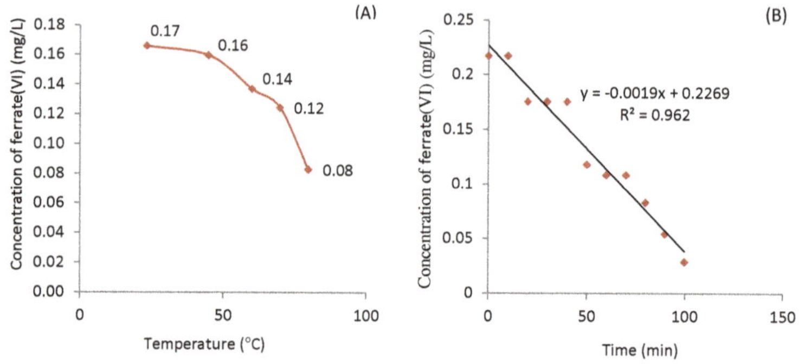

Figure 6. (A) Evaluation of ferrate(VI) degradation rate during 15 minutes under different temperatures and (B) evaluation of ferrate(VI) degradation under temperature of 23.5°C over time.

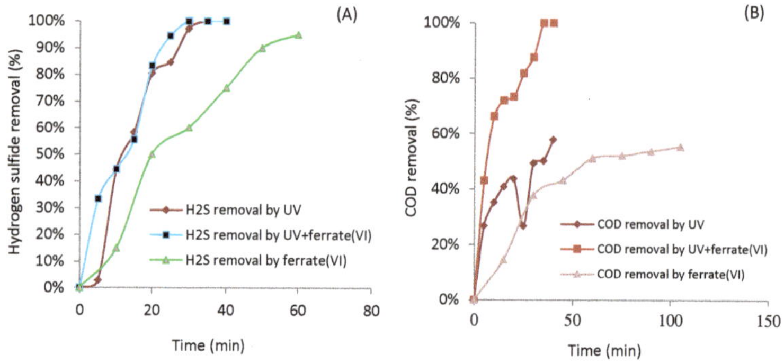

Figure 7. (A) Comparison of using UV, ferrate(VI) and their combination in the removal of hydrogen sulfide from municipal wastewater and (B) comparison of using UV, ferrate(VI) and their combination on COD removal from municipal wastewater.

Given that ferrate (VI) is a strong oxidizer, evaluating the effect of ferrate (VI) concentration on the removal of hydrogen sulfide and COD is highly important (4). As shown in Figure 4D, the removal efficiency of COD increases to a maximum value of about 60% when ferrate concentration reaches 1 mg/L. Nevertheless, further increase in oxidizer concentration has no significant effect on COD removal efficiency. This trend was not observed for hydrogen sulfide removal efficiency. The removal efficiency could reach about 100% when ferrate concentration increases to more than 1.5 mg/L. All results illustrated in Figure 4D were obtained for the residence time of 30 minutes.

Hydrogen sulfide removal and COD reduction

In the following sections, the results of the experiments pertaining to hydrogen sulfide and COD removals are discussed in terms of influencing parameters.

Hydraulic retention time

The contact time between an oxidizer and pollutants plays a critical role on the efficiency of pollutants' removal from wastewater (4). The results of this study demonstrate that hydrogen sulfide efficiency strongly depends on HRT (Figure 5A), and a large portion of this compound can be degraded after 50 minutes. Given that most damage caused by hydrogen sulfide occurs in wastewater collection systems, addition of adequate amount of ferrate (VI) to the starting point of the collection system can be considered as a promising solution for the problems of corrosion and malodor of the hydrogen sulfide.

pH

As demonstrated in Figure 4B, pH has an adverse effect on the removal efficiency. It implies that in acidic environments, the reaction rate of removal process is higher. This fact can be explained by examining the degradation rate of ferrate (VI). According to reaction 1, ferrate (VI) is produced during electrolysis of iron electrode in the presence of NaOH and distilled water.

$$Iron\ electrode \xrightarrow{Electrolysis+NaOH} Ferrate(VI) \quad (1)$$

Production of ferrate (VI) can generate a purple color in water. Ferrate (VI) in water can appear in two forms, including FeO_4^{-2} and $HFeO_4^-$. According to Eq. 3, the rate of ferrate consumption can be found by adding the consumption rates of these two species:

$$K[Ferrate] = K_1[HFeO_4^-] + K_2[HFeO_4^{2-}] \quad (3)$$

The rate constant for $HF_eO_4^-$ consumption (K_1=1.24×10^7 M/S) is dominant when compared to that of FeO_4^{2-} (K_2=8.41×10^2 M/S). Consequently, the reaction rate of $HFeO_4^-$ and hydrogen sulfide is greater than that of FeO_4^{2-} (17). Therefore, $HFeO_4^-$ is responsible for the removal of hydrogen sulfide rather than FeO_4^{2-}. In acidic environment, where pH is low, the dominant species is

$HFeO_4^-$, while in basic conditions ferrate is mainly in the form of FeO_4^{2-}. Because the rate constant of ferrate acidic species degradation is much higher than the basic species, ferrate (VI) is much more effective in acidic condition.

Temperature

As shown in Figure 5C and 5D, the removal efficiencies increase as temperature increases to 50°C. Nevertheless, the removal efficiencies are reduced for higher temperature. This fact can be justified by examining the results shown in Figure 4B which is the variation of ferrate (VI) concentration versus temperature in ferrate (VI) production section. Ferrate (VI) is gradually converted to ferric (III) due to self-degradation according to reaction 2 (4):

$$4Na_2FeO_4 + 10H_2O \rightarrow 8NaH + 4Fe(OH)_3 + 3O_2 \uparrow \quad (2)$$

The ferrate can be degraded due to both reactions with organic compounds and self-degradation. The rates of both reactions are accelerated by temperature. Nevertheless, as seen in Figures 4B and 6A, the self-degradation reaction speeds up significantly when the temperature exceeds 50°C. Thus, a significant portion of ferrate (VI) is consumed by self-degradation, and it decreases the contribution of ferrate to organic compounds degradation.

Ultraviolet radiation

UV radiation is a powerful means of degrading the molecules of organic compounds (15). In this study, the effect of UV radiation on the removal of hydrogen sulfide and COD from wastewater was investigated in the presence and absence of ferrate (VI) via a set of experiments. The results of these experiments are demonstrated in Figure 7A and 7B. As shown in Figure 7B, almost 100% of organic compounds was removed after 40 minutes when both UV and ferrate (VI) methods were used simultaneously. Nevertheless, when UV and ferrate (VI) methods were employed individually, 60% and 40% of hydrogen sulfide was removed, respectively, after the same time. Similar trend can be observed for the removal of hydrogen sulfide according to Figure 7A. These results show that UV and ferrate (VI) create synergy in the degradation process of organic compounds. Salari reported similar results for the removal of formaldehyde from synthetic wastewater by UV/ferrate (VI) (19). She reported that UV has a synergic effect on the removal of formaldehyde by ferrate (VI).

Conclusion

In this study, the removal of hydrogen sulfide and COD from municipal wastewater employing ferrate (VI) and UV was investigated. It was found that temperature, pH, HRT, ferrate (VI) concentration and UV radiation are effective factors in removing hydrogen sulfide and COD from municipal wastewater. The optimum condition for obtaining the highest hydrogen sulfide and COD removal is a pH value of less than 2, HRT of 60 minutes and a maximum temperature of 50°C. It was also concluded that employing UV radiation in addition to ferrate (VI) enhanced the removal efficiency. Using UV radiation and a combination of UV/ferrate (VI) to remove hydrogen sulfide shows that the combination of both methods is able to remove almost 100% of hydrogen sulfide in only 30 min, while it takes 60 minutes to obtain the same results using ferrate (VI) individually. Hydrogen sulfide is a serious issue in wastewater collection and treatment systems causing not only the production of offensive odor but also severe corrosive condition. The results of this study demonstrate that the combination of ferrate (VI) and UV radiation can be considered as a promising method to resolve the issue of hydrogen sulfide in wastewater collection systems.

Acknowledgments

This article was extracted from a Master of Science dissertation of Zeinab Eskandarin in the field of Chemical Engineering at Jami Institute of Technology, Isfahan, Iran. The authors appreciate Jami Institute of Technology for financial and logistic supports.

Competing interests

The authors declare that there is no conflict of interest that would prejudice the impartiality of this scientific work.

Authors' contributions

All authors of this study contributed equally for data collection, data analyses and manuscript writing.

References

1. Talaiekhozani A, Bagheri M, Goli A, Khoozani MR. An overview of principles of odor production, emission, and control methods in wastewater collection and treatment systems. J Environ Manage 2016; 170: 186-206. doi: 10.1016/j.jenvman.2016.01.021.
2. Talaiekhozani A, Masomi B, Hashemi SMJ. Evaluation of Gaseous Pollutants Emission Rate from Marvdasht Landfills. J Adv Med Sci Appl Technol 2016; 2(1): 162-75.
3. Fulazzaky MA, Talaiekhozani A, Majid MZA. Formaldehyde removal mechanisms in a biotrickling filter reactor. Ecol Eng 2016; 90: 77-81. doi: 10.1016/j.ecoleng.2016.01.064.
4. Talaiekhozani A, Bagheri M, Najafabadi NR, Borna E. Effect of nearly one hundred percent of municipal solid waste recycling in najafabad city on improving of its air quality. Journal of Air Pollution and Health 2016; 1(2): 111-22.
5. Kaempfer W, Berndt M. Polymer modified mortar with high resistance to acid to corrosion by biogenic sulfuric acid. the 9th International Congress on Polymers in Concrete; 19-21 October 1998; Bologna, Italy.

6. Sharma VK, Smith JO, Millero FJ. Ferrate (VI) oxidation of hydrogen sulfide. Environ Sci Technol 1997; 31(9): 2486-91. doi: 10.1021/es960755z.
7. Eskandari Z. Control of hydrogen sulfide and organic compounds in municipal wastewater by using ferrate (VI) [dissertation]. Isfahan, Iran: Jami Institute of Technology; 2016.
8. Tchobanoglous G, Burton FL. Wastewater Engineering. Management 1991; 7: 1-4.
9. Huber B, Herzog B, Drewes JE, Koch K, Müller E. Characterization of sulfur oxidizing bacteria related to biogenic sulfuric acid corrosion in sludge digesters. BMC Microbiol 2016; 16: 153. doi: 10.1186/s12866-016-0767-7.
10. Jiang Y, Goodwill JE, Tobiason JE, Reckhow DA. Impacts of ferrate oxidation on natural organic matter and disinfection byproduct precursors. Water Res 2016; 96: 114-25. doi: 10.1016/j.watres.2016.03.052.
11. Zyara AM, Torvinen E, Veijalainen A-M, Heinonen-Tanski H. The effect of chlorine and combined chlorine/UV treatment on coliphages in drinking water disinfection. J Water Health 2016: 14(4): 640-9. doi: 10.2166/wh.2016.144.
12. Jiang JQ, Lloyd B. Progress in the development and use of ferrate (VI) salt as an oxidant and coagulant for water and wastewater treatment. Water Res 2002; 36(6): 1397-408. doi: 10.1016/S0043-1354(01)00358-X.
13. Xie Y. Disinfection Byproducts in Drinking Water: Formation, Analysis, and Control. CRC Press; 2016.
14. Sharma VK. Potassium ferrate (VI): an environmentally friendly oxidant. Adv Environ Res 2002; 6(2): 143-56. doi: 10.1016/S1093-0191(01)00119-8.
15. Talaiekhozani A, Salari M, Talaei MR, Bagheri M, Eskandari Z. Formaldehyde removal from wastewater and air by using UV, ferrate (VI) and UV/ferrate (VI). J Environ Manage 2016; 184(Pt 2): 204-9. doi: 10.1016/j.jenvman.2016.09.084.
16. Salari M. Evaluation of formaldehyde removal from air by using ferrate (VI) in a chemical packed bed scrubber [dissertation]. Isfahan, Iran: Jami Institute of Technology; 2016.
17. Rush JD, Bielski BH. Decay of ferrate (V) in neutral and acidic solutions. A premix pulse radiolysis study. Inorg Chem 1994; 33(24): 5499-502. doi: 10.1021/ic00102a024.
18. American Public Health Association (APHA). Standard methods for the examination of water and wastewater. Washington, DC: APHA; 2005.
19. Salari M. Formaldehyde removal from wastewater and air by using UV, ferrate (VI) and UV/ferrate (VI) [dissertation]. Isfahan, Iran: Jami Institute of Technology; 2016.

17

Health risk assessment of citrus contaminated with heavy metals in Hamedan city, potential risk of Al and Cu

Oshin Rezaei Raja[1], Soheil Sobhanardakani[2*], Mehrdad Cheraghi[2]

[1]MSc Graduated, Department of the Environment, College of Basic Sciences, Hamedan Branch, Islamic Azad University, Hamedan, Iran
[2]Associate Professor, Department of the Environment, College of Basic Sciences, Hamedan Branch, Islamic Azad University, Hamedan, Iran

Abstract
Background: Fruits especially citrus species are an integral part of human diet. Contamination of foodstuffs by heavy and toxic metals via environmental pollution has become an inevitable challenge these days. Therefore, the effect of pollutants on food safety for human consumption is a global public concern. In this regards, this study was conducted for Al and Cu health risk assessment through the consumption of citrus species (orange, grapefruit, sweet lime and tangerine) in Hamedan city in 2015.
Methods: After collecting and preparing 4 samples from each citrus species with acid digestion method, the concentrations of Al and Cu were determined using inductively coupled plasma optical emission spectrometry (ICP-OES) with three replications. In addition, SPSS was employed to compare the mean concentrations of metals with maximum permissible limits (MPL) of the World Health Organization (WHO).
Results: The results showed that the mean concentrations of Al in citrus samples were 3.25±0.35 mg/kg and higher than WHO maximum permissible limits. The mean concentrations of Cu in citrus samples with 0.16±0.05 mg/kg are lower than WHO maximum permissible limits. Also, the computed health risk assessment revealed that there was no potential risk for children and adult by consuming the studied citrus.
Conclusion: Based on the results, consumption of citrus species has no adverse effect on the consumers' health, but as a result of the increased utilization of agricultural inputs (metal based fertilizers and pesticides, sewage sludge and wastewater) by farmers and orchardists, regular periodic monitoring of chemical pollutants content in foodstuffs are recommended for food safety.
Keywords: Citrus, Heavy metal toxicity, Health risk, Food safety, Hamedan

***Correspondence to:**
Soheil Sobhanardakani
Email: s_sobhan@iauh.ac.ir

Introduction

Vegetables and fruits, as they supply the important and essential components for human health such as minerals, vitamins, tartaric acid, citric acid, malic acid, mineral salts, cellulose and woody fibers, tannins, gums, pectin, volatile oils and coloring matters are an integral part of human diet. Consumption of these products on regular basis is vital in providing essential nutrients to the human body for promoting health (1-3). Nowadays, contamination of heavy metals in foods has become an unavoidable challenge (4). Heavy metals are important environmental pollutants, especially in regions with high anthropogenic pressure. Their presence in the environment particularly atmosphere and water, even in trace amounts, can lead to serious problems to all organisms. Heavy metal accumulation in soils and resulting in foodstuffs such as vegetables and fruits through irrigation with polluted water, metal based fertilizers and pesticides, emissions from industrial, transportation, harvesting process and storage of products can lead to adverse effects on food quality. So, heavy metals are among the major pollutants of food supply and may be considered as the most important problem to our environment (5-8).

Because heavy metals have long biological half-lives, non-biodegradable and persistent, they can be bioaccumulated via the soil-plant-food chains and the presence of high amount of heavy metals in the environment indicates a potential risk for human health and for the environment. Uptake of high amounts of heavy metals can lead to a number of serious adverse health effects such as depleting of some essential nutrients in the body, impaired psycho-social behaviors, decrease in immunological defenses, retardation in intrauterine growth, disabilities associated with undernourishment and a high spread of pharynx, larynx, esophagus, stomach and small intestine cancer (4,9). Although a number of heavy metals such as Cu, Fe,

Mn, Ni and Zn at lower quantities act as micronutrients, they become toxic at higher contents (10,11).

Potential health risk and food safety issues make this as one of the most serious health and environmental concerns (12). Therefore, over the past few decades, the increasing demand of food safety has motivated researches regarding the risk associated with the consumption of food products, contaminated by different types of environmental pollutants such as heavy and toxic metals and pesticides (13).

Aluminum is not considered to be an essential element in humans, however, it is prevalent in water, air, plants and consequently in all the food chain (7). Literally, Al is easily found at quantifiable concentrations in various tissues and biological fluids (14,15). It should be noted that high levels of Al can lead to a number of adverse health effects such as osteomalacia, colic, fatigue, dementia dialectica, anaemia and other blood disorders, neurodegenerative disorders, reduced renal function, dental caries, Parkinson and Alzheimer diseases along with kidney and liver dysfunctions, Al is considerably less toxic than other toxic metals (16-18).

Copper is one of most abundant trace elements with vitamin-like impact in human body and living systems ad it is found in a wide range of foods we eat such as nuts, several fruits and vegetables, red meat, shellfish, as well as in many vitamin supplements (19). Although small amount of Cu (50-120 mg) is found in the human body, it plays a critical role in different kinds of biochemical processes (7). Copper, an essential micronutrient for growing plants can be applied via organic and artificial fertilizers and it is important for healthy hormone secretion, nerve conduction, along with the growth of bones and connective tissue. A constant diet of Cu, even at entirely allowable limits, can break down the barrier that keeps undesirable toxins from entering the brain, and an increase in the production of beta-amyloid. On the contrary, critical doses of Cu can result to inflammation in the brain tissues, anorexia, fatigue, hair loss, acne, allergies, depression, premenstrual syndrome, migraines, anxiety, childhood hyperactivity, panic attacks, kidney and liver dysfunction, strokes, elevated cholesterol, adrenal hyperactivity and insufficiency, learning disorders, as well as autism and cancer (18-20).

Mausi et al reported that mango and orange fruits sold in Kenya posed no health risks to consumers based on their daily intake of metal levels, as the values were within PDTI standards of World Health Organization (WHO) (21). Chandorkar and Deota studied the health risk assessment of Ar, Cd, and Pb content of foods grown around the city of Vadodara (India) and noted that the study population was at risk of heavy metal toxicity through the consumption of food samples (22). Guerra et al studied the potential health risk assessment of Cd, Pb, Ni, Co and Cr in most regularly consumed vegetables in Brazil, and reported that the consumption of analyzed food samples can be considered safe and without adverse health effects to human (23).

Fruits as integral parts of human diet are widely consumed, they contribute a large fraction to the heavy metals intake, and therefore, strict control of these elements is recommended. Therefore, this study aims to investigate the health risk assessment of Al and Cu in four citrus species (grapefruit, sweet lime, orange, and tangerine) marketed in Hamedan city in 2015.

Methods

Sampling and sample preparation

In this study, 16 samples of popular citrus species (grapefruit, lemon, orange, and tangerine) were bought from different markets in Hamedan city so as to analyze their Al and Cu content. All specimens were washed with deionized water to eliminate airborne pollutants. In order to reduce the water content of fruit samples, the edible parts of the each sample were weighed and air-dried for a day. Then, in order to remove the sample moisture, all specimens were oven-dried at 70–80°C for 24 hours. Finally, dried samples were powdered using a grinder and then sieved through a Muslin cloth and packed into polyethylene bags (4).

Sample digestion

After weighing 0.5 g of each powdered citrus samples in three replicates for each sample and placed them in crucibles, the samples were digested with $HClO_4$ and HNO_3 (1:4) solution. The samples were left to cool and made up to a final volume of 25 mL with deionised water. Subsequently, the samples were well shaken and centrifuged at 3000 rpm to remove solid particles. The resulting homogenized samples were completely mixed before subsamples were taken for analysis in order to confirm the homogeneity of the mixture (4). Finally, the content of Al and Cu were analyzed in samples using the ICP-OES (Varian, 710-ES) with three replications.

Statistical analyses

Data were statistically analyzed utilizing one sample *t* test to compare the mean concentrations of Al and Cu with WHO maximum permissible limits using the SPSS statistical package version 20.

Human health risk assessment

Estimated average daily intakes (EADIs) of a metal in food and assumption of food consumption were used to investigate long term health risks to consumers. In this regard, for each type of exposure, the EADI was calculated by using Equation 1:

$$EADI = \frac{C \times F}{W \times D} \quad (1)$$

where C represents the concentration of metal in each commodity (mg/kg); *F* indicates the mean annual intake of food per person (the average daily intake of citrus species for adult and children is considered to be 0.01 kg per person per day); *D* indicates the number of days in a year (365 days); and *W* represents the mean body weight (70 kg for adult and 15 kg for children) (4,24).

The WHO has set values for toxicity, called acceptable daily intakes (ADIs) for a large number of chemicals, including some essential trace elements (25). The health risk indices were computed by dividing the EADI by the ADIs (mg/kg/day) established by FAO/WHO Codex Committee (24,26). The HI was computed by using Equation 2:

$$HI = \frac{EADI}{ADI} \quad (2)$$

When the HI >1; the food involved is considered a risk to the concerned consumers. When the HI <1, the food involved is considered as acceptable (no concern) to the concerned consumers (24).

Results

The contents of Al and Cu in the analyzed citrus samples are shown in Table 1. Data in Table 1 showed that the percentage of metals contamination of citrus samples were approximately 100%. The contents of Al in samples were higher than WHO maximum permissible limits (MPL) (0.20 mg/kg) (14). The highest content of Al was found in grapefruit (3.65 ± 0.82 mg/kg) while the lowest was found in tangerine (2.94 ± 0.37 mg/kg). In addition, the Cu contents of citrus samples were lower than WHO MPL (4.0 mg/kg) (27). The highest content of Cu was found in sweet lime (0.20 ± 0.18 mg/kg) and the lowest in tangerine (0.12 ± 0.03 mg/kg).

The systemic health risk assessment due to heavy metals encountered in citrus species is summarized in Table 2. The results showed that the calculated EADIs ranged between 1.55×10^{-5} to 2.20×10^{-3} mg/kg/day, while the hazard indices (HI) ranged from 3.87×10^{-4} to 2.36×10^{-2} for the analyzed metals indicating no direct hazard to human health, in spite of their presence in the food.

Discussion

Human exposure to heavy metals such as Al and Cu through environment or consumption of contaminated foods is known to be responsible for many human health problems (30,31).

The results of this study indicate that the mean concentrations of Al in citrus samples were 3.25 ± 0.35 mg/kg and were higher than WHO maximum permissible limits. In this regard, studies have shown that in plant origin foodstuffs, the total Al content in plant tissues is under the influence of Al in soil (32). Moreover, It was verified that Al accumulation in plants is a problem when they are grown in acid soils (33). Saracoglu et al determined Al content in apricot samples were purchased from Turkey ranged from 0.08 to 0.22 mg/kg (7).

The findings of this study show that the mean concentrations of Cu in citrus samples were 0.16 ± 0.05 mg/kg and lower than WHO maximum permissible limits. Researchers have determined Cu concentrations in various types of fruits from different countries in the world. Saracoglu et al determined Cu content in apricot samples purchased from Turkey, ranging from 0.92 to 6.49 mg/kg (7). Radwan and Salama surveyed the levels of Cu in various fruits sold in Egyptian markets, and reported that the Cu concentrations ranged from 0.90 to 25.90 mg/kg for Cu (5). Onianwa et al determined a mean Cu concentration of 2.13 mg/kg in some orange samples obtained from Nigerian markets (34). Parveen et al analyzed Cu content in some fruits marketed in Pakistan and reported that the mean concentrations of Cu in specimens were 1.61±1.94 mg/kg (35). Comparison of the results of this study with other studies is shown in Table 3.

As shown in Table 2, health index values of studied metals (Al and Cu) for children and adults are less than 1. Ac-

Table 1. Concentrations of heavy metals in the citrus samples (mg/kg)

Metal	Citrus species sample				Mean concentration ± SD
	Grapefruit	Sweet lime	Orange	Tangerine	
Al	3.65 ± 0.82	2.98 ± 1.12	3.43 ± 1.36	2.94 ± 0.37	3.25 ± 0.35
Cu	0.20 ± 0.08	0.20 ± 0.18	0.12 ± 0.09	0.12 ± 0.03	0.16 ± 0.05

Table 2. Acceptable and estimated daily intakes and health index for metals found in citrus samples

Metal	ADI (mg/kg/day) (14,28,29)	Mean Concentrations (mg kg^{-1})	EADI (mg/kg/day) (Children)	HI (Children)	EADI (mg/kg/day) (Adult)	HI (Adult)
Grapefruit						
Al	0.02	3.65	2.20×10^{-3}	1.10×10^{-1}	4.71×10^{-4}	2.36×10^{-2}
Cu	0.04	0.20	1.21×10^{-4}	3.01×10^{-3}	2.58×10^{-5}	6.46×10^{-4}
Sweet Lime						
Al	0.02	2.98	1.80×10^{-3}	8.99×10^{-2}	3.85×10^{-4}	1.92×10^{-2}
Cu	0.04	0.20	1.21×10^{-4}	3.01×10^{-3}	2.58×10^{-5}	6.46×10^{-4}
Orange						
Al	0.02	3.43	2.07×10^{-3}	1.03×10^{-1}	4.43×10^{-4}	2.21×10^{-2}
Cu	0.04	0.12	7.23×10^{-5}	1.81×10^{-3}	1.55×10^{-5}	3.87×10^{-4}
Tangerine						
Al	0.02	2.94	1.77×10^{-3}	8.86×10^{-2}	3.80×10^{-4}	1.90×10^{-2}
Cu	0.04	0.12	7.23×10^{-5}	1.81×10^{-3}	1.55×10^{-5}	3.87×10^{-4}

Table 3. Comparison of present mean values in fruit species with other studies result

Fruit Sample	Area	Metals (mg/kg)	
		Al	Cu
Grapefruit	Present sttudy	3.65	0.20
Sweet Lime		2.98	0.20
Orange		3.43	0.12
Tangerine		2.94	0.12
Grapefruit	Turkey[a]	-	5.12
Lemon		-	4.80
Finike orange		-	5.01
Valansiya orange		-	3.86
Yafa orange		-	4.28
Papaya	Ghana[b]	-	1.44
Water melon		-	1.90
Banana		-	3.23
Mango		-	1.24
Pear		-	1.48
Pineapple		-	4.23
Lima orange	Brazil[c]	-	46.80
Pera orange		-	35.60
Tahiti lime		-	69.00
Sweet lime		-	25.30
Orange	Nigeria[d]	-	29.70
Avocado		-	4.65
Pawpaw		-	6.50
Banana		-	10.20
Orange	Nigeria[e]	-	0.23
Avocado pear		-	3.10
Pawpaw		-	5.29
Pineapple		-	0.64

[a]Özcan et al(6); [b]Bempah et al (11); [c]de Moraes Barros et al (36); [d]Ihesinachi et al (37); [e]Ihesinachi and Eresiya (38).

cording to USEPA, the HI <1, implies that there will be no obvious health risks for consumers (39). The results of this study reveal that the average HI value was 0.01 for adults and 0.05 for children. So, it can be concluded that the Al and Cu in fruit samples do not present carcinogenic health risks to consumers. Nevertheless, the non-carcinogenic risks resulting from the utilization of citrus were greater for children than for adults.

Conclusion

This study investigates the levels and health risk assessment of Al and Cu in four commonly consumed citrus species (orange, grapefruit, sweet lime and tangerine) in Hamedan city, and can help in creating awareness of the detrimental effects caused by heavy metals. In line with the results of this study, it can be concluded that the percentage of metals contamination of citrus samples were about 100%. Furthermore, according to the results, no direct hazard to adult and children health was observed, due to the consumption of analyzed citrus species, despite the presence of Al and Cu in the samples.

Although controlled consumption of studied citrus species has no adverse effect on the consumers' health, due to the increased use of agricultural inputs, especially metal based fertilizers and pesticides, sewage sludge and wastewater by farmers and orchardists, regular periodic monitoring of chemical pollutants content specially heavy metals in foodstuffs are recommended for food safety.

Acknowledgments

The authors would like to appreciate the Hamedan Branch, Islamic Azad University for providing the necessary facilities to conduct and complete this study.

Competing interests

The authors declare that they have no competing interests.

Authors' contributions

All authors participated in the study design, literature search, writing of the manuscript, data acquisition, analysis, and interpretation. All authors critically reviewed, refined, and approved the manuscript.

References

1. Munzuroglu O, Karatas F, Geckil H. The vitamin and selenium contents of apricot fruit of different varieties cultivated in different geographical regions. Food Chem 2003; 83(2): 205-12. doi: 10.1016/S0308-8146(03)00064-5.
2. Zahoor A, Jaffar M, Saqib M. Elemental distribution in summer fruits of Pakistan. Nutr Food Sci 2003; 33(5): 203-7. doi: 10.1108/00346650310499712.
3. Dauchet L, Montaye M, Ruidavets JB, Arveiler D, Kee F, Bingham A, et al. Association between the frequency of fruit and vegetable consumption and cardiovascular disease in male smokers and non-smokers. Eur J Clin Nutr 2010; 64(6): 578-86. doi: 10.1038/ejcn.2010.46.
4. Orisakwe OE, Nduka JK, Amadi CN, Dike DO, Bede O. Heavy metals health risk assessment for population via consumption of food crops and fruits in Owerri, South Eastern, Nigeria. Chem Cent J 2012; 6(1): 77. doi: 10.1186/1752-153X-6-77.
5. Radwan MA, Salama AK. Market basket survey for some heavy metals in Egyptian fruits and vegetables. Food Chem Toxicol 2006; 44(8): 1273-8. doi: 10.1016/j.fct.2006.02.004.
6. Özcan MM, Harmankaya M, Gezgin S. Mineral and heavy metal contents of the outer and inner tissues of commonly used fruits. Environ Monit Assess 2012; 184(1): 313-20.doi: 10.1007/s10661-011-1969-y.
7. Saracoglu S, Tuzen M, Soylak M. Evaluation of trace element contents of dried apricot samples from Turkey. J Hazard Mater 2009; 167(1): 647-52. doi: 10.1016/j.jhazmat.2009.01.011.
8. Bıngöl M, Yentür G, Er B, Öktem AB. Determination of some heavy metal levels in soft drinks from Turkey using ICP-OES method. Czech J Food Sci 2010; 28(3): 213-6.
9. Arora M, Kiran B, Rani S, Rani A, Kaur B, Mittal N. Heavy metal accumulation in vegetables irrigated with water from different sources. Food Chem 2008; 111(4): 811-5. doi: 10.1016/j.foodchem.2008.04.049.
10. Al-Shawi AW, Dahl R. The determination of cadmium and six other heavy metals in nitrate/phosphate fertilizer solution by ion chromatography. Anal Chim Acta 1999;

391(1): 35-42. doi: 10.1016/S0003-2670(99)00200-7.

11. Bempah CK, Buah-Kwofie A, Osei-Tutu A, Denutsui D, Bentil N. Assessing potential dietary intake of heavy metals in some selected fruits and vegetables from Ghanaian markets. Elixir Pollut 2011; 39: 4921-6.
12. Singh A, Sharma RK, Agrawal M, Marshall FM. Risk assessment of heavy metal toxicity through contaminated vegetables from waste water irrigated area of Varanasi, India. Trop Ecol 2010; 51(2): 375-87.
13. D'Mello JPF. Food Safety: Contaminants and Toxins. Wallingford: CABI; 2003. p. 205-6.
14. Sato K, Suzuki I, Kubota H, Furusho N, Inoue T, Yasukouchi Y, et al. Estimation of daily aluminum intake in Japan based on food consumption inspection results: impact of food additives. Food Sci Nutr 2014; 2(4): 389-97. doi: 10.1002/fsn3.114.
15. Glynn AW. Concentration-dependent absorption of aluminum in rats exposed to labile aluminum in drinking water. J Toxicol Environ Health A 1999; 56(7): 501-12.doi: 10.1080/009841099157944.
16. Storey E, Masters CL. Amyloid, aluminium and the aetiology of Alzheimer's disease. Med J Aust 1995; 163(5): 256-9.
17. López FF, Cabrera C, Lorenzo ML, López MC. Aluminium content of drinking waters, fruit juices and soft drinks: contribution to dietary intake. Sci Total Environ 2002; 292(3): 205-13. doi: 10.1016/S0048-9697(01)01122-6.
18. Lokeshappa B, Shivpuri K, Tripathi V, Dikshit AK. Assessment of toxic metals in agricultural produce. Food Public Health 2012; 2(1): 24-9. doi: 10.5923/j.fph.20120201.05.
19. Itanna F. Metals in leafy vegetables grown in Addis Ababa and toxicological implications. Ethiopian J Health Dev 2002; 16(3): 295-302. doi: 10.4314/ejhd.v16i3.9797.
20. Ackah M, Anim AK, Zakaria N, Osei J, Saah-Nyarko E, Gyamfi ET, et al. Determination of some heavy metal levels in soft drinks on the Ghanaian market using atomic absorption spectrometry method. Environ Monit Assess 2014; 186(12): 8499-8507. doi: 10.1007/s10661-014-4019-8.
21. Mausi G, Simiyu G, Lutta S. Assessment of selected heavy metal concentrations in selected fresh fruits in Eldoret Town, Kenya. J Environ Earth Sci. 2014; 4(3): 1-8.
22. Chandorkar S, Deota P. Heavy metal content of foods and health risk assessment in the study population of Vadodara. Curr World Environ 2013; 8(2): 291-7. doi: 10.12944/CWE.8.2.15.
23. Guerra F, Ricardo Trevizam A, Muraoka T, Chaves Marcante N, Guidolin Canniatti-Brazaca S. Heavy metals in vegetables and potential risk for human health. Sci Agric 2012; 69(1): 54-60. doi: 10.1590/S0103-90162012000100008.
24. Omar WA, Zaghloul KH, Abdel-Khalek AA, Abo-Hegab S. Risk assessment and toxic effects of metal pollution in two cultured and wild fish species from highly degraded aquatic habitats. Arch Environ Contam Toxicol 2013; 65(4): 753-64. doi: 10.1007/s00244-013-9935-z.
25. Goldhaber SB. Trace element risk assessment: essentiality vs. toxicity. Regul Toxicol Pharmacol 2003; 38(2): 232-42. doi: 10.1016/S0273-2300(02)00020-X.
26. Food and Agriculture Organization of the United Nations/World Health Organization. Food Standards Programme. Codex Alimentarius Commission: Food Hygiene, Basic Texts (Third Edition). Rome; 2003.
27. Krejpcio Z, Sionkowski S, Bartela J. Safety of fresh fruits and juices available on the Polish market as determined by heavy metal residues. Polish J Environ Stud 2005; 14(6): 877-81.
28. Dehelean A, Magdas DA. Analysis of mineral and heavy metal content of some commercial fruit juices by inductively coupled plasma mass spectrometry. Sci World J 2013; 2013: 1-6. doi: 10.1155/2013/215423.
29. Fu QL, Liu Y, Li L, Achal V. A survey on the heavy metal contents in Chinese traditional egg products and their potential health risk assessment. Food Addit Contam B 2014; 7(2): 99-105. doi: 10.1080/19393210.2013.853106.
30. Singh R, Gautam N, Mishra A, Gupta R. Heavy metals and living systems: An overview. Ind J Pharmacol 2011; 43(3): 246-53. doi: 10.4103/0253-7613.81505.
31. Jackson LS, Knize MG, Morgan JN. Impact of Processing on Food Safety. New York: Springer Science & Business Media; 2012.
32. Gerhardsson L, Oskarsson A, Skerfving S. Acid precipitation—effects on trace elements and human health. Sci Total Environ 1994; 153(3): 237-45. doi: 10.1016/0048-9697(94)90203-8
33. Tyler G. Plant uptake of aluminium from calcareous soils. Experientia 1994; 50(7): 701-3. doi: 10.1007/BF01952878.
34. Onianwa P, Adeyemo A, Idowu O, Ogabiela E. Copper and zinc contents of Nigerian foods and estimates of the adult dietary intakes. Food Chem 2001; 72(1): 89-95. doi: 10.1016/S0308-8146(00)00214-4.
35. Parveen Z, Khuhro M, Rafiq N. Market basket survey for lead, cadmium, copper, chromium, nickel, and zinc in fruits and vegetables. Bull Environ Contam Toxicol 2003; 71(6): 1260-4. doi: 10.1007/s00128-003-8640-4.
36. de Moraes Barros HR, de Castro Ferreira TA, Genovese MI. Antioxidant capacity and mineral content of pulp and peel from commercial cultivars of citrus from Brazil. Food Chem 2012; 134(4): 1892-8. doi: 10.1016/j.foodchem.2012.03.090.
37. Kalagbor IA, Naifa PB, Umeh JN. Analysis of heavy metals in four fruits from Sii and Zaakpon communities in Khana, Rivers State. Int J Emerg Technol Adv Eng 2014; 4(5): 827-31.
38. Ihesinachi K, Eresiya D. Evaluation of heavy metals in orange, pineapple, avocado pear and pawpaw from a farm in Kaani, Bori, Rivers State Nigeria. Int Res J Publ Environ Health 2014; 1(4): 87-94.
39. Liang Q, Xue ZJ, Wang F, Sun ZM, Yang ZX, Liu SQ. Contamination and health risks from heavy metals in cultivated soil in Zhangjiakou City of Hebei Province, China. Environ Monit Assess 2015; 187: 754. doi: 10.1007/s10661-015-4955-y.

18

Increasing convergence rate in two-objective optimization of water distribution network with engineering judgment

Fariba Sherri[1], Amir Hossein Mahvi[2*], Abbas Toloie Eshlaghy[3], Amir Hessam Hassani[1]

[1]Department of Environmental Engineering, Faculty of Natural Resources and Environment, Science and Research Branch, Islamic Azad University, Tehran, Iran
[2]Department of Environmental Health Engineering, School of Public Health, Tehran University of Medical Sciences, Tehran, Iran
[3]Department of Industrial Management, Faculty of Management and Economic, Science and Research Branch, Islamic Azad University, Tehran, Iran

Abstract
Background: Water distribution networks (WDNs) are facilities that require massive investment and their optimization is very important. This study aimed to optimization and development of models for promoting WDNs with using engineering judgment. In this method, instead of controlling all system states, it is possible to search the optimal set of options based on engineering judgment and hydraulic and physical status of the system. Thus, the time to solve the optimization problem is greatly reduced, which is very important in widespread networks with many components. The case study was a WDN in western Tehran.
Methods: To reduce the calculation size and increase the convergence rate using engineering judgment, the parts of the network where there was no possibility of parallel piping was ignored. For other parts with a low pressure problem, parallel piping was defined. A FMGA and WaterGEMS hydraulic software were used to optimize the WDN. Cost minimization and pressure benefit maximization were the objective functions and the diameters of the pipes were considered to be the decision variables.
Results: The results of optimization the network showed that, the cost decreased 89.84% and the pressure in all nodes, except one node, reached within the standard range (26-60 mH_20). It included 2387 m of pipe with diameters of 100, 150, 200, 250, 350, 400 and 500 mm.
Conclusion: The results of optimization and modification of the network using engineering judgment confirm that the cost decreased significantly and the pressure level in all the nodes increased to above the allowable minimum pressure.
Keywords: Water, Judgment, Software, Pressure, Engineering

***Correspondence to:**
Amir Hossein Mahvi
Email: ahmahvi@yahoo.com

Introduction

A water distribution network (WDN) is an important component of citizen welfare (1). It handles the transfer of water from a reservoir to the consumer and consists of a water supply, consumers, distribution pipes, valves and pumps. This infrastructure provides the basic needs of a society (2,3). Researches are interested in the optimization of WDNs (4). and have used the evolutionary algorithms such as the genetic algorithm (GA) (5,6), ant colony optimization (ACO) (7,8), shuffled frog leaping algorithm (SFLA) (9,10), simulated annealing (SA) (11), honey-bee mating optimization (HBMO) (12), harmony search algorithm (HS) (13,14) and particle swarm optimization (PSO) (15,16). These algorithms do not require calculation of the derivative of the objective function, but the probability of finding a global optimum solution is more than the classic algorithms (17). The problems raised in WDNs are highly complex, so traditional optimization methods used to solve them are time-consuming and require a large amount of computation memory. This has decreased the efficiency of traditional optimization methods. Because WDNs are large-scale and have a large number of variables, evolutionary algorithms are preferred because they take less time to solve the problem (4). Even Fuzzy logic has been used for assessment of water quality in WDNs (18). Many real-world engineering problems are multi-purpose and usually conflict. To find an optimal response in such cases is not possible. In these cases, a set of solutions must be found that can establish the relative balance between different objectives. This set of solutions, known as the Pareto front, allows the designer engineer to choose one solution according to the specific conditions.

When faced with multi-objective optimization, classic optimization methods are inefficient for three major reasons. First, most cannot find several solutions in a single run. In addition, multiple applications of these methods do not guarantee the generation of widely different Pareto optimal solutions. Finally, most of them are not efficient in multi-objective optimizations that deal with discrete variables (19).

Seyoum et al used a penalty-free multi-objective evolutionary optimization to solve a real life network (20). Wang et al evaluated low and high level hybrid algorithm on optimization of WDNs. They compared the number of runs, convergence and diversity to detect the final solution (21). Tang et al used multi-objective analysis to reveal the benefits of water transfer between two reservoirs (22). In many of these methods, single-objective optimization is used to minimize the cost of pipes as the target. Some studies also have focused on the reliability and stochastic modeling of the demand (23,24).

Walski stressed the need to develop new models based on minimization of network costs and maximization of net profits (25). Halhal et al were the first ones who optimized WDN using a multi-objective GA. They minimized cost and maximized profit using a structured messy genetic algorithm (SMGA) to solve the optimization problems (26).

Walters et al used the SMGA to optimize the "anytown" distribution network (27). Prasad and Park optimized a two-loop reference network using a multi-objective GA. Their objective functions involved decreased the cost and increased the reliability of the network (19).

Although optimization techniques are more common, engineering judgment and experience are necessary along with them. Optimization techniques usually can be used as a backing instrument for decisions and provide a better result when combined (28). Iglesias-Ray et al combined engineering judgment and a model optimizer (pseudo-GA) to increase the energy efficiency and manage leakage in the WDNs. Their optimization problem had more than 70 000 decision variables and was divided into three parts. The first part was preliminary analysis to define the criteria for replacing pipes and pumps. The second was the use of an optimizer model and the third part was the correct adjustment of the control valves, pressure relief valves and pumps replacement. Here, a proper combination of optimization techniques and manual adjustments based on engineering judgment has been used to solve the general problems of water supply networks (29). Khedr and Tolson obtained a Pareto front with optimal quality by relying on engineering judgment without the need for complex calculations (28).

In the current study, a two-objective FMGA was used to obtain the best solution with the lowest cost and the most pressure benefit with the help of engineering judgment in a Pareto front from the set of non-dominant solutions to optimize a WDN. The FMGA performance was evaluated using Bentley WaterGEMS V8i hydraulic software for the design and modification of WDNs with the engineering judgment on a real WDN in western Tehran.

In this study, a new method for reducing the rate of convergence, optimization and development of models for promoting WDN has been presented. Based on the proposed method, instead of controlling all system states in the optimization model, it is possible to identify and search for the optimal set of options based on engineering judgment and hydraulic and physical status of the system. Thus, the time to solve the optimization problem is greatly reduced, which is very important in widespread networks with many components.

firstly, the nodes with pressure deficiency was determined. Then the parts of the network for which there was no possibility of parallel piping (physical limitations in urban areas for new piping such as narrow roads and intersections with underground facilities) was ignored and for the other parts with the low pressure problem and the upstream pipes with the lowest diameter were doubled. This is done because the maximum head loss is due to the smallest pipe diameter. Finally, the node pressure with the performed changes was checked to ensure the sufficiency of the pipe parallelizing. If there is still pressure deficiency in that point, the next upstream pipe is considered for parallelizing. It makes the decision space smaller and makes it easier to find a solution in a reasonable time. Based on the proposed method, instead of controlling all system states in the optimization model, it is possible to identify and search for the optimal set of options based on engineering judgment and hydraulic and physical status of the system. From the analysis of the results, it was observed that use of engineering judgment gave better results.

This study shows that relying on engineering judgment can produce reasonable of the Pareto front of multi objective WDN design problems. The engineering judgment solution generated a high-quality result and was able to identify a more realistic number of changes in the WDN with a very small computational budget and properly pressure in nodes.

Methods

Engineering judgment approach

In the engineering judgment approach, the operation is broken into two phases to obtain an appropriate practical Pareto front, minimize the cost and maximize the pressure profit. First, engineers must identify all practical and possible solutions. The most important part of this step, after determination of the objective functions, is to identify the decision variables. Hydraulic network analysis was utilized to detect nodes with low pressure for which parallel pipes can be used. The use of an optimizer algorithm is the second step in which the Pareto front is obtained by engineering judgment that provides cost-effective and efficient solutions. If there is more than one objective function, there will be a set of solutions so-called Pareto-optimal solution or non-dominated solution (30). These solutions are superior to others in the search space; however, one must recognize that optimization is an auxiliary tool for design engineers and the designer must select the proper solution using engineering judgment and expertise (31).

Optimization of water distribution network

In this study, WDN optimization is solved with objective functions of the cost minimization and the pressure benefit maximization. This is classified as multi-objective problems. The Eqs. 1 and 2 are objective functions used by Bentley WaterGEMS V8i software to optimization (32). Diameters of pipes and velocity in the pipes are considered as the decision variables and optimization constraint, respectively. Energy conservation and continuity laws are the main constraints which are always respected by the software Bentley WaterGEMS V8i.

$$Minimize\ F1 = \sum_{k=1}^{DP}(C_k(d_k)L_k) \tag{1}$$

$$Maximize\ F2 = \sum_{k=1}^{ND}[a\sum_{i=1}^{NJ}(\frac{JQ_{i,k}}{JQTotal_k})(\frac{P_{i,k}-P_{i,k}^{REF}}{P_{i,k}^{REF}})^b] \tag{2}$$

In the Eq. (1), *DP* represents the total number of design pipes in the network and $C_k(d_k)L_k$ is the cost of the pipe *k* with the length *L* and diameter *d*.

And in the Eq. 2,

ND: The number of design states

NJ: The number of nodes

$JQ_{i,k}$: The demand at the node i in alternative k

$JQTotal_k$: The total of demand in alternative k

$P_{i,k}$: The calculation pressure at the node i in alternative k

$P_{i,k}^{REF}$: The required minimum pressure determined by the designer

a, b: *a* is the linear parameter and *b* is the parameter which is less than 1 (about 0.5) (32).

Optimization algorithm

GA is an evolutionary algorithms based on natural selection. It was initially completed by Holland and then by Goldberg. First in the GA, multiple answers are randomly generated to the problem (an initial population) for which each answer is called a chromosome. Next, following the selection of the better chromosomes, the GA operators are composed and mutated. Finally, the current population composes with the new population resulted from the chromosomes composition and mutation. In fact, to create a new generation, the competence level is calculated according to the purpose. Those with more competence are more likely to produce the next generation. This algorithm continues until the desired limitation is satisfied.

GA as one of the most important evolutionary algorithms plays an important role in optimization of the WDNs. This algorithm has always faced a lot of changes and reforms. Fast messy GA is one of the GAs and in here is utilized as the optimizer model. It can evaluate chromosomes that are not equal in terms of gene strings and can increase the convergence rate by decreasing the chromosomes length and eliminating undesirable genes. FMGA acts as a filter to remove undesirable genes so that the population only contains short chromosomes with desirable genes. The filtering process continues until the length of all chromosomes decreases to a desirable length. This algorithm resists falling into a local minimum, which improves its ability to solve problems in a short time. FMGA is safer than standard GAs to be deceived and converges better to an optimal solution (33).

Optimal solutions in GA depend on factors such as population size, cut probability, splice probability, mutation probability and, most importantly, computational search space. Bentley WaterGEMS V8i software which works based on the fast messy GA was used to optimize the WDN in this study.

Assessment of the proposed Model

Case study

The case study was a WDN in Tehran, the capital of Iran. Given that the number of pipes and nodes of the network was huge, the network was simplified by deleting pipes with small diameters located as branches in the network. The simplified structure of the network consisted of 181 pipes, 138 nodes, a reservoir with a maximum water level of 1324.21 m and 10 pressure relief valves (PRV) (Figure 1) (34).

Table 1 lists the characteristics of the network pipes. It consisted of 45 019 m of steel and ductile iron pipes. The population and area covered by the reservoir contained 113378 peoples and 659.18 hectares, respectively. The maximum hourly demand (at 1 PM) was 682 L/s and the minimum hourly demand (at 4 AM) was 361 L/s.

The network available for the future (horizon plan) was examined. For the horizon plan, the maximum and minimum hourly demands were 1023.07 and 536.69 L/s, respectively. This network needed to be optimized to satisfy the minimum and maximum pressure in the nodes. Ductile iron pipes were used to improve the network.

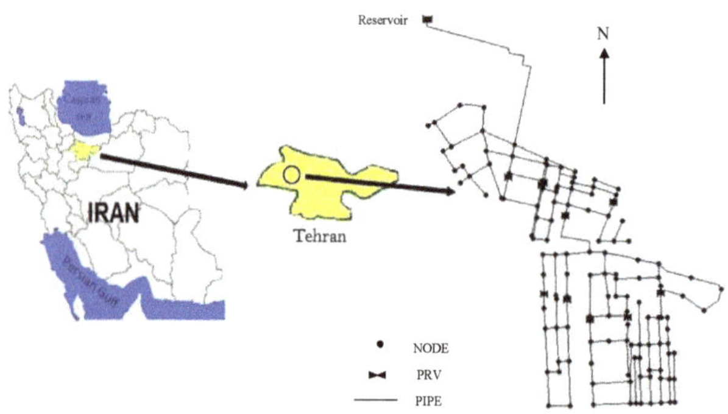

Figure 1. Layout of the WDS model (34).

Table 1. Characteristics of network pipes (34)

Material	Total length in the considered WDN (m)	Diameter (mm)
Ductile iron	835	60
	1170	80
	8052	100
	9681	150
	8718	200
	3263	250
	1836	300
	1438	350
	2290	400
	1130	500
	1714	700
Steel	7354	1200
	4513	900
Total	45019	

Nine commercial diameters were introduced to optimize the network (100, 150, 200, 250, 300, 350, 400, 500 and 600 mm at a cost of USD 30.53, 44.61, 60.79, 73.42, 92.76, 121.32, 145, 197.37 and 257.63 per unit length, respectively).

The network was examined under two conditions. In condition 1, parallel pipes were considered for all pipes except in the areas in which it was not possible and physical limitations. In condition 2, as is shown in Figure 2, parallel pipes were considered for pipes with low pressure problem and those that acted as pressure amplifiers in the mentioned pipes.

Results

After optimizing the maximum hourly demand at 1023.07 L/s, 20 solutions were obtained for each condition to optimize the network. These included new diameters for the parallel network pipes. In solution 16, 133 parallel pipes out of 167 predefined parallel pipes were selected. In solution 19, 13 out of 18 predefined parallel pipes were

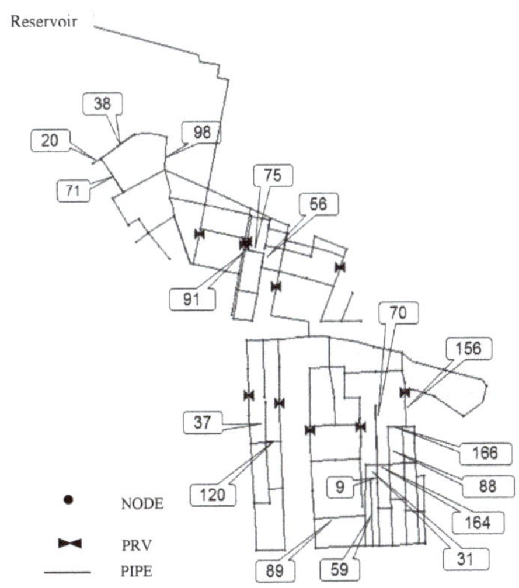

Figure 2. Layout of the WDS model in condition 2.

parallelized. In the creation of the scenario and its transfer to Bentley WaterGEMS, the effects of these diameters were imposed on the network.

Figures 3a and 3b show the Pareto front obtained for the cost versus pressure benefit for both conditions. Table 2 shows 20 solutions for minimizing the cost and maximizing the pressure benefit for both conditions using 167 parallel pipes (condition 1) and 18 parallel pipes (condition 2). As seen, almost at the same total benefit (total benefit = 168.843 for condition 1and 168.915 for condition 2). Solution 16 (condition 1) and solution 19 (condition 2) have the permitted pressure and the lowest cost (total cost of condition 1 = US$2 325 798 and of condition 2 = US$236 188). As observed, condition 2 is economically more cost-effective than condition 1, as the cost of condition 2 decreased 89.84% compared to condition 1.

After the network and economic analysis and the control of diameters to meet the minimum hourly demand (536.69 L/s), it was seen that condition 2, using parallel pipes in areas of low pressure, improved the pressure rate. Table 3 shows the pressure before and after optimization and reveals that the existing pipes could not meet future

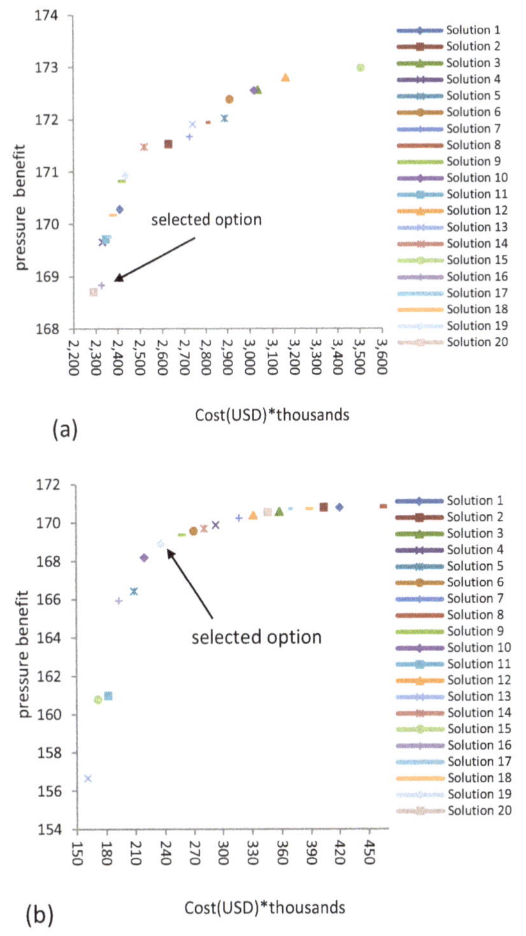

Figure 3. (a) Pareto front in condition 1 (b): Pareto front in condition 2

Table 2. Comparison of solutions in two conditions

Condition 1					Condition 2				
Solution	Total cost (USD)	Pressure benefit	Maximum pressure (mH_2O)	Minimum pressure (mH_2O)	Solution	Total cost (USD)	Pressure benefit	Maximum pressure (mH_2O)	Minimum pressure (mH_2O)
1	2409665	170.289	66.84	26.01	1	420504	170.801	66.72	26.05
2	2627280	171.54	66.82	26	2	403592	170.795	66.72	26.05
3	3035477	172.568	66.88	25.96	3	357456	170.577	66.72	26.04
4	2330004	169.662	67.07	25.91	4	293158	169.885	66.72	25.99
5	2889061	172.026	66.86	26.06	5	209102	166.435	66.72	25.82
6	2910663	172.389	66.77	26	6	269917	169.587	66.72	26.02
7	2724399	171.682	66.85	25.99	7	317362	170.258	66.72	25.99
8	2804143	171.952	66.75	25.96	8	462332	170.84	66.72	26.05
9	2417678	170.825	66.84	25.98	9	257790	169.376	66.72	25.85
10	3019325	172.558	66.89	25.98	10	219411	168.21	66.72	25.86
11	2343518	169.712	66.72	25.96	11	182057	160.983	66.72	24.93
12	3159449	172.806	66.89	25.9	12	331278	170.396	66.72	26.02
13	2738521	171.916	67.19	25.96	13	161244	156.679	66.72	24.17
14	2518845	171.489	66.84	25.96	14	280920	169.715	66.72	26.05
15	3505036	172.984	66.85	26.04	15	171257	160.788	66.72	24.96
16 (Selected option)	**2325798**	**168.843**	**66.85**	**26.06**	16	193443	165.939	66.72	25.43
17	2354281	169.759	66.79	25.84	17	367501	170.718	66.72	25.99
18	2377897	170.18	66.71	26.09	18	388651	170.732	66.72	26.05
19	2433348	170.928	66.84	25.95	**19 (Selected option)**	**236188**	**168.915**	**66.72**	**26.05**
20	2289175	168.717	66.85	26.05	20	345902	170.542	66.72	25.99
Control for minimum hourly demand			69.02	26.28	Control for minimum hourly demand			68.84	26.33

demand. This problem could be resolved using parallel pipes. According to the local standards (35), the network must be designed in such a way that the minimum and maximum pressure in the nodes to be satisfied. The allowable limit of pressure changes in the nodes is 26 to 60 mH_2O (in certain circumstances). As observed, at 4 nodes (123, 124, 125 and 126) the pressure is higher than the standard level. In condition 1, the pressure at the node 124 is 66.72 mH2O. Of course, the node of 124 was located in pressure zone 1 with a static pressure of 69.91 mH_2O. Such pressure is not unexpected because the nominal pressure of the pipes applied to the network was 10 atm, which makes the reported pressures acceptable.

Table 4 shows that the cost of condition 2 decreased 89.84% compared to condition 1.

Table 5 shows the nodes pressure percentage before and after optimization. As observed, the pressure percentage of less than 26 mH_2O has reached zero after optimization. But in some nodes, it is 2.9% and 0.72% in conditions 1 and 2, respectively.

Figure 4 shows that the pressure in the nodes is higher than the standard level in the two conditions (nodes 123, 124, 125 and 126 in condition 1 and 124 in condition 2). Figure 5 shows the diameter and amount of pipe to be used in the network for both conditions. Solution 16 in condition 1 includes 31304 m of pipe with diameters of 100, 150, 200, 250, 300, 350, 400, 500 and 600 mm and solution 19 in condition 2 includes 2387 m of pipe with diameters of 100, 150, 200, 250, 350, 400 and 500 mm.

Comparison of conditions 1 and 2 confirms that the implementation of solution 19 is more affordable.

Discussion

In this study, the FMGA and WaterGEMS hydraulic software were used to optimize a WDN in western Tehran with the help of engineering judgment. It was optimized using the objective functions of cost minimization and pressure benefit maximization under two conditions. Table 4 shows that the costs decreased 89.84% with the use of engineering judgment. Using this approach, for reducing calculation volume the parts of the network with no possibility of parallel piping (physical limitations such as narrow roads and intersections with underground facilities) was ignored. In parts experiencing low-pressure, parallel pipes were defined.

One-way analysis of variance (ANOVA) was used to compare the mean of the numbers obtained for the pressure variable in three options, including the existing pipes, condition1 and condition 2.

The mean of the pressure in the existing pipes has been 34.45 with the standard deviation of 16.88, in condition 1, 44.36 with the standard deviation of 9.60 and in condition 2, 41.55 with a standard deviation of 8.91. The results of analysis indicate that there is a significant difference between the three groups in the amount of pressure ($P<0.001$). Scheffe test analysis was used to find out what differences were observed among the groups.

It was observed that the mean pressure in the existing

Table 3. Comparison of nodes pressure before and after optimization in both conditions

Label	Existing pipe	Condition 1	Condition 2	Label	Existing pipe	Condition 1	Condition 2	Label	Existing pipe	Condition 1	Condition 2
		Solution 16	Solution 19			Solution 16	Solution 19			Solution 16	Solution 19
	Pressure (mH_2O)	Pressure (mH_2O)	Pressure (mH_2O)		Pressure (mH_2O)	Pressure (mH_2O)	Pressure (mH_2O)		Pressure (mH_2O)	Pressure (mH_2O)	Pressure (mH_2O)
1	27.7	27.7	27.7	47	29.9	37.33	39.32	93	47.99	48.01	48
2	-3.4	34.73	35.96	48	20.76	40.8	42.98	94	48.46	48.47	48.46
3	25.43	27.44	26.21	49	-9.58	38.07	36.3	95	56.13	56.35	56.12
4	26.01	26.06	26.05	50	37.77	38.25	38.18	96	47.47	48.85	47.49
5	25.85	26.07	26.08	51	37.48	38.23	38.12	97	40.9	50.78	41.34
6	26.15	26.12	26.13	52	35.86	40.56	35.97	98	50	50.27	49.99
7	28.32	27.24	28.49	53	13.34	44.23	41.77	99	50.09	50.42	50.08
8	31.58	31.69	31.58	54	25.99	40.41	40.75	100	57.16	57.45	57.18
9	24.45	27.08	26.71	55	44.32	44.74	44.2	101	38.86	52.38	38.88
10	24.93	27.15	26.77	56	35.04	37.43	40.95	102	58.2	58.51	58.24
11	27.64	27.64	27.64	57	34.86	37.45	40.54	103	58.87	58.95	58.94
12	30.1	31.35	30.85	58	13.09	45.19	41.97	104	44.29	52.91	44.58
13	-3.35	34.21	36.66	59	45.53	45.54	45.58	105	39.29	52.84	39.3
14	26.43	31.3	26.55	60	21.74	42.45	42.05	106	51.92	52.23	51.91
15	31.96	33.98	31.97	61	8.54	45.68	37.62	107	52.44	52.46	52.45
16	36.66	36.66	36.66	62	13.18	45.72	39.55	108	52.12	53.37	52.12
17	1.94	38.38	41.3	63	14.51	46.96	40.1	109	43.32	53.84	43.81
18	27.79	31.22	28.81	64	39.3	41.57	39.42	110	59.89	59.93	59.93
19	37.3	37.3	37.3	65	31.13	30.54	31.13	111	59.48	59.77	59.52
20	30.51	30.97	30.62	66	9.6	46.81	38.67	112	41.4	55.2	41.61
21	37.24	37.24	37.24	67	15.94	47.9	40.92	113	37.72	55.44	37.94
22	0.82	38.15	29.9	68	10.62	47.36	39.72	114	45.05	56.05	45.32
23	14.44	35.27	38.39	69	42.76	45.47	43.28	115	49.92	55.86	49.91
24	34.21	34.57	34.12	70	16.87	48.7	41.8	116	37.94	55.99	38.16
25	37.19	37.27	37.14	71	37.68	38.46	37.68	117	53.52	55.96	53.52
26	33.03	33.06	33.06	72	47.25	47.63	47.08	118	50.84	56.5	50.84
27	28.8	29.58	28.8	73	44.58	46.77	44.79	119	56.21	56.54	56.18
28	4.27	40.64	43.43	74	17.8	49.28	46.49	120	56.37	56.71	56.36
29	35.13	36.37	35.9	75	43.5	44.19	44.04	121	57.14	57.51	57.14
30	35.63	38.25	36.2	76	50.71	51.53	50.76	122	57.87	58.18	57.86
31	40.61	40.41	40.7	77	18.43	51.21	44.54	123	52.77	65.39	52.76
32	35.1	38.4	35.61	78	46.84	50.89	46.27	124	66.71	66.85	66.72
33	35.4	38.42	35.9	79	44.38	45.02	44.85	125	56.4	66.7	56.39
34	33.26	41.07	36.88	80	19.51	52.58	45.05	126	55.69	60.53	55.4
35	15.73	39.96	42.59	81	52.1	52.45	52.01	127	46.06	50.46	45.8
36	37.53	39.53	37.71	82	32.23	45.66	32.25	128	40.56	44.02	40.39
37	37.92	40.06	37.93	83	20.29	53.14	45.63	129	46.46	51.5	46
38	-11.45	36.26	34.35	84	52.88	53.03	52.87	130	37.29	37.35	37.26
39	17.15	43.86	39.46	85	45.11	46.53	45.1	131	50.57	50.93	50.47
40	4.88	41.69	34.45	86	28.61	47.03	28.83	132	44.85	45.76	45.24
41	10.69	44.33	40.95	87	45.26	46.73	45.29	133	48.01	48.18	47.96
42	33.5	42.61	37.96	88	53.75	53.84	53.78	134	33.25	38.09	35.34
43	16.34	44	39.26	89	54.09	54.08	54.11	135	32.9	35.79	32.91
44	-11.83	35.22	32.99	90	37.32	48.09	37.7	136	28.54	31.52	29.3
45	34.66	39.36	34.78	91	37.6	47.46	38.04	137	26.49	26.43	26.45
46	-9.68	37.37	35.14	92	47.74	47.74	47.74	138	56.61	56.63	56.64

pipes was significantly lower compared to conditions 1 and 2 ($P<0.001$). Also, the average pressure in condition 1 is 2.80 points more than the condition 2. But the difference between these conditions is not statistically significant ($P = 0.170$).

According to the results of analysis, due to the lack of significant difference in pressure between conditions 1 and 2, for choosing the ideal solution, the lowest cost and the number of nodes outside the normal range can be chosen as selection criteria.

As shown in Table 3, in condition 1, four nodes (nodes 124,123, 125 and 126), and in condition 2 only the node 124 are outside the normal range. Also, economically, condition 2 is more cost- effective than the condition

Table 4. Comparison of costs and total benefit in both optimizations

Parameters	Condition 1	Condition 2
Total cost (USD)	2325798	236188
Total benefit	168.843	168.915

1. In condition 1, there are nine pipe sizes with 331 304 m piping, but in condition 2 there are seven pipe sizes with 2387 m piping. According to Table 4, the cost of condition 2 dropped 89.84% than the condition 1. Thus, the condition 2 is preferable than the condition 1.

The results indicate that engineering knowledge and experience must be combined with optimization techniques to provide the best solution.

Iglesias-Rey et al employed engineering judgment and a model optimizer (pseudo-GA) to improve the energy efficiency and management of leakage in WDNs. The results showed that the total cost reached from €4 942 793 to €1 252 085. In fact it decreased approximately 3.9 times with a proper combination of optimization techniques and manual adjustments relying on using engineering judgment (29).

Khedr and Tolson (28) compared optimization techniques with an engineering judgment approach to design of WDN. They solved a simplified version of the Battle of

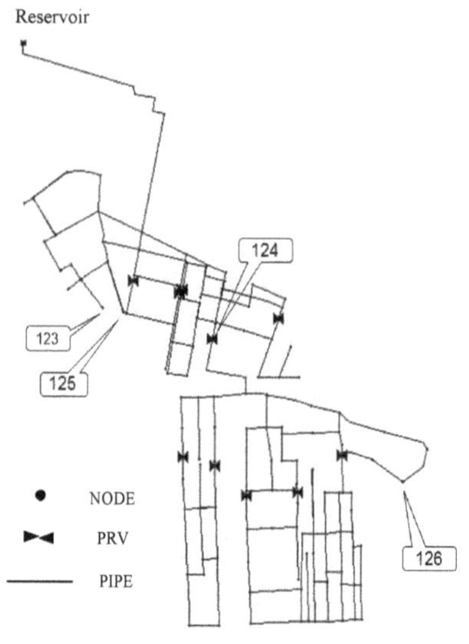

Figure 4. The nodes with the pressure higher than the standard level in two conditions.

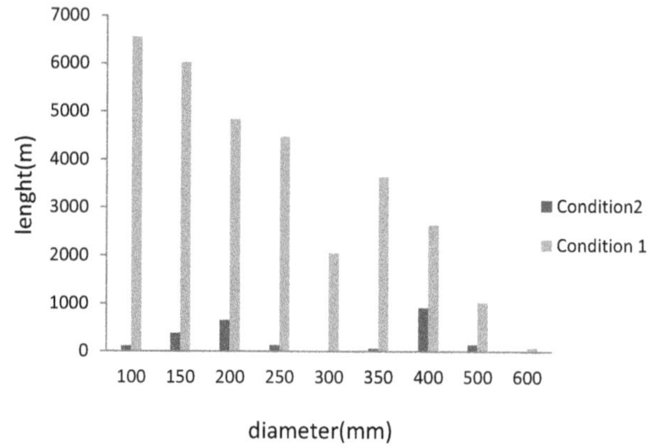

Figure 5. The diameter and amount of the proposed pipes in two conditions.

Algorithm settings for the two conditions are: Population size: 240, cut probability: 1.8, splice probability: 80, mutation probability: 0.9, Max.trials: 30000.

Background Leakage Assessment for Water Networks (BBLAWN) problem using both the engineering judgment approach and a multi objective optimization algorithm, and compared the corresponding Pareto fronts. Their study shows the engineering judgment solution generated a high-quality result to identify a more realistic number of changes in the WDN (about 10% of WDN) compared with BBLAWN teams based on global optimization algorithm (27).

Matthews et al showed the engineering judgment approach could provide a very good Pareto front. Moreover, this approach coupled with a heuristic global optimization algorithm can yield even better results than engineering judgment alone (36).

Conclusion

In engineering judgment approach, the operation was broken into two phases to obtain an appropriate practical Pareto front for minimizing the cost and maximizing the pressure benefit. First, engineers must identify all practical and possible solutions. The most important part of this step after determining the objective functions is to identify decision variables and decrease the search space. The second step includes using an optimizer algorithm by which the Pareto front is obtained with the help of engineering judgment to produce solutions which are cost-effective and efficient.

Table 5. Comparison of the nodes pressure percentage before and after the optimization of the network

Pressure limit (mH_2O)	Pressure (%) with the existing pipe	Pressure (%) The first condition	Pressure (%) The second condition
		Solution 16	Solution 19
<26	25.36	0	0
<40	35.51	36.23	47.83
<50	18.84	31.88	32.61
<60	19.57	28.99	18.84
<70	0.72	2.9	0.72

Many models of WDNs optimization perform designing based on minimizing the cost that leading to a pressure close to the minimum allowable pressure and decrease of the network performance. In this study, the multi-objective fast messy GA and Water GEMS V8i hydraulic software were used to optimize a real WDN in the western Tehran with the help of engineering judgment. This approach decreased the search space to produce the best possible solution with the lowest cost and the most pressure benefit using a Pareto front from a set of non-dominant solutions. To reduce the volume of computation the parts of the network for which there was no possibility of parallel piping was ignored and parallel piping was defined for other parts with low pressure. The results of optimization and modification of the network confirm that, with the help of engineering judgment, the cost decreased 89.84% in the real network and the pressure level in all the nodes increased to above the allowable minimum pressure.

Acknowledgments
This research is derived from a PhD thesis. The authors express their appreciation to the Tehran province water and wastewater co. for its cooperation in facilitating to obtain the data of network.

Competing interests
The authors declare that they have no competing interests.

Authors' contributions
All authors contributed equally and were involved in the study design, data collection, analysis, and interpretation of the data. All authors critically reviewed, refined, and approved the manuscript.

References
1. Scholten L, Scheidegger A, Reichert P, Maurer M. Combining expert knowledge and local data for improved service life modeling of water supply networks. Environ Model Softw 2013; 42: 1-16. doi: 10.1016/j.envsoft.2012.11.013.
2. Eslami A, Ghaffari M, Barikbin B, Fanaei F. Assessment of safety in drinking water supply system of Birjand city using World Health Organization's water safety plan. Environ Health Eng Manag 2018; 5(1): 39-47. doi: 10.15171/ehem.2018.06.
3. Nijhawan A, Jain P, Sargaonkar A, Labhasetwar PK. Implementation of water safety plan for a large-piped water supply system. Environ Monit Assess 2014; 186(9): 5547-60. doi: 10.1007/s10661-014-3802-x.
4. Shirzad A, Tabesh M, Heidarzadeh M. A new method for quasi-optimal design of water distribution networks. Water Resources Management 2015; 29(14): 5295-308. doi: 10.1007/s11269-015-1118-4.
5. Mambretti S, Orsi E. Optimizing pump operations in water supply networks through genetic algorithms. J Am Water Works Assoc 2016; 108(2): E119-E25. doi: 10.5942/jawwa.2016.108.0025.
6. Syahputra R. Distribution network optimization based on genetic algorithm. Journal of Electrical Technology UMY 2017; 1(1): 1-9.
7. Afshar A, Massoumi F, Afshar A, Marino MA. State of the art review of ant colony optimization applications in water resource management. Water Resources Management 2015; 29(11): 3891-904. doi: 10.1007/s11269-015-1016-9.
8. Dini M, Tabesh M. A New method for simultaneous calibration of demand pattern and Hazen-Williams coefficients in water distribution systems. Water Resources Management 2014; 28(7): 2021-34. doi: 10.1007/s11269-014-0592-4.
9. Mora-Melia D, Iglesias-Rey PL, Martínez-Solano FJ, Muñoz-Velasco P. The efficiency of setting parameters in a modified shuffled frog leaping algorithm applied to optimizing water distribution networks. Water 2016; 8(5): 182. doi: 10.3390/w8050182.
10. Roy P, Roy P, Chakrabarti A. Modified shuffled frog leaping algorithm with genetic algorithm crossover for solving economic load dispatch problem with valve-point effect. Appl Soft Comput 2013; 13(11): 4244-52. doi: 10.1016/j.asoc.2013.07.006.
11. Samora I, Franca MJ, Schleiss AJ, Ramos HM. Simulated annealing in optimization of energy production in a water supply network. Water Resources Management 2016; 30(4): 1533-47. doi: 10.1007/s11269-016-1238-5.
12. Ghajarnia N, Bozorg-Haddad O, Marino MA. Performance of a novel hybrid algorithm in the design of water networks. Water Management 2011; 164(4): 173-91. doi: 10.1680/wama.1000028.
13. Wang L, Pan QK, Tasgetiren MF. A hybrid harmony search algorithm for the blocking permutation flow shop scheduling problem. Comput Ind Eng 2011; 61(1): 76-83. doi: 10.1016/j.cie.2011.02.013.
14. Jung D, Kang D, Kim JH. Development of a hybrid harmony search for water distribution system design. KSCE Journal of Civil Engineering 2018; 22(4): 1506-14. doi: 10.1007/s12205-017-1864-3.
15. Ezzeldin R, Djebedjian B, Saafan T. Integer discrete particle swarm optimization of water distribution networks. Journal of Pipeline Systems Engineering and Practice 2014; 5(1): 04013013. doi: 10.1061/(ASCE)PS.1949-1204.0000154.
16. Surco DF, Vecchi TP, Ravagnani MA. Optimization of water distribution networks using a modified particle swarm optimization algorithm. Water Science and Technology: Water Supply 2017; 18(2): 660-78. doi: 10.2166/ws.2017.148.
17. Coelho B, Andrade-Campos A. Efficiency achievement in water supply systems—A review. Renewable and Sustainable Energy Reviews 2014; 30(C): 59-84. doi: 10.1016/j.rser.2013.09.010.
18. Gharibi H, Mahvi AH, Nabizadeh R, Arabalibeik H, Yunesian M, Sowlat MH. A novel approach in water quality assessment based on fuzzy logic. J Environ Manage 2012; 112: 87-95. doi: 10.1016/j.jenvman.2012.07.007.
19. Prasad TD, Park NS. Multiobjective genetic algorithms for design of water distribution networks. Journal of Water

Resources Planning and Management 2004; 130(1): 73-82. doi: 10.1061/(ASCE)0733-9496(2004)130:1(73).
20. Seyoum AG, Tanyimboh TT, Siew C. Practical application of penalty-free evolutionary multi-objective optimisation of water distribution systems. Water Utility Journal 2016; 12: 49-55.
21. Wang Q, Creaco E, Franchini M, Savic D, Kapelan Z. Comparing low and high-level hybrid algorithms on the two-objective optimal design of water distribution systems. Water Resources Management 2015; 29(1): 1-16. doi: 10.1007/s11269-014-0823-8.
22. Tang R, Zhang C, Ding W, Li Y, Zhou H. Use of multi-objective analysis to reveal the benefits of a water transfer project. Water Science and Technology: Water Supply 2016; 17(1): 259-66. doi: 10.2166/ws.2016.132.
23. Suribabu CR. Resilience-based optimal design of water distribution network. Appl Water Sci 2017; 7(7): 4055-66. doi: 10.1007/s13201-017-0560-2.
24. Sirsant S, Reddy MJ. Reliability-based design of water distribution networks considering mechanical failures. European Water 2017; 58: 407-14.
25. Walski TM. The wrong paradigm—why water distribution optimization doesn't work. Journal of Water Resources Planning and Management 2001; 127(4): 203-5. doi: 10.1061/(ASCE)0733-9496(2001)127:4(203).
26. Halhal D, Walters GA, Ouazar D, Savic DA. Water network rehabilitation with structured messy genetic algorithm. Journal of Water Resources Planning and Management 1997; 123(3): 137-46. doi: 10.1061/(ASCE)0733-9496(1997)123:3(137).
27. Walters GA, Halhal D, Savic D, Ouazar D. Improved design of "Anytown" distribution network using structured messy genetic algorithms. Urban Water 1999; 1(1): 23-38. doi: 10.1016/S1462-0758(99)00005-9.
28. Khedr A, Tolson BA. Comparing Optimization techniques with an engineering judgment approach to WDN design. Journal of Water Resources Planning and Management 2016; 142(5): C4015014. doi: 10.1061/(ASCE)WR.1943-5452.0000611.
29. Iglesias-Rey PL, Martinez-Solano FJ, Mora Melia D, Martinez-Solano PD. Combining engineering judgment and an optimization model to increase hydraulic and energy efficiency in water distribution networks. Journal of Water Resources Planning and Management 2016; 142(5): C4015012. doi: 10.1061/(ASCE)WR.1943-5452.0000605.
30. Hans AE. Multicriteria Optimization in Engineering and in the Sciences. New York: Springer; 1988.
31. Srinivas N, Deb K. Muiltiobjective optimization using nondominated sorting in genetic algorithms. Evol Comput 1994; 2(3): 221-48. doi: 10.1162/evco.1994.2.3.221.
32. Haestad Methods Solution Center. WaterGEMS For GIS user's guide. Watertown: Bentley Systems, Incorporated; 2005.
33. Goldberg DE, Deb K, Kargupta H, Harik G. Rapid Accurate Optimization of Difficult Problems Using Fast Messy Genetic Algorithms. Proceedings of the 5th International Conference on Genetic Algorithms; 1993; p. 56-64.
34. Sherri F, Mahvi AH, Toloie Eshlagy A, Hassani AH. A new approach in simultaneous calibration of Hazen-Williams coefficients and demand of nodes in of water distribution systems. Desalination Water Treat 2017; 74: 137-148. doi: 10.5004/dwt.2017.20617.
35. Islamic Republic of Iran Vice Presidency For Strategic Planning and Supervision. Design Criteria of Urban and Rural Water Supply and Distribution Systems. 2013. [In Persian].
36. Matthews JC, Selvakumar A, Sterling RL, Condit W. Innovative rehabilitation technology demonstration and evaluation program. Tunnelling and Underground Space Technology 2014; 39: 73-81. doi: 10.1016/j.tust.2012.02.003.

Study of the effectiveness of the third generation polyamide amine and polypropylene imine dendrimers in removal of reactive blue 19 dye from aqueous solutions

Sepideh Sadeghi[1,2], **Ghazal Raki**[1,2], **Asrin Amini**[1,2], **Nezamaddin Mengelizadeh**[1,2], **Mohammad Mehdi Amin**[1], **Majid Hashemi**[1,2,3*]

[1]Department of Environmental Health Engineering, School of Health, Isfahan University of Medical Sciences, Isfahan, Iran
[2]Student Research Committee, Isfahan University of Medical Sciences, Isfahan, Iran
[3]Environmental Health Engineering Research Center, School of Health, Kerman University of Medical Sciences, Kerman, Iran

Abstract

Background: Dye and colored materials cause health risks in water and therefore, must be removed from water supplies and wastewater. The aim of this study was to evaluate the effectiveness of the third generation poly(amidoamine) (PAMAM) and poly (propylene imine) dendrimers (PPI-G3) in the removal of reactive blue 19 (RB19) dye from aqueous solutions and determine the optimum conditions for the removal.

Methods: This study was performed in a laboratory and batch scale. In this study, synthetic wastewater was examined with three different concentrations of RB19 (25, 50, and 100 mg/L), different pHs (3, 7, and 10), various amounts of dendrimer (0.4, 0.8, 1.2, and 1.6 g/L), and at different times (15, 30, and 60 minutes) during the adsorption process. The remaining amount of dye was measured by spectrophotometer at 592 nm wavelength. Langmuir and Freundlich isotherms were also tested.

Results: The results showed that by increasing the reaction time and adsorbent dosage, the rate of dye removal increased while by increasing the initial dye concentration and pH, the dye removal efficiency was significantly decreased. In this study, with increase of pH from 3 to 10, dye removal efficiency at a concentration of 25 mg/L, decreased from 72% to 20% and 88% to 17% by PAMAM and PPI dendrimers, respectively. Excel software was used for data analysis.

Conclusion: Both adsorbents had a good dye removal efficiency, but PPI dendrimer was more effective in removing RB19. Adsorption data followed the Langmuir isotherm.

Keywords: Adsorption, Wastewater, Dendrimer, Polyamide amine, Polypropylene imine

***Correspondence to:**
Majid Hashemi
Email: mhashemi120@gmail.com

Introduction

The dye is one of the most important pollutants in industrial wastewater that is widely used in industrial sectors such as textile, leather, cosmetics, paper making, food processing and pharmaceutical industries (1,2). Reactive dyes are water-soluble and anionic that are largely used in the textile industry. Reactive blue 19 (RB19) with its special chemical structure is one of the anthraquinone dyes which are the most widely used in textile, after azo dyes. This dye has relatively low fixation efficiency on the fiber and thus, considerable amounts of waste can be imported to wastewater. Since this dye is stable in natural environments and highly resistant to light, chemical agents and biological degradation, discharging wastewater containing dye into the environment will cause irreversible and serious problems (3,4). Due to the low bio-decomposition of dyes, the conventional biological wastewater treatment processes are not effective in dye wastewater treatment. The colored wastewater is usually treated by physical or chemical methods such as coagulation and flocculation, precipitation, adsorption, membrane filtration, electrochemical techniques, ozonation and etc. The adsorption process is one of the physicochemical processes developed due to the low cost, ease of operation, flexibility, and simplicity for treatment of wastewater containing dye, organic materials and metals (5,6). One of the newest adsorbents is dendrimer. Dendrimer is a Greek term composed of two parts: "dendron" meaning "tree" and "meros" meaning "part" and implies the tree-like branched structure of this group of compounds (7). The structure of dendrimers is composed of 3 parts: An inner core, shell (generation) comprising repeating units, and terminal functional groups (8,9). Depending on the number of generations, dendrimers include a wide range of dimensions often in the nanometer range so that in the third generation (G3), their dimension is equal to insulin hormone (5). Dendrimer as a nanomaterial (7) and due to its large surface, has high and effective capability to remove the contaminants (10). The main features of dendrimers are having many active endgroups as well as the blank spaces between the branches of the internal connector that have the ability to attract and imprison foreign molecules such as water, dye and other materials by chemical bonding and encapsulation (9).

These polymers are of great importance in various fields such as medical and industrial applications and pharmaceuticals as a material to eliminate germs and in the field of water and wastewater treatment, to remove pollutants (11,12). Table 1 shows different types of dendrimers including polyamide amine-organosilicon (PAMAMOS), poly(amidoamine) (PAMAM), polypropylene imine (PPI), chiral, liquid crystal, Tecto, hybrid, multilingual and Micellar (9). Many studies have been done on the use of dendrimers to remove pollutants from the environment. Patel et al in a study examined the removal of metal ions Cu^{2+}, Ni^{2+} and Zn^{2+} of water using G1, G2, and G3 dendrimers with the number of hydroxyl groups 8, 32 and 128 as adsorbents (13). In another study, PPI and chitosan were evaluated as a biopolymer to remove reactive black 5 and reactive red 198 (5). The microparticles of polyamide imine-chitosan have been used as an adsorbent for the removal of reactive blue 21 (14). In this study, the G3 (PAMAM and PPI) dendrimers were used as the adsorbents for the removal of RB19 dye from wastewater.

Material and Methods

This study was conducted discontinuously (batch) on a laboratory scale. In this study, the third generation PAMAM and PPI dendrimers were purchased from the Institute of Tehran Paint, and their structural form is provided in Figure 1 (15). The dye used in this study, was purchased from Sigma-Aldrich and other chemicals used in the experiments, were obtained from Merck (Germany). Characteristics of RB19 dye are provided in Table 1 (16). In order to prepare the wastewater, the dye stock solution was created by solving the appropriate amount of RB19 powder in distilled water and then, different concentrations (25, 50, and 100 mg/L) were examined. In the present study, the influence of different adsorbent dosage (0.4, 0.8, 1.2, and 1.6 mg/L), pH (3, 7, and 10), initial dye concentration (25, 50, and 100 mg/L) and reaction time (15, 30, and 60 min) on dye removal was investigated. In order to mix and make a proper contact between the adsorbent and dye, a mechanical shaker at 250 rpm was used. The adsorbent was separated from the wastewater using membrane filter (0.45 micron in diameter). Sulfuric acid and sodium hydroxide 1N were used to adjust the pH. The remaining dye concentration was measured using a UV/Vis spectrophotometer (Optima SP-3000 Plus model, Japan) at 592 nm. The amount of dye adsorbed at equilibrium time (q_e) and dye removal efficiency (E) were respectively calculated using equations (1) and (2) (16).

Table 1. Characteristics of reactive blue 19 dye

Chemical Formula	$C_{22}H_{16}N_2Na_2O_{11}S_3$
Molecular Weight	626.5 g
Trade Name	Remazol Brilliant Blue R
Chemical Structure	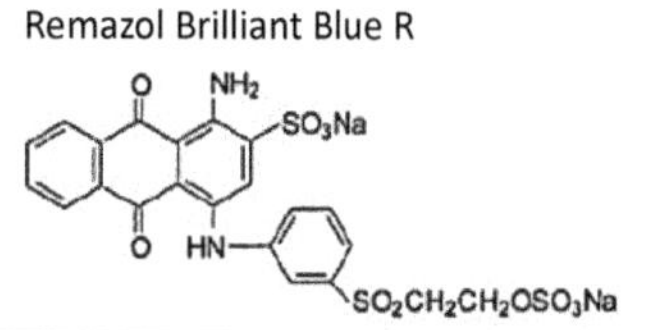

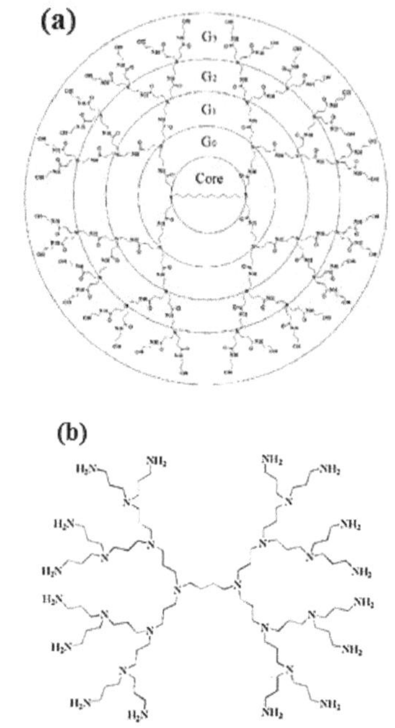

Figure 1. PAMAM (a) and PPI (b) dendrimers.

$$q_e = \frac{V}{M} \times (C_0 - C_e) \quad (1)$$

$$E = \frac{C_0 - C}{C_0} \times 100 \quad (2)$$

In these equations, C_0 and Ce respectively show the initial and final concentration of the dye (mg/L), V, volume of the solution (L), M, adsorbent mass (g), q_e, adsorption capacity at equilibrium time (mg/g), and E, dye removal efficiency (%) (17). Langmuir and Freundlich adsorption isotherms were examined to describe the relationship between the adsorbent and the colored solution. Langmuir and Freundlich isotherm models are described using equations (3) and (4):

$$\frac{C_e}{q_e} = \frac{1}{q_m b} + \frac{1}{q_m} C_e \quad (3)$$

$$\log q_e = \log k_f + \frac{1}{n} \ln C_e \quad (4)$$

Where q_m and b are Langmuir constants that are related to the maximum adsorption capacity (mg/g) and energy (L/mg). K_f and n, are Freundlich constants which show adsorbent capacity and intensity. Microsoft Excel software was used for data analysis and drawing charts.

Results

Determining the effect of pH and contact time on dye removal

The effect of pH on the removal of RB19 in the presence of adsorbent dose (0.4, 0.8, 1.2, and 1.6 g/L) in desired dye concentrations (25, 50, and 100 mg/L) was studied and the results are shown in Figure 2. As shown in this figure, the maximum dye removal efficiency was observed at pH 3. In the present study, the effect of the retention time (15, 30, and 60 minutes) as one of the independent variables in dye removal was also evaluated. The maximum adsorption for both PAMAM and PPI dendrimers took place 30 and 15 minutes respectively.

The effect of adsorbent dosage

Evaluating the effect of adsorbent dosage and determining the optimal dosage of adsorbent are the most important parameters that should be considered in the process of adsorption. The results of evaluation of the effect of adsorbent dosage on the removal of RB19 are shown in Figure 3. According to the findings of the study, by increasing adsorbent concentration from 0.4 to 1.6 g/L, dye removal efficiency by PAMAM and PPI dendrimers increased from 26% to 72% and from 35% to 88%, respectively. Thus, the optimal dosage of PAMAM and PPI dendrimers was 1.6 g/L because in the higher dosages, dye removal efficiency was constant or there was not a significant increase in dye removal.

The effect of dye concentration on the removal efficiency

To survey dye adsorption efficiency, the effect of dye concentration (25, 50, and 100 mg/L) on the removal efficiency was examined. Considering that adsorption at pH 3 had the highest efficiency, thus in this stage, adsorbent dosage 1.6 g/L and pH 3 were selected for dye adsorption. The results of this study are shown in Figure 4. As presented in this figure, by increasing concentrations from 25 to 100 mg/L, a significant reduction in the rate of dye removal was observed. So that by increasing dye concentration, dye removal efficiency decreased from 88% to 29% by PPI and also from 72% to 23.75% by PAMAM.

Evaluation of reactive blue 19 dye adsorption isotherm

Evaluation of experimental data with the Langmuir and Freundlich models and comparison of their correlation coefficients showed that RB19 dye adsorption obeyed the

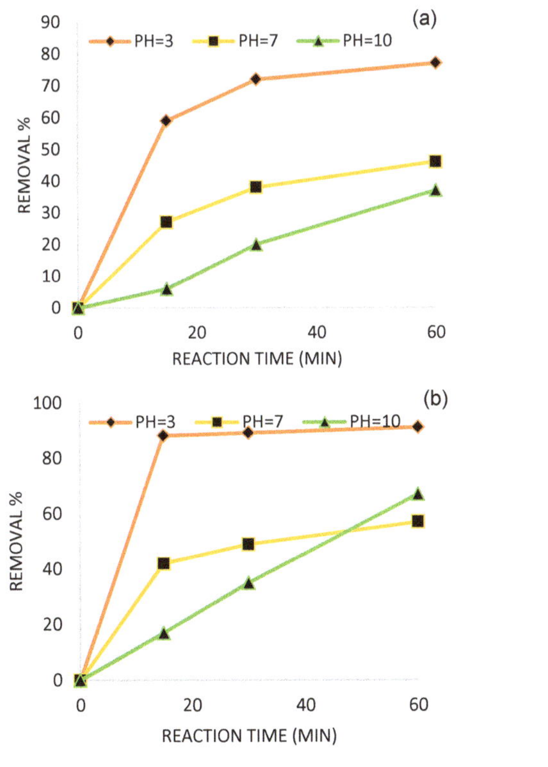

Figure 2. Effect of pH on dye removal efficiency by PAMAM (a) and PPI (b) dendrimers at different times. (dye concentration: 25 mg/L, dendrimer dosage: 1.6 g/L).

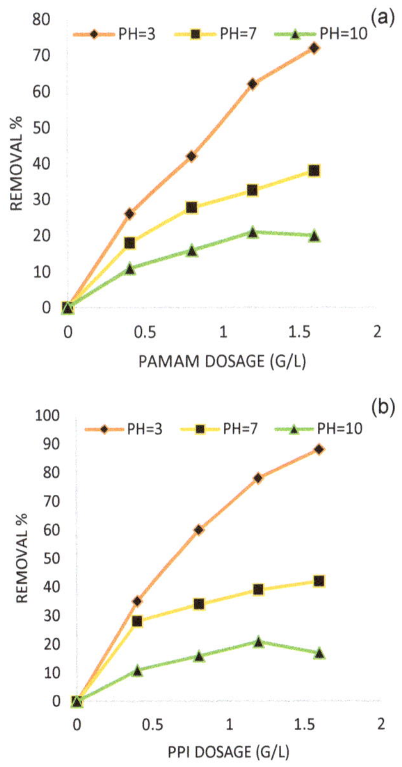

Figure 3. Effect of PAMAM (a) and PPI (b) dendrimers dosage on dye removal at different pHs (retention time: 30 min, dye concentration: 25 mg/L).

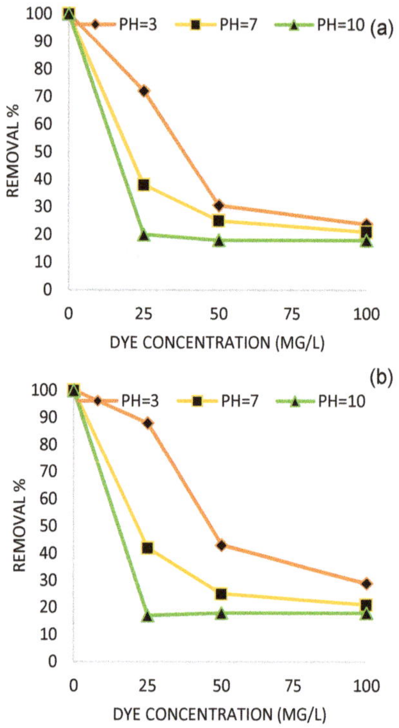

Figure 4. The effect of dye concentration on dye removal efficiency by PAMAM (a) and PPI (b) (retention time: 30 min, dendrimer dosage: 1.6 g/L).

Langmuir isotherm by the G3 PAMAM (R^2=0.94) and PPI (R^2=0.98) dendrimers (Figure 5). Maximum adsorption was 46.8 mg/g for PAMAM and 1029 mg/g for PPI. Values of parameters of Langmuir and Freundlich isotherms are presented in Table 2.

Table 2. Parameters of Langmuir and Freundlich models for RB19 dye adsorption by PPI and PAMAM dendrimers

Langmuir isotherm			Freundlich isotherm		
	PAMAM	PPI		PAMAM	PPI
Q_0 (mg/g)	46.8	1029	K_F	779.8	896.5
K_L	0.3049	0.233	n	3.18	4.29
R^2	0.9435	0.8965	R^2	0.84	0.89

Adsorption kinetics

To investigate the mechanism of adsorption of the RB19 on PPI and PAMAM, the pseudo-first- and second-order models were used. Table 3 shows the Kinetic parameters of the pseudo-first- and second-order models. The results showed that the R^2 values of the pseudo-second-order model (> 0.99) was higher than the pseudo-first-order (>0.90) for RB19 on PPI and PAMAM.

Discussion

Determining the effect of pH and contact time on dye removal

One of the most important effective factors in the adsorption process is pH that can affect adsorption capacity through affecting the degree of ionization and surface characteristics of adsorbent (18). The maximum dye removal efficiency was observed at acidic pH. In general, it can be stated that dye removal efficiency increases when the electrostatic attraction between the dendrimer with positively charged amino groups (NH_2) and anionic dye containing negatively charged sulfonic group, $R\text{-}SO_3^-$ happens. These groups produce hydrogen at acidic pH. Generally, an increase in the pH leads to a

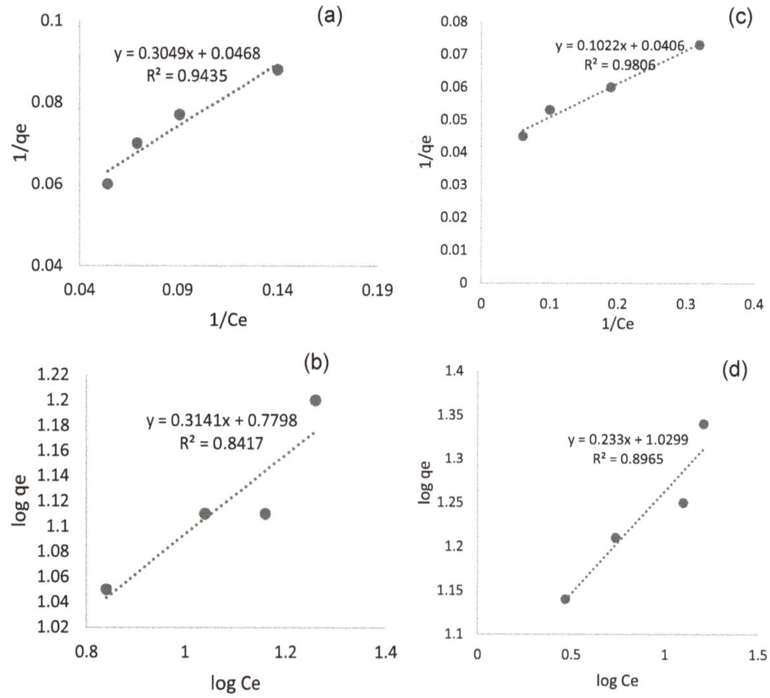

Figure 5. Adsorption isotherms of dye by PAMAM: Langmuir (a) and Freundlich (b) and by PPI: Langmuir (c) and Freundlich (d).

Table 3. The first-order and second-order adsorption rate constants, calculated adsorption capacity ($q_{e, cal}$) and experimental adsorption capacity (q_e) values of RB19 dye on the adsorbents

Adsorbent	$q_{e, exp}$ (mg/g)	First-order kinetic model			Second-order kinetic model		
		k_1 (1/min)	$q_{e, cal}$ (mg/g)	R^2	k_2 (g/mg/min)	$q_{e, cal}$ (mg/g)	R^2
PPI	15.31	0.0549	9.459	0.9062	0.045	15.037	0.9972
PAMAM	13.28	0.0457	9.351	0.9111	0.031	12.82	0.9918

decrease in the sites in the positively charged dendrimer and thus, a decrease in the dye removal efficiency (19,20). In general for anionic dyes, adsorption capacity increases by reducing the pH of the solution while it is reverse for cationic dyes (20). Almasian et al invetigated anionic dye removal by polyacrylonitrile and PAMAM. In this study, adsorption was higher at low pH and pH 2.1 was selected as the optimal pH (21). In the study of Salimpour Abkenar et al on anionic and cationic dye removal by PPI and flax, anionic dyes had better performance at pH 3 and cationic dyes, at alkaline pH (pH 11) (20). Yiyun and Jiepin used different generations of PAMAM dendrimers for removal of methane dyes, and the results showed that the removal of methylene blue and fuchsia acid was effective at high pH (22).

The maximum adsorption for both PAMAM and PPI dendrimers took place in the first 30 and 15 minutes, respectively. In fact, at the beginning of the adsorption reactions, vacant sites in the adsorbent effectively remove the dye, but after equilibrium, due to repulsive forces between molecules of adsorbent, removal efficiency reduces (23,24). In a study by Sadeghi-Kiakhani et al, G2 PPI dendrimer and chitosan were used in order to remove the reactive dye, in which by increasing the reaction time, the dye removal increased and the maximum dye removal efficiency was observed in the 30 minutes of the reaction (5). In a study by Amouei et al, removal of reactive orange 16 was evaluated by sunflower stalks. The results showed that by increasing the reaction time, color removal efficiency increased. In general, by increasing the contact time, the amount of adsorption to reach equilibrium time increased and then, remained almost constant (23).

The effect of adsorbent dosage

Evaluating the effect of adsorbent dosage and determining the optimal dosage of adsorbent are the most important parameters that should be considered in the process of adsorption. According to the findings of our study, by increasing adsorbent concentration until optimal dosage (1.6 g/L), dye removal efficiency by PAMAM and PPI dendrimers increased. It could be due to the increased surface area of the adsorbent and the number of available active sites for adsorption (14). In the study of Ghaneian et al on the effect of adsorption of the Russian knapweed flower powderon the removal of RB19, the results showed that by increasing the adsorbent dose from 0.2 to 0.6 g/100 mL, removal efficiency increased from 72% to 82% (16). In a study by Hayati et al, G2 PPI dendrimer was used for the removal of dyes from textile wastewater, and the results indicated that increasing concentrations of dendrimers leads to increased dye removal and the optimal dendrimer concentration was 1.8 g/L (19). In another study on the adsorption of reactive red 198 by titanium dioxide nanoparticles, by increasing the adsorbent concentration, the adsorption efficiency of dye increased and the optimal adsorbent dosage was reported 1 g/L (25), which are consistent with the results reported by Yiyun and Jiepin (22).

The effect of dye concentration on the removal efficiency

The PPI dendrimers can absorb dye molecules both the protonated functional groups (-NH_3^+, >NH^+-) and the cavities. In fact, due to saturation of adsorbent active sites by the dye molecules, an increase in the dye concentration can reduce adsorption capacity or dendrimer active surfaces (20). The results of Yiyun and Jiepin showed that in low dye concentrations, removal efficiency can be increased (22). In another study by Sadeghi-Kiakhani et al on removal of reactive black 5 and reactive red 198 dyes by polypropylene (PPI) dendrimers and chitosan as biopolymer, under the optimal amounts of process parameters, dye removal was reported 97% and 99% for reactive black 5 and reactive red 198, respectively (5). The effect of crap plant powder on the removal of RB19 dye showed that by increasing the initial concentration of dye, dye removal efficiency decreased in the research of Ghaneian et al (16).

Evaluation of reactive blue 19 dye adsorption isotherm

Adsorption isotherm studies have been carried out on 2 well-known Langmuir and Freundlich isotherm models. Langmuir isotherm model assumes monolayer adsorption onto a surface containing a finite number of adsorption sites of uniform strategies of adsorption with (with no interactions between the adsorbent molecules) no transmigration of adsorbate in the plane surface while Freundlich isotherm model assumes heterogeneous surface and multilayer adsorption (26). Evaluation of experimental data using the Langmuir and Freundlich models and comparison of their correlation coefficients showed that adsorption of RB19 dye obeys the Langmuir isotherm by the G3 PAMAM (R^2=0.94) and PPI (R^2=0.98) dendrimers (Figure 5).

Sayan et al examined thermodynamic adsorption in the removal of RB19 dye by activated carbon under ultrasonic irradiation. The results indicated that adsorption isotherm obeyed the Langmuir model and adsorption of RB19 was endothermic (26). In another study, nanocomposites including copper sulfide and PAMAM dendrimer were used to remove Isma acid fast yellow G dye, and the results indicated that adsorption obeyed Langmuir model and with pH increase, the rate of adsorption decreased (27).

Adsorption kinetics

The results showed that the R^2 values of the second-order kinetics (>0.99) was higher than the first-order kinetics (>0.90) for RB19 onto PPI and PAMAM. Moreover, the q_e value is calculated adsorption capacity for pseudo-second-order model was in agreement with the experimental q_e values than the pseudo-first-order model. Therefore, this study suggested that the second-order model well demonstrated the adsorption of RB19 kinetics. Similar result has been observed in the adsorption of reactive dye onto carbon nanotubes (28), nickel oxide (NiO) nanoparticles (29), palm shell powder, chitosan (30) and biomass (31).

Conclusion

Removal of RB19 dye from aqueous solutions by the third generation PAMAM and PPI dendrimers was investigated. Results of this study showed that by increasing pH, dye removal efficiency decreases and the highest dye removal efficiency was observed at pH 3. Also, by increasing the amount of adsorbent, dye removal efficiency increased. By increasing the initial dye concentration, dye removal efficiency decreased and by increasing the contact time, dye removal efficiency increased. In this study, the mechanism of adsorption of RB19 dye onto the third generation PAMAM and PPI dendrimers was fitted with Langmuir isotherm. According to the results, both adsorbents had good efficiency for dye removal, but PPI was more effective in removing RB19 dye. Adsorption data followed the Langmuir isotherm model. It seems that both dendrimers can be used for removal of different dyes in the textile industry as the environmental-friendly adsorbents.

Acknowledgments

The authors wish to thank Vice Chancellor of Research of Isfahan University of Medical Sciences (IUMS), Iran, for the financial support. Project code: 195093 and ethic code: IR.MUI.REC.1395.1.093.

Competing interests

The authors declare that they have no conflicts of interests.

Authors' contributions

All authors contributed equally to the study design, data collection, analysis, and interpretation. All authors critically reviewed, refined, and approved the manuscript.

References

1. Hadjltaief HB, Galvez ME, Zina MB, Da Costa P. TiO2/clay as a heterogeneous catalyst in photocatalytic/photochemical oxidation of anionic reactive blue 19. blue 19. Arab J Chem 2014;1-9. doi: 10.1016/j.arabjc.2014.11.006.
2. Malakootian M, Jafari Mansoorian H, Hosseini A, Khanjani N. Evaluating the efficacy of alumina/carbon nanotube hybrid adsorbents in removing Azo reactive red 198 and blue 19 dyes from aqueous solutions. Process Saf Environ Prot 2015; 96: 125-37. doi: 10.1016/j.psep.2015.05.002.
3. Maleki A, Mahvi AH, Rezaee R, Davari B. Removal of reactive blue 19 using natural and modified zeolites. Iranian Journal of Health and Environment 2013; 5(4): 519-30. [In Persian].
4. Khalque A, Ahammed SS, Khan SA, Awual R, Islam E, Rahman L. Utilizing Aman Husk for efficient removal of reactive dyes from industrial wastewater. Global Journal for Research Analysis 2018; 7(2): 663-5.
5. Sadeghi-Kiakhani M, Arami M, Gharanjig K. Dye removal from colored-textile wastewater using chitosan-PPI dendrimer hybrid as a biopolymer: optimization, kinetic, and isotherm studies. J Appl Polym Sci 2013; 127(4): 2607-19. doi: 10.1002/app.37615.
6. Srikantan C, Suraishkumar GK, Srivastava S. Effect of light on the kinetics and equilibrium of the textile dye (reactive red 120) adsorption by Helianthus annuus hairy roots. Bioresour Technol 2018; 257: 84-91. doi: 10.1016/j.biortech.2018.02.075.
7. Vunain E, Mishra AK, Mamba BB. Dendrimers, mesoporous silicas and chitosan-based nanosorbents for the removal of heavy-metal ions: a review. Int J Biol Macromol 2016; 86:570-86. doi: 10.1016/j.ijbiomac.2016.02.005
8. Malakootian M, Hashemi M, Toolabi A, Nasiri A. Investigation of nickel removal using poly(amidoamine) generation 4 dendrimer (PAMAM G4) from aqueous solutions. Journal of Engineering Research 2018; 6(2): 13-23.
9. Taghavi Pourian Azar N, Mutlu P, Khodadust R, Gunduz U. Poly(amidoamine) (PAMAM) Nanoparticles: Synthesis and Biomedical Applications. Hacettepe Journal of Biology and Chemistry 2013; 41(3): 289-99.
10. Sounthararajah DP, Loganathan P, Kandasamy J, Vigneswaran S. Adsorptive removal of heavy metals from water using sodium titanate nanofibres loaded onto GAC in fixed-bed columns. J Hazard Mater 2015; 287: 306-16. doi: 10.1016/j.jhazmat.2015.01.067.
11. Hayati B, Arami M, Maleki A, Pajootan E. Application of dendrimer/titania nanohybrid for the removal of phenol from contaminated wastewater. Desalination and Water Treatment 2016; 57(15): 6809-19. doi: 10.1080/19443994.2015.1012746.
12. Sajid M, Nazal MK, Ihsanullah I, Baig N, Osman AM. Removal of heavy metals and organic pollutants from water using dendritic polymers based adsorbents: a critical review. Sep Purif Technol 2018; 191:400-23. doi.org/10.1016/j.

seppur.2017.09.011.
13. Patel R, Patel H, Gajjar D, Patel PM. Removal of CU^{2+}, Ni^{2+} and Zn^{2+} metal ions from water by hydroxyl terminated s-triazine based dendrimer. Malaysian Polymer Journal 2014; 9(2): 70-7.
14. Wang P, Ma Q, Hu D, Wang L. Removal of reactive blue 21 onto magnetic chitosan microparticles functionalized with polyamidoamine dendrimers. React Funct Polym 2015; 91-92: 43-50. doi: 10.1016/j.reactfunctpolym.2015.04.007.
15. Lee S, Han SY, Lee TG, Chung G, Lee D, Oh HB. Observation of pronounced b*,y cleavages in the electron capture dissociation mass spectrometry of polyamidoamine (PAMAM) dendrimer ions with amide functionalities. J Am Soc Mass Spectrom 2006; 17(4): 536-43. doi: 10.1016/j.jasms.2005.12.004.
16. Ghaneian MT, Dehvari M, Jourabi Yazdi N, Mootab M, Jamshidi B. Evaluation of efficiency of Russian knapweed flower powder in removal of reactive blue 19 from synthetic textile wastewater. Journal of Rafsanjan University of Medical Sciences 2014; 12(10): 831-42. [In Persian].
17. Nadafi K, Nabizadeh Nodehi R, Jahangiri-Rad M. Removal of Reactive Blue 29 Dye from Water by Single-Wall Carbon Nanotubes. Iranian Journal Health and Enviroment 2011; 4(3): 359-68. [In Persian].
18. Ghaneian MT, Jasemizadeh T, Sahlabadi F, Miri M, Motabbefroee M. Survey thef efficiency of barberry stem powder in removal of the reactive Blue 19 from textile industrial wastewater. Journal of Rafsanjan University of Medical Sciences 2014; 13(7): 631-40. [In Persian].
19. Hayati B, Mahmoodi NM, Arami M, Mazaheri F. Dye removal from colored textile wastewater by poly(propylene imine) dendrimer: operational parameters and isotherm studies. CLEAN – Soil Air Water 2011; 39(7): 673-9. doi: 10.1002/clen.201000182.
20. Salimpour Abkenar S, Malek RM, Mazaheri F. Dye adsorption of cotton fabric grafted with PPI dendrimers: isotherm and kinetic studies. J Environ Manage 2015; 163: 53-61. doi: 10.1016/j.jenvman.2015.08.003.
21. Almasian A, Olya ME, Mahmoodi NM. Synthesis of polyacrylonitrile/polyamidoamine composite nanofibers using electrospinning technique and their dye removal capacity. J Taiwan Inst Chem Eng 2015; 49: 119-28. doi: 10.1016/j.jtice.2014.11.027.
22. Yiyun C, Jiepin Y. Effect of polyamidoamine dendrimers in decolorising triarylmethane dye effluent. Coloration Technology 2005; 121(2): 72-5. doi: 10.1111/j.1478-4408.2005.tb00254.x.
23. Amouei A, Asgharnia HA, Karimian K, Mahdavi Y, Balark D, Ghasemi SM. Optimization of dye reactive orange 16 (RO16) adsorption by modified sunflower stem using response surface method from aqueous solutions. Journal of Rafsanjan University of Medical Sciences 2016; 14(10): 813-26. [In Persian]
24. Mahmoodi NM, Hayati B, Arami M, Mazaheri F. Single and Binary system dye removal from colored textile wastewater by a dendrimer as a polymeric nanoarchitecture: equilibrium and kinetics. J Chem Eng Data 2010; 55(11): 4660-8. doi: 10.1021/je100248m.
25. Ghaneian MT, Ehrampush MH, Rahimi S, Ghanizadeh GH, AsKarshahi M. Absorbtion of reactive red 198 by nano titanium dioxide from of wastewater. Journal of Tolo Behdasht 2011; 9(4): 59-70. [In Persian].
26. Sayan E, Nuri Ata O, Esra Edecan M. Investigation of adsorption thermodynamics on removal of reactive blue 19 onto activated carbon under ultrasonic irradiation. J Chem Soc Pak 2014; 36(2): 232-8.
27. Abou Taleb MF, El-Trass A, El-Sigeny S. Synthesis of polyamidoamine dendrimer (PAMAM/CuS/AA) nanocomposite and its application in the removal of Isma acid fast yellow G Dye. Polym Adv Technol 2015; 26(8): 994-1002. doi: 10.1002/pat.3517.
28. Wu CH. Adsorption of reactive dye onto carbon nanotubes: equilibrium, kinetics and thermodynamics. J Hazard Mater 2007; 144(1-2): 93-100. doi: 10.1016/j.jhazmat.2006.09.083.
29. Monsef Khoshhesab Z, Ahmadi M. Removal of reactive blue 19 from aqueous solutions using NiO nanoparticles: equilibrium and kinetic studies. Desalination and Water Treatment 2016; 57(42): 20037-48. doi: 10.1080/19443994.2015.1101713.
30. Sreelatha G, Ageetha V, Parmar J, Padmaja P. Equilibrium and Kinetic studies on reactive dye adsorption using palm shell powder (an agrowaste) and chitosan. J Chem Eng Data 2010; 56(1): 35-42. doi: 10.1021/je1007263.
31. Mustafa MM, Jamal P, Alkhatib MF, Mahmod SS, Jimat DN, Ilyas NN. Panus tigrinus as a potential biomass source for reactive blue decolorization: isotherm and kinetic study. Electron J Biotechnol 2017; 26: 7-11. doi: 10.1016/j.ejbt.2016.12.001.

20

Effects of pistachio processing wastewater on treatment efficiency of urban wastewater using activated sludge process

Fatemeh Khademi[1], Kamyar Yaghmaeian[2], Mohammad Taheri[3], Mohammad Ali Hayatabadi[3], Alireza Nasiri[4], Mohammad Malakootian[4,5]*

[1]Islamic Azad University, Bandar Abbas Branch, Bandar Abbas, Iran
[2]Department of Environmental Health Engineering, School of Public Health, Tehran University of Medical Sciences, Tehran, Iran
[3]Kerman Water and Sewage Company, Kerman, Iran
[4]Environmental Health Engineering Research Center, Kerman University of Medical Sciences, Kerman, Iran
[5]Department of Environmental Health, School of Public Health, Kerman University of Medical Sciences, Kerman, Iran

Abstract
Background: In this study, the effect of wastewater produced by pistachio processing in processing terminals on wastewater treatment plants with activated sludge process was investigated.
Methods: The pilot of activated sludge was constructed at the laboratory scale similar to wastewater treatment plant in Kerman and it was investigated in different ratios of the wastewater of pistachio processing with urban wastewater and return sludge ratios of 50% and 75%. The best reactor efficiency in treating pistachio processing wastewater was obtained at a ratio of 5% with urban wastewater and return sludge of 50% plus 5 and 10% with urban wastewater and return sludge of 75%.
Results: The removal efficiency in the reactor in the aforementioned ratios was obtained to be 96.9%, 98%, and 96.2% for COD, 95.7%, 97.2%, and 93.3% for BOD_5, and 7.1%, 99%, and 96.97% for phenolic compounds, respectively. The removal efficiency of BOD_5 and COD reduced with the increase in the ratios of pistachio processing wastewater especially with the presence of phenolic compounds. The one-way analysis of variance (ANOVA) indicated that in the aforementioned ratios of pistachio processing wastewater with urban wastewater, the mean COD and BOD_5 in the output wastewater were significantly lower than the standards of Iranian environmental organization and guidelines of World Health Organization (WHO).
Conclusion: Activated sludge process is able to remove organic compounds of pistachio processing wastewater at a ratio of 5% and return sludge of 50% along with ratios of 5 and 10% and return sludge of 75% with urban wastewater with a high efficiency.
Keywords: Activated sludge process, Wastewater, Pistachio, Environment

***Correspondence to:**
Mohammad Malakootian
Email: m.malakootian@yahoo.com

Introduction

Iran is the largest exporter of pistachio with production of over 220 000 tons/year. Kerman province is considered as one of the major producers of pistachio and its best type in the world (1).

Pistachio processing in terminals produces annually large amounts of wastes and wastewater resulting from peeling, washing, and separation stages whose characteristics differ from those of urban wastewater (2,3). Research has shown that the green peel of pistachio is rich in phenolic compounds which are a group of planned secondary metabolite (4).

Phenolic compounds have high solubility in water, thus, they can be present in water resources. According to the classification by the United States Environmental Protection Agency (US EPA), phenolic compounds are regarded as primary contaminants which have a cumulative property for living creatures. Therefore, the wastewater of pistachio processing can cause serious problems due to loading high organic input into the receiving environment and wastewater treatment systems (5). Thus, it is necessary that these wastewaters are treated individually at the production site or if mixed with urban wastewater, they are pretreated appropriately (6). If there is no possibility for complete treatment, one can use adaptation method for pistachio wastewater with suitable

ratios, where it is discharged into the urban wastewater collection system such that it develops no problem for the biological treatment procedure. Khademi et al in Iran, investigated the quality and quantity of wastewater of pistachio processing terminals (7).
Electron oxidation process using graphite anode and examined electrocoagulation process using aluminum electrode planes investigated in the treatment of pistachio processing wastewater (8,9).
Furthermore, Bayar et al in Turkey, examined the effect of pH on the pretreatment of pistachio processing wastewater by electro-coagulation process (10).
As no research has been conducted so far on the adaptation of wastewater, the aim of this research was to investigate the possibility of discharging the pistachio processing wastewater into the urban wastewater network of urban wastewater treatment plants through activated sludge method.

Materials and Methods

This is an experimental and applied research, which was conducted in the pistachio harvesting season (September to October) during 2012 to 2014.
First, through field observations, the geographical position of all pistachio processing terminals located in Kerman city (20 units) was specified as census using GARMIN GPS76CSX device. To determine the quality of pistachio processing wastewater, sampling was done in 8 pistachio processing terminals and Samples were specified alphabetically (A-H). The quality of pistachio processing wastewater was determined in terms of BOD_5, COD, TSS, total Phenolic compounds and pH (11).
In the next stage, the quality and efficiency of wastewater treatment plant of Kerman city were determined. To do so, compound sampling was performed on raw input wastewater of Kerman treatment plant and the output wastewater of chlorination basin and the efficiency of the treatment plant was determined. To examine the effect of pistachio processing wastewater on urban wastewater, activated sludge pilot was designed, constructed, and operated at the laboratory scale similar to the wastewater treatment plant of Kerman, Iran (Figure 1).
The flow-diagram of the activated sludge pilot is shown in Figure 1, while the technical specifications of the reactor are provided in Table 1.
Next, the developed pilot was run using raw wastewater of Kerman city with a specific quality.
The efficiency of the activated sludge pilot in treating wastewater of pistachio processing G which had the worst quality, was determined. In the next stage, adaptation of wastewater of pistachio processing G with urban wastewater at return sludge ratios of 50% and 75% in the activated sludge pilot was conducted. This wastewater with different ratios (5%, 10%, 20%, 40%, 60%, and 80%) with Kerman raw wastewater whose quality had been already determined before entrance to the pilot, was examined. Storage tank was added to the activated sludge treatment system. The effect of different ratios of pistachio processing wastewater with urban wastewater on BOD_5, COD, total phenolic compounds of the output wastewater, mixed liquor suspended solids (MLSS) and sludge volume index (SVI) of activated sludge was determined at return

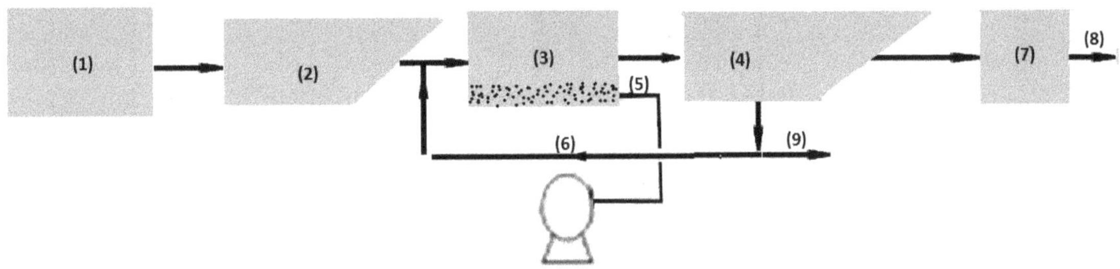

Figure 1. The flow diagram of the activated sludge pilot constructed for the research. 1) The tank for mixing pistachio and urban raw wastewater, 2) the primary sedimentation tank, 3) aeration tank 4) secondary sedimentation tank 5) air supply 6) return sludge 7) chlorination tank, 8) the output treated wastewater and 9) sludge discharge.

Table 1. The technical specifications of the activated sludge reactor design in the research

Unit	Shape	Length (cm)	Width (cm)	Height (cm)	Free Height (cm)	Volume (L)	Hydraulic Retention Time (h)
Primary sedimentation tank	Trapezoid	35	Surface (25) Bottom (10)	7	5	3	2
Aeration tank	Rectangular cube	40	25	10.5	5	10.5	7
Secondary sedimentation tank	Trapezoid	35	Surface (25) Bottom (10)	11	5	6	4

sludge ratios of 50 and 75% in three years, with the results being reported as mean.

The sampling methods and experiments were conducted according to the instruction of the *Standard Methods for the Examination of Water and Wastewater*, 20th edition (11). The concentration of phenolic compounds was determined according to Folin-Ciocalteu method (12). Absorption of the samples was measured by spectrophotometer at 765 nm versus control (12). This method was done for each of the standard solutions of gaelic acid followed by plotting the calibration curve of concentration against absorption. The information obtained from the measured parameters in the output wastewater was compared with Iranian environmental standard for surface and agricultural waters. Data were analyzed by SPSS 16.

Results

The geographical position of the 20 pistachio processing terminals in Kerman city is shown in Figure 2.

Investigation of the status of discharges of the wastewater of pistachio processing terminals in Kerman city indicated that 85% of these terminals discharge their wastewater into absorption wells, 10% to agricultural lands, and 5% do the wastewater treatment incompletely. The distance between the 8 selected pistachio processing terminals and urban wastewater lines was 10 m on average. The mean of the results obtained from determination of COD, BOD_5, total phenolic, total soluble solids (TSS), pH, and flow rate of raw wastewater of the selected pistachio processing terminals within the pistachio operation period is provided in Figure 3.

According to the obtained results, pistachio processing terminal G had the worst quality when compared with other terminals in terms of BOD_5, COD, TSS, total phenols and pH.

Examination of the quality of selected pistachio processing terminals during pistachio harvesting season in three years indicated that the mean concentration of the qualitative parameters of pistachio processing wastewater includes BOD_5= 6106 mg/L, COD = 21570 mg/L, pH = 5.6, TSS= 682 mg/L and phenolic compounds= 4154 mg/L. The high TSS removal efficiency showed an efficient settling process.

The results obtained from investigation of the quality of the input raw wastewater into the Kerman treatment plant and its output wastewater as mean in three years (with a total of 6 samples for every parameter) along with the efficiency of Kerman treatment plant in the mentioned years are provided in Table 2.

According to the obtained results, the qualitative parameters of Kerman wastewater activated sludge

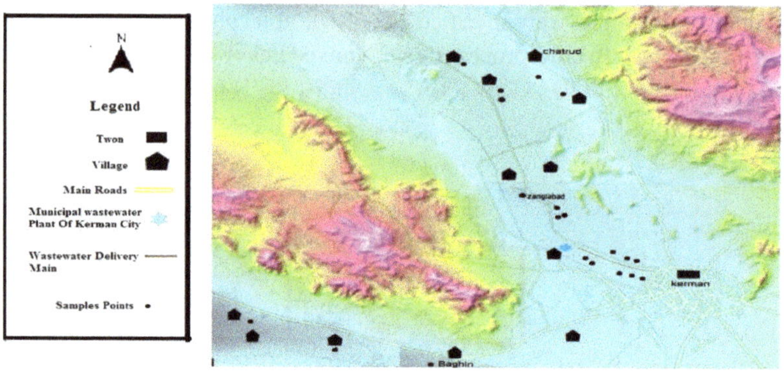

Figure 2. The geographical position of pistachio processing terminals in Kerman city.

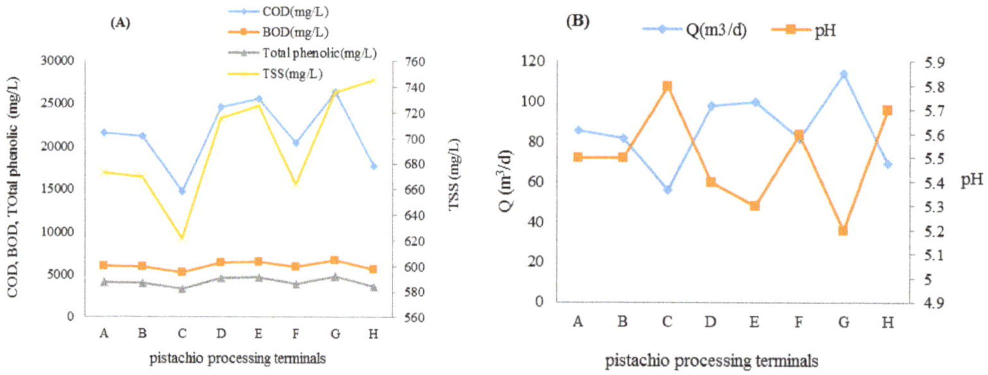

Figure 3. The mean values of BOD_5, COD, TSS, total phenolic compound (A), flow rate, pH (B) of the raw wastewater of pistachio processing terminals in Kerman city.

Table 2. The mean quality of the input raw wastewater and output wastewater of Kerman treatment plant in years 2012, 2013 and 2014 along with the efficiency of Kerman treatment plant

Parameters	BOD (mg/L)	COD (mg/L)	TSS (mg/L)	pH	Phenol (mg/L)
Input raw wastewater	254.1±7.8	434.5±10.1	509.8±4.8	7.4±0.1	>0.1
Output wastewater	42±3.1	68.3±2.7	52.2±3.2	7.5±0.1	>0.1
Efficiency of Kerman treatment plant (%)	83.4	84.3	89.7	-	-

treatment plant in Kerman city is within the standard limits of Iranian environmental standards in the wastewater output for agricultural and irrigation purposes. Further, enhancement of the treatment plant's output qualitative parameters was observed in three years.

The mean values of the results obtained from implementation and operation of the activated sludge pilot with the raw wastewater of Kerman city at return sludge ratios of 50% and 75% in three years in pistachio harvesting season are provided in Tables 3 and 4.

The maximum pilot removal efficiency of BOD_5 and COD during three years, over 7 days of stability for return sludge ratio of 50% and 75% was obtained to be 90.9% and 99.9% along with 92.1% and 92.1%, respectively.

Comparison of the mean values of the results obtained from the efficiency of Kerman city wastewater treatment plant with activated sludge pilot efficiency at return sludge ratios of 50% and 75% in removal of BOD_5 of the input raw wastewater in three years revealed that Kerman wastewater treatment plant with an input BOD of 256.3±3.2 and output of 42.6±2.9 mg/L has BOD removal efficiency of 83.3%. However, activated sludge pilot with the input BOD of 252.3 ± 2.2 mg/L and output of 22.8 ± 1.9 mg/L had a well efficiency of 90.9% with a return sludge ratio of 50% (Table 3). Furthermore, the efficiency of activated sludge pilot with the return sludge ratio of 75%, an input BOD of 260 ± 3.5 mg/L and output of 20.5 ± 1.9 mg/L, was obtained to be 92.1% (Table 4).

Comparison of the mean values of the results obtained from the efficiency of Kerman wastewater treatment plant with the efficiency of the activated sludge pilot at return sludge ratios of 50% and 75% in COD removal of the input raw wastewater in three years showed that Kerman city wastewater treatment plant with input COD of 434.5 ± 26.3 mg/L and output of 68.3 ± 10.3 mg/L had a COD removal efficiency of 84.3%. However, the activated sludge pilot with an input COD of 453.3 ± 3.1 mg/L and output of 39.3 ± 1.9 mg/L, had a removal efficiency of 90.9% with return sludge ratio of 50%. Further, the efficiency of activated sludge pilot with return sludge ratio of 75%, an input COD of 445.3 ± 3.9 mg/L and output of 35.1 ± 2 mg/L was obtained to be 92.1%. Increased

Table 3. The mean values of the results obtained from implementation and operation of activated sludge pilot with the raw wastewater of Kerman city at return sludge ratios of 50% in 3 years in pistachio harvesting season (early September until October)

Day	Q (mL/min)	HRT (h)	Input COD (mg/L)	Sewage COD (mg/L)	Removal efficiency (%)	Input BOD_5 (mg/L)	Sewage BOD_5 (mg/L)	Removal efficiency (%)
1	24	24	438.6±2.46	90.3±2.3	79.4	256±3.2	58.5±2.5	77.1
2	24	24	433±2.6	77.8±2.2	82	252±3.4	49.5±3.2	80.3
3	24	24	441±3	70.2±3	84	256.3±3.2	42.6±2.9	83.3
4	24	24	437±3	60.6±3.3	86.1	254±3	35.7±2.6	85.9
5	24	24	438±2.2	50.7±2.4	88.4	256.6±2.8	28.7±2.4	88.8
6	24	24	433±2.7	42.7±2.3	90.1	253±2.4	25.2±2.3	90
7	24	24	435.3±3.1	39.3±1.9	99.9	252.3±2.2	22.8±1.9	90.9

Table 4. The mean values of the results obtained from implementation and operation of activated sludge pilot with the raw wastewater of Kerman city at return sludge ratios of 75% in 3 years in pistachio harvesting season (early September until October)

Day	Q (mL/min)	HRT (h)	Input COD (mg/L)	Sewage COD (mg/L)	Removal efficiency (%)	Input BOD_5 (mg/L)	Sewage BOD_5 (mg/L)	Removal efficiency (%)
1	24	24	439.6±2.1	81.7±2.8	81.4	255.3±3	52.6±1.8	79.3
2	24	24	437.6±2.6	72±2.5	83.5	255±3.2	44.4±2.4	82.6
3	24	24	432.2±3.4	63.7±2.9	85.2	251±3	37.7±2.7	84.9
4	24	24	435.3±2.5	53.3±3.2	87.7	254±2.6	32±2.2	87.4
5	24	24	436.6±2.5	45.1±1.9	89.6	253.6±3.2	26.1±1.9	89.7
6	24	24	440.3±2	40.3±2.7	90.8	257±2.6	22.9±2.2	91
7	24	24	445.3±2.9	35.1±2	92.1	260±3.5	20.5±1.9	92.1

treatment during acidification was mainly a biological process, thus, the related enzymes played vital roles in the microbial metabolism. Sludge sedimentation is improved due to the release of intracellular polymers under anoxic and anaerobic conditions, which helps to bridge between biomass flocks to improve sedimentation.

According to the results obtained from comparison of efficiency of activated sludge pilot with the efficiency of Kerman wastewater treatment plant, the activated sludge pilot had a higher efficiency in COD, BOD_5 at return sludge ratios of 50% and 75%. In fact, all of the removed pollutants were assimilated and adsorbed on sludge.

The total mean of the results obtained from the investigation of the efficiency of activated sludge pilot in treating pistachio processing raw wastewater (processing terminal of pistachio G) of the two return sludge ratios of 50% and 75% in pistachio harvesting season from early September until October in three years are provided in Table 5.

The results obtained from determining the quality of raw wastewater of Kerman city and processing terminal of pistachio G before entrance to the pilot, have been mentioned as follows.

The mean concentration of BOD, COD, TSS, pH, and phenol of the raw wastewater of Kerman city in pistachio harvesting season in three years for entrance to the pilot was obtained to be 254.1±7.8, 434.5±10.1, 509.8±4.8, 7.4±0.1, and >0.1 mg/L, respectively.

The mean concentration of BOD, COD, TSS, pH, and total phenolic compounds of the processing terminal of pistachio G in three years for entrance to the pilot was obtained to be 6791±440, 26405.3±3266, 736±27, 5.23±0.25, and >4817.3±483 mg/L, respectively.

The mean results obtained from adaptation of the raw wastewater of processing pistachio terminal G with urban wastewater at return sludge ratios of 50% and 75% along with the efficiency of activated sludge pilot in removal of BOD_5, COD, and total phenolic compounds in pistachio harvesting season in three years are provided in Figure 4.

The mean of the results of effect of different ratios of pistachio processing wastewater and urban wastewater on MLSS and SVI of the activated sludge at return sludge ratios of 50 and 75% during three years is shown in Figure 5.

As can be observed in Figure 5, at ratios of 10%, 20%, 40%, 60%, 80%, and 100%, the pistachio processing wastewater with urban wastewater at return sludge ratio of 50%, due to decrease in the microorganisms of mixed liquid of the activated sludge, part of the input wastewater is outflowed as degraded from part of the aeration section of urban activated sludge. The increase in the SVI causes problem with sludge sedimentation. Further, decreased MLSS and

Table 5. The total mean of the results obtained from the investigation of the efficiency of activated sludge pilot in treating pistachio processing raw wastewater (processing terminal of pistachio G) at the two return sludge ratios of 50% and 75% in pistachio harvesting season from early September until October in 3 years

Pistachios raw sewage sludge with a ratio of 100% at return sludge ratios		50%	Efficiency (%)	75%	Efficiency (%)
BOD_5 (mg/L)	Input	6741±2.7	18.3	6747±3.7	21.7
	Output	5507±2		5282±3.3	
COD (mg/L)	Input	26368±3.8	23.7	26379±3.7	25.9
	Output	20110.6±2.8		19530.3±1.9	
Total phenolic (mg/L)	Input	4756±3.3	20	4772±3.9	23.2
	Output	3801.3±2.3		3662.1±2.6	

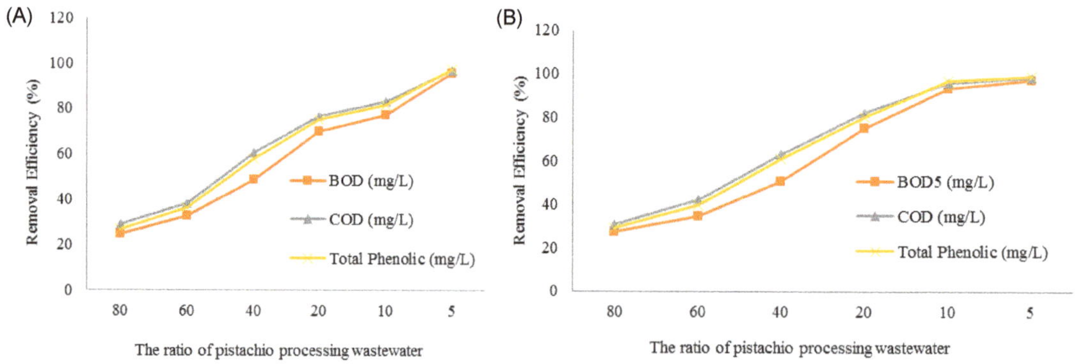

Figure 4. The mean results obtained from the adaptation of the raw wastewater of processing pistachio with urban wastewater at return sludge ratios of 50% (A) and 75% (B) along with the efficiency of activated sludge pilot in removal of BOD_5, COD, total phenolic in pistachio harvesting season in three years.

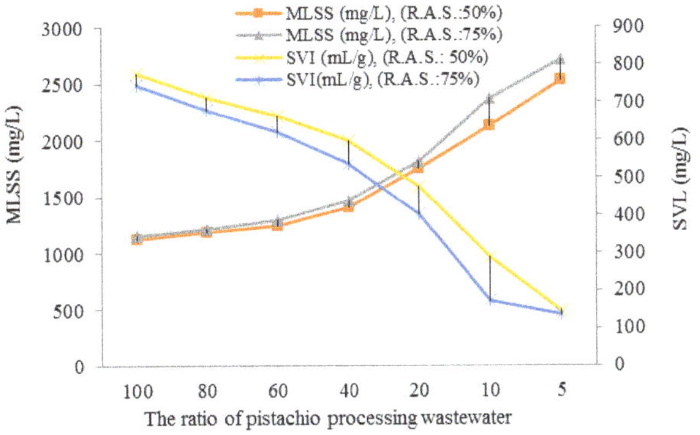

Figure 5. The mean of changes in MLSS and SVI of the activated sludge at different ratios of pistachio processing wastewater at return sludge ratios of 50% and 75% during pistachio harvesting season in 3 years..

sludge sedimentation were observed.

The standards of wastewaters output set by Iranian Environmental Protection Agency (13), US EPA (6), and the guidelines of World Health Organization (WHO) (14) in terms of allowable limits of BOD_5, COD, pH, and TSS, are provided in Table 6.

Discussion

The findings imply that the high ratio of COD and BOD of pistachio processing wastewater with primary contaminant qualitative characteristics (phenolic compounds) had a high toxicity for the environment (15-17).

The removal efficiency of COD and BOD_5 with pistachio processing wastewater ratios of 20%, 10%, 5%, 40%, 60%, 80%, and 100%, respectively, with urban wastewater and return sludge ratios of 50% and 75% revealed that at the ratio of 5% and return sludge of 50% and 75% along with 10% pistachio processing wastewater with the return sludge of 75%, it had a high efficiency in removing the parameters of BOD and COD. The presence of high nitrogen and carbon inorganic compounds in the pistachio terminal sewage system, with 75% sludge return ratio can be further removed. Because more sludge return rates, it increases the availability of organisms to nutrients which represents the production of interstitial organic compounds. On the other hand, the increase in the microbial mass of the system due to an increase in the organic matter and the increase in the cellular residence time with constant intake, will reduce the consumption rate of the substrate and increase the removal efficiency. Sawyer and Anderson in the United States managed to dilute the wastewater of an alcohol production unit up to 1% using urban wastewater collection system and a dripped two-stage filtration system with a return ratio of 1:3 and an organic loading of 0.95 kg BOD/m^3/d, they reduced the value of BOD from 485 mg/L to 25 mg/L (18). Burnett in the United States diluted the wastewater of an alcohol production unit with a COD of 90 000 mg/L and BOD of 25 000-300 000 mg/L up to 10% using urban wastewater and following neutralization, using a dripped filter with an organic loading of 4.8-6.1 kg COD/m^3/d and return ratio of 0.5, and achieved a COD removal of 40% (19). Costa Reis and Sant'Anna in Brazil removed COD from the wastewater of an alcohol unit diluted by up to 10% in a floating bed reactor with a retention time of 10-15 hours by up to 80% (20). The effect of dilution on wastewater in these studies showed congruence with pistachio processing wastewater research.

Galaieri et al in Tehran, evaluated the volume parameters of inoculation of the activated sludge, pH, and initial concentration in removing wastewater's phenol using discontinuous activated sludge system. They indicated that an aqueous solution containing 600 ppm phenol following 35 hours and 5 mL of inoculation volume of the activated sludge with a pH of 7, had the greatest phenol removal (21).

The best state for sludge sedimentation occurred at 5% of pistachio processing wastewater in the pilot of urban activated sludge with a return sludge ratio of 50%. With increase of the return sludge ratio from 50% to 75%, improvement of the performance of mixed liquid microorganisms and sedimentation of the sludge were observed. In this state, the ratio of pistachio processing wastewater treatment enhanced from 5% to 10%.

Nikazar et al in Tehran, compared the performance of sequencing batch reactor (SBR) for removing phenol at different concentrations. The greatest phenol removal efficiency was obtained to be 99.99% and the highest value of MLSS was 3801 mg/L in the reactor with a concentration of 150 mg/L (22). Elevation of the removal efficiency at the optimal concentration in these studies is in line with the present study.

At higher ratios of pistachio processing wastewater, the majority of microorganisms present in the reactor enter death phase, where the mentioned parameters of the

Table 6. The standard of Iranian Environmental Protection Agency, US Environmental Protection Agency and the guidelines of World Health Organization for the output of wastewaters

Entry	Parameter	Iranian Environmental Protection Agency (13)		Guidelines of WHO (14)	US EPA) (6)
		Discharges to surface water	Agriculture and irrigation use		
1	BOD_5 (mg/L)	30	100	100	30
2	COD (mg/L)	60	200	200	120
3	TSS (mg/L)	40	100	100	5
4	pH	6.5 - 8.5	6.5 - 8.5	6.5 - 8.5	6.5 - 8.4

output wastewater along with MLSS and SVI parameters increase and move towards unfavorable values. Further, the output wastewater of the reactor was not able to meet the standards of Iranian EPA along with the guidelines of WHO for discharging into surface waters and agricultural lands (13,14).

Takdastan et al in Tabriz, investigated the degree of degradation of para nitrophenol using aerobic biological method in an activated sludge batch reactor, such that by the injection of 100 mg/L para nitrophenol into the reactor, COD reduced from 772 to 193 mg/L, and the COD removal efficiency reached 75%. MLSS concentration increased slightly from 1290 to 1446 mg/L. The SOUR value was obtained to be 31 mgO_2/h.gVSS, SVI reduced to below 48 mL/g and the concentration of para nitrophenol in the output wastewater and the sludge were obtained as 63 and 0.09 mg/L, respectively. With injection of 150 mg/L para nitrophenol into the reactor, COD and the concentration of para nitrophenol of the output wastewater increased and the biological treatment system was disrupted (23). Shoukohi et al indicated that the Pearson coefficient between MLSS concentration and the phenol removal efficiency has a direct relationship, where its correlation has been significant with 100% confidence (24). According to their investigation, the effect of elevation of concentration on the reactor efficiency is in accordance with this study.

Conclusion

If properly managed and guided, activated sludge process is able to remove organic compounds from pistachio processing wastewater at a ratio of 5% and return sludge of 50% along with ratios of 5% and 10% and return sludge of 75% with urban wastewater with a high efficiency. However, at higher ratios, it does not have the required efficiency and the pistachio processing wastewater should be treated before being discharged into the urban wastewater system. Discharge of the wastewater of pistachio processing terminals without specific treatment or without using adaptation method to the urban wastewater by the activated sludge method is prohibited. It is suggested that at pistachio processing terminals close to wastewater collection system, provided that the aim is to use urban wastewater system, pistachio processing wastewater should be discharged through adaptation method with the city's wastewater collection system. In cases where there is no urban wastewater collection system around the pistachio processing terminal, independent wastewater treatment method is recommended. It is better to treat these wastewaters in accumulation complexes and through either biological or chemical methods.

Acknowledgments

This research is the result of a research plan which was financially supported by Water and Wastewater Engineering Company of Kerman province, and could not be realized without the assistance of this company. Hereby, the cooperation of the mentioned company is highly appreciated. The authors would like to gratitude Gholamabas Nekounam, MSc. of Health Faculty at Kerman University of Medical Sciences for his support and assistance.

Competing interests

The authors declare that they have no competing interests.

Authors' contributions

FK carried out the experiments. MM supervised the project, developed the theory and performed the computations. KY assisted to write the manuscript and analyze the results. AN assisted to write the paper. MT and MAH approved this project and assisted to fund of the project by Water and Wastewater Engineering Company of Kerman province.

References

1. Mehrnejad MR, Javanshah A. The Strategic Framework for Developing and Prooting Pistachio Research in Iran. Tehran: Jomhori; 2010. [In Persian].
2. Metcalf L, Eddy M, Tchobanoglous G, Burton FL, Stensel HD. Wastewater Engineering. 4th ed. New York: McGraw Hill Higher Education; 2003.
3. Eckenfelder WW Jr. Industrial Water Pollution Control. 3rd ed. New York: McGraw-Hill Companies; 2000.
4. Rajaei A, Barzegar M, Sahari MA. Investigation on antioxidative and antimicrobial activities of pistachio (pistachio vera) green hull extracts. Food Science and Technology 2011; 8(1): 111-20. [In Persian].
5. Goli AH, Barzegar M, Sahari MA. Antioxidant activity and total phenolic compounds of pistachio (*Pistachia vera*) hull extracts. Food Chem 2005; 92(3): 521-5. doi: 10.1016/j.foodchem.2004.08.020.
6. Norwegian Environment Agency. List of Priority Substances [cited 2018 Jun 1]. Available from: http://www.miljostatus.no/contentassets/37ad8a3f9ec343c29d72cef1b215560f/printpageandchildren.pdf.
7. Khademi F, Yaghmaeian K, Taheri M, Hayatabadi MA, Malakootian M. Qualitative and quantitative assessment of wastewater of pistachio processing terminals (case study: Kerman city). Toloo-e-Behdasht 2016; 14(6): 175-67. [In Persian].
8. Fıl BA, Boncukcuoglu R, Yilmaz AE, Bayar S. Electro-oxidation of pistachio processing industrial wastewater using graphite anode. Clean Soil Air Water 2014; 42(9): 1232-8. doi: 10.1002/clen.201300560.
9. Bayar S, Boncukcuoglu R, Yilmaz AE, Fil BA. Pre-treatment of pistachio processing industry wastewaters (PPIW) by

electrocoagulation using Al plate electrode. Sep Sci Technol 2014; 49(7): 1008-18. doi: 10.1080/01496395.2013.878847.
10. Bayar S, Yilmaz AE, Koksal Z, Boncukcuoglu R, Fil BA, Yilmaz MT. The effect of initial pH on pistachio processing industrial wastewater pre-treatment by electro coagulation method. International Conference on Sustainable Energy and Environmental Engineering; 2015 Oct 25-26; Bangkok.
11. American Public Health Association (APHA). Standard Methods for the Examination of Water & Wastewater. 21th ed. Washington DC USA: APHA; 2005.
12. Waterhouse AL. Current Protocols in food Analytical Chemistry: Determination of Total Phenolic. New York: John Wiley and Sons; 2002.
13. Department of Environment. Environmental standard. [cited 2018 Jun 2]. Available from: http://www.environment-lab.ir/environmental-standards. [In Persian].
14. WHO Scientific Group on Health Aspects of Use of Treated Wastewater for Agriculture and Aquaculture & World Health Organization. Health guidelines for the use of wastewater in agriculture and aquaculture: report of a WHO scientific group [meeting held in Geneva from 18 to 23 November 1987]. Geneva: World Health Organization; 1989.
15. Chen H, Yao J, Wang F, Zhou Y, Chen K, Zhuang R, et al. Toxicity of three phenolic compounds and their mixtures on the gram-positive bacteria Bacillus subtilis in the aquatic environment. Sci Total Environ 2010; 408(5): 1043-9. doi: 10.1016/j.scitotenv.2009.11.051.
16. Adeboye PT, Bettiga M, Olsson L. The chemical nature of phenolic compounds determines their toxicity and induces distinct physiological responses in Saccharomyces cerevisiae in lignocellulose hydrolysates. AMB Express 2014; 4: 46. doi: 10.1186/s13568-014-0046-7.
17. Machrafi Y, Prevost D, Beauchamp CJ. Toxicity of phenolic compounds extracted from bark residues of different ages. J Chem Ecol 2006; 32(12): 2595-615. doi: 10.1007/s10886-006-9157-1.
18. Sawyer CN, Anderson EJ. Anaerobic treatment of rum wastes. Water Sew Works 1949;96(3):112-4.
19. Burnett WE. Run distillery wastes- laboratory studies on aerobic treatment. Water Sewage Works 1973; 120: 107-11.
20. Costa Reis LG, Sant'Anna GL Jr. Aerobic treatment of concentrated wastewater in a submerged bed reactor. Water Res 1985; 19(11): 1341-5. doi: 10.1016/0043-1354(85)90298-2.
21. Galaieri H, Dollati F, Marandi R. Evaluation the parameters effect on phenol removal of artificial effluent using activated sludge. 6th National Conference & Exhibition on Environmental Engineering; 2012 Nov 17-21; Environment Faculty of Tehran University, Tehran; 2012. [In Persian].
22. Nikazar M, Solimani E, Alavi MR. Comparison the operation sequential batch reactors (SBR) for phenol removal in different concentrations. Second Environmental Engineering conference; 2008 May 20-21; Environment Faculty of Tehran University; Tehran; 2008. [In Persian].
23. Takdastan A, Mehrdadi N, Eslami A. Survey the rate biodegradable of Para nitrophenol by aerobic biological method in activated sludge batch reactor. Sixteenth National Conference on Environmental Health; 2013 Oct 1-3; Tabriz University of Medical Sciences, Tabriz; 2013. [In Persian].
24. Shoukohi R, Movahedian H, Jonidi-Jafari A, Hajian M. Survey the relationship between concentration MLSS, MLVSS and phenol removal efficiency in BF/AS system. 10th National Conference on Environmental Health; 2007 Oct-Nov 30-1; Hamadan University of Medical Sciences, Hamedan; 2007. [In Persian].

21

Assessing some heavy metals pollutions in sediments of the northern Persian Gulf (Bushehr province)

Mehdi Bibak[1*], Masoud Sattari[1,2], Ali Agharokh[3], Saeid Tahmasebi[4], Javid Imanpour Namin[1]

[1]Department of Fisheries, School of Natural Resources, University of Guilan, Sowmehsara, Iran
[2]Department of Marine Sciences, Caspian Sea Basin Research Center, University of Guilan, Sowmehsara, Iran
[3]Research Center for Persian Gulf, School of Persian Gulf, University of Persian Gulf, Bushehr, Iran
[4]Department of Statistics, School of Basic Sciences, University of Persian Gulf, Bushehr, Iran

Abstract
Background: Land and water pollution by heavy metals is a universal issue. Although the pollution affects all countries, but its range and severity vary hugely. The pollution of the marine environment by heavy metals is a worldwide problem. Marine sediments can be sensitive indicators for monitoring contaminants in aquatic environments.
Methods: The concentration of 10 elements (As, Cd, Cr, Cu, Al, Fe, Ni, Pb, Sb, and Zn) was determined in the sediments of four shoreline stations including Imam Hassan port, Ameri port, Bushehr port, and Nayband Bay at the west Persian Gulf from March to December 2017. The elements were measured by inductively coupled plasma mass spectrometry (ICP-MS). Data were analyzed using SPSS version 16.
Results: The contamination of the sediments was assessed based on the geoaccumulation index (Igeo) and enrichment factor (EF). Spearman correlation matrix was calculated between all the trace metals and major elements as well as corresponding sampling regions. Statistically significant inter-elemental correlations (e.g., Cr-Fe, Cr-Al, Cr-Ni, Cr-Zn, and Cr-Cu) were found between some metals. High EF levels for Fe, Al, and Pb suggest that metals in the sediments of the northern Persian Gulf could have originated from anthropogenic sources.
Conclusion: The contamination pattern of sediments is affected by factors such as sedimentation patterns, physical and chemical properties of the sediments. For example, sediments with fine-grained and high surface area-to-volume ratio can act as good absorbents for many pollutants.
Keywords: Heavy metals, Geologic sediment, Persian Gulf, Bushehr

***Correspondence to:**
Mehdi Bibak
Email:
mehdi.bibak65@yahoo.com

Introduction

The pollution of marine environment by heavy metals is a worldwide problem. Although the pollution affects all countries, but its range and severity vary hugely. In this regard, marine sediments can be sensitive indicators for monitoring contaminants in aquatic environments (1,2). The sediments are used as a stock for heavy metals and thus, need special investigation in projecting aquatic pollution research. Accumulation of heavy metals happen in upsaid sediments in aquatic environments by biological mechanisms. Sediment is a habitat and major nutrient source for aquatic organisms. These processes become toxic to sediment-dwelling organisms, resulting in death, reduced growth, reproductive impairment, and prevent species diversity (3,4). Heavy metals pollute marine sediments and other aquatic ecosystems, which happen due to industrialization (5,6).

Great quantities of pollutants, directly or indirectly, are discharged into the marine environment every day. In this regard, heavy metals are considered as serious pollutants of the aquatic environments due to their accumulative behavior (7,8). Heavy metals are one of the constituents that affect the marine ecosystems. Potential toxicity of heavy metals in an aquatic ecosystem is determined by their chemical form. Some elements are necessary for human health in trace amounts (Co, Cu, Cr, and Ni), while others are carcinogenic or toxic, mainly affecting the central nervous system (Hg, Pb, and As), the kidneys or liver (Hg, Pb, Cd, and Cu), or skin, bones and teeth (Ni, Cd, Cu, and Cr) (9). Heavy metals that have been recognized in the polluted environment include As, Cu, Cd, Pb, Cr, Ni, Hg, and Zn. The toxicity of trace metals

arises from their interference with an organism's uptake of essential metal ions such as sodium and calcium. For example, cadmium and zinc block the uptake of calcium ions, which are important for bone and teeth growth.
The interaction of some heavy metals with enzymes and their tendency to bind to the protein and other biological tissues also cause trace metal poisoning in organisms (10,11).
The Persian Gulf is a relatively shallow semi-enclosed sea with poor flushing characteristics. Thus, pollution discharges into the sea are diluted more slowly than into open marine ecosystems (12). The Persian Gulf has been facing many environmental challenges like biodiversity loss, industrial pollution, and harmful effluents (13). Additionally, environmental incidents, ship traffic, transportation, oil pollutants, and oil spills can cause different pollutants including heavy metals (14,15). Iran has the longest water joint border in the Persian Gulf and it has a way through the Strait of Hormuz and Makran Sea (Oman) to the Indian Ocean from the east. It leads to Arvand River in Khuzestan province from the west, which is composed of two Tigris and Euphrates rivers in Iraq as well as Karun River attachment in Iran (16).
The aim of this study was to determine the concentration of 10 metals (As, Cu, Cd, Pb, Cr, Ni, Zn, Fe, Al, and Sb) in sediments of the Persian Gulf.

Materials and Methods

Bushehr is located at 28° 55′ 19.84″ N and 50° 50′ 4.76″ E in southwestern Iran and around the northern of the Persian Gulf. In this study, 12 sampling sites in four regions of Imam Hassan port, Bushehr port, Ameri port, and Nayband Bay (Figure 1) were selected during March to December 2017.
Samples were obtained from the upper layer (5 cm depth) of sediments. Ten square meters of each area was sampled. The sediments were collected from the intertidal areas once low tides. Sediments were transferred to polyethylene containers and transported to the laboratory at 4°C. The samples were dried at 105°C for at least 24 hours (17). The samples were sieved through a 63 micron stainless steel sieve and agitated vigorously to produce homogeneity (18). One gram of each sample was digested in a mixture of concentrated HNO3 (65%) and HClO4 (60%) in the ratio of 4:1, first, at a low temperature (40°C) for 1 hour and then, at a high temperature (140°C) for 4 hours (19). The digested samples were then diluted to a certain capacity (25 mL) with double distilled water (DDW). The sample was then filtered through Whatman No. 40 filter paper. The heavy metals concentrations were determined by inductively coupled plasma mass spectrometry (ICP-MS) (scheme code was MMS-01: Detection limited were in ppm) in Zarazma Co.

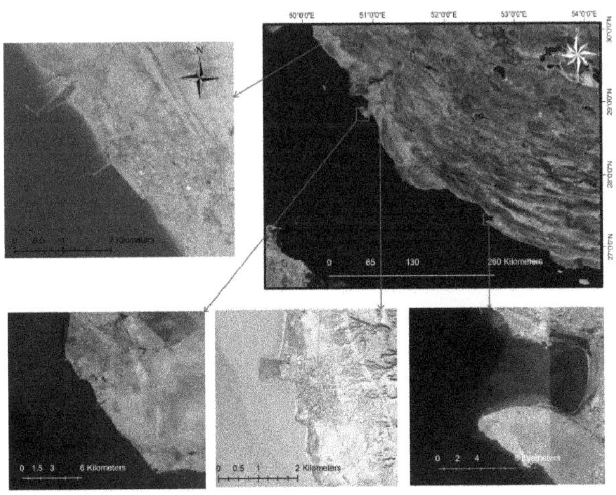

Figure 1. Sampling areas.

The geoaccumulation index (Igeo) is used to determine the level of contamination in sediments (20). For this purpose, the current concentrations is compared to pre-industrial levels using the following equation:

$$I_{geo} = \log_2 C_n / 1.5B_n$$

where Cn is the measured concentration of the element n in the pelitic sediment fraction (< 2 Am) and Bn is the geochemical background value in fossil argillaceous sediment (average shale). The quality of the sediments is divided into six classes by Igeo: unpolluted (Igeo < 0), unpolluted to moderately polluted (0 < Igeo < 1), moderately polluted (1 < Igeo < 2), moderately to strongly polluted (2 < Igeo < 3), strongly polluted (3 < Igeo < 4), strongly to extremely polluted (4 < Igeo < 5), and extremely polluted (5 < Igeo). Concentrations of the geochemical background are increased each time by the constant 1.5 in order to allow value fluctuations of a given substance in the environment as well as very small anthropogenic influences (21).
The enrichment factor (EF) is based on the standardization of a tested element against a reference one. According to contamination standards based on the EF, the following classes are obtained: EF<2 (deficiency to minimal enrichment), 2 ≤ EF < 5 (moderate enrichment), 5 ≤ EF < 20 (significant enrichment), 20 ≤ EF<40 (very high enrichment), and EF ≥ 40 (extremely high enrichment) (22). A reference element is the one characterized by low-frequency current variability. The most common reference elements are Sc, Mn, Ti, Al, and Fe (23). The EF is calculated using the following expression:

$$EF = [(Me)_s / (Fe)_s] / [(Me)_b / (Fe)_b]$$

where $(Me)_s$ and $(Me)_b$ are the concentrations of metal and normalizer, respectively. Also, $(Fe)_s$ and $(Fe)_b$ denote the sample and background sediments, respectively. Data were analyzed using SPSS version 16.

Results

Table 1 shows the concentrations of heavy metals in sediment samples collected from intertidal zone from four regions in the Bushehr province.
The obtained Igeo values for these four regions are shown

Table 1. Mean concentrations of each metal analyzed (mg/kg)

Metals	Region			
	Ameri Port	Bushehr Port	Imam Hassan Port	Nayband Bay
As	16.24±8.76	5.03±0.15	3.31±2.65	1.55±0.49
Cr	16.82±3.05	14.29±7.79	10.96±3.30	10.19±5.97
Cu	5.5±0.88	5.5±2.83	3.75±0.49	3.45±0.77
Pb	12.27±4.59	2.77±0.38	3.03±0.02	6.35±4.30
Zn	14.20±1.41	11.45±8.13	8.45±1.06	4.75±0.78
Al	3980±462.4	4189±5 3143.09	3828.5±688.01	695±173.2
Fe	8420±1824.3	5360±3309.26	4141±642.05	773.5±193.04
Ni	18.07±2	13.4±6.36	11.8±3.25	8.19±2.25
Sb	0.45±0.07	0.35±0.21	0.35±0.07	0.55±0.07
Cd	0.07±0.04	0.18±0.01	0.10±0.04	BDL*

*BDL: Below detection limit.

in Table 2.

Spearman correlation matrix was calculated between all the trace metals (and also major elements) and corresponding sampling regions (Table 3).

EF was used as an index to evaluate anthropogenic influences of heavy metals in sediments. When $0.5 \leq EF \leq 1.5$, the trace metals might have originated entirely from crustal materials or natural weathering processes, while when EF>1.5 suggested that a significant portion of trace metals is provided by other sources (24).

Discussion

In this study, sediment pollution was assessed using geoaccumulation index (Igeo) and EF. As can be seen in Table 1, metal levels in all areas of the sample sites in terms of frequency were Fe> Al> Cr> Ni> Zn> As> Pb> Cu> Sb> Cd. The relatively high concentrations of heavy metals during winter coincide principally with decreasing rate of organic matter decomposition, due to a low water temperature (25). The highest concentrations of Fe and Al were found in Ameri port while the lowest values of Cd were reported in Nayband Bay. Fe, Al, and Pb levels were found to be higher than the other metals in the four regions (Table 2). Approximately, 42% of the samples were included in Class 0. A small percentage of the samples was included in Class 2, 3, and 4 for Al, Fe, Cd, and Pb. There were significant inter-elemental correlations (e.g., Cr-Fe, Cr-Al, Cr-Ni, Cr-Zn, and Cr-Cu). Inverse correlations

Table 2. Index of geoaccumulation for sampling regions

Region	Class								
	Fe	Al	Cd	Pb	Ni	As	Zn	Cu	Sb
Imam Hassan	0	2	0	0	0	1	0	0	1
Bushehr	3	3	2	2	0	1	0	0	1
Ameri	4	3	0	3	0	1	0	0	1
Nayband	3	2	0	2	0	0	0	0	1

Table 3. The Spearman correlation matrix between element levels in the sediment of the Persian Gulf

	Cr	Al	Fe	Ni	As	Cu	Pb	Zn	Sb	Cd
Cr	1									
Al	.934	1								
Fe	1.000**	.927	1							
Ni	.997**	.901	.998**	1						
As	.857	.639	.865	.891	1					
Cu	.947	.999**	.940	.917	.667	1				
Pb	.131	-.233	.150	.213	.557	-.195	1			
Zn	.996**	.961*	.995**	.986**	.816	.971	.046	1		
Sb	-.724	-.923	-.711	-.664	-.310	-.907	.589	-.780	1	
Cd	.317	.570	.303	.253	.072	.546	-.715	.381	-.756	1

*Correlation is significant at the 0.05 (2-tailed); ** Correlation is significant at the 0.01 (2-tailed).

Table 4. Enrichment factor in sampling regions

Metals Region	As	Cr	Cu	Pb	Zn	Al	Fe	Ni	Sb	Cd
Imam Hassan	EF<2	EF<2	EF<2	EF<2	EF<2	EF<2	EF<2	EF<2	EF<2	EF<2
Bushehr port	EF<2	EF<2	EF<2	EF<2	EF<2	2<EF<5	2<EF<5	EF<2	EF<2	EF<2
Ameri port	EF<2	EF<2	EF<2	2<EF<5	EF<2	2<EF<5	2<EF<5	EF<2	EF<2	EF<2
Nayband Bay	EF<2	EF<2	EF<2	EF<2	EF<2	EF<2	EF<2	EF<2	EF<2	EF<2

were observed between Sb and Cd with almost all other elements, suggesting a positive strong significant correlation among Cr, Al, Fe, Ni, and Cu, which were clustered in the first class. Sb had a negative correlation with other elements in the first class. High EF level for Fe, Al, and Pb demonstrated that the levels of metals in sediments of the Persian Gulf could have originated from anthropogenic sources (Table 4). According to these results, the presence of these metals in the same groups might reflect a similar behavior or suggest common bio-originated sources.

The highest mean concentrations of Pb (12.27 mg/kg), Ni (18.07 mg/kg) and As (16.24 mg/kg) were reported in the Ameri port. This concentration was similar to the results of Ganugapenta et al (26) and Zhuang & Gao (27) but lower than that of El Tokhi et al (21). The pollution in the Ameri port was higher than other ports, due to the increase in the amount of agricultural wastewater, industrial wastewater, oil pollution and shipping in this port compared to the other ports.

Conclusion

Oil extraction and recent regional wars in the Persian Gulf have released millions of oil barrels, including detectable trace metals, to the sea. Seawater desalination plants in the Persian Gulf with a capacity of millions cubic meters freshwater per day are another sources of heavy metal pollution in the area. In addition to pollutants, the contamination pattern of sediments is affected by factors such as sedimentation patterns, physical and chemical properties of the sediments. For example, sediments with fine-grained and high surface area-to-volume ratio can act as good absorbents for many pollutants.

Acknowledgements

The authors wish to thank all who assisted in conducting this research.

Competing interests

The authors declared no competing interests.

Authors' contributions

All authors contributed in all aspects of this research like performing surveys and experiments, preparing the figures and tables, designing the experiments, analyzing the data graphical elaborations, writing the manuscript, and providing critical revision of the paper.

References

1. Pekey H, Karakas D, Ayberk S, Tolun L, Bakoglu M. Ecological risk assessment using trace elements from surface sediments of Izmit Bay (Northeastern Marmara Sea) Turkey. Mar Pollut Bull 2004; 48(9-10): 946-53. doi: 10.1016/j.marpolbul.2003.11.023.
2. Sinem Atgin R, El-Agha O, Zararsız A, Kocataş A, Parlak H, Tuncel G. Investigation of the sediment pollution in Izmir Bay: trace elements. Spectrochim Acta Part B At Spectrosc 2000; 55(7): 1151-64. doi: 10.1016/S0584-8547(00)00231-7.
3. Rubio B, Nombela MA, Vilas F. Geochemistry of major and trace elements in sediments of the Ria de Vigo (NW Spain) an assessment of metal pollution. Marine Pollution Bulletin 2000; 40(11): 968-80.
4. Reimann C, de Caritat P. Intrinsic flaws of element enrichment factors (EFs) in environmental geochemistry. Environ Sci Techno 2000; 134(24): 5084-91.
5. Yang H, Rose NL. Distribution of mercury in six lake sediment cores across the UK. Sci Total Environ 2003; 304(1-3): 391-404. doi: 10.1016/s0048-9697(02)00584-3.
6. Capillo G, Silvestro S, Sanfilippo M, Fiorino E, Giangrosso G, Ferrantelli V, et al. Assessment of electrolytes and metals profile of the Faro Lake (Capo Peloro Lagoon, Sicily, Italy) and its impact on *Mytilus galloprovincialis*. Chem Biodivers 2018; 15(5): e1800044. doi: 10.1002/cbdv.201800044.
7. Abdolahpur Monikh F, Maryamabadi A, Savari A, Ghanemi K. Heavy metals' concentration in sediment, shrimp and two fish species from the northwest Persian Gulf. Toxicol Ind Health 2015; 31(6): 554-65. doi: 10.1177/0748233713475498.
8. Forouhar Vajargah M, Mohamadi Yalsuyi A, Hedayati A, Faggio C. Histopathological lesions and toxicity in common carp (Cyprinus carpio L. 1758) induced by copper nanoparticles. Microsc Res Tech 2018; 81(7): 724-9. doi: 10.1002/jemt.23028.
9. Olubunmi FE, Olorunsola OE. Evaluation of the status of heavy metal pollution of sediment of Agbabu bitumen deposit area, Nigeria. Eur J Sci Res 2010; 41(3): 373-82.
10. Torre A, Trischitta F, Faggio C. Effect of CdCl2 on regulatory volume decrease (RVD) in *Mytilus galloprovincialis* digestive cells. Toxicol In Vitro 2013; 27(4): 1260-6. doi:

10.1016/j.tiv.2013.02.017.
11. Savorelli F, Manfra L, Croppo M, Tornambe A, Palazzi D, Canepa S, et al. Fitness Evaluation of Ruditapes philippinarum Exposed to Ni. Biol Trace Elem Res 2017; 177(2): 384-93. doi: 10.1007/s12011-016-0885-y.
12. Burgos-Aceves MA, Cohen A, Smith Y, Faggio C. MicroRNAs and their role on fish oxidative stress during xenobiotic environmental exposures. Ecotoxicol Environ Saf 2018; 148: 995-1000. doi: 10.1016/j.ecoenv.2017.12.001.
13. Fernandez JM, Hayes PJ, Smith MR. Comprehensive Analytical Chemistry. Elsevier; 1992. p. 345-47.
14. Khoshnood Z, Mokhlesi A, khshnood R. Bioaccumulation of some heavy metals and histopathological alterations in liver of Euryglossa orientalis and Psettodes erumei along North Coast of the Persian Gulf. Afr J Biotechnol 2010; 9(41): 6966-72. doi: 10.5897/AJB10.944.
15. Sabzalizadeh S, Dehghan Madise S. Assessment of heavy metals pollution in the sediments of Lifeh-Busaf (northwest of Persian Gulf) using Geo-accumulation Index. Iranian Scientific Fisheries Journal 2010; 19(3): 51-60. [In Persian].
16. Pourang N, Nikouyan A, Dennis JH. Trace element concentrations in fish, surficial sediments and water from northern part of the Persian Gulf. Environ Monit Assess 2005; 109(1-3): 293-316. doi: 10.1007/s10661-005-6287-9.
17. Tatina M, Oryan S, Gharibkhani M. Surveying the amount of heavy metals (Ni, Pb,Cd & V) accumulation derived from oil pollution on the muscle tissue of Pelates quadrilineatus from the Persian Gulf. Mar Biol 2009; 1(1): 28-39. [In Persian].
18. Dehghan Chenari A, Lak R. Effects of heavy metals from petroleum pollutions on carbonate surface sediments of the Persian Gulf. Int J Environ Res 2014; 8(2): 297-304. doi: 10.22059/ijer.2014.720.
19. Ismail A, Badri MA, Ramlan MN. The background levels of heavy metal concentration in sediments of the west coast of Peninsular Malaysia. Sci Total Environ 1993; 134 (Suppl 1): 315-23. doi: 10.1016/S0048-9697(05)80032-4.
20. Muller G. Index of geoaccumulation in sediments of the Rhine River. GeoJournal 1969; 2: 108-18.
21. El Tokhi M, Amin BM, Alaabed SA. Environmental assessment of heavy metals contamination of bottom sediments of Oman Gulf, United Arab Emirates. J Pollut Eff Cont 2017; 5(4): 203. doi: 10.4172/2375-4397.1000203.
22. Tanner PA, Leong LS, Pan SM. Contamination of heavy metals in marine sediment cores from Victoria Harbour, Hong Kong. Marine Pollution Bulletin 2000; 40(9): 769-79.
23. Zhuang W, Gao XL.Acid-volatile sulfide and simultaneously extracted metals in surface sediments of the southwestern coastal Laizhou Bay, Bohai Sea: concentrations, spatial distributions and the indication of heavy metal pollution status. Mar Pollut Bull 2013; 76(1-2): 128-38.
24. Taylor SR, McLennan SM. The geochemical evolution of the continental crust. Rev Geophys 1995; 33(2): 241-65. doi: 10.1029/95RG00262.
25. Bazzi AO. Heavy metals in seawater, sediments and marine organisms in the Gulf of Chabahar, Oman Sea. J Oceanogr Mar Sci 2014; 5(3): 20-9. doi: 10.5897/JOMS2014.0110.
26. Ganugapenta S, Nadimikeri J, Chinnapolla SRRB, Ballari L, Madiga R, Nirmala K, et al. Assessment of heavy metal pollution from the sediment of Tupilipalem Coast, southeast coast of India. International Journal of Sediment Research 2018; 33(3): 294-302. doi: 10.1016/j.ijsrc.2018.02.004.
27. Zhuang W, Gao X. Integrated assessment of heavy metal pollution in the surface sediments of the Laizhou Bay and the coastal waters of the Zhangzi Island, China: comparison among typical marine sediment quality indices. PLoS One 2014; 9(4): e94145. doi: 10.1371/journal.pone.0094145.

Synthesis of 2,4-dinitrophenylhydrazine loaded sodium dodecyl sulfate-coated magnetite nanoparticles for adsorption of Hg(II) ions from an aqueous solution

Soheil Sobhanardakani[1*], Raziyeh Zandipak[2], Lobat Taghavi[3]

[1]Associate Professor, Department of the Environment, College of Basic Sciences, Hamedan Branch, Islamic Azad University, Hamedan, Iran
[2]Msc, Young Researchers & Elite Club, Hamedan Branch, Islamic Azad University, Hamedan, Iran
[3]Assistant Professor, Department of the Environmental Pollution, College of Environment and Energy, Science and Research Branch, Islamic Azad University, Tehran, Iran

Abstract
Background: The rapid increase in agricultural and industrial development has made heavy metal pollution a serious environmental problem and public health threat; therefore, removal of heavy metals from water is important. The current study prepared DNPH@SDS@Fe_3O_4 nanoparticles as a novel and effective adsorbent for removal of Hg(II) ions from an aqueous solution.
Methods: A selective adsorbent for Hg(II) was synthesized by coating Fe_3O_4 nanoparticles with sodium dodecyl sulfate which was further functionalized with 2,4-dinitrophenylhydrazine (2,4-DNPH). The synthesized nanoparticles were characterized by Fourier transform infrared spectroscopy (FTIR), x-ray diffraction (XRD), scanning electron microscopy (SEM) and SEM–EDXSt. The effects of pH, dose of adsorbent and shaking time on adsorption capacity were investigated. The kinetics and equilibrium of adsorption of the metal ions were thoroughly studied.
Results: SEM showed that the size of the nanoparticles was 20 to 35 nm. The maximum adsorption capacity for Hg(II) was 164.0 mg g^{-1} for an adsorbent dose of 0.04 g at pH 7.0, 25°C and the initial metal concentration was 25 mg L^{-1},which was greater than for most adsorbents previously examined for Hg(II) adsorption. Adsorption experimental data showed good correlation with the pseudo-second-order model and Langmuir isotherm model.
Conclusion: The results indicated that the DNPH@SDS@Fe_3O_4 nanoparticles are an efficient adsorbent for removal of heavy metal from wastewater.
Keywords: Adsorption, Mercury, Magnetite nanoparticles, Wastewater, Kinetics

***Correspondence to:**
Soheil Sobhanardakani
Email: s_sobhan@iauh.ac.ir

Introduction

Recent industrial activity has increased the presence of hazardous heavy metals such as mercury (Hg) ions in wastewater (1). Hg (II) is hazardous to health and the environment and is extreme toxic, even at low levels. This element biomagnifies in the food chain and bioaccumulates in organisms (2). It exists in the environment in the form of inorganic mercury as metallic, mercurous (Hg_2^{2+}), mercuric (Hg^{2+}); and organic mercury containing methyl, ethyl and phenyl groups (3). The potential sources of Hg(II) ions in wastewater includes fungicides, the electronics industry, nonferrous metal smelting, metals used in manufacturing, paints, pigments and batteries (4).

Mercury can cause neurodegenerative diseases such as amyotrophic lateral sclerosis, Alzheimer disease and Parkinson disease. Elemental mercury and inorganic compounds damage the immune system and kidneys and methyl mercury can cause cardiovascular disease and nervous system disorders (5). The World Health Organization (WHO) sets the permissible limit of Hg(II) for drinking water at 0.001 mg L^{-1}.

Adsorption, coagulation-flocculation, ion-exchange, membrane filtration, chemical precipitation, flotation and electrochemical have all been employed for metal ion removal from aqueous solution; however, all these methods except adsorption are inefficient for the removal of some metal ions and incur high operational costs (6,7). Adsorption has been demonstrated to be the most efficient method for removal of heavy metal ions from

aqueous solutions (8,9).
Adsorbents such as chitosan, iron oxides, activated carbon and polymeric materials have been applied to remove heavy metal from wastewater; however, these adsorbents have shown low adsorption capacities or low rates of adsorption (10). Recently, magnetite nanoparticles (Fe_3O_4) have been tested as an efficient adsorbent. They offer a high surface area, magnetic response and the unique advantage of easy separation under external magnetic fields (11,12). To further facilitate the adsorption affinity, surface modification through physical coatings and covalent bonding has been explored to enable specific metal complexation (13).
The present work coated magnetite nanoparticles with sodium dodecyl sulfate (SDS), an anionic surfactant. Next, 2,4-dinitrophenylhydrazine (2,4-DNPH), an important reagent which can act as an electron pair donor to react with most hard and intermediate cations, was chemically immobilized on the surface of the SDS-coated magnetite nanoparticles and the product was used for adsorption of Hg(II) ions from an aqueous solution. The effects of pH, adsorbent dose and contact time on the adsorption efficiency were studied. The adsorption mechanisms are discussed in detail. The adsorption kinetic and isotherm were also investigated.

Methods

Apparatus and reagents

Hg(II) ion concentration was measured by cold vapor atomic absorption spectroscopy (CVAAS) using a direct mercury analyzer (DMA-80). Adjustments in pH were carried out using a pH meter (Metrohm; model 713; Switzerland). The adsorbent (DNPH@SDS@Fe_3O_4 nanoparticles) was characterized using scanning electron microscopy (SEM; Hitachi-S4160; Japan) and x-ray diffraction (XRD; Italstructure; ADP2000; Italy). Fourier transform infrared spectroscopy (FTIR) at 400 to 4000 cm^{-1} in KBr were recorded on a FTIR spectrometer (PerkinElmer; spectrum 100). The porosity and specific surface area were defined by N_2 adsorption–desorption porosimetry (77 K) using a porosimeter (MicrotacBEL; Japan).
All chemicals and reagents used in this work were of analytical grade and were purchased from Merck (Germany). Double-distilled water was used throughout the study. A stock solution of Hg(II) (1000 mg L^{-1}) was prepared by dissolving the appropriate amount of $Hg(NO_3)_2{\cdot}6H_2O$ in double-distilled water.

Synthesisof Fe_3O_4 nanoparticles

The magnetite nanoparticles were synthesized using the conventional co-precipitation method. Briefly, $FeCl_3.6H_2O$ (11.68 g) and $FeCl_2.4H_2O$ (4.30 g) were dissolved in 200 mL of double-distilled water under nitrogen gas in an ultrasonic bath at 85°C for a few minutes to produce smaller, more homogenous particles. Next, 20 mL of 30% NH_3 was added dropwise to the solution. The color of the suspension turned black immediately. The precipitated Fe_3O_4 nanoparticles were washed with ddH_2O and once with 0.02 mol L^{-1} NaCl and were separated using an external magnetic field (14).

Synthesis of magnetite modified with 2,4-DNHD sodium dodecyl sulfate

For this purpose, 2.0 g of magnetite nanoparticles were diffused in 50 mL of water and mixed with 100 mg of SDS. Next, 20 mL of a solution of 0.90 g 2,4-DNPH in concentrated hydrogen chloride + acetonitril was added. The suspension was stirred at 60°C for 3 hours and then the solvent was evaporated and the remaining produce was washed, air-dried, and stored in a closed bottle for subsequent use.

Point of zero charge of the DNPH@SDS@Fe_3O_4 nanoparticles (pH_{PZC})

The point of zero charge (PZC) is a characteristic of metal oxides (hydroxides) and is of fundamental importance to surface science. It relates to adsorption and describes the condition at which the electrical charge density on a surface is zero. The pH_{PZC} of the DNPH@SDS@Fe_3O_4 nanoparticles was determined in degassed 0.01 mol L^{-1} $NaNO_3$ solution at 20°C. Aliquots of 30 ml of 0.01 mol L^{-1} $NaNO_3$ were mixed with 30 mg of DNPH@SDS@Fe_3O_4 nanoparticles in several beakers. The pH of the solutions were adjusted to 2.0 to 9.0 using solutions of 0.01 mol L^{-1} of HNO_3 and/or NaOH as appropriate. The initial pH of the solutions were recorded and the beakers were covered with parafilm and shaken for 24 hours. The final pH values were recorded. The pH_{pzc} was recorded as the point at which $pH_{initial} = pH_{final}$.

Adsorption kinetics and equilibrium testing

Adsorption kinetics testing was carried out by adding 0.04 g of DNPH@SDS@Fe_3O_4 nanoparticles to three initial concentrations of Hg(II) (25, 50, 90 mg L^{-1}), which was then shaken at 150 rpm and 25°C. At specified time intervals, a strong permanent magnet used to separate the synthesized adsorbent. The concentration of the residual Hg(II) in the solution was measured by CVAAS using a DMA-80. The amount of metal adsorbed at each time interval per unit mass of the adsorbent q_t (mg g^{-1}) was calculated as:

$$q_t = \frac{(C_0 - C_t)V}{W} \quad (1)$$

where C_0 (mg L^{-1}) denotes the initial liquid-phase concentration of the metal, C_t (mg L^{-1}) denotes the concentration at time t, V (l) denotes the volume of the solution and W (g) denotes the mass of the adsorbent. Adsorption equilibrium testing was carried out by adding 0.04 g of DNPH@SDS@Fe_3O_4 nanoparticles to 15 ml of initial concentrations Hg(II) solution (25 to 420 mg L^{-1}). The samples were shaken for 24 hours to reach equilibrium, then the solution was analyzed for residual metal (Hg(II)) content using CVAAS on a DMA-80. The number of adspecies at equilibrium, q_e (mg g^{-1}), were computed as:

$$q_e = \frac{(C_0 - C_e)V}{W} \quad (2)$$

where C_e (mg L^{-1}) denotes the metal concentration at equilibrium.

Results

Characterization of modified magnetite nanoparticles

In order to better understand the adsorption behavior of DNPH@SDS@Fe_3O_4 nanoparticles toward the metal ions, their structural properties were characterized. The XRD pattern of DNPH@SDS@Fe_3O_4 nanoparticles is shown in Figure 1. The SEM and SEM-EDX results of the Fe_3O_4 nanoparticlesare shown in Figure 2. Figure 3 shows the FTIR spectra of the DNPH@SDS@Fe_3O_4 nanoparticles.

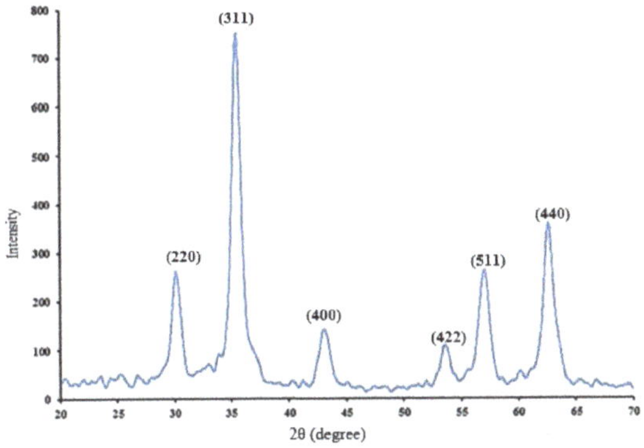

Figure 1. XRD patterns of Fe_3O_4 nanoparticles.

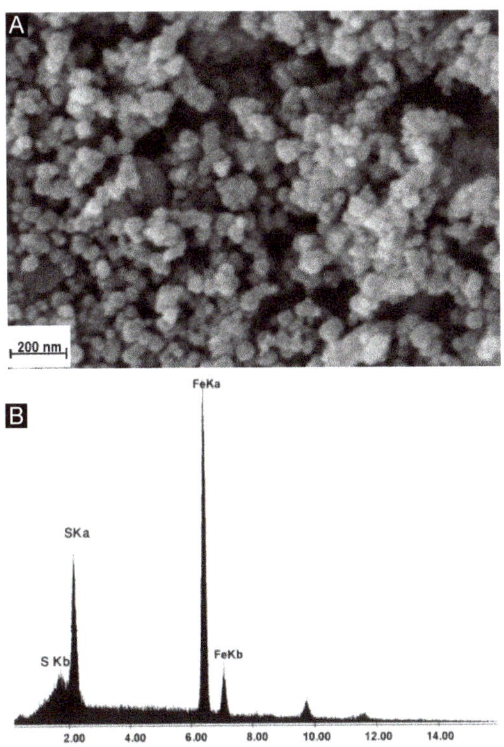

Figure 2. (A) SEM image of DNPH@SDS@Fe_3O_4 nanoparticles; (B) SEM-EDX spectrum of SDS@Fe_3O_4 nanoparticles.

Effect of solution pH

The PZC (pH_{pzc}) of the DNPH@SDS@Fe_3O_4 adsorbent was 5.0 (Figure 4A). The influence of pH on Hg(II) adsorption onto DNPH@SDS@Fe_3O_4 was studied at pH values ranging from 1.0 to 10 (Figure 4B). The maximum adsorption of Hg(II) onto DNPH@SDS@Fe_3O_4 occurred at about pH 7.0.

Effect of adsorbent dose

The relation between adsorbent dose and removal of Hg(II) ions by DNPH@SDS@Fe_3O_4 nanoparticles is shown in Figure 5. The percentage of Hg(II) removed increased as the dose of adsorbent increased from 0.01 to 0.04 g; thus, 0.04 g of adsorbent dose was selected as the optimum dose for the remaining experiments. Increasing the dose of

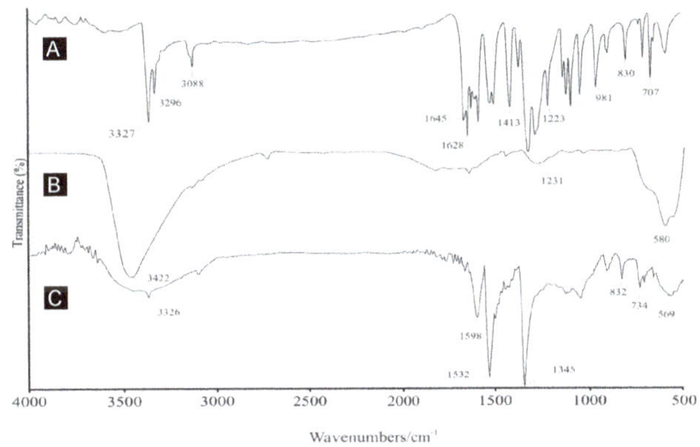

Figure 3. FTIR spectrum of: (A) 2,4-DNPH; (B) magnetite nanoparticles; (C) DNPH@SDS@Fe_3O_4 nanoparticles

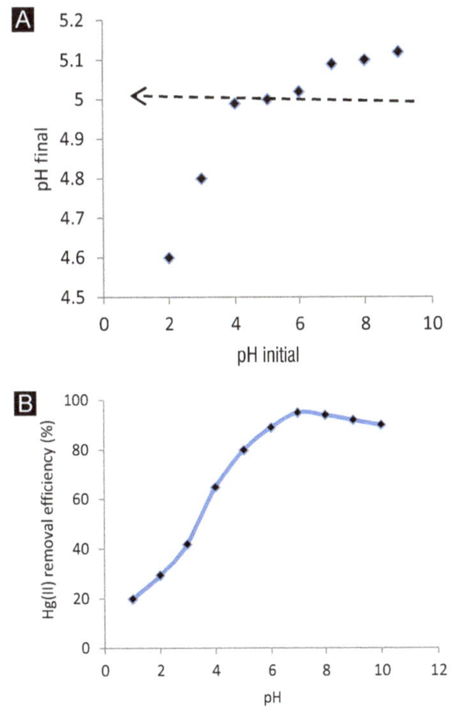

Figure 4. (A) PZC of DNPH@SDS@Fe_3O_4 nanoparticles; (B) effect of pH on the removal of Hg(II) ions (C_0= 25 mg l^{-1}, contact time = 45 min, dose of adsorbent = 0.04 g, temperature = 25°C).

adsorbent beyond this point did not appreciably increase Hg(II) adsorption.

Effect of shaking time and kinetics

The effect of contact time was determined at three concentrations of Hg(II) solution (5, 50 and 90 mg L^{-1}). The removal of Hg(II) from solution by DNPH@SDS@ Fe_3O_4 as a function of contact time is shown in Figure 6. It is clear that the adsorption capacity of Hg(II) onto the DNPH@SDS@Fe_3O_4 nanoparticles increased as the contact time increased. Three kinetic models were tested to determine the adsorption kinetics of Hg(II) ions onto DNPH@SDS@Fe_3O_4 nanoparticles (Figure 7). The kinetic parameters are listed in Table 1.

Equilibrium

Freundlich and Langmuir equilibrium isotherms were used to determine the adsorption mechanism of DNPH@ SDS@Fe_3O_4 for Hg(II) ions. The equilibrium isotherm for adsorption of Hg(II) ions onto the magnetite nanoparticles at pH 7.0 and 25°C is shown in Figure 8. The adsorption constants obtained from the isotherms are listed in Table 2.

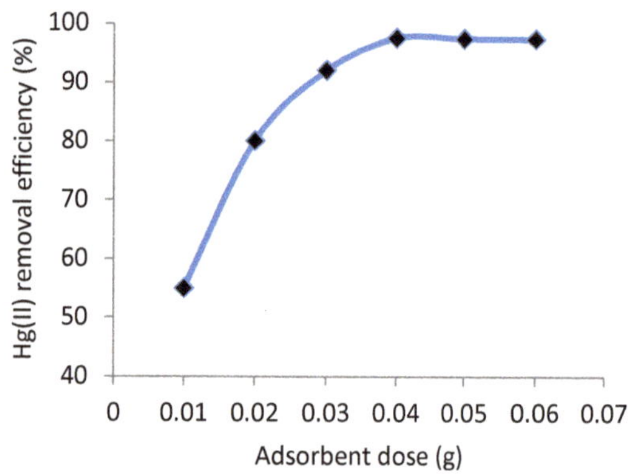

Figure 5. Effect of dose of DNPH@SDS@Fe_3O_4 nanoparticles on removal of Hg(II) ions (C_0= 25 mg l^{-1}, initial pH = 7.0, contact time = 45 min, temperature = 25°C).

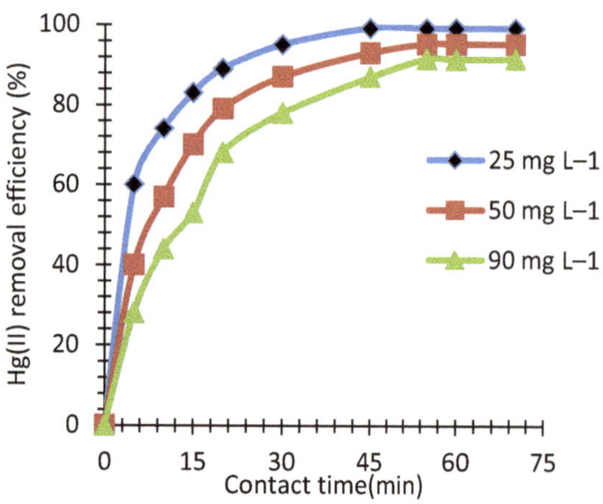

Figure 6. Effect of contact time on removal of Hg(II) ions (C_0= 25, 50 and 90 mg l^{-1}, initial pH = 7.0, dose of adsorbent = 0.04 g, temperature = 25°C).

Discussion

As shown in Figure 1, the XRD pattern of the magnetite is consistent with the six diffraction peaks at (220), (311), (400), (422), (511) and (440) recorded by the Joint Committee on Powder Diffraction Standards (JCPDS card, file No. 79-0418) indexed to the cubic spinel phase of Fe_3O_4. The XRD pattern confirms that the synthesized adsorbents are Fe_3O_4. The SEM image of the nanoparticles in Figure 2A shows an average diameter for DNPH@ SDS@Fe_3O_4 nanoparticles of 20 to 35 nm with a spherical shape.

Typical SEM-EDX elemental analysis of SDS@Fe_3O_4 nanoparticles is shown in Figure 2B and shows the peak areas of Fe and S from which the atomic ratio of Fe/S was calculated to be 96.66/3.34 = 28.9. This represents the surface SDS coverage ratio of the adsorbent. An adjacent monolayer of the surface coating is the likely source of incompleteness of surface protonation and the existence of spatial resistance for the surface coating reaction.

The specific surface area was determined using the BET

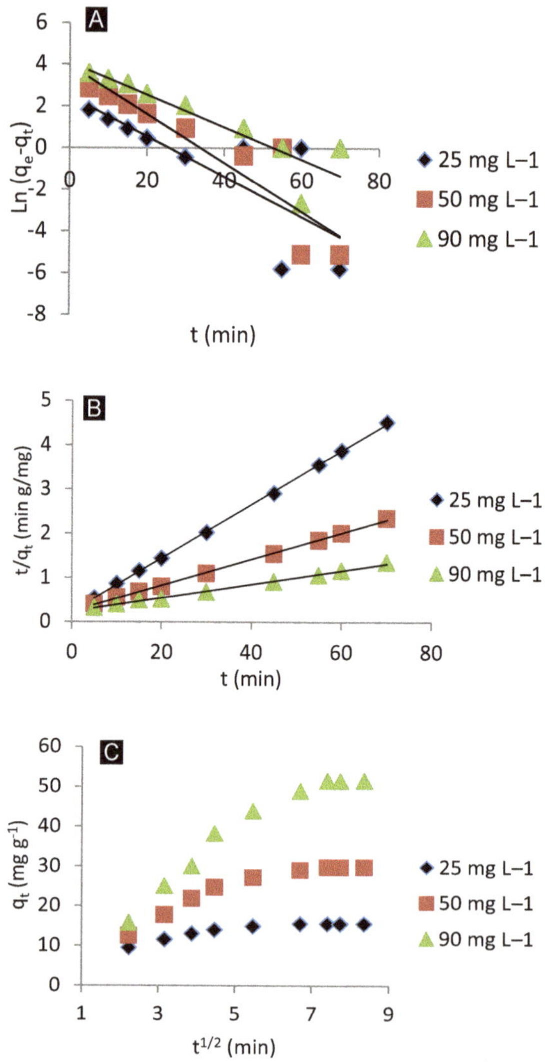

Figure 7. Adsorption of Hg(II) ions onto DNPH@SDS@Fe_3O_4 nanoparticles at 25°C: (A) pseudo-first-order kinetic; (B) pseudo-second-order kinetic; (C) intraparticle diffusion.

Table 1. Kinetic model parameters for adsorption of Hg(II) ions onto adsorbent at 25°C

C_0 (mg L^{-1})	q_e exp(mg g^{-1})	Pseudo-first-order			Pseudo-second-order			Intraparticle diffusion		
		q_{e1}(mg g^{-1})	k_1 (min^{-1})	R^2	q_{e2} (mg g^{-1})	k_2 (g mg^{-1} min^{-1})	R^2	K_{id} (mg g^{-1} min$^{-0.5}$)	A	R^2
25	15.69	11.88	0.096	0.618	16.47	0.016	0.999	0.909	8.851	0.836
50	30.28	52.24	0.116	0.830	33.78	0.003	0.998	2.652	10.14	0.877
90	60.22	59.38	0.077	0.808	64.10	0.001	0.998	5.855	7.433	0.932

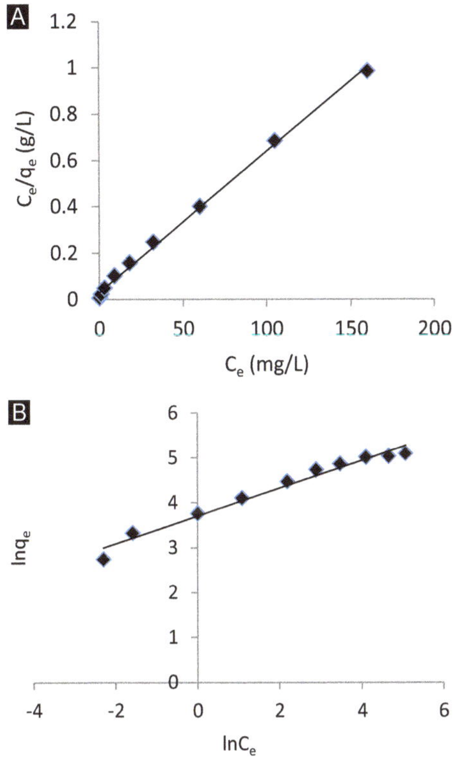

Figure 8. Equilibrium isotherms for adsorption of Hg(II) ions onto DNPH@SDS@Fe_3O_4 at 25°C: (A) Langmuir; (B) Freundlich.

Table 2. Isotherm parameters of adsorption of Hg(II) ions onto DNPH@SDS@Fe_3O_4 at 25°C

Langmuir			Freundlich		
b (L mg^{-1})	q_m (mg g^{-1})	R^2	K_f (mg$^{1-(1/n)}$ l$^{1/n}$ g^{-1})	n	R^2
0.206	164.0	0.996	40.77	3.237	0.972

equation applied to the adsorption data (15,16). The results showed that the average specific surface areas of the Fe_3O_4 and DNPH@SDS@Fe_3O_4 nanoparticles were 99.3 and 75.5 m^2 g^{-1}, respectively. It can be concluded from these values that this type adsorbent comprises nanoparticles with relatively large specific surface areas. The decrease in surface area of the adsorbent as compared to the magnetite nanoparticles may be the result of aggregation after modification with DNPH.

The FTIR spectra of the 2,4-DNPH, Fe_3O_4 nanoparticles and adsorbent are shown in Figures 3A, 3B and 3C, respectively. The Fe_3O_4 nanoparticles are evident from the occurrence of a strong absorption band in the FTIR spectrum which encompasses the characteristic wave number of 580 cm^{-1}. This pattern corresponds to Fe–Ovibrations in Fe_3O_4 (Figure 3B). After immobilization of 2,4-DNPH onto magnetite nanoparticles, a visible broadband of modified Fe_3O_4 nanoparticles appears in the 3200–3500 cm^{-1} region caused by stretching vibrations of N–H or OH groups with varying degrees of hydrogen bonding. The absorption spectrum shows absorption bands (cm^{-1}) at 1345, 1532 and 1598 that correspond to the bending vibrations of the N–H group (Figure 3C). Comparison of these characteristic spectral bands indicates that the surface-modified Fe_3O_4 nanoparticles contain the -NH-functional group.

Aqueous solution pH is an important controlling parameter in heavy metal adsorption. Figure 4B shows that an increase in pH from 2.0 to 7.0 increased adsorption of Hg(II) from 20% to 95%. The effect of solution pH could be explained by the ionization and surface charge of the adsorbate. The surface of the DNPH@SDS@Fe_3O_4 nanoparticles is covered with carboxyl, hydroxyl, and amino groups that vary in form at different pH values. At low pH (pH $<pH_{pzc}$), the groups on the surface are positively charged by protonation that causes electrostatic repulsion with the positively charged Hg(II) ions. As the pH increased (pH $>pH_{pzc}$), the number of negatively charged sites increased and the concentration of H_3O^+ decreased, weakening the competition between the positively charged H_3O^+ and mercury for the same functional groups, improving Hg(II) ion removal efficiency. Hg(II) speciation also favors adsorption at pH 7.0 due to the presence of Hg(II) species. Above pH 7.0, adsorption gradually decreased due to formation of $Hg(OH)_2$, which tends to precipitate at higher pH values. Similar behavior has been reported for Hg(II) adsorption on resin-loaded magnetic ß-cyclodextrin beads (3).

Figure 5 shows that an increase in adsorbent dose rapidly increased removal efficiency of Hg(II). Hg(II) adsorption increased from 55% at a dose of 0.01 g to 97.5% at a dose of 0.04 g of DNPH@SDS@Fe_3O_4 nanoparticles and did not further increase because the number of sites available for metal ion adsorption was limited. The increase in adsorption efficiency with an increase in adsorbent dose is likely the result of the increased number of binding sites available for the target pollutant (Hg(II)). It can be concluded that the optimum dose (0.04 g) of adsorbent provided sufficient adsorption sites for 97.5% adsorption efficiency under the experimental conditions. A value of 0.04 g of DNPH@SDS@Fe_3O_4 was thus selected as the optimum dose. These results are in agreement with the results of previous studies (17).

The adsorption of Hg(II) onto DNPH@SDS@Fe_3O_4 nanoparticles at different initial concentrations was studied as a function of contact time to determine the adsorption equilibrium time. Initially, a large number of empty surface sites are available for adsorption that increase the rate of transfer and diffusion. After this stage, the appearance of strong repulsive forces and saturation of the pores and vacant surface sites with Hg(II) molecules on in solid and bulk phases decreases the migration and diffusion rates. Figure 6 shows that the contact time required for the Hg(II) solution at 25, 50 and 90 mg L^{-1} to reach equilibrium was 45, 55 and 55 minutes, respectively. The removal rate significantly decreased with an increase in initial Hg(II) concentrations from 25 to 90 mg L^{-1}.

To evaluate the kinetics of adsorption, the experimental results were compared with those predicted by the pseudo-first-order, pseudo-second-order and the intraparticle diffusion models. The pseudo-first-order model was described by Lagergren (18) as:

$$\ln(q_e - q_t) = \ln(q_e) - \frac{k_1 t}{2303} \qquad (3)$$

where: q_e (mg g^{-1}) and q_t (mg g^{-1}) denote the adsorption capacity at equilibrium and at time t (min), respectively and k_1 (min^{-1}) denotes the rate constant of the pseudo-first-order kinetic model. The linear form of the pseudo-second-order kinetic model can be expressed as (18):

$$\frac{t}{q_t} = \frac{1}{k_2 q_e^2} + \frac{t}{q_e} \qquad (4)$$

where: q_t (mg g^{-1}) denotes the adsorption uptake at time t (min) and k_2 (g $mg^{-1}min^{-1}$) denotes the rate constant of the pseudo-second order kinetic model.

The intraparticle diffusion model proposed by Weber and Morris was used to identify the mechanism involved in adsorption as (19):

$$q_t = k_{id} t^{0.5} + A \qquad (5)$$

where K_{id} (mg $g^{-1}min^{-0.5}$) denotes the intraparticle diffusion rate constant and A (mg g^{-1}) is a constant denoting the thickness of the boundary layer.

The kinetic constants and correlation coefficient (R^2) of Hg(II) adsorption onto DNPH@SDS@Fe_3O_4 are presented in Table 1. As shown, R^2 indicates a better fit for the pseudo-second-order model to the experimental data than the other models (Figure 7). The low k_2 values (0.016, 0.003 and 0.001 g mg^{-1} min^{-1}) indicate that the rate of Hg(II) adsorption is rapid. Moreover, the q_e values for the pseudo-second-order model were similar to those for the experimental q_e (Table 1). It is reasonable to conclude that the rate limiting step during adsorption is chemisorption involving valence forces through the sharing or exchange of electrons between the metal ions and the adsorbent.

Figure 7C indicates that the data points are related in three stages. First, the adsorption rate is high, which can be imputed to external surface adsorption during which the Hg(II) diffuses through the solution to the external surface of the adsorbent or the boundary layer diffusion of the solute molecules. In the second stage, adsorption slows during the intraparticle diffusion as the rate-controlling step. The third stage is the final equilibrium phase during which intraparticle diffusion slows and levels off either because the remaining metal content is extremely low in solution or maximum adsorption has been attained.

Adsorption isotherm models explain the relationship between adsorbents and adsorbates (Hg(II)) at equilibrium. The adsorption isotherms of Hg(II) onto the DNPH@SDS@Fe_3O_4 nanoparticles were tested at a pH of 7.0 at different initial Hg(II) concentrations (25 to 420 mg L^{-1}). The Langmuir (20) and Freundlich (21) equations are shown in Eqs. (6) and (7), respectively:

$$\frac{C_e}{q_e} = \frac{C_e}{q_m} + \frac{1}{q_m b_1} \qquad (6)$$

$$\ln q_e = \frac{1}{n}\ln C_e + \ln k_f \qquad (7)$$

where c_e (mg L^{-1}) denotes the equilibrium concentration of adsorbate ions in solution, q_e (mg g^{-1}) denotes the equilibrium adsorption capacity of the adsorbent, q_m (mg g^{-1}) denotes the maximum adsorption capacity of adsorbent for monolayer coverage, b (L mg^{-1}) is a constant related to adsorption free energy, K_f ($mg^{1-(1/n)}$ $L^{1/n}$ g^{-1}) is a constant related to adsorption capacity, and n is an empirical parameter related to adsorption.

The results indicate that the linear correlation coefficient for the Langmuir model (R^2= 0.996) was higher than the Freundlich model (R^2= 0.972), but the Langmuir model fitted the adsorption information better than the Freundlich model. It confirms that the adsorption occurs in a monolayer and that the adsorption of each molecule has equal activation energy. It is clear that adsorption occurs on a homogeneous surface. The maximum adsorption capacity using the Langmuir model was 164.0 mg g^{-1} at 25°C.

Hg(II) adsorption capacity was compared with the results of other adsorbents from previous studies. The adsorption capacities of thiol-modified Fe_3O_4@SiO_2 was 148.8 (22), resin-loaded magnetic ß-cyclodextrin beads was 88.4 (3), grapheme oxide was 80.7 (5) and chitosan was 52.6 (23); however, a higher adsorption capacity and better removal efficiency were recorded for the DNPH@SDS@Fe_3O_4 nanoparticles prepared in this work.

Conclusion

DNPH@SDS@Fe_3O_4 nanoparticles were synthesized and used for adsorption of Hg(II) ions from aqueous solutions. The BET surface area of the adsorbent was 75.5 m^2 g^{-1}, which can facilitate diffusion and adsorption of Hg(II) onto the adsorbent. An acceptable adsorption efficiency for Hg(II) was observed as the initial pH of the solution varied from 1.0 to 7.0. The mechanism for Hg(II) adsorption can be explained by the electrostatic interaction of the negatively charged surfaces of DNPH@SDS@Fe_3O_4 nanoparticles and Hg(II) species. Hg(II) adsorption followed pseudo-second-order kinetics. Equilibrium data were well-described by the Langmuir isotherm, suggesting that adsorption occurred mainly in a monolayer. The results show that DNPH@SDS@Fe_3O_4 nanoparticles are an economic adsorbent for removal of Hg(II) ions from aqueous solution.

Acknowledgments
The authors are grateful to the Hamedan Branch, Islamic Azad University for providing the facilities to conduct and complete this study.

Competing interests
The authors declare that they have no competing interests.

Author contributions
All authors participated in the study design, literature search, writing of the manuscript, and data acquisition, analysis, and interpretation. All authors critically reviewed, refined, and approved the manuscript.

References

1. Cegłowski M, Schroeder G. Preparation of porous resin with Schiff base chelating groups for removal of heavy metal ions from aqueous solutions. Chem Eng J 2015; 263: 402-11. doi: 10.1016/j.cej.2014.11.047.
2. Cui L, Wang Y, Gao L, Hu L, Yan L, Wei Q, et al. EDTA functionalized magnetic graphene oxide for removal of Pb (II), Hg (II) and Cu (II) in water treatment: adsorption mechanism and separation property. Chem Eng J 2015; 281: 1-10. doi: 10.1016/j.cej.2015.06.043.
3. Cui L, Wang Y, Gao L, Hu L, Wei Q, Du B. Removal of Hg(II) from aqueous solution by resin loaded magnetic β-cyclodextrin bead and graphene oxide sheet: synthesis, adsorption mechanism and separation properties. J Colloid Interface Sci 2015; 456: 42-9. doi: 10.1016/j.jcis.2015.06.007.
4. Wang Z, Xu J, Hu Y, Zhao H, Zhou J, Liu Y, et al. Functional nanomaterials: Study on aqueous Hg(II) adsorption by magnetic Fe_3O_4@SiO_2-SH nanoparticles. Journal of the Taiwan Institute of Chemical Engineers 2016; 60: 394-402. doi: 10.1016/j.jtice.2015.10.041.
5. Fan Q, Yang Y, Hao Y, Zhao X, Feng Y. Preparation of three-dimensional PANI/GO for the separation of Hg(II) from aqueous solution. J Mol Liq 2015; 212: 557-62. doi: 10.1016/j.molliq.2015.10.008.
6. Al-Zoubi H, Ibrahim KA, Abu-Sbeih KA. Removal of heavy metals from wastewater by economical polymeric collectors using dissolved air flotation process. J Water Process Eng 2015; 8: 19-27. doi: 10.1016/j.jwpe.2015.08.002.
7. Zhang Z, Zhang H, Zhu L, Zhang Q, Zhu W. Hierarchical porous Ca $(BO_2)_2$ microspheres: hydrothermal–thermal conversion synthesis and their applications in heavy metal ions adsorption and solvent-free oxidation of benzyl alcohol. Chem Eng J 2016; 283: 1273-84. doi: 10.1016/j.cej.2015.08.073.
8. Sobhanardakani S, Parvizimosaed H, Olyaie E. Heavy metals removal from wastewaters using organic solid waste—rice husk. Environ Sci Pollut Res Int 2013; 20(8): 5265-71. doi: 10.1007/s11356-013-1516-1.
9. Zandipak R, Sobhanardakani S. Synthesis of $NiFe_2O_4$ nanoparticles for removal of anionic dyes from aqueous solution. Desalin Water Treat 2015; 57(24): 11348–60. doi: 10.1080/19443994.2015.1050701.
10. Sun Y, Yue Q, Gao B, Gao Y, Xu X, Li Q, et al. Adsorption and cosorption of ciprofloxacin and Ni(II) on activated carbon-mechanism study. J Taiwan Inst Chem Eng 2014; 45(2): 681-8. doi: 10.1016/j.jtice.2013.05.013.
11. Afkhami A, Moosavi R. Adsorptive removal of Congo red, a carcinogenic textile dye, from aqueous solutions by maghemite nanoparticles. J Hazard Mater 2010; 174(1-3): 398-403. doi: 10.1016/j.jhazmat.2009.09.066.
12. Lai L, Xie Q, Chi L, Gu W, Wu D. Adsorption of phosphate from water by easily separable Fe_3O_4@SiO_2 core/shell magnetic nanoparticles functionalized with hydrous lanthanum oxide. J Colloid Interface Sci 2016; 465: 76-82. doi: 10.1016/j.jcis.2015.11.043.
13. Wang J, Zheng S, Shao Y, Liu J, Xu Z, Zhu D. Amino-functionalized Fe_3O_4@ SiO_2 core–shell magnetic nanomaterial as a novel adsorbent for aqueous heavy metals removal. J Colloid Interface Sci 2010; 349(1): 293-9. doi: 10.1016/j.jcis.2010.05.010.
14. Sobhanardakani S, Zandipak R. 2,4-Dinitrophenylhydrazine functionalized sodium dodecyl sulfate-coated magnetite nanoparticles for effective removal of Cd (II) and Ni (II) ions from water samples. Environ Monit Assess 2015; 187(7): 412. doi: 10.1007/s10661-015-4635-y.
15. Brunauer S, Emmett PH, Teller E. Adsorption of gases in multimolecular layers. J Am Chem Soc 1938; 60(2): 309-19. doi: 10.1021/ja01269a023.
16. Venkatesha T, Viswanatha R, Nayaka YA, Chethana B. Kinetics and thermodynamics of reactive and vat dyes adsorption on MgO nanoparticles. Chem Eng J 2012; 198-199: 1-10. doi: 10.1016/j.cej.2012.05.071.
17. Shan C, Ma Z, Tong M, Ni J. Removal of Hg(II) by poly (1-vinylimidazole)-grafted Fe_3O_4@SiO_2 magnetic nanoparticles. Water Res 2015; 69: 252-60. doi: 10.1016/j.watres.2014.11.030.
18. Azizian S. Kinetic models of sorption: a theoretical analysis. J Colloid Interface Sci 2004; 276(1): 47-52. doi: 10.1016/j.jcis.2004.03.048.
19. Ghaedi M, Biyareh MN, Kokhdan SN, Shamsaldini S, Sahraei R, Daneshfar A, et al. Comparison of the efficiency of palladium and silver nanoparticles loaded on activated carbon and zinc oxide nanorods loaded on activated carbon as new adsorbents for removal of Congo red from aqueous solution: Kinetic and isotherm study. Mater Sci Eng C 2012; 32(4): 725-34. doi: 10.1016/j.msec.2012.01.015.
20. Langmuir I. The adsorption of gases on plane surfaces of glass, mica and platinum. J Am Chem Soc 1918; 40(9): 1361-403. doi: 10.1021/ja02242a004.
21. Freundlich H, Heller W. The adsorption of cis-and trans-azobenzene. J Am Chem Soc 1939; 61(8): 2228-30. doi: 10.1021/ja01877a071.
22. Zhang S, Zhang Y, Liu J, Xu Q, Xiao H, Wang X, et al. Thiol modified Fe_3O_4@ SiO_2 as a robust, high effective, and recycling magnetic sorbent for mercury removal. Chem Eng J 2013; 226: 30-8. doi: 10.1016/j.cej.2013.04.060.
23. Sobhanardakani S, Zandipak R, Javanshir-Khoei A, Hosseini SM, Moslemi M, Delfieh P. Removal of Hg (II) and Cd (II) Ions from aqueous solution using chitosan: kinetics and equilibrium studies. Iran J Health Sci 2015; 3(2): 21-30. doi: 10.7508/ijhs.2015.02.004.

Evaluation of drinking water quality indices

Nematollah Jafarzadeh[1,2], Maryam Ravanbakhsh[3,4*], Kambis Ahmadi Angali[5], Ahmad Zare Javid[6,7], Darioush Ranjbar Vakil Abadi[8], Shirin Ardeshirzadeh[9]

[1]Professor of Environmental Health Engineering, Department of Environmental Health Engineering, School of Public Health, Ahvaz Jundishapur University of Medical Sciences, Ahvaz, Iran

[2]Professor of Environmental Health Engineering, Environmental Technologies Research Center, Ahvaz Jundishapur University of Medical Sciences, Ahvaz, Iran

[3]PhD Student of Environmental Health Engineering, Student Research Committee, Ahvaz Jundishapur University of Medical Sciences, Ahvaz, Iran

[4]PhD Student of Environmental Health Engineering, Department of Environmental Health Engineering, School of Public Health, Ahvaz Jundishapur University of Medical Sciences, Ahvaz, Iran

[5]Associated Professor of Biostatistics, Department of Biostatistics, Ahvaz Jundishapur University of Medical Sciences, Ahvaz, Iran

[6]Assistant Professor of Nutrition Sciences, Nutrition and Metabolic Diseases Research Center, Ahvaz Jundishapur University of Medical Sciences, Ahvaz, Iran

[7]Assistant Professor of Nutrition sciences, Hyperlipidemia Research Center, Ahvaz Jundishapur University of Medical Sciences, Ahvaz, Iran

[8]Assistant Professor of Environmental Health Engineering, Department of Environmental Health Engineering, School of Public Health, Bushehr University of Medical Sciences, Bushehr, Iran

[9]Medical Practice, Observer at Cleveland Clinic Foundation, Family Medicine Department, Ohio, USA

Abstract

Background: Internal corrosion and the formation of scale in water distribution pipes are the most important problems for an urban water distribution system. Physical, chemical, or biological factors can lead to these two processes. Internal corrosion and scale formation can impact health, economy, and aesthetics. This study assessed the physicochemical quality parameters and evaluated the potential for corrosion and scale formation in drinking water at the distribution systems of 5 selected cities in Bushehr province (Kangan, Dashtestan, Dashti, Bushehr, and Ganaveh) from 2009-2012.

Methods: This study was carried out based on laboratory data collected from monthly samplings of tap water in the Water and Wastewater Company of Bushehr province during the years 2009-2012. Internal corrosion and scale formation rates were calculated using the Ryznar, Langelier, Aggressive, and Puckorius indices.

Results: The results of the Ryznar, Puckorius, Aggressive and Langelier indices indicated that the drinking water in the 5 selected cities of Bushehr province was corrosive. Moreover, the majority of parameters used to determine water quality exceeded Iran's national standards.

Conclusion: It is concluded that there is problem of water corrosion and scaling in drinking water of distribution systems in selected cities.

Keywords: Bushehr, Water corrosiveness, Scale formation, Saturation index

***Correspondence to:**
Maryam Ravanbakhsh
Email: mapseh@gmail.com

Introduction

A sufficient quantity of drinking water with an appropriate quality is vital for human life. Social, agricultural, and industrial developments have led to the contamination of water resources and reduced water quality. In addition, population growth and increasing living standards in communities have increased the need for water (1). The World Health Organization (WHO) considers the provision of clean water an indicator of health development in developing countries (2). Therefore, an urban water supply system should provide an adequate amount and desirable quality of water (2). The quality of urban drinking water is affected by its chemical stability, which can cause corrosion and scale formation in water pipes.

According to the ISO (8044), corrosion is defined as the chemical and physical interactions between metal and its surrounding environment affecting the characteristics of metal and causing damage to it, its surrounding environment, and its technical system (3). Generally, this environment involves either a liquid medium, such as water flow in pipes, or a solid medium, such as soil contact with buried pipes (4). Water's potential for corrosion is affected by physical, chemical, and biological factors. Physical factors include flow rate and temperature; chemical factors include alkalinity, hardness, conductivity, dissolved oxygen, presence of sulfate and chloride ions, dissolved gases, and pH (5,6). Biological corrosion, which is generally created by iron and sulfate reducing bacteria, is resistant to high amounts of chlorine in a distribution system (6). White et al in 2011 indicated that bio-films grown in the distribution system can impact corrosion in a drinking water distribution system. Moreover, they introduced some other microorganisms influencing corrosion, which can grow in heavy metal-rich environments (7). Scale formation can harm treatment plant equipment and fixtures. Scaling is problematic as it forms inside the pipes and reduces the area available for carrying water. Corrosion can lead to economic, health, and aesthetic problems. Corrosion in distribution pipes can cause perforation, the loss of large volumes of water, and it can reduce the space available for carrying water (8). Water with high corrosion potential can dissolve the initial materials in metal pipes and fittings and cause health problems. Corrosion may also allow heavy metals such as cadmium, copper, lead, and other pollutants to enter the water. Industry spends billions of dollars because of corrosion. According to 2002 reports, the expenses associated with corrosion in Australia, the United Kingdom, Japan, and some other countries were several times greater than the respective country's gross national product (9). There are no exact statistics available regarding the expenses caused by corrosion and scaling in Iran; however, it has been indicated that greater than 30% of the distributed water is lost because of corrosion in the distribution systems (10).

In Iran, similar studies have been conducted to evaluate the stability of drinking water. Ebrahimi et al in 2012 investigated the potential for corrosion and scale formation in the drinking water distribution system of Koohdasht city using indices of corrosion and scale formation. Their results indicated that the drinking water in Koohdasht tends to be corrosive (11). Similarly, Jaafarzadeh et al found that the drinking water in Ahwaz city is corrosive and contains high concentrations of metals (exceeding EPA standards) including lead, iron, cadmium, copper, and manganese (12). Furthermore in 2012, Malakootian and Fatehizadeh evaluated the quality of drinking water in Bardsir city, Iran using the Langelier, Ryzener, and Aggressive indices and found that it is scaling (5). Paseban et al assessed the drinking water distribution system of Bojnoord, Iran in 2012 and found that the drinking water distribution system in Bojnoord had a scaling tendency. They suggested that scaling may lead to reductions in flow and pressure in the water system (13). In another study in 2011, Mokhtari et al indicated that the drinking water in Ardabil city was slightly corrosive (14). Zazouli et al found similar results in Yasooj city (15).

According to the statistics obtained from the Water and Wastewater Company in Bushehr province, the amount of water produced for the province in the first 6 months of 2011 was 1 118 507 cubic meters from underground reservoirs and 37 761 285 cubic meters from surface water, of which the total amount of 23 146 177 cubic meters was consumed. Considering the high levels of water lost from the distribution system, the evaluation of reasons for such losses seems to be essential. The water pipes used in the distribution networks of these cities are galvanized metal with a lifetime of usually about 20 years. One important factor involved in water loss in distribution systems is corrosion. Therefore, it is necessary to study all factors involved in the corrosion potential of water resources and its trend in distribution systems. The potential of water to corrode and scale is characterized by its stability. Instability and low quality water in pipes can lead to corrosion and scaling in them (16,17).

Methods

This study aimed to determine the corrosion and scaling potential of drinking water in 5 selected cities in Bushehr province (Bushehr, Kangan, Dashtestan, Ganaveh, and Dashti) using the Ryzener, Puckorius, Aggressive, and Langelier water stability indices.

Bushehr has a latitude of 27°14′N and a longitude of 50°6′, 52°58′E from the Greenwich meridian and is surrounded by Khuzestan, Kohkiloye, and Boyer Ahmad provinces from the north, the Persian Gulf and Hormozgan province from the south and west, and Fars province from the east. It has an area of about 27 653 km^2 and population of over 886 000 people. According to political divisions, this province consists of nine cities, 5 of which (Kangan, Dashtestan, Dashti, Bushehr, and Ganaveh) were selected for the present study. The geographic location of this province is shown in Figure 1. The main sources of drinking water in this province are rivers, springs, groundwater wells, and the Kowsar Dam reservoir.

Data used in this study was obtained from the Bushehr Water and Wastewater Company and included water quality parameters such as the concentrations of anions (chloride, sulfate, ammonium, nitrate, nitrite, fluoride), cations (calcium, magnesium, iron), total hardness, total alkalinity, pH, and total dissolved solids (TDS) that were determined regularly from drinking water distribution systems in the selected cities. Samples were taken in Bushehr, Dashtestan, and Dashti from 2010 to 2011, in Kangan from 2008 to 2012, and in Ganaveh from late 2007 to late 2010. The total number of samples used in this study was 228. Processes including sterilization of containers, sampling, and transportation and storage of samples in the lab were performed according to standard procedures (18).

Statistical analysis was performed using SPSS software (version 16), and the mean results were compared with Iranian National and EPA standards. A single sample

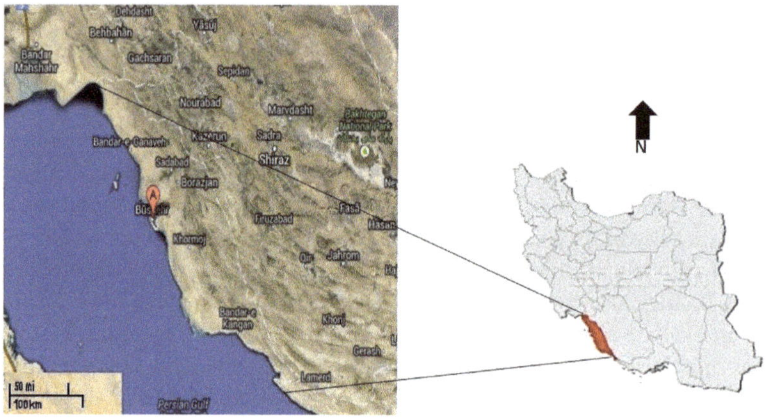

Figure 1. Geographical situation of Bushehr province.

t test was used to compare the means of water quality parameters with the standard limitations, applying $P < 0.05$ as the minimum level of significance.

To determine the corrosion and scaling potential of the water distribution systems in the 5 selected cities, the Langelier (LI), Ryzener (RI), Puckorius (PI) and corrosivity (AI) indices were used with equations 1 to 6.

$pHs = 9.3 + A + B - (C + D)$ (11) Eq. 1

$A = (Log10\ (TDS)-1)/10$ (14)

$B = -13/12Log10\ (°C + 273) + 34.5$ (14)

$C = Log10\ (Ca2+\ as\ CaCO3)- 0.4$ (14)

$D = Log10\ (Total\ Alkalinity\ as\ CaCO3\ )$ Eq. 2

$RI = 2\ pHs - pH$ (11) Eq. 3

$LI= pHs - pH$ (11) Eq. 4

$PI = 2\ pHs - pHeq$ (11) Eq. 5

$pHeq = 1.465\ Log\ (Total\ Alkalinity\ as\ mg/l)+ 4.54$ (11) Eq. 6

Results

The results (Tables 1 and 2) of the Langelier saturation index showed that the drinking water in Ganaveh (0.38) and Bushehr (0.18) was unstable and had a slight scaling tendency. The drinking water of Dashtestan (-0.062), Dashti (-0.23), and Kangan (-1.09) was shown to be corrosive. In comparison, the water of Ganaveh had the highest potential for scaling, and that of Kangan and Dashtestan had the maximum and minimum corrosion potential, respectively.

Table 1. The interpretation of indices (11)

Saturation Index	Ranges	Interpretation
	> 0	Water is scale forming (non-aggressive)
LI	< 0	Water is not scale forming (aggressive)
	= 0	Water is balanced
	< 4	Water is scale forming (non-aggressive)
	5-6	Water is relatively scale forming
RI	6-6.5	Water is stable
	6.5 -7	Water is aggressive
	> 8	Water is aggressive strongly
	<10	Water is aggressive strongly
AI	10-12	Water is aggressive
	>12	Water is non-aggressive
PI	<6	Water is aggressive
	>6	Water is scale forming (non-aggressive)

The results of Aggressive index indicated that the water quality of Bushehr (12.21) and Ganaveh (12.3) cities were sever aggressive and Dashti (11.75), Dashtestan (11.91), and Kangan (10.96) were low aggressive.

Results of the Ryznar stability index indicated that the aggressiveness of the drinking water was low in Bushehr (7.01) and Ganaveh (7.04), medium in Dashtestan (7.51) and Dashti (7.78), and severe in Kangan (9.78).

According to the Puckorius Scaling Index (PSI), the 5 selected cities had a tendency for corrosion. The highest tendency was observed in Kangan (11.21).

Table 3 summarizes the results of the water quality parameter analysis. The results showed that the mean concentrations of water quality parameters in some cases were higher than Iran's national allowable limits. In Kangan, there were no significant differences between mean concentrations of No^{-2} and SO^{-4} and the Iranian national standard. TDS, total hardness, Mg^{+2}, and Cl^{-} concentrations showed positive significant differences with Iranian national standards, because they exceeded the national allowable limitations. The parameters of F^{-}, NO^{-3}, NH_3, Ca^{+2}, Fe^{+2}, and pH had no significant differences with national standards.

Assessing the water quality parameters in the selected cities revealed that the mean concentration of TDS in Kanghan and Dashtestan exceeded the national allowable limits. Total hardness of water in Kanghan, Dashtestan, Dashti, and Bushehr also exceeded national limits. The water in Kangan had the highest chloride concentration among the selected cities which was significantly different from national standards in that it exceeded national limits. The results in Kangan and Dashti revealed that the mean concentrations of total hardness and Mg^{+2} had positive significant differences with Iranian national standards. The evaluation of total water hardness in Bushehr and Ganaveh cities showed a positive significant difference with national standards. Ca^{+2} concentration in Ganaveh had no significant difference with national standards. The rest of the assessed data showed no significant differences

Table 2. Interpretation of water quality Indices in selected cities

Index	Selected city	Kangan	Dashtestan	Dashti	Bushehr	Genaveh
	n	110	57	14	20	27
LSI	Mean± SD	-1.09±0.7	-0.062±0.9	-0.23±0.99	0.18±0.44	0.38±1.25
	(min-max)	(-3.14-0.39)	(-4.89-1.30)	(-2.56-.58)	(-.82-0.9)	(-2.4-2.9)
	Interpretation	Aggressive	Aggressive	Aggressive	Scale formation	Scale formation
AI	Mean± SD	10.96±0.78	11.91±0.92	11.75±1.03	12.21±0.45	12.3±1.2
	(min-max)	(8.7-12.4)	(6.98-13.3)	(9.31-12.6)	(11.18-12.94)	(9.57-14.94)
	Interpretation	Median aggressive	Low aggressive	Low aggressive	Not aggressive	Not Aggressive
RI	Mean± SD	9.78±1.27	7.51±1.7	7.78±1.70	7.01±0.77	7.04±1.74
	(min-max)	(7.41-13.77)	(5.41-16.98)	(6.64-11.93)	(5.7-9.16)	(4.12-11.26)
	Interpretation	Median aggressive	Median aggressive	Median aggressive	Low aggressive	Low aggressive
PSI	Mean± SD	11.21±1.4	7.23±2.1	7.63±1.8	6.62±1.3	7.39±1.46
	(min-max)	(9.02-16.29)	(5.32-18.5)	(5.82-11.84)	()4.75-10.56	(5.37-11.46)
	Interpretation	Sever aggressive	Aggressive	Aggressive	Aggressive	Aggressive

Table 3. Analysis of physicochemical quality of drinking water in 5 selected cities in Bushehr province

Parameter	City					Iranian national standard[b] (19)	EPA (19)	Unit
	Kanghan n = 110	Dashtestan n =57	Dashtie n =14	Bushehr n =20	Ghenaveh n =27			
TDS	2170.32±745.9 (55-9260)	2785.9±369.2 (9-1538)	951.79±525.9 (50-1797)	1438.95±1514.8 (220-7650)	552.85±344.6 (100-1080)	1500	500	mg/L
P value	0.013[a]	0.000[a]	0.002[a]	0.859[a]	0.000[a]	–	–	–
Total hardness	915.87±944.3 (12-3812)	587.6±315.9 (4-1600)	579.1±315.94 (4-1600)	583±163.3 (220-1100)	454.1±218.84 (60-720)	500	350	mg/L as $CaCO_3$
P value	0.000[a]	0.041	0.375	0.035	0.286	–	–	–
Total Alkalinity	1.00±0.16 (1-1)	2.11±0.345 (1-2)	2.05±0.250 (2-2)	2.18±0.367 (1-3)	1.99±0.63 (1-2)	–	–	mg/L as $CaCO_3$
P value	-	-	-	-	-	–	–	–
Fluoride	0.86±0.74 (0-6)	0.81±0.6 (0-3)	0.57±0.5 (0-1)	0.53±0.31 (0-1)	0.56±0.353 (0-1)	1.7	2	mg/L
P value	0.000[a]	0.000[a]	0.000[a]	0.000[a]	0.000[a]	–	–	–
Chloride	883.33±1678.5 (25-7460)	206.55±138 (11-994)	207.76±91.7 (40-390)	193.96±95.9 (89-532)	193.96±95.8 (4-216)	400	250	mg/L
P value	0.003[a]	0.000[a]	0.000[a]	0.000[a]	0.000[a]	–	–	–
Sulfate	370.12±307.4 (8-2350)	240.63±186.6 (0-1360)	130.25±63.6 (20-260)	194.08±143.4 (72-464)	294.35±173.8 (25-672)	400	250	mg/L
P value	0.310	0.000	0.0000	0.000	0.004	–	–	–
Nitrite	2.62±3.95 (0-24)	0.01±0.38 (0-0)	0.1±0.021 (0-0)	0.02±0.089 (0-0)	0.0±0.004 (0-0)	3	1	mg/L
P value	0.312	0.000[a]	0.000[a]	0.000[a]	0.000[a]	–	–	–
Nitrate	6.09±7 (0.67-)	38.11±26.8 (0-32)	24.7±87.6 (0-329)	1.65±2.73 (0-10)	2.06±4.28 (0-21)	50	10	mg/L
P value	0.000[a]	0.739	0.299	0.000[a]	0.000[a]	–	–	–
Ammonium	0.22±0.1 (0-0)	0.01±0.024 (0-0)	0.03±0.05 (0-0)	0.6±0.223 (0-1)	0.13±0.47 (0-176)	1.5	–	mg/L
P value	0.000[a]	0.000[a]	0.000[a]	0.000[a]	0.000[a]	–	–	–
Calcium	114.3±102.8 (3-401)	131.95±69 (0-321)	154.24±91.75 (3-361)	183.1±189.2 (22-1041)	134.45±108.2 (0-1041)	250	–	mg/L
P Value	0.000[a]	0.00[0a]	0.02[a]	0.00[0a]	0.77	–	–	–
Magnesium	156.63±183.3 (1-720)	65.77±52.42 (0-249)	84.97±120.04 (2-478)	37.78±20 (13-102)	32.45±19.95 (13-102)	50	–	mg/L
P value	0.000	0.027	0.295	0.013	0.000	–	–	–
Iron	0.13±0.280 (0-2)	0.04±0.088 (0-0)	0.05±0.116 (0-1)	0.07±0.2 (0-1)	0.05±0.14 (0-1)	0.3	0.3	mg/L
P value	0.000[a]	0.000[a]	0.000[a]	0.000[a]	0.000[a]	–	–	–
pH	7.19±1.51 (6-8)	7.39±0.280 (7-8)	7.34±0.314 (7-8)	7.38±0.317 (7-8)	7.80±0.852 (6-10)	8.5	9	–
P value	0.000[a]	0.000[a]	0.000[a]	0.000[a]	0.000[a]	–	–	–

[a] One sample test used to mean comparison between cities data and Iranian national standards, applying $P < 0.05$ as the minimum level of significance.

[b] Maximum allowable limits: the maximum ions concentrations in the water that Its continued consumption will not be hazardous for human health with weight of 76 kg and daily water consumption of 2 L.

with national standards.

Discussion

The present study has shown that the drinking water of distribution systems in 4 of the studied cities had a tendency to corrode, and in Bushehr the tendency of scale formation was predominant. pH is an important parameter for predicting corrosion. Normally, a higher pH indicates a lower corrosion rate. A high pH contributes to protective film formation by reducing the film-free corrosion rate and increasing the precipitation rate. pH influences $FeCO_3$, and, under certain circumstances, increasing pH decreases $FeCO_3$ solubility and increases super saturation (20). According to the current findings, the concentrations of totally dissolved solids, chloride, and sulfate ions in drinking water are high in most of the cities studied, and this can lead to increased corrosion tendency in water. pH and total alkalinity can determine the distribution of CO_2, HCO_3, and CO_3^{2-} fractions in water. Species distribution depends on pH value, temperature, and ionic strength. This finding is consistent with the findings of other similar studies (21, 22).

Some studies have found that low pH (23) and high concentrations of chloride and sulfate ions (24) can lead to water being more aggressive. Husband et al showed that perforation of a copper pipe is related to cold and chlorinated waters with high pH and sulfate and low alkalinity (25). Moreover, a high correlation was found between the severity of corrosion in metal pipes and the ratio of sulfate and chloride ions to bicarbonate (26,27). These results suggest that high metal concentrations associated with low bicarbonate concentrations in water may increase its corrosion tendency. Furthermore, a low concentration of bicarbonate ions can increase the corrosivity of water, and high concentrations of dissolved solids and oxygen can increase the severity of corrosion. Dissolved solids may lead to increased water conductivity and result in increasing corrosion potential, which would be intensified with the predominance of sulfate and chloride ions and would be reduced with the predominance of bicarbonate ions (23).

In the present study, Kanghan had the most aggressive water among all the studied cities, followed by Dashtestan, Dashti, and Ganaveh. According to the results (Tables 1 and 3), the drinking water in Kanghan had the highest concentrations of chloride, sulfate, nitrite, ammonium, iron, magnesium, fluoride, and total hardness. Moreover, the lowest pH and alkalinity and the highest concentration of TDS which may contribute to causing severe corrosivity of drinking water were observed in this city. Ammonia is responsible for the rapid consumption of oxygen and chlorine in water (28). Furthermore, low alkalinity and high ionic concentrations of sulfate, chloride, iron, magnesium, and total hardness could accelerate the corrosivity of water (29). It has been suggested that high concentrations of iron in drinking water may result in a severe corrosion potential for the water distribution network of Kangan (17,30). A high TDS concentration is usually associated with high concentrations of ions which increases the conductivity of water. The water's ability to conduct a corrosive current has a direct relationship with conductivity (31). In contrast, the water quality indices showed that Bushehr had the best water quality parameters and a mild tendency to scaling formation. Compared with Iran's national standards, the parameter values were in accordance with allowable limits, suggesting a more desirable water quality. It is suggested that the intensity of a lower corrosion potential can be attributed to a better quality of water in the pipes of a distribution system. Similarly, the other cities which have a higher water quality may have less aggressive drinking water. According to the findings of this study and other similar studies, the drinking water in some provinces of Iran does not have appropriate chemical stability; it seems that there are corrosion and scale formation problems in the distribution systems of these provinces. Other studies have been carried out in Shahroud (32), Kian (33), Kerman (2), Shahrood (34), Zanjan (3), Ahvaz (35), Kashan (36) and other cities to assess the quality of drinking water using aggressive and scaling indices. According to these studies, the drinking water in Shahroud, Kian, Ardebil, Zanjan, Ahvaz, and Kashan has a tendency to be aggressive, and in Kerman, there is a scaling potential. Selvam et al (37) evaluated the water quality index of groundwater resources around the coastal city of Tuticorin in southern India and concluded that the groundwater quality was impaired by man-made activities. Ravikumar et al studied the water quality index to determine the quality of surface water of Mallathahalli Lake in India and showed that the lake water fell into the poor water category (38).

Conclusion

According to the findings of the present study, the majority of water quality parameters of the drinking water in distribution pipes of the 5 studied cities exceeded Iranian national and EPA standards. Therefore, it seems there is a problem with the usage of water in some specific industries. Moreover, the drinking water in the selected cities was not chemically stable, and thus it seems there is a problem of water corrosion and scaling in the drinking water distribution systems. Corrosion byproducts and leakage of heavy metals and iron in drinking water pipes are common results of aggressive waters. Therefore, it is recommended that corrosion and scaling indices be used in future studies to attain the information needed by water operators to predict the extent of corrosion and metal leakage in water distribution networks and to enable the continuous monitoring of water quality.

Acknowledgments

This research was supported by the Deputy of Research and Technology, Ahvaz Jundishapur University of Medical Sciences. The authors hereby express their gratitude.

Competing interests

The authors declare that they have no competing interests.

Authors' contributions

All authors contributed equally and were involved in designing the study, data collection, and article approval.

References

1. Samaei MR, Ebrahimy A, Ehrampoosh MH, Talebi P, Khalili MH, Morovati R. A study of the physical and chemical quality of potable water in yazd. Journal of Toloo-e-Behdasht 2008; 6(2): 50-7. [In Persian].
2. Malakootian M, Fatehizadeh A, Meydani E. Investigation of corrosion potential and precipitation tendency of drinking water in the Kerman distribution system. Toloo-e-Behdasht 2012; 11(3): 1-10. [In Persian].
3. Sarin P, Snoeyink VL, Bebee J, Jim KK, Beckett MA, Kriven WM, et al. Iron release from corroded iron pipes in drinking water distribution systems: effect of dissolved oxygen. Water Res 2004; 38(5): 1259-69.
4. Teimouri F, Sadeghi M, Drees F, Hashemi H, Shakeri K, Rezaei S. Survey of corrosion or scaling potential of resources, storage and distribution of waste supply system in Kian by using Langelier and Ryznar indexes. Journal of Health System Research 2012; 8(1): 78-84. [In Persian].
5. Malakootian M, Momeni J. Quality survey of drinking water in Bardsir, Iran 2009-2010. Scientific Journal of Rafsanjan University 2012; 11(4): 403-10. [In Persian].
6. Mahvi AH, Dindarlou K, Jamali HA, Alipour V. Corrosion and scaling in Bandar Abbas Pipe water network. Hormozgan Medical Journal 2011; 14(4): 335-40. [In Persian].
7. White C, Tancos M, Lytle DA. Microbial community profile of a lead service line removed from a drinking water distribution system. Appl Environ Microbiol 2011; 77(15): 5557-61. doi: 10.1128/AEM.02446-10.
8. Dietrich AM, Glindemann D, Pizarro F, Gidi V, Olivares M, Araya M, et al. Health and aesthetic impacts of copper corrosion on drinking water. Water Sci Technol 2004; 49(2): 55-62.
9. Koch GH, Brongers MP, Thompson NG. Cost of Corrosion Study Unveiled. Houston: NACE International; 2002.
10. Ghaneian M, Ehrampoush M, Ghanizadeh G, Amrollahi M. Survey of corrosion and precipitation potential in dual water distribution system in Kharanagh district of Yazd province. Toloo-e-Behdasht 2008; 7(3-4): 65-72. [In Persian].
11. Ebrahimi A, Kamarehie B, Asgari G, Seid Mohammadi A, Roshanaei G. Drinking water corrosivity and sediment in the distribution network of Kuhdasht, Iran. Health System Research 2012; 8(3): 480-6. [In Persian].
12. Jaafarzadeh Haghighi Fard N, Maraghie S, Marashi S, Mobad P. Assessment of bacterial, chemical and physical parameters in water treatment intake of Ahvaz. Journal of Jundishapour University of Medical Sciences 2008; 6(2): 227-34. [In Persian].
13. Motesaddi Zarandi S, Paseban A, Atamaleki A, Ahmadabadi M, Yanegh O, Ghorbanpoor R, et al. Corrosion and scaling potential of Bojnurd drinking. Journal of North Khorasan University of Medical Sciences 2015; 6: 913-24.
14. Mokhtari SA, Aalighadri M, Hazrati S, Sadeghi H, Gharari N, Ghorbani L. Evaluation of corrosion and precipitation potential in Ardebil drinking water distribution system by using Langelier & Ryznar indexes. Journal of Health 2010; 1(1): 14-23. [In Persian].
15. Zazouli M, Barafrashtehpour M, Sedaghat F, Mahdavi Y. Assessment of scale formation and corrosion of drinking water supplies in Yasuj (Iran) in 2012. Journal of Mazandaran University of Medical Sciences 2013; 22(2):100-8. [In Persian].
16. Rabbani D, Miranzadeh MB, Motlagh AA. Study for determination of industrial water corrosivity in Kashan Fajre Sepahan Galvanizing Mills during 2005-2006 Iran. Pak J Biol Sci 2008; 11(1): 131-4.
17. Savari J, Jaafarzadeh Haghighi Fard N, Hassani AH, Shams Khoramabadi G. Heavy metals leakage and corrosion potential in Ahvaz drinking water distribution network. Water and Wastewater 2007; 18(4): 16-24. [In Persian].
18. APHA. Standard Methods for the examination of water and wasfewater. 20th ed. Washington DC, USA: American Public Health Association/American Water Works Association/Water Environment Federation; 1998.
19. Shabankareh Fard E, Hayati R, Dobaradaran S. Evaluation of physical, chemical and microbial quality of distribution network drinkingwater in Bushehr, Iran. Iranian South Medical Journal 2015; 17(6): 1223-35.
20. Al-Rawajfeh AE, Al-Shamaileh EM. Inhibition of corrosion in steel water pipes by ammonium pyrrolidine dithiocarbamate (APDTC). Desalination 2007; 206(1–3): 169-78. doi: 10.1016/j.desal.2006.02.065.
21. Al-Rawajfeh AE, Al-Shamaileh EM. Assessment of tap water resources quality and its potential of scale formation and corrosivity in Tafila province, South Jordan. Desalination 2007; 206(1): 322-32. doi: 10.1016/j.desal.2006.01.039.
22. Atasoy AD, Yesilnacar MI. Effect of high sulfate concentration on the corrosivity: a case study from groundwater in Harran Plain, Turkey. Environ Monit Assess 2010; 166(1-4): 595-607. doi: 10.1007/s10661-009-1026-2.
23. Schock M. Internal corrosion and deposition control. In: American Water Works Association. Water Quality and Treatment: A Handbook of Community Water Supplies. New York: McGraw-Hil; 1999. p. 1-17.
24. Agatemor C, Okolo PO. Studies of corrosion tendency of drinking water in the distribution system at the University of Benin. Environmentalist 2008; 28(4): 379-84. doi: 10.1007/s10669-007-9152-2.
25. Husband PS, Boxall JB. Asset deterioration and discolouration in water distribution systems. Water Res 2011; 45(1): 113-24. doi: 10.1016/j.watres.2010.08.021.
26. Zhang Z, Stout JE, Yu VL, Vidic R. Effect of pipe corrosion scales on chlorine dioxide consumption in drinking water distribution systems. Water Res 2008; 42(1-2): 129-36. doi: 10.1016/j.watres.2007.07.054.
27. Vermilyea DA. Stress corrosion cracking of iron and nickel base alloys in sulfate solutions at 289 C. Corrosion 1973; 29(11): 442-50. doi: 10.5006/0010-9312-29.11.442.
28. Nawrocki J, Raczyk-Stanislawiak U, Swietlik J, Olejnik A, Sroka MJ. Corrosion in a distribution system: Steady water and its composition. Water Res 2010; 44(6): 1863-72. doi: 10.1016/j.watres.2009.12.020.
29. Pierre RR. Corrosion Basics: An Introduction. 2nd ed. Houston TX: NACE; 2006.

30. Sarin P, Snoeyink VL, Bebee J, Jim KK, Beckett MA, Kriven WM, et al. Iron release from corroded iron pipes in drinking water distribution systems: effect of dissolved oxygen. Water Res 2004; 38(5): 1259-69. doi: 10.1016/j.watres.2003.11.022.
31. Roberge PR. Handbook of Corrosion Engineering. New York: McGraw-Hill; 1999.
32. Nazemie S, Fadaie H. Survey of corrosion and precipitation potential in water distribution system in Shahrod city using water corrosivity index. 16th National Environmental Health Conferences; Oct 1-3, 2013. Tabriz: Tabriz University of Medical Sciences; 2013. [In Persian].
33. Singley JE, Beaudet BA, Markey PH, DeBerry DW, Kidwell JR, Malish DA. Corrosion Prevention and Control in Water Treatment and Supply Systems. Park Ridge: Noyes Publications; 1985.
34. Azari A, Nazemi S, Kakavandi B, Rastgar A. Survey of scaling and corrosion potential in drinking water resources of Shahrood city by using stability indexes in 2013. Journal of Sabzevar University of Medical Sciences 2016; 22(6): 944-54. [In Persian].
35. Savari J, Jaafarzadeh Haghighifard N, Hassani AH, Shams Khoramabadi GH. Heavy metals leakage and corrosion potential in Ahvaz drinking water distribution network. Water Wastewater J 2008; 18(4): 16-24.
36. Rabbani D, Miranzadeh MB, Paravar A. Evaluating the corrosive and scale-forming indices of water in the villages under the coverage of Kashan Rural Water and Wastewater Company During 2007-9. Feyz 2011; 15(4): 382-8. [In Persian].
37. Selvam S, Manimaran G, Sivasubramanian P, Balasubramanian N, Seshunarayana T. GIS-based evaluation of water quality index of groundwater resources around Tuticorin coastal city, south India. Environ Earth Sci 2014; 71(6): 2847-67. doi: 10.1007/s12665-013-2662-y.
38. Ravikumar P, Aneesul Mehmood M, Somashekar RK. Water quality index to determine the surface water quality of Sankey tank and Mallathahalli lake, Bangalore urban district, Karnataka, India. Appl Water Sci 2013; 3(1): 247-61. doi: 10.1007/s13201-013-0077-2.

24

Survey of water quality in Moradbeik river basis on WQI index by GIS

Mohammad Taghi Samadi[1], Shahram Sadeghi[2], Alireza Rahmani[3], Mohammad Hossien Saghi[4*]

[1]Associate Professor, Department of Environmental Health Engineering, School of Public Health and Center for Health Research, Hamadan University of Medical Sciences, Hamadan, Iran

[2]MSc Student of Environmental Health Engineering, Student Research Committee, Department of Environmental Health Engineering, School of Public Health, Kurdistan University of Medical Sciences, Sanandaj, Iran

[3]Professor, Department of Environmental Health Engineering, School of Public Health and Center for Health Research, Hamadan University of Medical Sciences, Hamadan, Iran

[4]Ph.D student of Environmental Health Engineering, Department of Environmental Health Engineering, School of Public Health, Shahid Beheshti University of Medical Sciences, Tehran, Iran

Abstract

Background: Survey of pollution and evaluation of water quality in rivers with Oregon Water Quality Index (OWQI) and GIS are effective tools for management of the impact of environmental water resources. The information in calculating the WQI of Moradbeik river allowed us to take our tests results and make a scientific conclusion about the quality of water. GIS can be a powerful tool for developing solutions for water resources problems for assessing water quality, determining water availability, preventing flooding, understanding the natural environment, and managing water resources on a local or regional scale.

Methods: The WQI of Moradbeik river consists of nine tests: Fecal Coliform (FC), Biochemical Oxygen Demand (BOD_5), Nitrates (NO_3), Total Phosphate (PO_4), pH, temperature, Dissolved Oxygen (DO), turbidity, and Total Solid (TS). Water quality of Moradbeik river was investigated for 12 months. Concentrations of these nine variables were normalized on a scale from 0 to 100 and translated into statements of water quality (excellent, good, regular, fair, and poor). Also this data were analyzed with WQI index, and then river basis on water quality was zoning by GIS.

Results: The average of WQI was 61.62, which corresponded to "medium" quality water at the sampling point 1 (best station) and decreased to around 26.41 (bad quality) at sampling point 6. The association between sampling points and water quality indexes was statistically significant ($P<0.05$).

Conclusion: Based on physical, chemical and biological agent monitoring and also with control of water quality indexes of these points, we observed wastewater and other river pollutants.

Keywords: Water quality evaluation, Water quality index, GIS, Moradbeik

***Correspondence to:**
Mohammad Hossien Saghi
Email: saghi9@gmail.com

Introduction

Different regions of the world are being faced by different types of problems associated with the occurrence, use, and control of water resources, which may endanger the sustainable development of these resources. The quality of surface waters is a very sensitive issue. Anthropogenic influences as well as natural processes degrade surface waters and impair their use for drinking, industry, agriculture, recreation, and other purposes (1,2). The Water Quality Index (WQI) has been considered to give a criteria for surface water classification based on the use of standard parameters for water characterization (3,4). So chemical, physical, and biological constituents are quantified in all rivers of the world. The problem is the quantity of analysis required and the cost to accomplish them. In order to resolve this problem, regulatory agencies have been creating and using a general index as a management tool. One of the oldest of these tools is the WQI developed by the National Sanitation Foundation (NSF) of United States which is based on the analysis of nine parameters: Fecal Coliform (FC), Biochemical Oxygen Demand (BOD_5), Nitrates (NO_3), Total Phosphate (PO_4), pH, temperature, Dissolved Oxygen (DO), turbidity, and Total Solid (TS). Its output ranges from 0 to 100, where 100 represents perfect water quality conditions while zero indicates water that is not suitable for the intended use without further treatment. This index is a mathematical instrument used to transform large quantities of water characterization data into a single number, which represents the water quality level. The use of WQI is a simple practice, which allows adequate classification of water quality (5,6). The general WQI was developed by Brown *et al* and improved by Deininger for the NSF (7). The Scottish Development Department (SDD) in collaboration with regional institutions for river quality preservation, Solway Purification Board (Solway RPB), the Tweed Purification (Tweed RPB) carried out an extensive research in order to evaluate the river quality in Scotland according to an NSF study (8). Horton suggested that various water quality data could be integrated into an overall index. Over the years many indices have been calculated, each for its own purpose as follows: Dalkey, Liebman, Prati *et al*, O'Connor, Harkins, Walski and Parker, Inhaber, and Service for Rhine Pollution Monitoring (RPM) (8). The purpose of the Oregon Water Quality Index (OWQI) was to improve understanding of water quality issues by integrating complex data and generating a score that describes water quality status and evaluates water quality trends. Although some information is lost when integrating multiple water quality variables, this loss is outweighed by the gain in understanding of water quality issues by the public and policy makers. Improved understanding is very important to water resource managers in terms of increased support for water resource improvement efforts. The science of water quality has improved markedly since the introduction of OWQI in the 1970s (8). OWQI was improved in 1995 to reflect advances in the knowledge of water quality and designing water quality indices. The primary purpose of this report is to describe the historical basis and define the improved design of OWQI. Improved OWQI is widely used and maintained by Oregon Department of Environmental Quality (DEQ). OWQI has been used to report water quality status and trends in Oregon to state legislators and other water resource policy makers via presentations and to the public through reports accessible on the internet. OWQI has been used as a supplement to more traditional reporting formats, such as Water Quality Status Assessment (305(b)) Report (9) and the annual McKenzie Watershed Water Quality Report (9). The quality of water is equally important as its quantity. Remote sensing and GIS are effective tools for water quality mapping and land cover mapping essential for monitoring, modeling, and environmental change detection. Geographic Information System (GIS) can be a powerful tool for developing solutions for water resources problems in assessing water quality, determining water availability, preventing flooding, understanding the natural environment, and managing water resources on a local or regional scale (10,11). Therefore, the aim of this study was to determine the WQI of stream reaches exposed to point and non-point discharges of flow through trout farm and diffuse discharges of rural community and agricultural, then this river zoning by GIS.

Methods

Description of the watershed investigated

Moradbeik river drains a 30 km^2 catchment in south of Hamedan. The population relies on Moradbeik river as a source of domestic, industrial, and irrigation water; habitat for fish and other wildlife; and a place to recreate. The river also receives highly treated and untreated municipal and industrial wastewater from the urban population. The catchment is illustrated in (Table 1). The main river length is 18 km.

Procedure for watershed sampling

Collection vials, sample stabilization, and transportation to the laboratory as well as sample storage were done considering the recommendations of standard methods (12) and Environmental Protection Agency (EPA) (13). Samples were taken from at least 40 cm under the water surface and whenever it was possible, at the middle of the stream. Samples were never taken when it was raining, but only at least 72 h after the rain had stopped, so that the river had returned to its usual flow condition. Sampling for watersheds was carried out over 12 months, covering all stations. The samples were taken approximately every month and were transported to the laboratory for quantification of the other parameters after the determination of field parameters.

Sample collection and analysis

Temperature, pH (standard method 4500), and Dissolved Oxygen (DO) (standard method 4500) values of water samples were measured in the field using a thermometer and a portable pH meter (HACH, sension 1) and DO meter (HACH, 6), respectively. Samples for Biochemical Oxygen Demand (BOD) (standard method 5210) were collected in 250-ml-glass bottles which were closed under the water. Polythene bottles were soaked in 1.5 m nitric acid for 4 days and washed thoroughly with distilled and deionized water before use. Water samples were collected in polythene bottles. BOD was determined after incubation for 5 days in tightly stoppered bottles in the dark at 20 °C and determination of the oxygen consumed (12). The total suspended solids and total dissolved solids were separated by filtering the water through 0.45 mm paper filter and determined according to standard procedures (12). Conductivities were measured at 25 °C directly usy ing a digital conductivity meter (HACH, session 5). Nitrate (NO_3^-) (standard method 4500) was determined by spectrophotometric method at 220 nm (UV-1700 Pharma Spect Shimadzo) Inorganic phosphates (P-PO_4^{-3}) (4500) were estimated by the phosphomolybdate method photometrically after conversion of orthophosphates by being digested with persulfate. Turbidity was quantified using

Table 1. Description of the sampling points

Point	Altitude (m)	Geographic coordinates
1	2080	34°44.820 48°30.261
2	1960	34°45.314 48°30.619
3	1940	34°45.593 48°30.666
4	1900	34°46.769 48°30.702
5	1778	34°48.981 48°31.116
6	1740	34°51.497 48°30.959

a turbidimetry HACH (2100N). Fecal Coliforms were assayed according to the membrane filter technique (colony/100 mL).

Water Quality Index (WQI) calculations

WQI is a mathematical instrument used to transform large quantities of water quality data into a single number which summarizes different quality parameters. WQI is an index of water quality for a particular use. Mathematically, the index is an arithmetic weighing of normalized water quality measurements. The normalizations, as well as the weighings are different for different water uses (5). WQI used in this paper was calculated in three different ways and based on parameters: the first one produced a NFSWQI based on the parameters proposed by Brown and Forsythe (14) including: biochemical oxygen demand, dissolved oxygen, total fecal coliform, pH, temperature, total nitrate, total phosphorus, total solids, and turbidity. This WQI is calculated using the following equation:

$$WQI = \sum_{i=1}^{n} w_i q_i$$

where "WQI" is a number between 0 and 100 to indicate the water quality index; "q_i" is water quality score of parameter, "a" number between 0 and 100, obtained from the respective "curve average of quality" variation, as a function of concentration nor measurement; "n" the number of parameter used to calculated WQI, and "w_i" the weighing factor of parameter, a number between 0 and 1.

Results

Water quality of Moradbeik river was investigated for 12 months. The index was applied to six monitoring points in the aquatic bodies. Concentrations of these nine variables were normalized on a scale from 0 to 100 and translated into statements of water quality (excellent, good, regular, fair, and poor) (for example Figure 1A-B). The value of NSFWQI in 12 months is shown in (Figure 2).

The association between month and NSFWQI is not statistically significant (P>0.05; Figure 3). The association between station and NSFWQI was statistically significant (P<0.05; Figure 4).

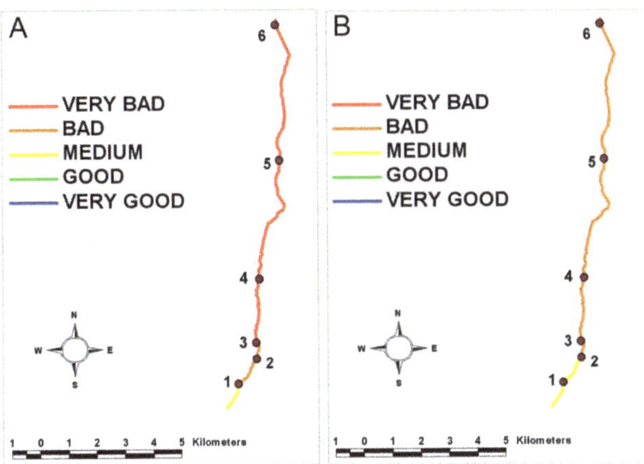

Figure 1. Water quality index map. A) October and November. B) September.

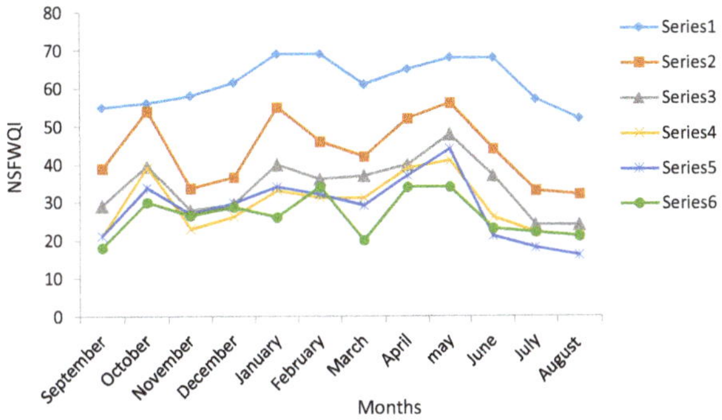

Figure 2. Value of NSFWQI in 12 months

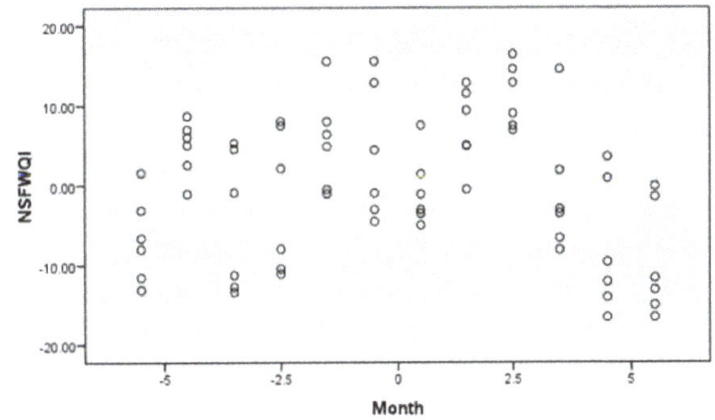

Figure 3. Association between month and NSFWQI

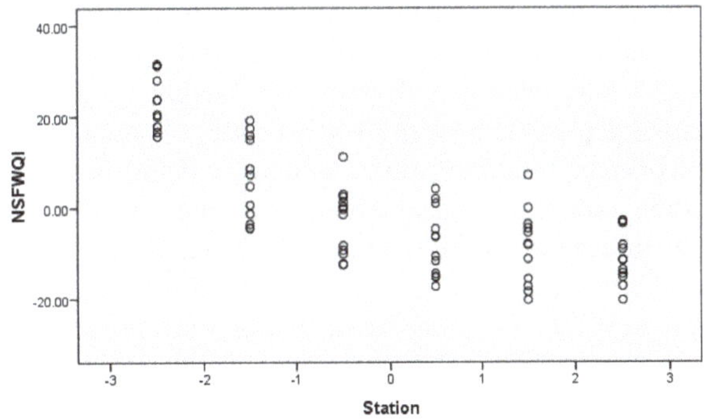

Figure 4. Association between stations and NSFWQI

Discussion

Gazzaz *et al* (15) surveyed the artificial neural network modeling of the water quality index for Kinta river (Malaysia) in 2012 using water quality variables as predictors. The result of a survey by Koçer and Sevgili (16) in 2014 showed that WQI_{min} calculated using NH^+_4-N and TON is a useful and easily applicable methodology in the assessment of the impacts of trout farm effluents on the stream water quality. Modified pollution indices were used to evaluate the pollution status in the middle section of the Lower Seyhan River Basin by employing a GIS software (ArcGIS 9.3) for data processing, estimations, and evaluations. Air Quality Index (AQI) and WQI were utilized to evaluate air and water pollution levels, respectively (12). A survey by Ma *et al* (17) in 2013 with subject of "a modified water quality index for intensive shrimp ponds of Litopenaeus vannamei" has shown modified WQI based on the varifactors was applied to evaluate the water quality in shrimp culture ponds. The result revealed that the overall water quality in the shrimp ponds was mainly excellent during the early period and deteriorated in the mid to late period. The average of NSFWQI was 61.62, which corresponds to "medium" quality water at sampling point 1 (16), the best station, and decreased to around 6.41 (bad quality) at sampling point 6. The demonstrated monitoring of Moradbeik watersheds was slightly affected by the section of the river within this town. The association between month and NSFWQI was not statistically significant (P>0.05; Figure 3). The association between station and NSFWQI was statistically significant (P<0.05; Figure 4). The results of this study are consistent with Gazzaz *et al*'s study. The WQI predictions of this model had a significant, positive, and very high correlation (r= 0.977, P<0.01) with the measured WQI values, implying that the model predictions explain around 95.4% of the variation in the measured WQI values (15). WQI_{min} calculation with NH_4^+-N, Total Organic Nitrogen (TON), soluble reactive phosphorus, and total organic phosphorus, which were selected using the principal component analysis findings meaningfully, classified the sampling points. Further reduction of parameters to NH_4^+-N and TON in WQI_{min} calculation achieved a similar successful classification of the sampling points (16). Sampling point 1 were more suitable than other sampling points for drinking uses. This station's water with pH adjustment and primary treatment can be used for drinking, but other stations are not potable and must be post treated. Results indicated that for the available data and time frame considered in the study, air and water qualities generally were in good conditions (low pollution), yet, precautions could still be taken for improvement (16,18). WQI evaluates water quality synthetically; furthermore, it reveals the outcome when some of the other variables deteriorate significantly. Thus, this study illustrates the necessity and usefulness of multivariate statistical techniques for interpreting large and complex data sets regarding water quality in shrimp culture pond. Furthermore, the evaluation results revealed that the modified WQI can be used as a tool for determining water quality. There is no index for significant statistical relationship between the months of the study sample (P>0.05).

Conclusion

Distribution of pollution, river pollution, and low rainfall in the months to months is more rain. Pollution of rivers in the rainy months also cannot generally undo. Based on physical, chemical, and biological agent monitoring and also with control of water quality indexes in these points, we observed that wastewater and other forms of pollution can arrive the river. Results also indicated the need for improvement of monitoring network for better assessment of the environmental quality in the whole basin (17). In general, GIS tools were very helpful in the development of the indices. There are many reasons for pollution levels in rivers, being more rural, urban, and industrial sewage discharge than the annual rainfall.

Acknowledgements

The authors are most grateful to the laboratory staff of the Department of Environmental Health Engineering, School of Public Health, Hamadan University of Medical Sciences, Iran, for their financial support and collaboration in this research.

Competing interests

The authors declare that they have no competing interests.

Authors' contributions

MTS, MHS, and AR conceived and designed the study. SHS and MHS performed the literature search and wrote the manuscript. All authors participated in the data acquisition, analysis and interpretation. All authors critically reviewed, refined and approved the manuscript.

References

1. Carpenter SR, Caraco NF, Correll DL, Howarth RW, Sharpley AN, Smith VH. Nonpoint pollution of surface waters with phosphorus and nitrogen. Ecol Appl 1998; 8(3): 559-68.
2. Jarvie HP, Whitton BA, Neal C. Nitrogen and phosphorus in east coast British rivers: speciation, sources and biological significance. Sci Total Environ 1998; 210-211: 79-109.
3. Couillard D, Lefebvre Y. Analysis of water quality indices. J Env Manage 1985; 21(2): 161–79.
4. Sánchez E, Colmenarejo MF, Vicente J, Rubio A, García MG, Travieso L, *et al.* Use of the water quality index and dissolved oxygen deficit as simple indicators of watersheds pollution. Ecol Indic 2007; 7(2): 315–28.
5. Chang NB, Chen HW, Ning SK. Identification of river water quality using the fuzzy synthetic evaluation approach. J Environ Manage 2001; 63(3): 293–305.
6. Simões FS, Moreira AB, Bisinoti MC, Gimenez SM, Yabe MJ. Water quality index as a simple indicator of aquaculture effects on aquatic bodies. Ecol Indic 2008; 8(5): 476-84.
7. He LM, He ZL. Water quality prediction of marine recreational beaches receiving watershed baseflow and stormwater runoff in southern California. Water Res 2008; 42(10-11): 2563-73.
8. Dunnette DA. A geographically variable water quality

index used in Oregon. Journal of the Water Pollution Control Federation 1979; 51(1): 53-61.
9. Cude CG. Oregon water quality index a tool for evaluating water quality management effectiveness. J Am Water Resour Assoc 2001; 37(1): 125-37.
10. Skidmore AK, Bijker W, Schmidt K, Kumar L. Use of remote sensing and GIS for sustainable land management. ITC Journal 1997; 3(4): 302-15.
11. Asadi SS, Vuppala P, Reddy MA. Remote sensing and gis techniques for evaluation of groundwater quality in municipal corporation of Hyderabad (Zone-V), India. Int J Environ Res Public Health 2007; 4(1): 45-52.
12. Clesceri LS, Greenberg AE, Eaton AD. Standard methods for the examination of water and wastewater. 20th edition. APHA American Public Health Association;1998
13. Walsh P, Wheeler W. Water quality index aggregation and cost benefit analysis. U.S. Environmental Protection Agency National Center for Environmental Economics 1200 Pennsylvania Avenue. July, 2012. Available from: http://www.epa.gov/economics
14. Brown MB, Forsythe AB. Robust tests for the equality of variances. Journal of the American Statistical Association 1974; 69(346): 364-7.
15. Gazzaz NM, Yusoff MK, Aris AZ, Juahir H, Ramli MF. Artificial neural network modeling of the water quality index for Kinta River(Malaysia) using water quality variables as predictors. Mar Pollut Bull 2012; 64(11): 2409-20.
16. Koçer MA, Sevgili H. Parameters selection for water quality index in the assessment of the environmental impacts of land-based trout farms. Ecol Indic 2014; 36: 672-81.
17. Ma Z, Song X, Wan R, Gao L. A modified water quality index for intensive shrimp ponds of Litopenaeus vannamei. Ecol Indic 2013; 24: 287-93.
18. Golge M, Yenilmez F, Aksoy A. Development of pollution indices for the middle section of the Lower Seyhan Basin (Turkey). Ecol Indic 2013; 29: 6-17.

Evaluation the concentration of mercury, zinc, arsenic, lead and cobalt in the Ilam city water supply network and resources

Ahmad Reza Yazdanbakhsh[1], **Mohammad Hasan Abasi**[2], **Zeinab Gholami**[3], **Moayed Avazpour**[4,5*]

[1]Associate Professor of Environmental Health Engineering, Department of Environmental Health Engineering, School of Public Health, Shahid Beheshti University of Medical Sciences, Tehran, Iran.
[2]BSc of Environmental Health, Ilam University of Medical Sciences, Ilam, Iran.
[3]MSc of Environmental Health Engineering, Student Research Committee, School of Public Health, Tehran University of Medical Sciences, Tehran, Iran
[4]PhD Student Research Office of Environmental Health, Department of Environmental Health Engineering, School of Public Health, Shahid Beheshti University of Medical Sciences, Tehran, Iran
[5]Lecturer of Environmental Health, Department of Environmental Health Engineering, Ilam University of Medical Science, Ilam, Iran

Abstract

Background: The presence of heavy metals in water resources above threshold levels can be toxic and carcinogenic for consumers. This study determined the concentrations of heavy metals in the drinking water distribution network and resources of the city of Ilam in Iran.

Methods: In this cross-sectional study from 6 sources of water supply and also, different parts of the water supply system of Ilam city, samples were collected based on standard sampling methods. The samples were tested with a BRAIC atomic absorption spectrophotometer. The data was analyzed using nonparametric Mann-Whitney test.

Results: The concentration of zinc in all water sources of the city of Ilam was higher than WHO guidelines and Iranian standard 1053. Contamination by cobalt, arsenic and lead from Ilam dam, Pich-e Ashoori well and Haft Cheshmeh well was higher than national and international standards. The amount of cobalt and mercury at Ilam dam was significantly different from the levels at other sources ($P < 0.05$).

Conclusion: The use of pesticides in the agricultural sector, contamination of water by human waste and aged and worn water pipes are the likely sources of the increased concentrations of heavy metals, especially lead and arsenic. Because there is a cumulative effect from these metals, appropriate measures are necessary by the relevant agencies to address this problem.

Keywords: Water resources, Distribution network, Heavy metals, Ilam

***Correspondence to:**
Moayed Avazpour
Email: m_f_1859@yahoo.com

Introduction

Population growth has increased the use of water for consumption, agriculture, industry and other uses. This not only affects the quantity of water resources, but also reduces their quality (1,2). Today, in addition to microbial contamination, chemical contamination is a major safety issue related to water.

Contamination of water resources by heavy metals is a serious environmental problem (3,4). High levels of heavy metals in the water have adverse effects on human health (5). Metals enter water resources either naturally or through contamination. Rock and soil in contact with water are the largest natural source of contamination by heavy metals (6). Major sources of industrial pollution, such as mining, and some fertilizers can lead to disposal of untreated or semi-refined wastewater containing heavy metals (7,8).

Heavy metals do not degrade; they gradually precipitate and accumulate in fatty tissues, muscles, bones and joints of creatures (9). Mercury, lead, arsenic and metals that exhibit high electronegativity, such as nickel and cobalt, have a strong affinity for amino groups and sulfhydryl groups. Enzymes are destroyed by these metals and lose power. In addition, metals enter and disrupt human metabolism (10,11).

Lead, zinc, cadmium, arsenic, cobalt and mercury are among the most dangerous poisons (12). Arsenic is not a heavy metal, but a metalloid (13). The main target of arsenic is the brain. Skin cancer can also be an effect of chronic exposure to arsenic (14). Lead can cause mental retardation, behavioral disorders, memory problems and mood changes. The most important effects of lead is impairment of the neurological development of children (15). Zinc can cause lung, prostate and kidney cancer. Cobalt can cause skin and cellular damage. Mercury can impair the senses of touch, sight and hearing (16).

Many studies have examined the role of heavy metals on human health in various aquatic environments. Babaei et al (14) in Kashmir, Ghaemi et al in Tehran (17), Mohammedan et al in Zanjan (18), Baptista et al in Brazil (19), Wang et al in China (20) and Zorer et al in Turkey (21) have all conducted such studies. The drinking water in Ilam, Iran derives from various dams and well in the region. The present study evaluated the quality of the drinking water in the distribution network resources for Ilam for the presence of heavy metals (mercury, zinc, arsenic, lead and cobalt) and compared the percentages with national standards and international guidelines.

Materials and Methods

The present research is a cross-sectional study conducted in 2015. After competing the necessary surveys and library research and identifying the resources and water distribution networks of Ilam, the appropriate sample size was determined. To ensure coverage of Ilam for sampling, water reservoirs and the distribution networks associated with each storage tank were selected as sampling stations. A total of six water sources were determined for Ilam: Gham Gerdalan (Ilam) dam, Pich-e Ashoori well, Haft Cheshmeh well, the Ghoch Ali wells, Naghlieh well and Gol Gol spring.

Sampling was conducted monthly over the course of four months. At each sampling time, 30 samples (6 source of drinking water and 4 samples from any source distribution network) were collected. Throughout the sampling period, 120 samples were collected and sent to the laboratory. Sampling, storage and fixing of the samples was performed according to standard methods (22). The samples were transferred to the Laboratory of Pharmacology to determine the levels of mercury, zinc, arsenic, lead and cobalt. Samples were tested using an atomic absorption spectrophotometer (BRAIC; model WFX-130; China). To determine the concentration of these metals in the distribution network resources, the data was analyzed at a significance level of $P = 0.05$. Because non-normal data was used, the non-parametric Mann-Whitney test was applied in SPSS software. The results were compared with national standards (standard 1053 for Iranian water) and international standards (WHO and USEPA; Table 1).

Table 1. Average concentration of heavy metals, according to national and international standards

Heavy metals	National and international standards		
	WHO	EPA	Standard 1053 of Iran
Lead (mg/L)	0.01	0.015	0.05
Arsenic (mg/L)	0.01	-	0.005
Mercury (mg/L)	0.001	-	0.006
Zinc (mg/L)	5	5	3
Cobalt (mg/L)	-	0.005	0.005

Results

Table 2 lists the mean, standard deviation (SD) and minimum and maximum concentrations of mercury in drinking water sources Ilam. The results show that from the six main sources of drinking water in Ilam, only Ilam dam and Gol Gol spring show traces of mercury. No other source was found to contain mercury. The mercury concentration in the drinking water sources of Ilam was lower than the average concentration of heavy elements in accordance with national and international standards.

Table 3 shows that the concentration of zinc in all water sources in the city of Ilam was higher than national and international standards. The maximum amount of zinc recorded was 6.0440 mg/L from Naghlieh well. The minimum concentration was recorded for Gham Gerdalan dam.

The results showed that the arsenic levels in water resources of Ilam dam and Gol Gol spring were higher than the national and international standards, but all other sources recorded values below these standards. The lowest amount of arsenic was observed in Naghlieh well (0.0038 mg/L; Table 4).

Table 5 shows that the concentrations of lead at Ilam dam and Gol Gol spring were higher than the national and international standards. All other sources of drinking water for Ilam recorded concentrations below these standards. The lowest amounts of lead were observed in Haft Cheshmeh well (0.0075 mg/L) and Naghlieh well (0.0078 mg/L).

Table 6 shows the amount of cobalt in drinking water sources of Ilam. The results showed that the maximum concentration of cobalt was found in Naghlieh well (0.042 mg/L) and Ilam dam (0.014 mg/L). Cobalt concentrations

Table 2. Mean, SD, minimum and maximum concentrations of mercury in drinking water of Ilam

Water resources	Hg(mg L^{-1})				
	N	Mean	SD	Minimum	Maximum
Gham Gerdalan dam	20	0. 000605	0.0001938	0.0003	0.0009
Pich Ashoori well	20	0.000	0.000	0.000	0.000
Haft Cheshmeh well	20	0.000	0.000	0.000	0.000
Ghoch Ali wells	20	0.000	0.000	0.000	0.000
Naghlieh well	20	0.000	0.000	0.000	0.000
Gol Gol	20	0.000215	0.0000233	0.000	0.0009

Abbreviation: SD, Standard Deviation.

Table 3. Mean, SD, minimum and maximum concentrations of zinc in drinking water of Ilam

Water resources	Zn (mg/L)				
	N	Mean	SD	Minimum	Maximum
Gham Gerdalan dam	20	4.4050	1.02211	2.90	5.80
Pich Ashoori well	20	5.0950	0.20641	4.80	5.50
Haft Cheshmeh well	20	5.0755	0.49318	4.70	5.71
Ghoch Ali wells	20	4.6925	0.39633	4.00	5.60
Naghlieh well	20	6.0440	0.68015	4.50	5.80
Gol Gol	20	4.7950	0.62275	3.60	5.60

Abbreviation: SD, Standard Deviation.

Table 4. Mean, SD, minimum and maximum concentrations of arsenic in drinking water of Ilam

Water resources	As (mg/L)				
	N	Mean	SD	Minimum	Maximum
Gham Gerdalan dam	20	0.05075	0.00715	0.0350	0.0620
Pich Ashoori well	20	0.0095	0.00153	0.0070	0.0120
Haft Cheshmeh well	20	0.0062	0.00034	0.0058	0.0070
Ghoch Ali wells	20	0.0051	0.00074	0.0035	0.0061
Naghlieh well	20	0.0038	0.00020	0.0300	0.0042
Gol Gol	20	0.0507	0.007293	0.0035	0.0360

Abbreviation: SD, Standard Deviation.

Table 5. Mean, SD, minimum and maximum concentrations of lead in drinking water of Ilam

Water resources	Pb (mg/L)				
	N	Mean	SD	Minimum	Maximum
Gham Gerdalan dam	20	0.06770	0.01067	0.0500	0.0850
Pich Ashoori well	20	0.01688	0.00116	0.0150	0.0186
Haft Cheshmeh well	20	0.00750	0.00082	0.0060	0.0089
Ghoch Ali wells	20	0.01591	0.00353	0.0100	0.0220
Naghlieh well	20	0.00789	0.00091	0.0060	0.0089
Gol Gol	20	0.06640	0.01682	0.0450	0.0870

Abbreviation: SD, Standard Deviation.

Table 6. Mean, SD, minimum and maximum concentrations of cobalt in drinking water of Ilam

Water resources	Co (mg/L)				
	N	Mean	SD	Minimum	Maximum
Gham Gerdalan dam	20	0.01412	0.00407	0.0090	0.0200
Pich Ashoori well	20	0.00890	0.00083	0.0070	0.0100
Haft Cheshmeh well	20	0.00189	0.00016	0.0013	0.0022
Ghoch Ali wells	20	0.00045	0.00008	0.0003	0.0006
Naghlieh well	20	0.04230	0.00435	0.0350	0.0500
Gol Gol	20	0.00037	0.00016	0.0001	0.0007

Abbreviation: SD, Standard Deviation.

were also recorded at the Ghoch Ali wells.

Comparison the results of heavy elements present in all water sources are shown in Table 7. The results show that there was a significant difference between the amount of mercury found at Ilam dam and other sources of drinking water ($P<0.05$). There was no significance difference ($P>0.05$) between the concentration of zinc found at Ilam dam, Ghoch Ali wells and Gol Gol spring, but compared to the other sources, there was a significant difference ($P<0.05$). The arsenic concentrations at Ilam dam and Gol Gol spring showed no significant difference, but these values were significantly different from the amount found in the other sources ($P<0.05$). There was no significant difference ($P>0.05$) between of the concentrations of lead at Ilam dam and Gol Gol spring, but these values were significantly different when compared with the other sources ($P<0.05$). The concentration of cobalt at Ilam dam was significantly different from those at the other sources ($P<0.05$).

Figure 1 shows the differences in zinc concentrations for the different sources of drinking water of Ilam. Figure 2 shows the differences in the concentrations of other heavy elements.

Discussion

The use of heavy metals in industry has caused these metals to enter the environment in various ways. Heavy metals play essential roles in living bodies for building

Table 7. Comparison of heavy metal concentrations in drinking water of Ilam

Water resources		*P* value				
		Hg	Zn	As	Pb	Co
Gam Gerdalan dam	Pich Ashoori well	< 0.001	0.004	< 0.001	< 0.001	< 0.001
	Haft Cheshmeh well	< 0.001	0.005	< 0.001	< 0.001	< 0.001
	Ghoch Ali wells	< 0.001	0.626	< 0.001	< 0.001	< 0.001
	Naghlieh well	< 0.001	0.009	< 0.001	< 0.001	< 0.001
	Gol Gol	< 0.001	0.286	1.0	0.996	< 0.001
Pich Ashoori well	Haft Cheshmeh well	1.0	1.0	0.999	0.007	< 0.001
	Ghoch Ali wells	1.0	0.253	0.998	0.999	< 0.001
	Naghlieh well	1.0	1.0	0.991	0.011	< 0.001
	Gol Gol	< 0.001	0.581	< 0.001	< 0.001	< 0.001
Haft Cheshmeh well	Ghoch Ali wells	1.0	0.305	1.0	0.01	0.434
	Naghlieh well	1.0	1.0	1.0	1.0	< 0.001
	Gol Gol	< 0.001	0.650	< 0.001	< 0.001	0.372
Ghoch Ali wells	Naghlieh well	1.0	0.402	1.0	0.032	< 0.001
	Gol Gol	< 0.001	0.994	< 0.001	< 0.001	1.0
Naghlieh well	Gol Gol	< 0.001	0.755	< 0.001	< 0.001	< 0.001

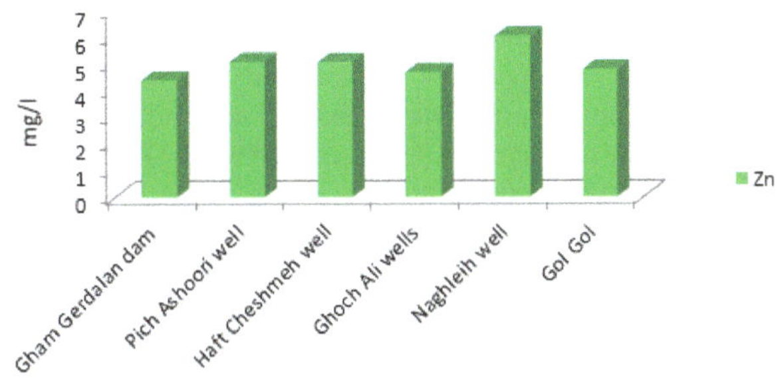

Figure 1. Concentrations of zinc in sources of drinking water of Ilam.

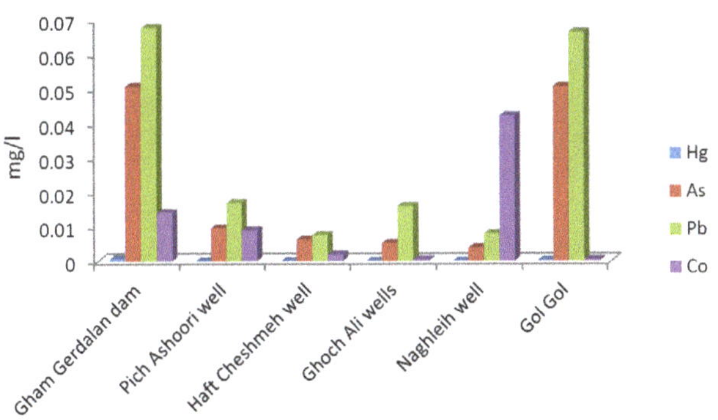

Figure 2. Concentrations of mercury, arsenic, lead and cobalt in sources of drinking water of Ilam.

molecules and acting as coenzymes. Some metals within a specific desirable amount are essential to life. The amount of metals in different ecosystems is influenced by different factors. They may occur through in aquatic systems as the results of natural factors, such as soil erosion and flooding, or by artificial factors, including human activity (23).

The results indicate that the contamination of Ilam dam and Gol Gol spring with arsenic, lead and cobalt is higher than at other sources. It appears that pollution by heavy metal from these two sources originated from human and animal waste, agricultural activity, feeder, cattle and poultry farms, asphalt factories, stone crushers, greenhouse complexes and landfill in the area through which the river basin flows into Ilam dam and Gol Gol spring. Pollution of the drinking water with mercury was lower than the national and international standards. The maximum concentration of zinc pollution was recorded at Naghlieh well (6.044 mg/L).

Orhan Gendoz et al evaluated arsenic levels in the Simav plain in Turkey and reported that the average concentration of arsenic was 0.099 mg/L and the maximum was 0.561 mg/L (24). Mosaferi et al studied arsenic levels in water resources in Hashtrood in Iran and found that arsenic in the drinking water of 50 villages was higher than the national standard (25). In the current study, the concentrations of arsenic in at Gam Gerdalan dam and Gol Gol spring were high and above national and international standards. At the other resources, it was lower than national international standards.

The difference in arsenic levels could be due to the use of herbicides in agricultural land in the dam basin. A survey in 2013 by Ho et al on surface and internal contamination of estuarine sediment of the Kam River in Haiphong in Vietnam with heavy metals found that the development of industrial and agricultural activity in the region had significantly increased pollution by heavy metals such as lead, zinc and arsenic (26).

Nahid et al examined heavy metals in the drinking water in different parts of the city of Tehran in Iran and concluded that the concentration of lead in some areas was higher than USEPA standards (27). Mohammedan et al examined pollution of water wells in Zanjan and showed that lead concentrations in the water resources in Zanjan were above WHO standards. In the present study, the amount of lead in the water supply at Ilam dam was 0.0679 mg/L and at Gol Gol spring was 0.0661 mg/L, which are above national and international standards.

High levels of lead can be of natural or human origin. Lead particles from gasoline combustion directly or indirectly land on the surface of the soil and plants and ultimately enter water systems. The existence of high levels of lead in the water supply at Ilam dam, Gol Gol spring and the water distribution network could have originated from the entry of lead from aging pipes and fittings used in water distribution systems themselves.

Savari et al investigated heavy metals in water resources in the city of Ahvaz in Iran and reported the amount of zinc to be 3.18 mg/L. This is inconsistent with the results of the present study which found that the concentrations of zinc in all water resources of Ilam was higher than that amount. High levels of elements like cobalt and zinc in the wells of Ilam were caused by the shallowness of the wells and the type of soil.

Conclusion

The results show that the sources of contamination of Ilam dam and Gol Gol spring with heavy metals were higher than national standards and USEPA and WHO guidelines. Zinc concentrations in all water sources were higher than the standard level; thus, it is necessary for authorities to initiate necessary action to decrease these amounts. The high levels of lead at Ilam dam and Gol Gol spring also require measures such as use of lead-free pipes and replacement of the aging pipe network.

Because of the importance of heavy metal pollution and the increase in the population and industry, it is necessary that sources of drinking water in Ilam should be monitored continuously to assure that they adhere to lawful standards. It is recommended that all industries, manufacturing units and contaminated human and natural resources should be examined in the catchment area to determine the main causes of water contamination at each station area. Failure to monitor and control these resources will exacerbate the influence of heavy metals threatening the welfare of the urban population of Ilam. It is recommended that water from some sources that contain higher concentrations of these metals should be mixed with other water sources having lower concentrations to reduce the amount consumed and prevent the accumulation of heavy metals in the food chain and the ecosystem.

Acknowledgements

This research was supported by the Deputy of Research and Technology of Ilam University of Medical Sciences. It was conducted according to a plan approved by the Student Research Committee (code 938045/150). The authors would like to thank the University for its support.

Competing interests

The authors declare that we have no competing interests.

Author contribution

All authors contributed equally and participated in the collection, analysis and interpretation of the data. All authors critically reviewed, refined and approved the manuscript.

References

1. Buragohain M, Bhuyan B, Sarma HP. Seasonal variations of lead, arsenic, cadmium and aluminium contamination of groundwater in Dhemaji district, Assam, India. Environ Monit Assess 2010; 170(1-4): 345-51. doi: 10.1007/s10661-009-1237-6.
2. Rabani M, Jafarabadi Ashtiani A, Mehrdadsharif A. The measurement of heavy metal (Ni.Pb.Hg) pollution in sediments of the Persian Gulf/operational area Assaluyeh. Journal of Environmental Science and Technology 2008; 9(3): 23-31. [In Persian].
3. Shahriari A. Determination of heavy metals (Cd, Cr, Pb, Ni) in edible tissues of Lutjans Coccineus and Tigeratooh Croaker in the Persian Gulf-2003. J Gorgan Univ Med Sci. 2005; 7(2): 65-7. [In Persian].
4. Lesmana SO, Febriana N, Soetaredjo FE, Sunarso J, Ismadji S. Studies on potential applications of biomass for the separation of heavy metals from water and wastewater. Biochemical Engineering Journal 2009; 44(1): 19-41. doi: 10.1016/j.bej.2008.12.009.
5. Hou D, He J, Lü C, Ren L, Fan Q, Wang J, et al. Distribution characteristics and potential ecological risk assessment of heavy metals (Cu, Pb, Zn, Cd) in water and sediments from Lake Dalinouer, China. Ecotoxicol Environ Saf 2013; 93: 135-44. doi: 10.1016/j.ecoenv.2013.03.012.
6. Smith AH, Steinmaus CM. Health effects of arsenic and chromium in drinking water: recent human findings. Annu Rev Public Health 2009; 30: 107-22. doi: 10.1146/annurev.publhealth.031308.100143.
7. Brahman KD, Kazi TG, Afridi HI, Naseem S, Arain SS, Ullah N. Evaluation of high levels of fluoride, arsenic species and other physicochemical parameters in underground water of two sub districts of Tharparkar, Pakistan: a multivariate study. Water Res 2013; 47(3): 1005-20. doi: 10.1016/j.watres.2012.10.042.
8. Rajaei Q, Pourkhabbaz AR, Hesari Motlagh S. Assessment of heavy metals health risk of groundwater in Ali Abad Katoul Plian. Journal of North Khorasan University of Medical Sciences 2012; 4(2): 155-62. [In Persian].
9. Sacmacı S, Kartal S, Sacmacı M. Determination of Cr (III), Fe (III), Ni (II), Pb (II) AND Zn (II) ions by FAAS in environmental samples after separation and preconcentration by solvent extraction using a triketone reagent. Fresenius Environmental Bulletin 2012; 21(6): 1563-70.
10. Buschmann J, Berg M, Stengel C, Winkel L, Sampson ML, Trang PTK, et al. Contamination of drinking water resources in the Mekong delta floodplains: arsenic and other trace metals pose serious health risks to the population. Environ Int 2008; 34(6): 756-64. doi: 10.1016/j.envint.2007.12.025.
11. Kumar M, Kumar A, Singh S, Kumar R, Walia TP. Uranium content measurement in drinking water samples using track etch technique. Radiat Meas 2003; 36(1-6): 479-81. doi: 10.1016/S1350-4487(03)00176-8.
12. Argos M, Kalra T, Pierce BL, Chen Y, Parvez F, Islam T, et al. A prospective study of arsenic exposure from drinking

water and incidence of skin lesions in Bangladesh. Am J Epidemiol 2011; 174(2): 185-94. doi: 10.1093/aje/kwr062.

13. Rosado JL, Ronquillo D, Kordas K, Rojas O, Alatorre J, Lopez P, et al. Arsenic exposure and cognitive performance in Mexican schoolchildren. Environ Health Perspect 2007; 115(9): 1371-5.
14. Babaee Y, Alavi Moghadam MR, Ghassemzadeh F, Arbab Zavar MH. Arsenic contamination of groundwater in the Kashmar Koohsorkh. Journal of Environmental Science and Technology 2008; 10(3): 29-36. [In Persian].
15. Farias SS, Casa VA, Vázquez C, Ferpozzi L, Pucci GN, Cohen IM. Natural contamination with arsenic and other trace elements in ground waters of Argentine Pampean Plain. Sci Total Environ 2003; 309(1-3): 187-99. doi: 10.1016/S0048-9697(03)00056-1.
16. Elinge CM, Itodo AU, Peni IJ, Birnin-Yauri UA, Mbongo AN. Assessment of heavy metals concentrations in bore-hole waters in Aliero Community of Kebbi State. Advances in Applied Science Research 2011; 2(4): 279-82.
17. Ghaemi P, Rustami hozori S, Ghaemi A. Determination of copper and lead concentrations in drinking water in Tehran. Journal of Environmental Studies 2004; 30(36): 27-32. [In Persian].
18. Mohammadian M, Nouri J, Afshari N, Nassiri J, Nourani M. Investigation of heavy metals concentrations in the water wells close to Zanjan zinc and lead smelting plant. Iranian Journal Of Health And Environment 2008; 1(1): 51-6. [In Persian].
19. Baptista Neto JA, Smith BJ, McAllister JJ. Heavy metal concentrations in surface sediments in a nearshore environment, Jurujuba Sound, Southeast Brazil. Environ Pollut 2000; 109(1): 1-9.
20. Wang H, Wang CX, Wang ZJ, Cao ZH. Fractionation of heavy metals in surface sediments of Taihu Lake, East China. Environ Geochem Health 2004; 26(2): 303-9. doi: 10.1023/B:EGAH.0000039594.19432.80.
21. Zorer OS, Ceylan H, Dogru M. Determination of heavy metals and comparison to gross radioactivity concentration in Soil and Sediment Samples of the Bendimahi River Basin (Van, Turkey). Water Air Soil Pollut 2009; 196(1-4): 75-87. doi: 10.1007/s11270-008-9758-0.
22. American Public Health Association, American Water Works Association, Water Environment Federation. Standard Methods for the Examination of Water and Wastewater. Washington: APHA, AWWA, WEF; 1999.
23. khorasani N, shaygan J, Karimi Shahri N. a survey of heavy metal concentration in the upper sediment layers of Bandar Abbas coasts. Iranian Journal Natural Research 2005; 58(4): 861-9. [In Persian].
24. Gunduz O, Simsek C, Hasozbek A. Arsenic pollution in the ground water of simav plain, Turkey: its impact on water quality and human health. Water Air Soil Pollut 2010; 205(4): 43-62. doi: 10.1007/s11270-009-0055-3.
25. Mosaferi M,Taghipour H, Hassani AM, Borghei M, Kamali Z, Ghadirzadeh A. Study of arsenic presence in drinking water sources: a case study. Iranian Journal of Health and Environment 2008; 1(1): 19-28. [In Persian].
26. Ho HH, Swennen R, Cappuyns V, Vassilieva E, Neyens G, Rajabali M, et al. Assessment on pollution by heavy metals and arsenic based on surficial and core sediments in the Cam River-mouth, Haiphong province, Vietnam. Soil and Sediment Contamination: An International Journal 2013; 22(4): 415-32. doi: 10.1080/15320383.2013.733445.
27. Nahid P, Moslehi P. Heavy metals concentrations on drinking water in different aeras of Tehran as ppb and methods of remal them. Iranian Journal of Food Science And Technology 2008; 5(1): 29-35. [In Persian].

Effects of sewage sludge and its biochar on soybean yield in fine-textured loess soil

Hamed Fathi Dokht[1*], Seyed Alireza Movahedi Naeini[2], Esmaeil Dordipour[2], Liz Wollsene De Jong[3], Ebrahim Hezarjaribi[4]

[1]PhD Student of Soil Science, Department of Soil Sciences, Gorgan University of Agricultural Sciences and Natural Resources, Gorgan, Iran
[2]Associate Professor of Soil Science, Department of Soil Sciences, Gorgan University of Agricultural Sciences and Natural Resources, Gorgan, Iran
[3]Professor of Soil Science, Department Agroecology, Aarhus University, Aarhus, Denmark
[4]Research Instructor of Plant Breeding, Golestan Agricultural and Natural Resources Research Center, Gorgan, Iran

Abstract
Background: The application of biochar as a soil amendment has achieved popularity and has been found to improve considerably soil nutrient status and plant productivity in low fertile soils.
Methods: This study used two different methods to apply sewage sludge and sewage sludge biochar with three replicates in a completely randomized block design. The experiment included (1) control, (2) mulch biochar (20 t/ha), (3) incorporated biochar (20 t/ha), (4) mulch sewage sludge (45 t/ha), and (5) incorporated sewage sludge (45 t/ha). This study aimed to investigate the potential benefits of sewage sludge and its biochar as mulch and incorporated into the top-soil for soil fertility, growth, yield, and shoot nutrient concentration of soybean and the effects of treatments on dry bulk density, water content, and mechanical resistance.
Results: The results showed higher values of height, biological yield, grain yield, and number of pods in the biochar mulch treatments (116.5, 10627.01, 4180, and 71.2, respectively). Moreover, it was determined that the addition of biochar and sewage sludge to soil resulted in increased shoot nitrogen (N), phosphorous (P), and potassium (K) concentrations, soil P, soil K, and total soil N (TN) compared with the control plots. Results from this study also showed that biochar and sewage sludge had positive impacts on the physical properties of soil. Bulk density and soil mechanical resistance were decreased, and volumetric water content was increased.
Conclusion: Heating sludge removed all pathogens. It is recommended that short- and long-term studies be conducted in the region to investigate further increases in the combined biochar and sewage sludge application rates.
Keywords: Nitrogen (N), Phosphorous (P), Potassium (K), Pathogens

***Correspondence to:**
Hamed Fathi Dokht
Email: hamed_fathi_dokht@yahoo.com

Introduction

Soybean (*Glycine max* L. Merrill) originated in China where it has been cultivated since the 11th century. It was first grown in Iran in 1939. Soybean production in Iran has steadily increased from 6000 tons in 1970 to 1.5 million tons in 2007 (1). The use of organic and inorganic fertilizers can increase soil fertility. The main disadvantages of chemical fertilizers are their low availability to poor farmers (2) and soil degradation problems (3) resulting from the use of chemical fertilizers in consecutive years. Organic fertilizers are capable of modifying nutrient efficiency. Emphasis has to be placed on the use of organic fertilizer and microbial inoculants to ensure that agriculture remains a viable part of a healthy and sustainable ecosystem. Sewage sludge is a residual material from wastewater treatment process. Iran currently produces more than 25 million tons of sewage sludge (moisture of 80%) in wastewater treatment plants, thereby creating a great environmental responsibility for plant owners and local officials (4). Biochar is a solid material gained from the thermo-chemical alteration (pyrolysis) of biomass in an oxygen-limited condition. It can be used for various purposes, such as a factor for soil enhancement, an enhancer of fertilizer use efficiency, protection against certain environmental pollutants, and as a way to reduce greenhouse gases (GHG) (5). The pyrolysis process is suitable for sludge management compared with the currently-used land filling, incineration, or direct agricultural utilization methods (5). The pyrolysis process can reduce the volume of sewage sludge, eradicate pathogens, and convert organic matter (OM) into bio-fuel, bio-oil, and biochar (6). Products of biochar from the sewage sludge produced through pyrolysis are rich in elemental carbon and nutrients and have plenty of exchangeable cations and surface adsorption sites which are very useful in eliminating pollutants from wastewater (7) and improving soil fertilization (8). Hence, biochar produced from sewage sludge through pyrolysis are of high interest. Biochar is charred OM considered for use as a soil modification. It is a fine-grained charcoal intended for use in both ameliorating the physical, chemical, and biological properties of soil and sequestering soil carbon (9). Products generated in the pyrolysis process include biochar, gas, and liquid. Differences in ratios are caused by the type of OM and temperature (10). Though different biochars share the basic characteristics (e.g., high C content), all of them have different particular characteristics related to the properties of the raw organic materials (feedstock) and the pyrolysis parameters used for production (11). Biochar has the valuable ability to retain water and nutrients. Moreover, the material offers a conductive habitat for soil microorganisms and, thus, can contribute to soil fertility (12). Biochar can also be used to improve water retention in farmlands, reduce nitrous oxide emissions, balance soil acidity, and increase soil organic carbon (13). Sewage sludge and its biochar have the potential for use as a fertilizer in agricultural production to resolve the needs of plants. Total and available phosphorous (P) and potassium (K) concentrations were increased by the pyrolysis process, although those of total and available N were decreased. Adding sewage sludge biochar to non-fertile and contaminated soils increased the fresh matter weight of Chinese cabbage (14). In addition, organic waste applied as mulch or incorporated in the topsoil affected the physical properties of soil and plant growth (15), which, in turn, contributed to higher crop yield.

Iran's agriculture makes heavy use of farm chemicals which can adversely affect soil quality, the environment, and human health while increasing production costs. However, sewage sludge (biomass) can be converted on the farm to produce biochar, which can contribute to improving soil quality and increasing yield. In the long term, this method of soil amelioration can contribute to more sustainable farming. The goals of this study were: 1) to assess the potential benefits of sewage sledge and its biochar as mulch incorporated into topsoil for soil fertility, growth, and yield of soybean; and 2) to determine the impact of sewage sludge and its biochar as mulch incorporated into topsoil on dry bulk density, water content, and mechanical resistance.

Material and Methods

Description of the study area

Field trials were conducted at Gorgan University of Agricultural Sciences and Natural Resources, Golestan province, Iran (approx. 37° 45' N, 54° 30' E; altitude 13 m A.S.L.). The average annual rainfall there was 607 mm and mean air temperatures were 13.5°C from 2005 to 2015. Sewage sludge and sewage sludge biochar were applied on June 3, 2014.

Collection of soil samples

Composite soil samples were taken from a depth of 0-30 cm before the experiments. The soil samples were taken to the laboratory, air-dried, and passed through a 2-mm sieve. pH and electrical conductivity (EC) were determined in a 1:2.5 (g/mL) soil: water ratio using a pH meter (Mettler Toledo Delta 320) (16) and EC meter (DDS-307A) (17), respectively. Cation exchange capacity (CEC) was determined using $NH_4OAc/NaNH_4OAc$ at 7.0 (18). OM was determined with the Walkley-Black method (19). Calcium carbonate equivalent (CCE) was measured by acid neutralization (20). Total nitrogen (TN) was determined with the Kjeldhahl method, and available phosphorous (P) according to Olsen method (21). The available potassium (AK) was extracted from saturated paste by vacuum pump and analyzed by flame emission spectrophotometry. Finally, soil texture was determined using the Bouyoucos method (22). All analyses were performed in parallel and in triplicate, and the results are listed in Table 1.

Sewage sludge preparation and characteristics

Sludge used for the pyrolysis treatment in this study was collected from the Kordkuy sewage sludge treatment plant (Golestan province, Iran). Biochar was produced by pyrolyzing the biomass at various temperatures under O_2-limited conditions. Dried sewage sludge was collected from Gorgan, and the sewage sludge was pyrolyzed in a closed ceramic pot under O_2-limited conditions at 350°C for 3 hours (23,24). The pH, EC, and CEC of the biochar and sewage sludge were determined and analyzed with the same methods as mentioned above for soil testing.

The yield of biochar was determined as follows:

$$\text{Biochar yield (\%)} = (W_2/W_1)*100 \quad (1)$$

where W_1 is the dry weight of the sewage sludge sample prior to pyrolysis and W_2 is the biochar weight. The biochar yield was calculated by subtracting the chemical reagent weight from the biochar mass produced.

Ash was separated by placing the biochar or sewage sludge sample in a nickel crucible and heating it at 700°C for 2 hours under air (25). The ash content was calculated as follows:

$$\text{Ash content (\%)} = (\text{M Ash/M Biochar or sewage sludge})*100 \quad (2)$$

where M Ash was the mass of ash and M Biochar or sewage sludge was the mass of biochar or sewage sludge.

Table 1. Main characteristics of studied soils

pH	EC (dS m^{-1})	CEC (cmol^{+}kg^{-1})	Clay g kg^{-1}	Silt g kg^{-1}	Sand g kg^{-1}	OM (%)	TN (%)	K mg/kg	P mg/kg
7.8	0.88	20	339	461	200	1.6	0.09	245	9.8

Table 2. Basic characterization of sewage sludge and biochar pyrolyzed at 350°C

Treatments	Moisture (%)	Yield (%)	Hydrophobicity	Ash (%)	pH	EC dSm^{-1}	CEC $cmol^{(+)}(kg^{-1})$	BET surface area (m^2/g)
Sewage sludge	84.1	-	-	44.45	6.3	4.2	23.1	1.1
Biochar (350°C)	-	48.1	Hydrophilic	60.2	7.9	2.4	21.2	34.2

Finally, the surface area of the biochar and sewage sludge was measured using the Brunauer, Emmett and Teller (BET) method that measured N2 gas sorption (0.162 nm^2) at 77 K. Approximately 200 mg of ground biochar was outgassed at 120°C for 16.5 hours and then analyzed on an Autosorb-1 Surface Area Analyzer (Quantachrome Instruments). Five data points with relative pressures of 0.05–0.3 were used to calculate the surface area (26). Hydrophobicity was measured using the ethanol droplet test (27). All analyses were performed in triplicate and in parallel, and results are listed in Table 2.

Moisture content

Based on ASTM D2216-98 (28), the moisture content was determined by oven drying sewage sludge at 110°C until constant weight.

Determination of heavy metal content in sludge and biochar

Aqua regia (mixture of HF, $HClO_4$, HNO_3, and H_2SO_4 acids in the same ratio) was used to determine the total content of heavy metals (29). The heavy metal content of the digestate was then measured with an atomic absorption spectrophotometer equipped with a graphite furnace (Analytik Jena, AAS6 Vario). This experiment determined the Cd, Cu, Ni, Pb, Cr, Co, and As contents of sludge and biochar.

Sewage sludge and sewage sludge biochar agronomic properties

Total carbon (c), nitrogen (N), hydrogen (H), and oxygen (O) contents of sewage sludge and biochar were determined using an EA3000 Element Analyzer (Euro Vector, Italy). Total phosphorus (TP) and available phosphorous (AP) contents were determined by molybdenum antimony colorimetry. Total potassium (TK) and AK contents were determined by flame emission spectrometry. Hydrolysable nitrogen (HN) content was determined by alkaline hydrolysis and distillation.

Preparation of experimental plots

The experiment was a completely randomized block design with three replicates conducted on 2 m × 3 m experimental plots ploughed to a depth of 30 cm. The soil surface was pulverized and raked even. Levees were constructed to separate the plots. In this experiment, treatment 1 (T1) was the untreated control, treatment 2 (T2) was soil with mulch biochar at 20 t/ha, treatment 3 (T3) was soil with incorporated sewage sludge biochar at 20 t/ha, treatment 4 (T4) was soil with mulch sewage sludge (dry weight) at 45 t/ha, and treatment 5 (T5) was soil with incorporated sewage sludge (dry weight) at 45 t/ha (Figure 1). Each treatment consisted of 5 sub-treatments of chemical fertilizer (ST): ST1 (0 kg/ha), ST2 (N (100 kg/ha)), ST3 (N (100 kg/ha) and P (50 kg/ha)), ST4 (N (100 kg/ha) and K (300 kg/ha)), and ST5 (N (100 kg/ha), K (300 kg/ha), and P (50 kg/ha)) (Figure 1). Fertilizer sources for N, P, and K were urea, triple superphosphate, and K sulfate.

The land was ploughed and disked before any treatments were applied. For the mulch treatments, the sewage sludge biochar and sewage sludge were applied uniformly over the land instantly after sowing manually. Where the sewage sludge biochar and sewage sludge were to be incorporated (T3 and T5), it was initially applied uniformly as mulch and then incorporated to a depth of 0.2 m manually using a spade before drilling. To determine the appropriate amount of biochar application rates, previous studies were reviewed. The application amount used in previous studies varied widely from below 5 t/ha to over 100 t/ha. Results of previous studies indicated that a rate in the range of 5 t/ha to 20 t/ha is appropriate for agricultural crops, though the appropriate rate depends on various factors such as primary soil characteristics and type of crop being grown. The application of sludge was 2.25 times greater than that of biochar based on the sludge-to-biochar conversion yield.

Crop cultivation

The soybean variety used in this experiment was D.P.X3589. Each plot was sub-divided into rows with an inter-row distance of 30 cm and 15 cm between

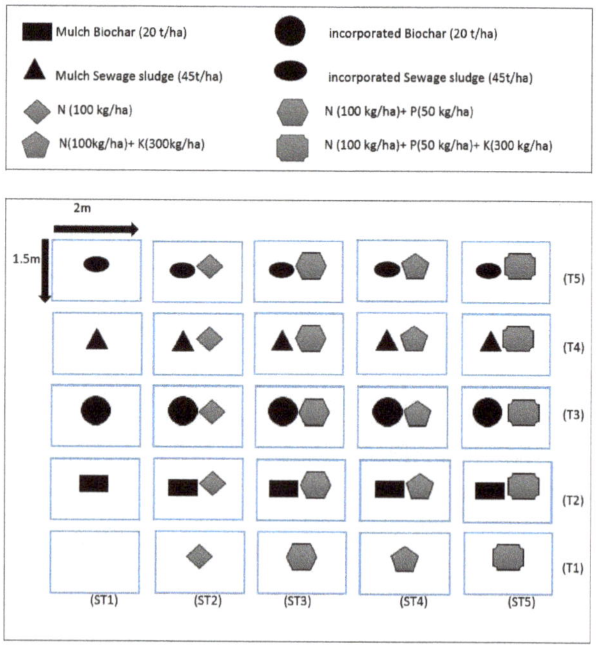

Figure 1. Field trial layout showing treatments

planting holes. Six seeds were planted in each hole. After seedling emergence, 3 seedlings were removed, leaving the strongest three remaining in each hole. Weeds were regularly removed by hand-weeding after planting. Based on soil moisture content, the irrigation interval was generally 20 to 25 days in the first stages of plant growth; in summer, irrigation intervals of 10-15 days were applied.

Measurements

Plant height per plant (cm): Plant height was measured from the base to the top with the aid of a measuring tape at the end of cropping.

Number of pods per plant: All pods of the plants were harvested and counted.

Grain yield (kg/ha) and biomass (kg/ha): At maturity, all pods were harvested, all grains were recovered after shelling, and then they were weighed using top load balance (with a basin-like top and calibrated in kg) per plot. The grain yield per hectare was determined as follows:

Grain Yield (kg/ha) and biomass (kg/ha) = (Grain Yield (kg/ha) or biomass (kg/ha)* 10000 m^2)/Plot Area (m^2)

Soil physical properties: Volumetric water content was determined gravimetrically on soil cores for depths of 0-0.08 m at 6 times (M1, M2,..., M6) by oven drying. On the same dates, mechanical resistance was measured in the field with a portable cone penetrometer (S1, S2,..., S6). Dry bulk density samples were collected using a standard core sampler with a sampling volume of 100.8 cm 3 at 6 times (Bd1, Bd2,..., Bd6) for each treatment and all replicates according to the core method. Samples were collected at depths of 0–08 cm and oven dried for approximately 24 hours at 105°C to a constant mass. Soil bulk density was determined (equation 2) by dividing the known mass of soil by the soil volume it occupied:

$$\rho b = \text{soil mass / volume occupied by soil sample} \quad (3)$$

where soil was measured in grams and volume in cm^3.

Tissue nutrient concentration: N, P, and K contents in plant shoot material at flowering were determined using a standard digest technique (30). N was measured using Kjeltec, P using a colorimetric method, and K by flame photometry. Plant shoot samples were selected at random from the center of each plot between two adjacent rows on July 10, 2015 in three replicates per plot and mixed.

Soil chemical properties: Available P (from a sodium bicarbonate extractant), K (using a 1N NH_4OAc extractant), and soil TN (using a K chloride extractant) contents were determined at harvest. Soil samples were selected at random from the center of each plot between 2 adjacent rows at depths of 0 to 0.2 m on October 20, 2014 in 3 replicates per plot and mixed.

Statistical analysis

The data collected from this experiment was analyzed and compared for statistical differences using the Analysis of Variance (ANOVA) method and Duncan's new multiple range test (DMRT), and the software package Statistical Package for the Social Sciences (SAS) was used to process the analyses.

Results and Discussion

Sewage sludge and biochar characterization

The sewage sludge and biochar properties, including yield, ash content, pH, CEC, moisture, hydrophobicity, EC, and BET, are summarized in Table 2. The yield of biochar produced at 350°C was 48.1%, which was in accordance with the result obtained by Méndez et al (24). The pyrolysis process increased the percentage of ash content, as reported by others (31,32). The percentage of ash content was increased from 59.45% for sewage sludge to 64.2% for biochar. Pyrolysis also increased pH as results of polymerization/condensation reactions and acidic surface groups released during pyrolysis showed (24). The EC, a factor used to evaluate the content of total dissolved salts in a sample, was 4.2 dSm^{-1} for sewage sludge; this figure declined with pyrolysis to 2.4 dSm^{-1} for biochar at 350°C (24). The CEC value decreased with the pyrolysis temperature from 23.1 cmol $^{(+)}kg^{-1}$ for sewage sludge to 21.2 for biochar at 350°C. These results are in agreement with those obtained by Méndez et al (24). The declines in CEC with the pyrolysis process may be due to functional groups released within the pyrolysis process (24). SSA, one of the main factors in evaluating adsorption, is involved in the distribution of pore structure (33). The results showed that the BET surface area of biochar increased during the pyrolysis process, as reported by others (24,34). The BET surface area of the biochar obtained at 350°C (34.2 m^2/g) was larger than that of sewage sludge (1.1 m^2/g).

Heavy metal contents of sewage sludge and biochar

Table 3 presents the heavy metal contents of sewage sludge and biochar. Within wastewater treatments, nearly 50% to 80% of the heavy metal content present in wastewater is preserved in the sewage sludge; this clarifies the high concentration of particular metals in sewage sludge (35). In this study, the heavy metal content was not as high and did not exceed the EPA thresholds for sewage sludge land usage (values in bold letters in Table 3) (36).

Table 3. Heavy metal analysis of the sewage sludge and biochars

Treatment	Heavy metal (mg/kg)						
	Cd	Cu	Ni	Pb	Pb	As	Co
Sewage sludge	2.1	54.2	61	91	28.3	BDL	BDL
Biochar (350°C)	3.8	98	112	158	52	BDL	BDL
US EPA, 1993	**85**	**4300**	**420**	**840**	**3000**	**75**	**50**

Abbreviation: BDl, below detection limit.

Table 4. Agronomic properties of sewage sludge and sewage sludge biochar

Treatment	C (%)	O (%)	H (%)	N (%)	TP g/kg	TK g/kg	HN g/kg	AP g/kg	AK g/kg
Dried sewage sludge	25.2	19.48	5.1	5.7	8.5	7	0.18	0.3	1.8
Biochar (350°C)	26.4	8.30	1.1	3.9	14	12.2	0.05	1.1	2.1

Sewage sludge and sewage sludge biochar agronomic properties

Pyrolysis process factors like temperature, residence time, and heating rate can affect the quality and quantity specifications of biochar (37). In the current study, sewage sludge was changed at a relatively low pyrolysis temperature (350°C), allowing for biochar that is suitable for agricultural uses (38). The elemental content changes in C, H, O, and N are also listed in Table 4. As expected in the pyrolysis process for the biochar, the C content increased, while the H, N, and O contents decreased. Temperature can increase the carbon content, possibly due to the increasing degree of carbonization. However, the decreases in O and H elements could be attributed to the breaking of weaker bonds in the biochar structure and the highly carbonaceous materials yielded with increased temperature. This fact was also suggested by results obtained by other authors (33). The contents of major plant nutrients N, P, and K were determined to assess whether biochar was a favorable fertilizer for agricultural land. N and HN (0.05%) contents of biochar were lower than those of dried sewage sludge on the grounds of the volatilization of N within pyrolysis. N can be removed through the losses of NH_4-N and NO_3-N fractions as well as volatile matter containing N groups at 200°C (9,39). Nonetheless, the accessibility of N in biochar remains unclarified (33). The TP and K contents of biochar were increased by 64.5% and 74.1%, respectively.

1000-grain weight, grain yield, biological yield, plant height, and number of pods per plant

As shown in Tables 5 and 6, the sewage sludge, its biochar and chemical fertilizer had a significant effect on 1000-grain weight, grain yield, biological yield, plant height, and number of pods per plant. The highest values of 1000-grain weight, grain yield, biological yield, plant height, and number of pods per plant were found in mulch biochar (T2) (230.1 g, 4182 kg ha^{-1}, 10627 kg ha^{-1}, 116.53 cm, and 71.2, respectively), and the lowest values were observed in the control treatment (T1) (155.4 g, 2086 kg ha^{-1}, 8117 kg ha^{-1}, 92.2 cm, and 36.2, respectively). In the chemical fertilizer treatment (ST), the highest values of 1000-grain weight, grain yield, biological yield, plant height, and number of pods per plant were found in NPK (ST5) (220.76 g, 3821.47 kg ha^{-1}, 10421.2 kg ha^{-1}, 114.4 cm, and 66.13, respectively), and the lowest values were observed in the control treatment (ST1) (201.82 g, 3148.7 kg ha^{-1}, 9343.6 kg ha^{-1}, 103.3 cm, and 57.64, respectively). Yamika and Ikawati reported that the values of 1000-grain weight, grain yield, biological yield, plant height, and number of pods per plant were increased in soybean with the addition of biochar in soils (40). Moreover, Yooyen et al reported that the combination of inorganic and organic fertilizers (0, 0.5, and 1 t ha^{-1}) increased seed yield up to 3.5 t ha^{-1}, which was closely similar to this experiment (41). Navin et al also showed that adding biochar to soil can increase soybean yield (42). Falodun et al (43)and Liang et al (44) showed that plant height was increased by organic and inorganic fertilizers.

Effects of biochar and sewage sludge on some soil physical properties

Biochar and sewage sludge have the potential to modify the physical condition of soil, including factors such as surface area, pore size distribution, bulk density, water-holding capacity, and soil mechanical resistance (45).

Dry bulk density for depths of 0.0-0.08 m during the growing season

Differences in dry bulk density among treatments

Table 5. Effects of treatment applications on average soybean yield

Treatment	1000-grain weight (g)	Number of pods per plant	Plant height (cm)	Grain yield (kg/ha)	Biological yield (kg/ha)
Factor (A) (biochar and sewage sludge T1, ...,T5)					
Control (T1)	155.41^{c}	36.2^{c}	92.2^{c}	2086.47^{c}	8117.62^{d}
Biochar mulch (T2)	230.1^{a}	71.2^{a}	116.53^{a}	4182.73^{a}	10627.01^{a}
Biochar incorporation (T3)	227.6^{ab}	70.73^{a}	115.53^{a}	4055.8^{a}	10528.6^{a}
Sewage sludge mulch(T4)	225.98^{ab}	66.2^{b}	112.46^{b}	3714.00^{b}	10160.8^{b}
Sewage sludge incorporation (T5)	220.44^{b}	65.13^{b}	111.87^{b}	3612.73^{b}	9979.3^{c}
Factor (B) (chemical fertilizer ST1, ...,ST5)					
NPK (ST5)	220.76^{a}	66.13^{a}	114.4^{a}	3822.93^{a}	10421.2^{a}
NK (ST4)	216.51^{a}	63.46^{b}	110^{b}	3689.53^{b}	10001.8^{b}
NP (ST3)	211.73^{b}	62^{bc}	112.86^{a}	3580.67^{c}	9938.51^{bc}
N (ST2)	208.84^{b}	60.4^{c}	108^{b}	3390.33^{d}	9753.2^{c}
Control (ST1)	201.82^{c}	57.64^{d}	103.3^{c}	3147.87^{e}	9343.6^{d}

Table 6. Analysis of Variance on average soybean yield

Sources of variances	Degree of freedom	Means of squares				
		1000-grain weight (g)	Number of pods per plant	Plant height (cm)	Grain yield (kg/ha)	Biological yield (kg/ha)
Replications	2	463.1	0.89	3.72	47902.45	25080
Factor A (biochar and sewage sludge)	4	15166.9**	3202.32**	1497.44**	10505373**	15908765.6**
Error A	8	120.3	10.81	9.78	20450.6	37355.8
Factor B (chemical fertilizer)	4	789.7**	158.58**	283.6**	1045790.2**	2303761.5**
Error B	-	39.7	12.09	8.32	26055	67272
A ×B	16	18.08	1.34	17.01	14212.5	104051.1
CV	-	2.9	5.6	2.62	4.57	2.62

**Significant at the 0.01 level; *Significant at the 0.05 level.

were significant as determined by ANOVA for the 6 measurement dates (Bd1, Bd2,..., Bd6). Dry bulk density values within 0.0 to 0.08 m with mulch applications (T2 and T4) and incorporation treatments (T3 and T5) were still lower than the control (T1) during the growing season (at all measurement dates), irrespective of the statistics (Figure 2). Chan and Xu reported that bulk density in top soil was decreased with the addition of biochar to soils (11). Moreover, Jones et al (46) showed that OM reduced bulk density by soil dilation and improvement in aggregate stability. The decrease in bulk density in the biochar and OM-amended soils might be caused by the physical dilution effects, as reported by Jones et al, who showed that increasing OM content can significantly decrease bulk density. In addition, the application of biochar decreased soil dry bulk density (45). It can lead to alterations in soil aggregate sizes, as shown by Jones et al (46) and MacRae and Mehuys (45). No difference in soil dry bulk density was seen among sub-treatments.

Volumetric soil water content for depths of 0-0.08 m during growing season

ANOVA results showed differences between treatments for the volumetric water content in soil within the growing season were significant. The biochar mulch application (T2) significantly increased the volumetric water content in soil relative to other treatments during the growing season. The water content of soil for mulch biochar (T2), incorporation biochar(T3), incorporation sewage sludge (T5), and mulch sewage sludge (T4) were significantly greater than the control(T1) (Figure 3). Relative to incorporation treatments (T3 and T5), the effect of mulch treatments (T2 and T4) on soil volumetric water content became more pronounced at the growing season, especially when the evaporation potential was high. Asai et al (47) and Chan et al (48) indicated that applying biochar to soil can improve water-holding capacity and water retention which may be a factor in both yield improvement and plant nutrient availability. Jones et al showed that sewage sludge, because of its high OM content, significantly increased the volumetric water content of soil in treatments (46). Moreover, soil water content rates among sub-treatments were not different.

Soil Mechanical Resistance during Growing Season

Differences among treatments for soil mechanical resistance were significant for all measurements as shown by analysis of variance. Soil mechanical resistance in both incorporation (T3 and T5) and mulch (T2 and T4) treatments was significantly less than the control (T1) during the growing season (Figure 4). Soil mechanical resistance with biochar mulch (T2) was significantly less than the other treatments at the end of the growing season. Gantzer et al reported a decrease in soil

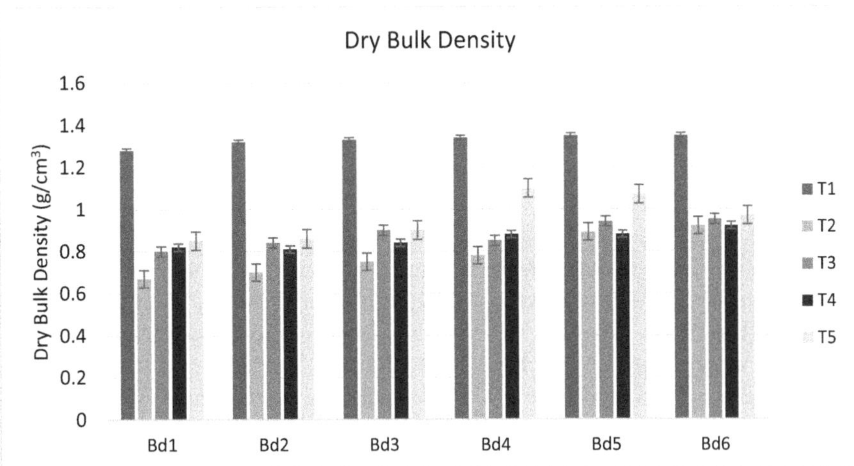

Figure 2. Influence of biochar and sewage sludge on dry bulk density.

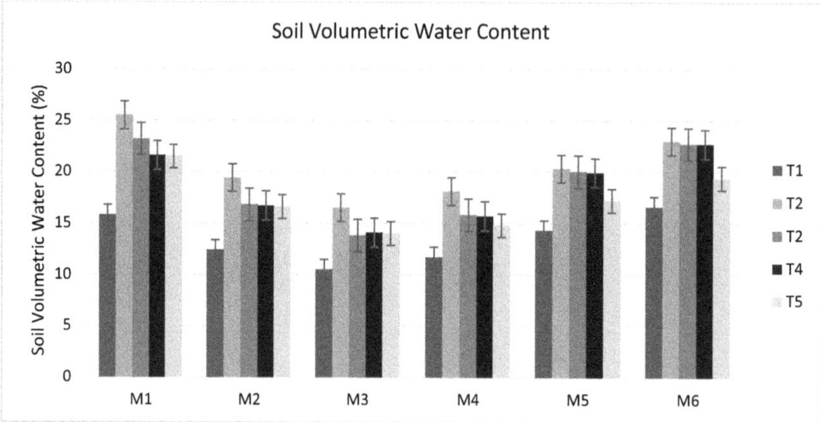

Figure 3. Influence of biochar and sewage sludge on soil volumetric water content.

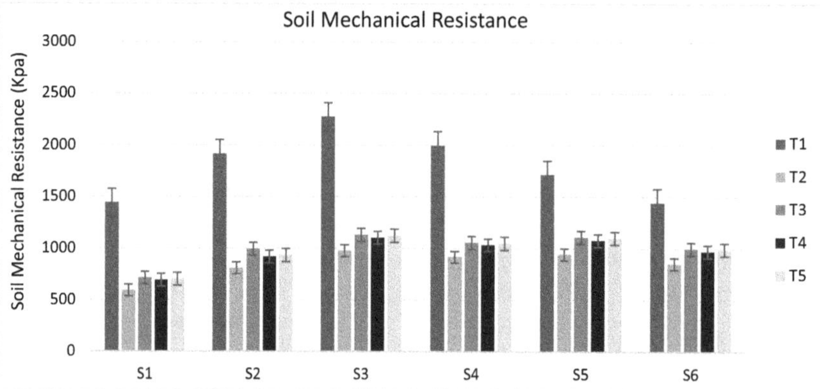

Figure 4. Influence of biochar and sewage sludge on soil mechanical resistance.

mechanical resistance with the addition of biochar (49), and Mukherjee and Lal also noted a decrease in soil mechanical resistance with the incorporation of OM to the soil (50). Furthermore, soil mechanical resistance rates among sub-treatments were not different. Yao et al showed that the application of biochar can improve the physical qualities of soil in terms of bulk density, soil mechanical resistance, and water holding capacity (51). Decreases in soil mechanical resistance could be attributed to the amelioration of the aggregation and the decrease in bulk density (51).

Effects of treatment applications on post-cropping soil chemical analysis

This study demonstrated that the addition of mulch sewage sludge (T4) and incorporation sewage sludge (T5) as well as their biochars (T2 and T3) increased total soil nitrogen (TN) compared to the control (T1) due to the high content of N in sewage sludge. Lehman and Joseph reported an increase in TN with the addition of sewage sludge (9). The sewage sludge treatments (T4 and T5) showed a higher increase in TN compared to the biochar treatments (T2 and T3). TN contents of biochar were lower than those of sewage sludge, which could be attributed to the volatilization of N within pyrolysis. N can be removed through the losses of NH_4-N and NO_3-N fractions, like volatile matter containing N groups, at 200°C (9,39). Some researchers showed that the application of biochar can increase the uptake and efficiency of N fertilizers in soil as a result of reducing N leaching and also by affecting the soil N cycle (52). The results of this study showed that the addition of N chemical fertilizer to soil can significantly increase TN compared to the control plot.

The results of the addition of mulch sewage sludge (T4) and incorporated sewage sludge (T5) as well as their biochars (T2 and T3) increased the available phosphorus in soil compared to the control (T1). Lehman and Joseph reported an increase in soil phosphorus with the addition of sewage sludge (9). Several studies reported an increase in AP after the application of biochar (53- 55). The biochar treatments (T2 and T3) showed a higher increase in soil phosphorus compared to the sewage sludge treatments (T4 and T5). Biochar can increase nutrient pools in soil due to its high surface area and can increase essential nutrient contents in soil. Furthermore, the results of this study showed that the addition of NP (ST3) and NPK(ST5) chemical fertilizer in soil can increase soil phosphorus significantly compared to the NK (ST4), N (ST2), and control (ST1) plots (Tables 7 and 8).

The effects of treatments on the AK of soil were significant. The results showed that the addition of mulch sewage sludge (T4) and incorporated sewage sludge (T5) as well

Table 7. Effects of treatment applications on post-cropping soil chemical analysis

Treatment	N (%)	P (mg kg^{-1})	K (mg kg^{-1})
Factor (A)			
Control (T1)	0.106^{c}	11.4^{c}	390.3^{c}
Biochar mulch (T2)	0.118^{b}	23.5^{a}	466.6^{a}
Biochar incorporation (T3)	0.116^{b}	23.69^{a}	443^{b}
Sewage sludge mulch (T4)	0.158^{a}	16.9^{b}	466.5^{a}
Sewage sludge incorporation (T5)	0.155^{a}	16.5^{b}	455.5^{a}
Factor (B)			
NPK (ST5)	0.125^{bc}	22.9^{a}	468.6^{a}
NK (ST4)	0.143^{a}	15.05^{b}	472.8^{a}
NP (ST3)	0.142^{a}	22.72^{a}	434.2^{b}
N (ST2)	0.129^{ab}	15.57^{b}	426.5^{b}
Control (ST1)	0.113^{c}	15.9^{b}	420^{b}

Table 8. Analysis of variance post-cropping soil chemical analysis

Sources of variances	df	Means of squares		
		N	P	K
Replications	2	0.00030	1.81	375.8
Factor A (biochar and sewage sludge)	4	0.0088**	408.44**	15129.6**
Error A	8	0.00013	2.59	186.18
Factor B (chemical fertilizer)	4	0.0022**	241.8**	9039.5**
Error B	-	0.0039	2.58	383
A ×B	16	0.000061	1.44	70.53
CV	-	15.26	8.7	4.4

**Significant at the 0.01 level; *Significant at the 0.05 level.

Table 9. Effects of treatment applications on shoot nutrient concentration

Treatment	N (%)	P (mg kg^{-1})	K (mg kg^{-1})
Factor (A)			
Control (T1)	2.08^{c}	0.17^{c}	1.18^{d}
Biochar mulch (T2)	3.86^{b}	0.46^{a}	2.34^{b}
Biochar incorporation (T3)	3.76^{b}	0.45^{a}	2.27^{c}
Sewage sludge mulch (T4)	4.57^{a}	0.34^{b}	2.4^{a}
Sewage sludge incorporation (T5)	4.38^{a}	0.32^{b}	2.34^{b}
Factor (B)			
NPK (ST5)	3.83^{b}	0.409^{a}	2.51^{a}
NK (ST4)	4.06^{a}	0.298^{b}	2.54^{a}
NP (ST3)	3.82^{b}	0.428^{a}	1.83^{b}
N (ST2)	4.05^{a}	0.319^{b}	1.84^{b}
Control (ST1)	2.89^{c}	0.304^{b}	1.81^{b}

Table 10. Analysis of variance for shoot nutrient concentration

Sources of variances	df	Means of squares		
		N	P	K
Replications	2	0.083	0.0021	0.066
Factor A (biochar and sewage sludge)	4	14.6**	0.212**	4.05**
Error A	8	0.1	0.0019	0.0046
Factor B (chemical fertilizer)	4	3.47**	0.057**	2.188**
Error B	-	0.04	0.0017	0.0192
A ×B	16	0.047	0.00212	0.0168
CV	-	5.38	11.84	6.5

**Significant at the 0.01 level; *Significant at the 0.05 level.

as their biochars (T2 and T3) increased soil AK compared to the control (T1). The sewage sludge treatments (T4 and T5) showed higher increases in soil AK compared to the biochar treatments (T2 and T3). Bierdman and Harpole reported an increase in soil AK with the addition of biochar (56). In addition, the results of this study showed that the addition of NK (ST4) and NPK (ST5) chemical fertilizer in soil can increase soil AK significantly ($P <$ 0.05) compared to the NP (ST3), N (ST2), and control (ST1) plots (Tables 7 and 8).

TN, soil available P, and soil AK were higher with mulch treatments (T2 and T4) than the incorporation treatments (T3 and T5) (Tables 7 and 8). Some researchers have found that biochar applications can result in increased contents of total, extractable, and/or exchangeable soil nutrients (56-60).

Effect of biochar and sewage sludge applications on shoot nutrient concentration

This study demonstrated that the addition of mulch sewage sludge (T4) and incorporation sewage sludge (T5) as well as their biochars (T2 and T3) increased N concentration in shoots compared to the control (T1). The sewage sludge treatments (T4 and T5) showed higher increased N concentrations in shoots compared to the biochar treatments (T2 and T3). Results of this study showed that the addition of N (ST2, ST3, ST4, and ST5) chemical fertilizer to soil increased the N concentration in shoots significantly compared to the control plot (ST1) (Tables 9 and 10). Chan et al and Taghizadeh-Toosi et al reported similar effects for biochar on N uptake. One way biochar may affect soil nutrients is by reducing leaching losses (54,55). Nutrients can be adsorbed to biochar's surface, which slows leaching (11,61).

According to the results of the present study, P concentration in shoots was significantly increased by the application of mulch sewage sludge (T4) and incorporation sewage sludge (T5) as well as their biochars (T2 and T3) compared to the control (T1). The biochar treatments (T2 and T3) showed higher increased P concentration in shoots compared to the sewage sludge treatments (T4 and T5). The results of this study showed that the addition of PN (ST3) and NPK (ST5) chemical fertilizers to soil can increase the P concentration in shoots significantly compared to the NK (ST4), N (ST2), and control plots (ST1) (Tables 9 and 10). Lehmann and Rondon indicated an increase in P concentration in plants with the application of biochar (62). The increase in P concentration as a consequence of the application of biochar may be attributed to the high P content in sewage sludge biochar. The highest soil P concentration was seen in biochar-amended soils (Table 4).

Potassium concentration in shoots was also significantly increased by the addition of mulch sewage sludge (T4) and incorporation sewage sludge (T5) as well as their biochars (T2 and T3) compared to the control (T1). The sewage sludge treatments (T4 and T5) showed higher increases in K concentration in shoots than the biochar treatments

(T2 and T3). Moreover, the addition of NK (ST4) and NPK (ST5) chemical fertilizers to soil increased the K concentration in shoots significantly compared to the NP (ST3), N (ST2), and control plots (ST1) (Tables 9 and 10). Van Zwieten et al reported an increase in concentration of plant K with the addition of biochar (63). The adhesion of soil colloidal particles to each other depends upon the net attractive forces between them, called soil aggregation. This property is very important for soil structure. Soil with a good structure provides favorable conditions for the movement of water and nutrients and their absorption by plants (64).

Biochar cost estimates and correlations

The costs related to biochar depend on various parameters, e.g., the cost of the feedstock, feedstock collection and transportation, production technology and temperature, production scale, distribution and handling. The difference in costs to produce biochar depends on raw materials like green waste, poultry litter, and wastewater sewage. Lehmann and Rondon presented an estimate of \$50-200/ton of biochar (62). The cost of biochars produced from sewage sludge is negligible, as currently these wastes generally have a cost associated with their exposure. Results of the current study indicated that production costs may be as low as \$40/ton. Correlations between biological yield and K concentration in shoots, P concentration in shoots, N concentration in shoots, mechanical resistance at 6 times (S1, S2,..., S6), and volumetric water contents at 6 times (M1, M2,..., M6) were obtained. A significant positive correlation with biological yield was observed in K concentration in shoots ($P<0.0001$, 0.84), P concentration in shoots ($P<0.0001$, 0.81), N concentration in shoots ($P<0.0001$, 0.74) and volumetric water contents (M2) ($P<0.0001$, 0.72). Soils with high SSA and fine clay particles have significant effects on mechanical resistance to the root growth by cementing different sizes of soil particles. Dry bulk density and soil water content can control the soil mechanical resistance (64). In the sub-humid climate of northern Iran, in addition to the cementing effects of high quantity fine clays in high specific surface soils, slow saturation and incomplete swelling of soils may induce greater cohesion and mechanical resistance (64). The correlation between soil K concentration and plant K uptake was not significant (65).

Conclusion

This study showed that sewage sludge and sewage sludge biochar have the potential to increase growth and yield of soybean. To-date, research has clearly demonstrated that biochar improves plant productivity and yield and soil fertility through several mechanisms. Physical aspects change with biochar, which can improve soil water-holding capacity. The biochar physical structure, its high porosity, and the initial associated high surface area result in soil fertility advantages over the direct addition of nutrients. Biochar applications can also increase nutrient retention and availability. This study also found that plant tissue K, P, and N concentrations and soil P, N, and K increased with the application of biochar sewage sludge and sewage sludge. These nutrients may be directly introduced to the soil through labile organic compounds associated with biochar sewage sludge and become available as these compounds weather. This paper also reports that biochar and sewage sludge amendment had positive impacts on soil physical properties; for example, biochar and sewage sludge decreased the bulk density of the soil. Biochar and sewage sludge treatments led to increased soil water content and decreased soil mechanical resistance. Having perfect knowledge about the circumambient environment is of high priority for researchers and agriculture policy makers. In soils containing silt clay loam texture, illite minerals, and with high SSA, Heating sludge removed all pathogens. It is recommended that short- and long-term studies be conducted in the region with further increases in the rates of combined biochar and sewage sludge application to determine the optimum level of application required for optimum crop production.

Acknowledgments

This research was funded by Gorgan University of Agricultural Sciences and Natural Resources and Golestan Water and Wastewater Company (Iran).

Competing interests

The authors affirm that this article is their original work and that they have no conflicts of interest to declare.

Author's contributions

All authors were involved in all stages of the article. On behalf of the co-authors, the corresponding author bears full responsibility for this submission.

References

1. Shurtleff W, Aoyagi A. History of soybeans and soy foods in the Middle East; 2007. [cited 2015 June 12]. Available from: http://www.soyinfocenter.com/HSS/middle_east.php.
2. Garrity DP. Agroforestry and the achievement of the Millennium Development Goals. Agrofor Syst 2004; 61(1): 5-17.
3. Khaim S, Chowdhury MAH, Saha BK. Organic and inorganic fertilization on the yield and quality of soybean. Journal of the Bangladesh Agricultural University 2013; 11(1): 23-8. doi: 10.3329/jbau.v11i1.18199.
4. Fahiminia M, Farzadkia M, Jafari Mansoorian H, Majidi G, Mirzabeygi Arhany M, Khanjani N, et al. The status of wastewater management in Shokuhieh industrial park (A case study of Qom province). Environ Health Eng Manage J 2015; 2(4): 165-71.
5. Hwang IH, Ouchi Y, Matsuto T. Characteristics of leachate from pyrolysis residue of sewage sludge. Chemosphere 2007; 68(10): 1913-19. doi: 10.1016/j.chemosphere.2007.02.060.

6. Domínguez A, Menéndez JA, Inguanzo M, Pís JJ. Production of bio-fuels by high temperature pyrolysis of sewage sludge using conventional and microwave heating. Bioresurse Technol 2006; 97(10): 1185-93. doi: 10.1016/j.biortech.2005.05.011
7. Smith KM, Fowler GD, Pullket S, Graham NJ. Sewage sludge-based adsorbents: a review of their production, properties and use in water treatment applications. Water Res 2009; 43(10): 2569-94. doi: 10.1016/j.watres.2009.02.038.
8. Hossain MK, Strezov V, Chan KY, Nelson PF. Agronomic properties of wastewater sludge biochar and bioavailability of metals in production of cherry tomato (*Lycopersicon esculentum*). Chemosphere 2010; 78(9): 1167-71. doi: 10.1016/j.chemosphere.2010.01.009.
9. Lehman J, Joseph S, eds. Biochar for Environmental Management: Science, Technology and Implementation. 2nd ed. London: Taylor & Francis Ltd; 2009.
10. Verheijen F, Jeffery S, Bastos AC, van der Velde M, Diafas I. Biochar Application to Soils A Critical Scientific Review of Effects on Soil Properties, Processes and Functions. Italy: European Commission, Joint Research Centre, Institute for Environment and Sustainability; 2010.
11. Chan KY, Xu Z. Biochar nutrient properties and their enhancement. In: Lehmann J, Joseph S, eds. Biochar for Environmental Management: Science and Technology. London: Earthscan; 2009.
12. Kookana RS, Sarmah AK, Van Zwieten L, Krull E, Singh B. 3 Biochar application to soil: agronomic and environmental benefits and unintended consequences. In: Advances in Agronomy San Diego. San Diego: Elsevier; 2011. p. 103-43.
13. Zhang A, Bian R, Pan G, Cui L, Hussain Q, Li L, et al. Effects of biochar amendment on soil quality, crop yield and greenhouse gas emission in a Chinese rice paddy: a field study of 2 consecutive rice growing cycles. Field Crops Res 2012; 127: 153-160. doi:10.1016/j.fcr.2011.11.020.
14. Liu T, Liu B, Zhang W. Nutrients and heavy metals in biochar produced by sewage sludge pyrolysis: its application in soil amendment. Pol J Environ Stud 2014; 23(1): 271-5.
15. Amini S, Movahedi Naeini SA, Mashayekhi K. Effects of paper-mill sludge as a mulch versus topsoil incorporation on potassium uptake and the grain yield of rain-fed wheat in a high specific surface loess soil with illite dominance in clay fraction. Appl Environ Soil Sci 2012; 2012: 624824. doi: 10.1155/2012/624824.
16. Thomas GW, Sparks DL, Page AL, Helmke PA, Loeppert RH, Soltanpour, et al. Soil pH and soil acidity. Methods of soil analysis. Part 3-chemical methods. Madison, WI: Soil Science Society of America; 1996.
17. Rhoades JD. Salinity: electrical conductivity and total dissolved solids. In: Sparks JD, Sparks DL, Page AL, Helmke PA, Loeppert RH, Soltanpour MA, et al, eds. Methods of Soil Analysis, Part 3 - Chemical Methods. Madison, WI: Soil Science Society of America; 1996. p. 417–435.
18. Sumner M E, Miller W P. Cation Exchange Capacity and Exchange Coefficients. Madison, USA: Soil Science of America and American Society Agronomy; 1996.
19. Nelson DW, Sommers L. Total Carbon, Organic Carbon, and Organic Matter. Madison, USA: ASA-SSSA; 1982.
20. Richards LA. Diagnosis and improvement of saline and alkali soils. Soil Sci 1954; 78(2): 154.
21. Olsen SR, Cole CV, Frank S,Dean LA. Estimation of Available phosphorus in Soils by Extraction with Sodium Bicarbonate. Washington DC: United States Department of Agriculture; 1954.
22. Bouyoucos GJ. Hydrometer method improved for making particle size analyses of soils. Agron J 1962; 54(5): 464-5. doi:10.2134/agronj1962.00021962005400050028x.
23. Lu H, Zhang W, Wang S, Zhuang L, Yang Y, Qiu R. Characterization of sewage sludge-derived biochars from different feedstocks and pyrolysis temperatures. J Anal Appl Pyrolysis 2013; 102: 137-43.
24. Méndez A, Terradillos M, Gascó G. Physicochemical and agronomic properties of biochar from sewage sludge pyrolysed at different temperatures. J Anal Appl Pyrolysis 2013; 102: 124-30. doi: 10.1016/j.jaap.2013.03.006.
25. Yuan JH, Xu RK, Zhang H. The forms of alkalis in the biochar produced from crop residues at different temperatures. Bioresour Technol 2011; 102(3): 3488-97. doi: 10.1016/j.biortech.2010.11.018.
26. Mukome FN, Zhang X, Silva LC, Six J, Parikh SJ. Use of chemical and physical characteristics to investigate trends in biochar feedstocks. J Agric Food Chem 2013; 61(9): 2196-204. doi: 10.1021/jf3049142.
27. De Jonge LW, Moldrup P, Jacobsen OH. Soil-water content dependency of water repellency in soils. Soil Sci 2007; 172(8): 577-88. doi: 10.1097/SS.0b013e318065c090.
28. American Society for Testing and Materials International (ASTM International). Standard Test Methods for Laboratory Determination of Water (Moisture) Content of Soil and Rock by Mass/ASTM D2216 - 10. Philadelphia: ASTM International; 2010. doi: 10.1520/D2216-10.
29. Tessier A, Campbell PG, Bisson M. Sequential extraction procedure for the speciation of particulate trace metals. Anal Chem 1979; 51(7): 844-51. doi: 10.1021/ac50043a017.
30. Westerman RL. Soil Testing and Plant Analysis. Madison,Wis, USA: Soil Science Society of America; 2005.
31. Zielińska A, Oleszczuk P, Charmas B, Skubiszewska-Zięba J, Pasieczna-Patkowska S. Effect of sewage sludge properties on the biochar characteristic. J Anal Appl Pyrolysis 2015; 112: 201-213. doi: 10.1016/j.jaap.2015.01.025.
32. Oh TK, Choi B, Shinogi Y, Chikushi J. Characterization of biochar derived from three types of biomass. Journal of the Faculty of Agriculture, Kyushu University 2012; 57(1): 61-6.
33. Song XD, Xue XY, Chen DZ, He PJ, Dai XH. Application of biochar from sewage sludge to plant cultivation: influence of pyrolysis temperature and biochar-to-soil ratio on yield and heavy metal accumulation. Chemosphere 2014; 109: 213-20. doi: 10.1016/j.chemosphere.2014.01.070.
34. Lester JN, Sterritt RM, Kirk PWW. Significance and behavior of heavy metals in waste water treatment processes II. Sludge treatment and disposal. Sci Total Environ 1983; 30: 45-83. doi: 10.1016/0048-9697(83)90003-7.
35. US EPA. Standards for the use or disposal of sewage sludge. Federal Register 1993; 58(32):9248-415.
36. Agrafioti E, Bouras G, Kalderis D, Diamadopoulos E. Biochar production by sewage sludge pyrolysis. J Anal Appl Pyrolysis 2013; 101: 72-78. doi: 10.1016/j.jaap.2013.02.010.
37. Shinogi Y, Yoshida H, Koizumi T, Yamaoka M, Saito T. Basic characteristics of low-temperature carbon products from waste sludge. Adv Environ Res 2003; 7(3): 661-665.
38. Gaskin JW, Steiner C, Harris K, Das KC, Bibens B. Effect of low-temperature pyrolysis conditions on biochar for agricultural use. Transactions of the Asabe 2008; 51(6): 2061-9.
39. Agboola K, Moses SA. Effect of biochar and cowdung on nodulation, growth and yield of soybean (*Glycine max* L.

Merrill). Int J Agric Biosci 2015; 4(4): 154-60.
40. Yamika WSD, Ikawati KR. Combination inorganic and organic fertilizer increased yield production of soybean in rain-field Malang, Indonesia. American-Eurasian Journal of Sustainable Agriculture 2012; 6(1): 14-17.
41. Yooyen J, Wijitkosum S, Sriburi T. Increasing yield of soybean by adding biochar. J Environ Res Dev 2015; 9(4): 1066-74.
42. Navin K, Singh VK, Thakur RB, Kumar N. Effect of level and time of N application on the performance of winter rice. J Appl Biol 1996; 6(1-2): 48-53.
43. Falodun EJ, Osaigbovo AU, Remison SU. The effect of packaged organic and inorganic fertilizers on the growth and yield of soyabean (*Glycine max*). African Journal of General Agriculture 2010; 6(3): 169-75.
44. Liang B, Lehmann J, Solomon D, Kinyangi J, Grossman J, O'neill B, et al . Black carbon increases cation exchange capacity in soils. Soil Sci Soc Am J 2006; 70(5): 1719-1730.
45. MacRae RJ, Mehuys GR. The effect of green manuring on the physical properties of temperate-area soils. In: Stewart BA, ed. Advances in Soil Science. New York: Springer; 1985. p. 71-94.
46. Jones DL, Edwards-Jones G, Murphy DV. Biochar mediated alterations in herbicide breakdown and leaching in soil. Soil Biol Biochem 2011; 43(4): 804-13. doi: 10.1016/j.soilbio.2010.12.015.
47. Asai H, Samson BK, Stephan HM, Songyikhangsuthor K, Homma K, Kiyono Y, et al. Biochar amendment techniques for upland rice production in Northern Laos: 1. Soil physical properties, leaf SPAD and grain yield. Field Crops Res 2009; 111(1-2): 81-4. doi: 10.1016/j.fcr.2008.10.008.
48. Chan KY, Van Zwieten L, Meszaros I, Downie A, Joseph S. Agronomic values of greenwaste biochar as a soil amendment. Australian Journal of Soil Research 2008; 45(8): 629-34.
49. Gantzer CJ, Buyanovsky GA, Alberts EE, Remley PA. Effects of soybean and corn residue decomposition on soil strength and splash detachment. Soil Sci Soc Am J 1987; 51(1): 202-6. doi: 10.2136/sssaj1987.03615995005100010042x.
50. Mukherjee A, Lal R. Biochar impacts on soil physical properties and greenhouse gas emissions. Agronomy 2013; 3(2): 313-39. doi: 10.3390/agronomy3020313.
51. Yao Y, Gao B, Zhang M, Inyang M, Zimmerman AR. Effect of biochar amendment on sorption and leaching of nitrate, ammonium, and phosphate in a sandy soil. Chemosphere 2012; 89(11): 1467-71. doi: 10.1016/j.chemosphere.2012.06.002.
52. Nigussie A, Kissi E, Misganaw M, Ambaw G. Effect of biochar application on soil properties and nutrient uptake of lettuces (Lactuca sativa) grown in chromium polluted soils. American-Eurasian Journal of Agriculture and Environmental Science 2012; 12(3): 369-76.
53. Van Zwieten L, Kimber S, Morris S, Chan KY, Downie A, Rust J, et al. Effects of biochar from slow pyrolysis of papermill waste on agronomic performance and soil fertility. Plant Soil 2010; 327(1-2): 235-46.
54. Zwieten Van L, Kimber S, Morris S, Chan KY, Downie A, Rust J. Effects of biochar from slow pyrolysis of papermill waste on agronomic performance and soil fertility. Plant Soil 2010; 327(1-2): 235-46.
55. Taghizadeh-Toosi A, Clough TJ, Sherlock RR, Condron, LM. Biochar adsorbed ammonia is bioavailable. Plant Soil 2012; 350(1-2): 57-69. doi: 10.1007/s11104-011-0870-3.
56. Biederman LA, Harpole WS. Biochar and its effects on plant productivity and nutrient cycling: a meta-analysis. GCB Bioenergy 2013; 5(2): 202-14. doi:10.1111/gcbb.12037.
57. Dempster DN, Gleeson DB, Solaiman Z, Jones DL, Murphy DV. Decreased soil microbial biomass and nitrogen mineralisation with eucalyptus biochar addition to a coarse textured soil. Plant Soil 2012; 354(1-2): 311-24. doi: 10.1007/s11104-011-1067-5.
58. Dempster DN, Jones DL, Murphy DV. Organic nitrogen mineralization in two contrasting agro-ecosystems is unchanged by biochar addition. Soil Biol Biochem 2012; 48: 47-50. doi: 10.1016/j.soilbio.2012.01.013.
59. Nelson NO, Agudelo SC, Yuan W, Gan J. Nitrogen and phosphorus availability in biochar-amended soils. Soil Sci 2011; 176(5): 218-26. doi: 10.1097/SS.0b013e3182171eac.
60. Beck DA, Johnson GR, Spolek GA. Amending greenroof soil with biochar to affect runoff water quantity and quality. Environ Pollut 2011; 159(8-9): 2111-8. doi: 10.1016/j.envpol.2011.01.022.
61. Lehmann J. A handful of carbon. Nature 2007; 447(7141): 143-4. doi: 10.1038/447143a.
62. Lehmann J, Rondon M. Bio-char soil management on highly weathered soils in the humid tropics. In: Uphoff N, ed. Biological Approaches to Sustainable Soil Systems. Boca Raton, FL: CRC Press; 2006. p. 517-30.
63. Van Zwieten L, Meszaros I, Downie A, Chan YK, Joseph S. Soil health: can the cane industry use a bit of 'black magic'. Aust Canegrower 2008; 17: 10-11.
64. Borselli L, Carnicelli S, Ferrari GA, Pagliai M, Lucamante G. Effect of gypsum on hydrological, mechanical and porosity properties of a kaolinitic crusting soil. Soil Technol 1996; 9(1–2): 39-54. doi: 10.1016/0933-3630(95)00034-8.
65. Amini S, Movahedi Naeini SAR. Effects of paper mill sludge application on physical properties of an illitic loess slowly swelling soil with high specific surface area and wheat yield in a temperate climate. J Agric Sci 2013; 5(1): 295-313. doi: 10.5539/jas.v5n1p295.

Efficiency of electrical coagulation process using aluminum electrodes for municipal wastewater treatment

Samad Gholami Yengejeh[1], Hossein Jafari Mansoorian[2,3], Gharib Majidi[1], Ahmad Reza Yari[4*], Narges Khanjani[5]

[1]MSc of Environmental Health Engineering, Department of Environmental Health Engineering, School of Public Health, Qom University of Medical Sciences, Qom, Iran
[2]Lecturer of Environmental Health Engineering Research Center, Department of Environmental Health, School of Public Health, Kerman University of Medical Science, Kerman, Iran
[3]Young Researchers and Elite Club, Hamedan Branch, Islamic Azad University, Hamedan, Iran
[4]Assistant Professor of Research Center for Environmental Pollutants, Qom University of Medical Sciences, Qom, Iran
[5]Associate Professor of Environmental Health Engineering Research Center, Department of Environmental Health Engineering, School of Public Health, Kerman University of Medical Sciences, Kerman, Iran

Abstract
Background: The reuse of treated municipal wastewater is an important source of water for different purposes. This study evaluated the efficiency of the electrocoagulation process in removing turbidity, total suspended solids (TSS), chemical oxygen demand (COD), nitrate, and phosphate from wastewater at the treatment facility in Karaj, Iran.
Methods: This experimental study was performed at a pilot scale and in a batch system. A 4-liter tank made from safety glass with 4 plate electrodes made from aluminum was unipolarly connected to a direct current power supply with a parallel arrangement. Wastewater samples were taken from the influent at the Karaj wastewater treatment facility. Rates of turbidity, TSS, COD, nitrate, and phosphate removal under different conditions were determined.
Results: The highest efficiency of COD, TSS, nitrate, turbidity, and phosphate elimination was achieved at a voltage of 30 volts and a reaction time of 30 minutes. The rates were 88.43%, 87.39%, 100%, 80.52%, and 82.69%, respectively.
Conclusion: Based on the results of this study, electrocoagulation is an appropriate method for use in removing nitrate, phosphate, COD, turbidity, and TSS from wastewater.
Keywords: Electrical coagulation process, Wastewater treatment

***Correspondence to:**
Ahmad Reza Yari
Email: yari1ahr@gmail.com

Introduction

Treated municipal wastewater is a valuable source of water for different purposes such as irrigating agricultural land and green spaces, breeding fish, and artificially recharging underground wastewater (1). Wastewater carries various pollutants that can be classified into the following four groups based on their characteristics: total suspended solids (TSS), non-organic particles, organic particles, and microorganisms (2).

Among the parameters that are important for evaluating the performance of wastewater facilities are the chemical oxygen demand (COD) and nitrogen and phosphorus compounds in the effluent of such treatment units. Organic decomposable materials are always measured according to biological oxygen demand (BOD) and COD. If these materials enter the environment untreated, their biological stability can lead to decreased oxygen resources, anaerobic circumstances, and the dissemination of bad odors as from hydrogen sulfide (3,4).

The phosphate compounds in wastewater become dissolved orthophosphates after hydrolysis and biological degradation which is used in water resources by microorganisms, especially algae. Phosphorus can lead to increased algae growth and eutrophication in bodies of water. Therefore, controlling the growth of algae and preventing eutrophication in water resources by removing phosphorus from municipal wastewater can be important factors for saving the quality of limited fresh water resources (5).

Electrocoagulation is the coagulation of material onsite using electrolysis with soluble anode electrodes which are usually made from Fe or Al (6). The electrochemical oxidation of Fe or Al at the anode makes metal cathodes, but oxidation at the cathode usually makes hydrogen. The electrochemical process includes many chemical and physical phenomena such as electrical discharge, anode oxidation, cathode reduction, coagulation, electrophoresis migration, and surface adsorption. Moreover, in electrocoagulation, the solution does not saturate with anions and its salt content does not increase as with chemical sedimentation (7).

The use of Al electrodes produced reactions in this study as listed below (6):

$$Al(OH)_2^+ + H_2O \rightarrow Al(OH)_3 + H^+ \quad (1)$$

$$Al(OH)_3 + H_2O \rightarrow Al(OH)_4^- + H^+ \quad (2)$$

$$Al^{3+} + H_2O \rightarrow Al(OH)^{2+} + H^+ \quad (3)$$

$$Al(OH)^{2+} + H_2O \rightarrow Al(OH)_2^+ + H^+ \quad (4)$$

In this process, metal hydroxides are made and combined with pollutants. Bigger clots are formed, and as air bubbles rise to the surface of the reactor, the lighter particles remain floating. Thus, by using the two phenomena of floating and sedimentation, pollutants are removed from the environment (8). Therefore, the mechanism of removal in the electrocoagulation method is the particle instability method which includes electrical double-layer compression, surface adsorption and electrical neutralization, sweeping coagulation, and bridging between particles (9). Meanwhile, the metal hydroxides in the aqua phase remain as a colloid suspension and are able to remove pollutants from wastewater by complexing or electrostatic force (surface adsorption), which leads to coagulation. In the mechanism of forming surface complexes, pollutants as chemicals will make complexes with metal hydrated cations; through sedimentation or floating on the surface of the solution, which leads to the formation of sludge, they will remove pollutants (10). Electrocoagulation has been used successfully to treat the wastewater of industries such as plating, textiles, olive oil production, laundry, tannery, paper and pulp, fermentation, and slaughtering (6,7).

Wang et al evaluated turbidity removal from wastewater at 20 volts and reached a 98% efficiency (11). Merzouk et al investigated the removal of suspended solids from textile wastewater by using the electrical-coagulation and flotation process and achieved a removal efficiency under optimum conditions (11.55 mA/cm^2 current intensity, 1 cm space between electrodes, and 10 minutes of contact time) of 86.5% (12).

Among the advantages of electrical coagulation are the capabilities of adjusting to different situations, selecting, automatic regulation, and compatibility with environmental conditions, simple equipment, high speed, no needing for specific chemicals, small sludge production volumes, short contact time, simple operations, and the capability of treating a wide range of pollutants (7). This study evaluated the efficiency of electrocoagulation using aluminum electrodes for treating municipal wastewater from the Karaj wastewater treatment plant.

Material and Methods

Sampling and analysis

In this study, combined sampling was performed on the influent of the Karaj wastewater facility. Samples were transferred to the lab in 50-liter polyethylene containers. To prevent any change in quality before analysis, the samples were stirred again. All experiments were done according to the standard methods for the examination of water and wastewater (13). The methods used to analyze each parameter are shown in Table 1.

Electrocoagulation cell

This study was done experimentally in a plexiglass batch reactor with the dimensions of 16×16×16 cm and a volume of 4 liters which was resistant to acid. Four unipolar electrodes made from aluminum sheets with the dimensions of 14×14 cm and a thickness of 0.2 cm were installed in the reactor and connected to the feeding source unipolarly and parallelly. This pattern consumes less energy in comparison to the series method. In this experiment, the electrical digital converter Matrix Ltd (MPS-30051) which can convert urban electricity to direct electrical current was used. The electricity entering the electrical converter was an alternating current, and the exiting electricity was direct. The Rogale multimeter was used to measure the electrical current. Cell contents were stirred using a Heidolph MR 3001 stirring device at 200 rpm. The electrodes were placed 2 cm away from the base of the reactor so that stirring could be done efficiently.

Electrocoagulation procedure and calculations

The efficiency of the electrocoagulation process in removing turbidity, TSS, COD, nitrate, and phosphate from wastewater from the Karaj facility was evaluated under the circumstances shown in Table 2.

Table 1. Methods used to analyze each parameter

Pollutant	Unit	Experimental number	Method	Model	Reference
Nitrate	mg/L	SM4110-B	Spectrophotometer	HACH America	(13)
Phosphate	mg/L	SM4110-B	Spectrophotometer	HACH America	(13)
COD	mg/L	SM5220-D	Spectrophotometer	HACH America	(13)
TSS	mg/L	2540	Total solids dried at 103-105 c	-	(13)
Turbidity	NTU	SM2130-B	Nephlometry	HACH America	(13)

Based on the results of other studies, technical and economic constraints, and the aim of the researcher to deal with a real-time situation, this study evaluated a short contact time with high voltage (contact time: 15 minutes; voltage: 45 V), average voltage times with average voltage (contact times: 20 and 30 minutes; voltage: 30 V), and a long contact time with low voltage (contact time: 45 minutes; voltage: 15 V).

Samples (50 mL) were taken from the middle of the reactor. After the end of each experiment, the electrodes were scratched with water, washed with diluted hydrogen chloride, and weighed after washing with distilled water. In Figure 1, the schematic view of the electro-coagulation reactor used in this research is shown.

The electrical energy used is an important economic parameter in the electrocoagulation process; it is calculated using the following formula (14):

$$P = (U.i.t)/V \qquad (5)$$

where P is the amount of electricity used (kwh/L), U is the electrical voltage (V), I is the electrical intensity (A), t is the contact time (h), and V is the volume of wastewater or water (L) for each setting up of the device (14). Also, the number of electrodes used at the end of each experiment was determined by subtracting the initial and final weights.

Results

The quality indices of the municipal wastewater sampled can be seen in Table 3.

The present study evaluated the efficiency of the electrocoagulation process using aluminum electrodes in removing COD, TSS, nitrate, turbidity and phosphate from raw municipal wastewater in 3 different situations 1) contact times of 45 and 60 minutes with the potential difference of 15 V and electrical currency of 1.5 A; 2) contact times of 20 and 30 minutes with the potential difference of 30 V and electrical currency of 3.4 A; 3) contact time of 15 minutes with the potential difference of 40 V and electrical currency of 5.3 A.

Table 2. Experimental conditions

Contact time (min)	Potential difference (V)	Electrical currency (A)
15	45	5.3
20	30	3.4
30	30	3.4
45	15	1.5
60	15	1.5

The elimination efficiency of nitrate, phosphate, COD, turbidity, and TSS by electrocoagulation is shown in Table 4. As can be seen, with increases in voltage and contact time, the removal rates of nitrate, phosphate, COD, turbidity, and TSSs increased as well. The maximum removal efficiency rates of COD, TSS, nitrate, turbidity, and phosphate were seen under conditions of 30 V and a reaction time of 30 minutes and were 88.43%, 87.39%, 100%, 80.52%, and 82.69%, respectively.

Figure 2 presents the number of electrodes used (g/m^3) in different voltages. The maximum electrode usage was 2.62 g/m^3 at 45 V and a contact time of 60 minutes. Figure 3 presents the electrical energy (kWh/L) consumed in different times and voltages during the electrocoagulation process. The maximum amount of electricity used was 0.095 kWh/L at 60 minutes and 45 V.

Discussion

The maximum efficiency for removing turbidity, TSS and COD occurred in 30 minutes of contact time and at 30 V; the minimum efficiency for removing turbidity occurred at 15 V of potential difference and in a contact time of 45 minutes. As Table 4 shows, at specific reaction times and with increases in electricity voltage, the efficiency of removal increased. Rahmani and Samarghandi studied the efficiency of the electrocoagulation method in removing COD from wastewater; they concluded that by increasing the voltage, the percentage of COD removed increased as well (15). Arbabi et al investigated the electrocoagulation process for treating wastewater from a baker's yeast plant. The COD and turbidity removal efficiency rates turbidity increased with the promotion of current density (16). The results of this study corresponded to those of Rahmani and Samarghandi and those of Arbabi et al. Increasing current density produced larger and heavier flocs. The number of bubbles decreased, but their density increased. As a result, lightweight particles were removed by the bubbles (17).

Based on the results of this study, it can be concluded that

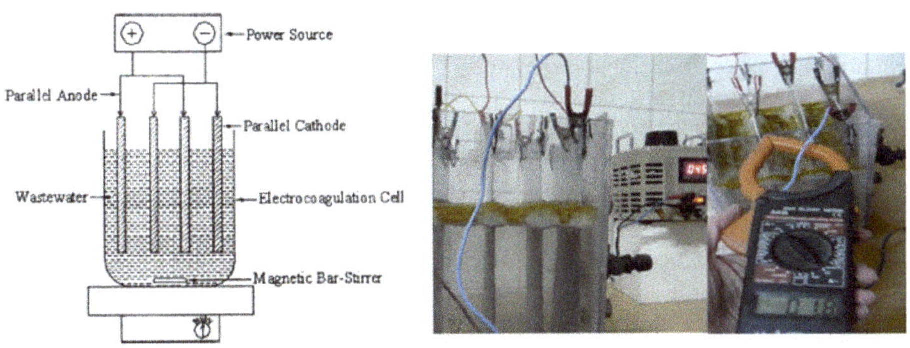

a b

Figure 1. Bench-scale EC reactor with monopolar electrodes in parallel connection (a); image of electrocoagulation (b).

Table 3. Quality indices of sampled wastewater

Parameter	Unit	Samples number	Minimum	Maximum	Mean	SD
Suspended solids	mgL^{-1}	9	225	232	228.3	±3.5
Turbidity	NTU	9	152	165	157.5	±6.57
Phosphate	mgL^{-1}	9	6.52	6.76	6.64	±0.12
Nitrate	mgL^{-1}	9	5.9	6.5	6.22	±0.3
COD	mgL^{-1}	9	415	441	428	±13
pH	-	9	6.3	8.1	7.2	±0.9

Table 4. Effects of potential difference and reaction time on elimination of nitrate, phosphate, COD, turbidity, and suspended solid particles

Pollutant	Reaction time (min)	Electrical currency (A)	Potential difference (V)	Mean concentration of wastewater entering electrocoagulation	Mean concentration of wastewater exiting electrocoagulation	Mean percent of elimination
	15	3.4	30	6.3	1.2	80.95
	20	5.3	45	6.2	1.16	81.29
Nitrate	30	3.4	30	6.5	0	100
	45	1.5	15	5.9	1.1	81.35
	60	1.5	15	6.1	0.5	91.80
	15	3.4	30	6.64	1.79	73.04
	20	5.3	45	6.6	1.56	76.36
Phosphate	30	3.4	30	6.76	1.17	82.69
	45	1.5	15	6.52	1.73	73.46
	60	1.5	15	6.56	1.97	69.97
	15	3.4	30	437	93	78.71
	20	5.3	45	430	76	82.32
COD	30	3.4	30	441	51	88.43
	45	1.5	15	415	131	68.43
	60	1.5	15	421	121	71.25
	15	3.4	30	159	52.96	66.69
	20	5.3	45	158	42.83	72.89
Turbidity	30	3.4	30	163	32.56	80.52
	45	1.5	15	152	74.5	50.99
	60	1.5	15	155	71.73	53.72
	15	3.4	30	231	59.66	74.17
	20	5.3	45	225	41	81.78
TSS	30	3.4	30	230	29	87.39
	45	1.5	15	229	96.33	57.93
	60	1.5	15	226	85.66	62.1

increasing retention time to 30 minutes increases removal efficiency, but no significant increase in removal efficiency will result when retention time is longer than 30 minutes. Other studies have shown that at a contact time of less than 15 minutes (between 10 V to 40 V), removal efficiency is low, because coagulants do not form; a contact time of more than 60 minutes is not wise because of economic reasons, the formation of many clots, and the possibility of their breakage. Therefore, the appropriate time interval for the electrocoagulation process is between 15 to 60 minutes. Merzouk et al in 2009 (12), Parsa et al in 2011 (18), and El-Ashtoukhy et al in 2010 (19) confirm the above results. In this study, the maximum efficiency for nitrate removal happened at a contact time of 30 minutes and 30 V; the least efficiency occurred at 30 V and a contact time of 20 minutes. In a study done by Bazrafshan et al, the maximum removal of nitrate from the solution by iron electrodes (96%) happened at 40 volts, 60 minutes of contact time, and an initial concentration of 5 mg/L (20). Their study showed that as contact time and potential difference increase, removal rates increase as well, which is similar to the present study. In another study done by Razavi et al on laundry wastewater using electrocoagulation, results showed that using this method with aluminum electrodes can lead to a nitrate removal efficiency of 68% (21).

In the current study, the maximum efficiency in removing phosphate happened at 30 minutes and 30 V, and the minimum efficiency happened at 15 V and 60 minutes of contact time. According to these results, increases in voltage have a stronger effect in comparison to reaction time in increasing efficiency. In the study of Takdastan et al about eliminating phosphate, in 90 minutes with 30 V, the efficiency was 99%. They further showed that increases in voltage and reaction time caused increased efficiency (22). Valero and colleagues' study in 2011 (23) confirmed these results.

One of the most important criteria in choosing an appropriate process for removing pollutants is its cost-efficiency; therefore, after each series of experiments with

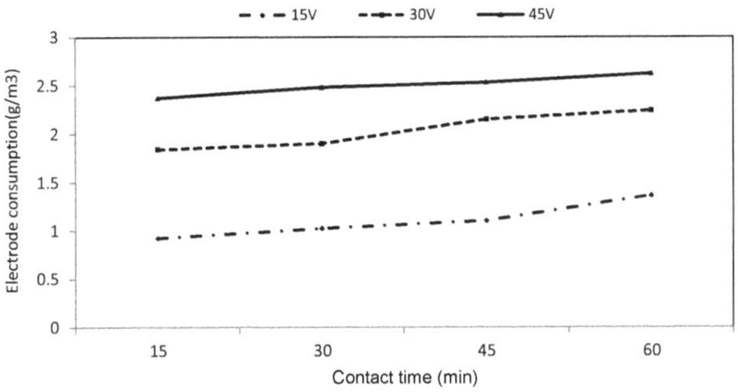

Figure 2. Number of electrodes used at different voltages and contact times.

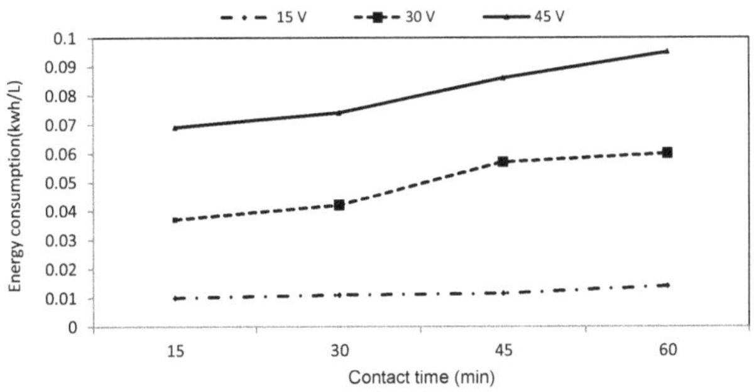

Figure 3. Energy used at different voltages and contact times.

the electrocoagulation process, the aluminum electrodes were weighed and the number of electrodes used under different circumstances was determined. As voltage increased in the process, the number of electrodes used also increased. The maximum amount of decrease in electrode weights happened at 45 V, and the minimum amount happened at 15 V (Figure 2). The results of Adhoum and Monser on removing phenol compounds using electrocoagulation showed that even though removal efficiency increases with increases in voltage, electrode and energy consumption increases as well (24).
The minimum and maximum amounts of energy consumed were 15 V and 45 V, respectively. As can be seen in Figure 3, energy consumption is directly correlated with voltage; as voltage increased, electricity consumption increased as well. Increased voltage is related to increased speed of the production of electrolysis products, such as hydrogen in the cathode, and increases in reduction conditions which lead to increased elimination efficiency (25). This finding is in line with those of Pletcher (26) and Malakootian et al (17). Decreases in electrical intensity also occur with increases in electrolysis time; this is related to the surface absorption–reabsorption reaction and blocking currency. This result is in line with those of Walsh (27) and Mahvi et al (28).

Conclusion

This study evaluated the capabilities of the electrocoagulation method with aluminum electrodes for removing COD, TSS, nitrate, turbidity, and phosphate from municipal wastewater. The results showed that the efficiency of removing pollutants is directly related to increased voltage and contact time. The efficiency rates of removing COD, TSS, nitrate, turbidity, and phosphate from municipal wastewater using electrocoagulation with aluminum electrodes at 30 V and 30 minutes of contact time were 88.43%, 87.39%, 100%, 80.52%, and 82.69%, respectively. The results of this research can be used by municipal wastewater facilities.

Authors' contributions

AY and SGY designed the study. GM and HJM performed the literature search and wrote the manuscript. NK carried out significant improvement in English. All authors participated in data acquisition, analysis, and interpretation. All authors critically reviewed, refined, and approved the manuscript.

Competing interests

The authors declare that they have no competing interests.

Acknowledgments

The authors wish to thank the Faculty of Health at the Qom University of Medical Sciences for cooperating in this study.

References

1. Amin MM, Hashemi H, Ebrahimi A, Bina B, Movahhedian Attar H, Jaberi A, et al. Using combined processes of filtration and ultraviolet irradiation for effluent disinfection of Isfahan north wastewater treatment plant in pilot scale. Water and Wastewater 2011; 22(2): 71-7. [In Persian].
2. Nabizadeh R, Nemati R, Aslani H. Effect of potassium permanganate on effectiveness of chitosan in removing of turbidity and fecal coliforms from wastewater. Water and Wastewater 2012; 23(3): 95-102. [In Persian].
3. Tchobanoglous G, Louis F, Burton H, Stensel D. Wastewater Engineering: Treatment and Reuse. 4th ed. NewYork, NY: McGraw-Hill; 2004.
4. Nemerow NL, Agardy FJ, Salvato JA. Environmental Engineering. 5th ed. New Jersey, NJ: John Wiley & Sons; 2003.
5. Hormozi Nejad M, Takdastan A, Jaafarzadeh Haghighifard N, Ahmadi Mogadam M, Mengelizadeh N. Removal of phosphorus, organic matter and suspended solids by poly aluminum chloride from municipal wastewater (case study: Ahvaz west wastewater treatment plant). Jundishapur J Health Sci 2008; 4(3): 1-13. [In Persian].
6. Jafari Mansoorian H, Mahvi AH, Jonidi Jafari A. Removal of lead and zinc from battery industry wastewater using electrocoagulation process: Influence of direct and alternating current by using iron and stainless steel rod electrodes. Sep Purif Technol 2014; 135: 165-75. doi: 10.1016/j.seppur.2014.08.012.
7. Jafari Mansoorian H, Rajabizadeh A, Bazrafshan E, Mahvi AH. Practical assessment of electrocoagulation process in removing nickel metal from aqueous solutions using iron-

rod electrodes. Desalination and Water Treatment 2012; 44 (1–3): 29-35. doi: 10.1080/19443994.2012.691708.

8. Emamjomeh MM, Sivakumar M. Review of pollutants removed by electrocoagulation and electrocoagulation/ flotation process. J Environ Manage 2009; 90(5): 1663-79. doi: 10.1016/j.jenvman.2008.12.011.
9. Malakootian M, Jafarzadeh Haghighi Fard N, Ahmadian M, Loloei M. Influence of fenton process on treatability of Kerman city solid waste leachate. Iran J Health Environ 2010; 3(2): 123-34. [In Persian].
10. Kashefialasl M, Khosravi M, Marandi R, Seyyedi K. Treatment of dye solution containing colored index acid yellow 36 by electrocoagulation using iron electrodes. Int J Environ Sci Tech 2006; 2(4): 365-71.
11. Wang CT, Chou WL, Kuo YM. Removal of COD from laundry wastewater by electro coagulation/electro flotation. J Hazard Mater 2009; 164(1): 81-86. doi: 10.1016/j.jhazmat.2008.07.122.
12. Merzouk B, Gourich B, Sekki A, Madani K, Chibane M. Removal turbidity and separation of heavy metals using electrocoagulation–electroflotation technique A case study. J Hazard Mater 2009; 164(1): 215-22. doi: 10.1016/j.jhazmat.2008.07.144.
13. American Water Work Association, APHA. Standard Methods for the Examination of Water and Wastewater. 22th ed. .Washington, DC: APHA; 2012.
14. Martínez SS, Bahenab CL. Chlorbromuron urea herbicide removal by electro-Fenton reaction in aqueous effluents. Water Res 2009; 43(1): 33-40. doi: 10.1016/j.watres.2008.09.036.
15. Rahmani A, Samarghandi MR. Electrochemical removal of COD from effluents. Water and Wastewater 2007; 18(4): 9-15. [In Persian].
16. Arbabi M, Shafiei S, Sedehi M, Mazaheri-Shoorabi E. Investigation electrocoagulation process by using iron and stainless steel electrodes for baker's yeast wastewater treatment. J Shahrekord Univ Med Sci 2014; 16(5): 1-12. [In Persian].
17. Malakootian M, Mansoorian HJ, Moosazadeh M. Performance evaluation of electrocoagulation process using iron-rod electrodes for removing hardness from drinking water. Desalination 2010; 255(1-3): 67–71. doi: 10.1016/j.desal.2010.01.015.
18. Basiri Parsa J, Rezaei Vahidian H. Removal of acid brown 14 in aqueous media by electrocoagulation: optimization parameters and minimizing of energy consumption. Desalination 2011; 278(1–3): 295-302. doi: 10.1016/j.desal.2011.05.040.
19. El-Ashtoukhy ES, Zewail TM, Amin NK. Removal of heavy metal ions from aqueous solution by electrocoagulation using a horizontal expanded Al anode. Desalination and Water Treatment 2010; 20(1-3): 72-9. doi: 10.5004/dwt.2010.1127.
20. Bazrafshan E, Soory MM, Mostafapoor F, Mansoorian HJ, Paseban A. Application of aerated electrochemical process to nitrate removal from aqueous. J North Khorasan Univ Med Sci 2011; 3(4): 25-33. [In Persian].
21. Razavi M, Saeedi M, Jabbari E. The effect of operating conditions on simultaneous removal of phosphate, nitrate and COD from laundry wastewater by electrocoagulation using aluminum electrodes. J Environ Stud 2012; 38(3): 75-84. [In Persion].
22. Takdastan A, Azimi A, Salari Z. The use of electrocoagulation process for removal of turbidity, COD, detergent and phosphorus from carwash effluent. Water and Wastewater. 2011; 22(3): 19-25. [In Persian].
23. Valero D, Ortiz M, García V, Expósito E, Montiel V, Aldaz A. Electrocoagulation of wastewater from almond industry. Chemosphere 2011; 84(9): 1290-5. doi: 10.1016/j.chemosphere.2011.05.032.
24. Adhoum N, Monser L. Decolourization and removal of phenolic compounds from olive mill wastewater by electro coagulation. Chem Eng Technol 2004; 43(10): 1281-7. doi: 10.1016/j.cep.2003.12.001.
25. Koparal AS, Ogutveren UB. Removal of nitrate from water by electroreduction and electro coagulation. J Hazard Mater 2002; 89(1): 83-94. doi: 10.1016/S0304-3894(01)00301-6.
26. Pletcher D. Guide to electrochemical technology for synthesis, separation and pollution control. [cited 20 Nov 2016] Available from: https://electrosynthesis.com/wp-content/uploads/2015/11/guide-to-electrochemical-technology.pdf.
27. Walsh FC. Electrochemical technology for environmental treatment and clean energy conversion. Pure Appl Chem 2001; 73(12): 1819-37. doi: 10.1351/pac200173121819.
28. Mahvi AH, Mansoorian HJ, Rajabizadeh A. Performance of electrocoagulation process for removal of sulphate ion from aqueous environments using plate aluminum electrodes. J Qom Univ Med Sci 2009; 4(3): 21-8. [Persian].

Groundwater quality assessment using the water quality pollution indices in Toyserkan Plain

Soheil Sobhanardakani[1*], Lobat Taghavi[2], Behzad Shahmoradi[3], Amin Jahangard[4]

[1]Associate Professor, Department of the Environment, College of Basic Sciences, Hamedan Branch, Islamic Azad University, Hamedan, Iran
[2]Assistant Professor, Department of the Environmental Pollution, College of Environment and Energy, Science and Research Branch, Islamic Azad University, Tehran, Iran
[3]Assistant Professor, Kurdistan Environmental Health Research Center, Kurdistan University of Medical Sciences, Sanandaj, Iran
[4] MSc, Young Researchers & Elite Club, Hamedan Branch, Islamic Azad University, Hamedan, Iran

Abstract
Background: Iran is located within the dry and semi dry regions, thus almost 90% of the required water is secured via the use of groundwater. Owing to the increasing pollution of water resources, this study was performed to evaluate water quality pollution indices for heavy metals (As, Zn, Pb and Cu) contamination monitoring in Toyserkan Plain during spring and summer in 2012.
Methods: A total of 20 ground water wells were chosen randomly. The samples were filtered (0.45 μm) and kept cool in polyethylene bottles. Samples were taken for the analysis of metals, the former was acidified with HNO_3 to pH lower than 2. Metal concentrations were determined using ICP-OES.
Results: The results revealed that the mean values of contamination index (C_d), heavy metal pollution index (HPI) and heavy metal evaluation index (HEI) in samples for spring season were -2.81, 9.74 and 1.20, respectively and were -2.67, 9.51 and 1.32, respectively in samples for summer season and this indicates low contamination levels. Comparing the mean concentrations of the evaluated metals with WHO permissible limits demonstrated a significant difference ($P<0.05$). Thus, the mean concentrations of the metals were significantly lower than the permissible limits.
Conclusion: Although the heavy metal pollution of the ground water in Toyserkan Plain is not higher than permissible limits, the irregular and long-term utilization of agricultural inputs, use of wastewater and sewage sludge in agriculture, over utilization of organic fertilizers and establishment of pollutant industries can threaten the ground water, and cause irreversible damages in this area.
Keywords: Heavy metals, Groundwater, Water quality pollution indices, Toyserkan Plain

***Correspondence to:**
Soheil Sobhanardakani
Email: s_sobhan@iauh.ac.ir

Introduction

Urbanization, industrialization, agriculture and exploitation of natural resources, are basic activities associated with living in contemporary societies that imposed pollutant loads especially toxic metals into natural cycles like soil, water and air cycles (1,2). Nowadays, globally, heavy metals have been taken into consideration owing to their toxicity, ability to accumulate in the biota and adverse health effects even at low concentrations (3-5). One of the most important properties of heavy metals, which differentiate them from other toxic pollutants, is that they are not easily biodegradable in the environment (2,6). Apart from the potential toxicity of heavy metals to living organisms, these pollutants are stable in the environment and tend to accumulate in the tissues of plants and animals (7). Water cycle due to the dynamism and its constant nature is more exposed to pollution when compared to the soil and atmosphere cycles (5). Water resources are very important for mankind existence and economical development. People around the globe have utilized groundwater as a source of drinking water, and even today more than 50% the world's population depends on groundwater for survival (8,9). So, the contamination of these resources by heavy metals is a serious ecological problem.

Metals like Cu, Fe and Zn, are essential metals since they play important roles in biological systems, while As, Hg and Pb are toxic, even in trace amounts (10,11). It should be noted that the essential metals can also produce toxic effects at high concentrations (11,12).

Arsenic is a widely distributed metalloid, occurring in the biosphere. The combustion of fossil fuel for production of energy and smelting of non-ferrous metals are two significant processes that lead to the contamination of the environment, especially the source of atmospheric pollution for this element. Other sources of As are arsenical pesticides, chemicals such as chromated copper arsenate used for the manufacturing of wood preservatives (13).

Zinc is an essential structural and functional element in biological systems which often catalyzes reactions, binds to substrates by favoring various reactions via the mediation of redox or oxidation–reduction reactions, through reversible changes in the oxidation state of the metal ions. Of course, Zn harms some physiological activities such as breathing and causes other diseases (14,15).

Lead may damage the kidney and cause symptoms of chronic toxicity, such as poor reproductive capacity, impaired organ function, tumors, blood pressure and hepatic abnormalities (16). In addition, Pb can also affect brain activity by interfering with neurotransmitter release and synapse formation. Exposure to Pb via the accidental ingestion of Pb paint, inhalation of traffic exhaust fumes and the consumption of Pb-contaminated food can cause the reduction of IQ, learning disabilities, hyperactivity, slow growth, impaired hearing and antisocial behaviors (17-19).

Copper is one of the most abundant trace elements with vitamin-like impact in human body and living systems and is found in a wide range of foods eaten by humans such as nuts, many fruits and vegetables, red meat, shellfish, and in many vitamin supplements. Despite the small amount of Cu (50-120 mg) found in the human body, it plays a crucial role in different kinds of biochemical processes (20). Copper which is an essential micronutrient for growing plants should be supplied via organic and artificial fertilizers for healthy hormone secretion, nerve conduction, and the growth of bones and connective tissue. A constant diet of Cu, even at entirely allowable limits, can break down the barrier that keeps undesirable toxins from entering the brain, and an increase in the production of beta-amyloid. On the other hand, critical doses of Cu can cause inflammation in the brain tissues, anorexia, fatigue, hair loss, acne, allergies, depression, premenstrual syndrome, migraines, anxiety, childhood hyperactivity, panic attacks, kidney and liver dysfunction, strokes elevated cholesterol, adrenal hyperactivity and insufficiency, learning disorders, autism and cancer (21).

Quality indices employ a series of reproducible judgments to compile the effects of all the pollution parameters and are a useful and relatively easy method to evaluate the composite influence of overall pollution (22). For heavy metal contamination assessment in water resources, several methods were developed. The Contamination index, the Heavy metal potential index and the heavy metal evaluation index (HEI) are pollution indices which help in assessing the present level of pollution (23).

Iran is located within the dry and semi dry regions, thus almost 90% of the required water is secured via the use of groundwater (24). Owing to the geological structure of Toyserkan township especially minerals containing As, Zn, Pb, and Cu (25), and also rapid agricultural growth in Toyserkan Plain, discharge of heavy and toxic metals into groundwater resources of this region via overutilization of agricultural inputs like chemical and organic fertilizers especially phosphorus fertilizers, zinc sulfate and metal-containing pesticides are on the increase. Therefore, this study was conducted for the assessment of water quality pollution indices for heavy metals (As, Zn, Pb and Cu) contamination monitoring in Toyserkan Plain during spring and summer in 2012.

Methods

Study area

The study was carried out in Toyserkan township in Hamedan province which is located in the western part of Iran. The area of Toyserkan Plain is 792 km^2. Drinking water for residents of the Toyserkan township is supplied from 1243 wells, 400 springs and 220 aqueducts (25).

Sampling and sample analysis

In this study, based on the Cochran's sample size formula, groundwater samples were collected from 20 wells with depth of 10 to 90 m and discharge on the range of 1 to 6 L/s basis of different land use pattern, including agricultural and residential areas from open and tube wells to assess the heavy metal contamination during spring and summer seasons in 2012. Figure 1 illustrates the sampling stations in the study area. The samples were collected in acid washed 200 mL polyethylene bottles to prevent unpredictable changes in characteristic as per standard procedures (26). The collected samples were filtered (Whatman no. 42), preserved with 6N HNO_3 (Suprapur Merck, Germany) and kept at a temperature of 4°C for further analysis (26,27). Concentrations of heavy metals (As, Zn, Pb and Cu) in water samples were determined employing ICP-OES (Varian, 710-ES, Australia).

Valuation methods

Three documented methods evaluated in this study are the contamination index (C_d), heavy metal pollution index (HPI) and HEI developed or proposed by Backman et al, Prasad and Bose, and Edet and Offiong, respectively (27-29).

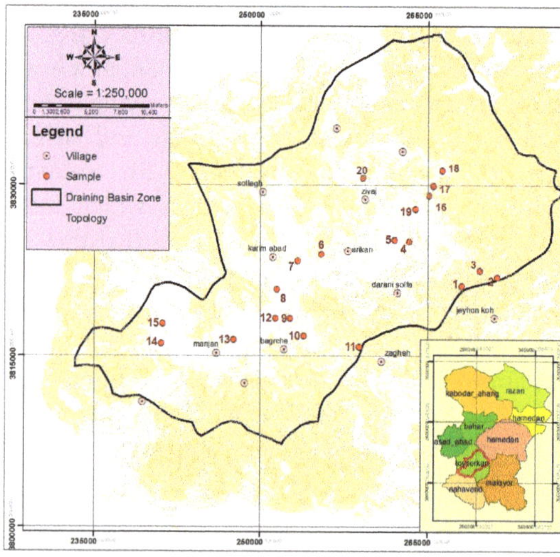

Figure 1. Map of sampling stations.

The contamination index

In this method, the quality of water is evaluated by the calculation of the degree of contamination. The C_d is computed separately for each sample of water analyzed, as a sum of the contamination factors of individual components exceeding the upper permissible value. Hence, the C_d summarizes the combined effects of several quality parameters considered harmful to household water. The contamination index is calculated from Equations 1 and 2:

$$Cd = \sum_{i=1}^{n} C_{fi} \quad (1)$$

where

$$C_{fi} = \frac{CA_i}{CN_i} - 1 \quad (2)$$

C_{fi} indicates contamination factor for the *i*th component, C_{Ai} indicates analytical value for the *i*th component, C_{Ni} indicates upper permissible concentration of the *i*th component (N denotes the 'normative value') (2,27).

The resultant C_d value identifies areas of varying contamination levels which are grouped into three categories as follows: $C_d < 1$ (low), $C_d = 1\text{-}3$ (medium) and $C_d > 3$ (high) (5,27).

The upper permissible concentration value (C_{Ni}) was taken as the maximum admissible concentration (MAC).

Heavy metal pollution index

The HPI represents the total water quality with respect to heavy metals and based on weighted arithmetic quality mean method. The HPI is developed in two steps. The first is by establishing a rating scale for each selected parameter giving weightage and second is by selecting the pollution parameter on which the index is to be based. The rating system is an arbitrarily value between 0 to 1 and its selection depends on the importance of individual quality considerations in a comparative way or it can be evaluated by making values inversely proportional to the recommended standard for the corresponding parameter (2,27,30,31). In computing the HPI, Prasad and Bose (29) considered unit weightage as a value inversely proportional to the recommended standard (S_i) of the corresponding parameter as proposed by Reddy (32).

The HPI model is presented in Equation 3 (31):

$$HPI = \frac{\sum_{i=1}^{n} W_i Q_i}{\sum_{i=1}^{n} W_i} \quad (3)$$

where Q_i= the sub-index of the *i*th parameter, W_i= the unit weightage of the *i*th parameter and n= the number of parameters considered. The sub-index of the parameter is calculated by Equation 4:

$$Q_i = \sum_{i=1}^{n} \frac{\{M_i(-)I_i\}}{(S_i - I_i)} \quad (4)$$

where M_i indicates the monitored value of heavy metal of *i*th parameter, I_i indicates the ideal value of the *i*th parameter and S_i indicates the standard value of the *i*th parameter. The sign (−) indicates numerical difference of the two values, ignoring the algebraic sign. Low heavy metal pollution (HPI < 100), heavy metal pollution on the threshold risk (HPI = 100) and high heavy metal pollution (critical pollution index) (HPI > 100). If the samples have HPI values greater than 100, water is not potable (2,27,29,31).

In computing the HPI for the present study, As, Zn, Pb and Cu were utilized. The weightage (W_i) was taken as the inverse of MAC, S_i the World Health Organization (WHO) standard for drinking water and I_i the guide value for the chosen element (Table 1).

Heavy metal evaluation index

HEI focuses on heavy metals in water samples for estimating the water quality (33). The water quality index is classified into three categories which include HEI <400 (low heavy metals), 400 <HEI < 800 (moderate to heavy metals) and HEI> 800 (high heavy metals). The index is calculated from Equation 5 (5):

$$HEI = \sum_{i=1}^{n} \frac{H_c}{H_{mac}} \quad (5)$$

where H_c is the monitored value of the *i*th parameter and H_{mac} the MAC of the *i*th parameter (27,33).

Results

The results of the heavy metal concentrations in ground water samples for spring and summer seasons are given in Tables 2 and 3. Also the correlation matrix between elements for spring and summer seasons are presented in Table 4.

The results indicate that As, Zn, Pb and Cu concentrations in groundwater samples collected from Toyserkan Plain in spring season ranged from 0.08 to 7.48, 0.12 to 15.64, 0.09 to 5.50 and 0.89 to 13.58 μg/L, respectively.

The results indicate that As, Zn, Pb and Cu concentrations in groundwater samples collected from Toyserkan Plain in summer season ranged from 0.57 to 7.21, 0.41 to 16.42, 0.19 to 4.46 and 6.54 to 15.76 μg/L, respectively.

Correlation at 5% level of significance (P > 0.05) demonstrated no significant correlation between the following pairs: As and Zn, Pb, Cu; Zn and Pb; Cu and Pb and Cu in water samples for spring and summer seasons.

The computed C_d, HPI and HEI values for each location, correlation between index values and concentration of metal and correlation between different indices values for spring and summer seasons are given in Tables 5 to 7, respectively.

The computed C_d demonstrates that the mean values in

Table 1. Standard used for the indices computation (27)

	W	S	I	MAC
As	0.02	50	10	50
Zn	0.0002	5000	3000	5000
Pb	0.70	100	10	1.50
Cu	0.001	1000	2000	1000

Abbreviations: W; weightage (1/MAC); S, standard permissible in ppb; I, highest permissible in ppb; MAC, maximum admissible concentration/ upper permissible.

Table 2. Concentration of As, Zn, Pb and Cu (μg L^{-1}) in groundwater samples collected from Toyserkan Plain in spring season

Station	As	Zn	Pb	Cu
1	7.48	3.75	1.13	7.78
2	4.20	1.80	0.10	8.49
3	6.21	10.27	0.09	7.22
4	0.40	2.57	1.61	0.89
5	3.28	6.05	1.06	6.67
6	2.06	3.41	0.17	9.93
7	5.62	7.42	4.38	13.58
8	1.79	15.64	1.70	11.51
9	6.44	11.20	5.50	9.16
10	3.56	3.61	1.12	10.44
11	4.62	1.62	1.25	9.08
12	0.08	1.05	1.27	10.34
13	7.04	0.40	0.31	10.25
14	0.35	0.17	3.96	11.83
15	1.18	1.17	2.43	11.86
16	3.81	1.05	2.37	12.06
17	2.96	0.12	0.52	5.57
18	3.50	0.29	2.80	5.83
19	3.78	3.20	1.10	5.80
20	5.20	2.14	0.42	3.56
Mean	3.68 ± 2.24	4.24 ± 3.85	1.66 ± 1.51	8.59 ± 3.19

Table 3. Concentration of As, Zn, Pb and Cu (μg/L) in groundwater samples collected from Toyserkan Plain in summer season

Station	As	Zn	Pb	Cu
1	0.57	4.75	1.55	12.60
2	3.23	8.55	2.39	6.54
3	5.06	8.83	1.39	13.14
4	5.83	3.23	1.78	11.65
5	3.45	11.07	1.78	11.13
6	7.08	4.18	0.46	11.93
7	0.76	10.24	1.64	12.91
8	3.91	9.08	1.05	13.25
9	7.21	16.42	1.34	12.72
10	5.75	9.83	0.19	12.74
11	7.11	3.73	4.46	11.42
12	3.37	0.94	3.00	15.76
13	3.89	8.54	1.98	12.25
14	5.26	2.66	1.98	12.72
15	2.10	3.53	2.58	12.15
16	3.86	0.41	1.53	13.02
17	3.23	0.87	1.43	13.39
18	3.22	1.74	1.16	13.30
19	3.06	1.28	1.49	13.31
20	1.86	16.03	3.74	13.83
Mean	3.99 ± 1.93	6.30 ± 4.89	1.85 ± 1.01	12.49 ± 1.71

spring and summer seasons were -2.80 and -2.67, respectively, and indicate low contamination. The computed HPI shows that the mean values in spring and summer seasons were 9.74 and 9.51, respectively and are lower than 100, the critical value for drinking water for all the locations. Moreover, the computed HEI shows that the mean values in spring and summer seasons were 1.20 and 1.32, respectively and indicate low heavy metal pollution.

A comparison between the indices and heavy metal concentration demonstrates significant correlation with Pb for spring and summer samples (Table 6). This indicates that Pb is the main contributory parameter. Moreover, the correlation between C_d, HPI and HEI is significant (Table 7). Therefore, the three existing methods; the Contamination index, the HPI and the HEI provide same results.

Discussion

In order to assess the groundwater resources of Toyserkan Plain, groundwater samples were collected from 20 wells. Four elements including As, Zn, Pb and Cu in the samples were measured and used in calculating C_d, HPI and HEI indices. The results demonstrated that the mean concentration of metals in samples in spring season (μg L^{-1}) were 3.68 ± 2.24 for As, 4.24 ± 3.85 for Zn, 1.66 ± 1.51 for Pb and 8.59 ± 3.19 for Cu, respectively (Table 2). In addition, the mean concentration of metals in samples in summer season (μg L^{-1}) were 3.99 ± 1.93 for As, 6.30 ± 4.89

Table 4. Correlation matrix between elements

	As	Zn	Pb	Cu
	Spring			
As		0.219	-0.380	-0.009
Zn			0.230	0.163
Pb				0.378
	Summer			
As		0.030	-0.125	-0.115
Zn			0.018	-0.161
Pb				-0.061

Table 5. Evaluation indices

Station	Spring			Summer		
	C_d	HPI	HEI	C_d	HPI	HEI
1	-3.09	10.06	0.91	-2.94	10.08	1.06
2	-3.84	11.40	0.16	-2.33	9.00	1.67
3	-3.81	11.27	0.19	-2.96	9.94	1.04
4	-2.92	10.02	1.08	-2.67	9.36	1.31
5	-3.22	10.43	0.78	-2.73	9.52	1.27
6	-3.83	11.47	0.16	-3.54	10.80	0.46
7	-0.95	6.68	3.25	-2.88	9.97	1.12
8	-2.82	9.83	1.18	-3.21	10.38	0.79
9	-0.19	5.42	3.81	-2.95	9.84	1.05
10	-3.17	10.34	0.83	-3.74	11.19	0.26
11	-3.06	10.12	0.93	-0.87	6.49	3.13
12	-3.14	10.41	0.86	-1.92	8.32	2.08
13	-3.64	10.97	0.36	-2.59	9.39	1.41
14	-1.34	7.50	2.66	-2.56	9.30	1.44
15	-2.34	9.10	1.66	-2.22	8.86	1.77
16	-2.33	8.97	1.67	-2.89	9.84	1.11
17	-3.59	11.02	0.41	-2.97	10.02	1.03
18	-2.06	8.53	1.94	-3.15	10.32	0.85
19	-3.18	10.34	0.81	-2.93	9.97	1.07
20	-3.61	10.98	0.39	-1.45	7.63	2.55
Mean	-2.81	9.74	1.20	-2.67	9.51	1.32

Table 6. Correlation between index values and concentration of metals

Parameter	C_d		HPI		HEI	
	r	*P*	*r*	*P*	*r*	*P*
Spring						
As	0.007	0.978	-0.056	0.815	0.015	0.948
Zn	0.240	0.308	-0.250	0.287	0.244	0.299
Pb	0.999[a]	0.000	-0.996[a]	0.000	0.998[a]	0.000
Cu	0.381	0.097	-0.376	0.102	0.390	0.089
Summer						
As	-0.067	0.779	-0.003	0.990	-0.068	0.777
Zn	0.020	0.993	-0.022	0.926	0.021	0.929
Pb	0.998[a]	0.000	-0.991[a]	0.000	0.998[a]	0.000
Cu	-0.068	0.777	0.080	0.737	-0.067	0.780

[a] Correlation is significant at the 0.01 level (2-tailed).

Table 7. Correlation between different indices values

	r	*P*
Spring		
C_d vs. HPI	-0.999[a]	0.000
C_d vs. HEI	0.999[a]	0.000
HPI vs. HEI	-0.998[a]	0.000
Summer		
C_d vs. HPI	-0.997[a]	0.000
C_d vs. HEI	1.000[a]	0.000
HPI vs. HEI	-0.997[a]	0.000

[a] Correlation is significant at the 0.01 level (2-tailed).

for Zn, 1.85±1.01 for Pb and 12.49±1.71 for Cu, respectively (Table 3). The higher values of metals concentration in summer season compared to spring season may be due to reduced aquifer recharge and reduced dilution of pollutants as a result of increase in water temperature during the summer season.

Based on the results, heavy metal pollution is not observed in any cases. According to Table 5, the values of C_d, HPI and HEI indices were found in the range of -3.84 to -0.19, 5.42 to 11.47 and 0.16 to 3.81, respectively, in spring season. The highest values of C_d, HPI and HEI were found in the sample collected from stations number 9, 6 and 9, respectively. Furthermore, during summer season, the values of same indices are increased up to -3.74 to -0.87, 6.49 to 11.19 and 0.26 to 3.13, respectively. The highest values of C_d, HPI and HEI in this season were found in the sample collected from station number 11, 10 and 11, respectively. The higher values of C_d, HPI and HEI may be due to agricultural activities during spring and summer seasons. The C_d, HPI and HEI values of the samples within study area are found below the critical pollution index (100), and this indicates that the water is not critically polluted with respect to studied heavy metals and is suitable for drinking.

Nevertheless, the values of these three indices in groundwater collected from Toyserkan Plain are totally below the critical values, but severe precautions such as managing the use of agricultural inputs (chemical fertilizers and metal-containing pesticides), use of wastewater and sewage sludge in agriculture, over utilization of organic fertilizers and establishment of pollutant industries must be taken into consideration in this area. In this regard, Sobhanardakani (34) evaluated water quality pollution indices for groundwater resources of Ghahavand Plain in Hamedan province and reported that mean values of C_d, HPI and HEI in samples for spring 2012 were -2.27, 9.01 and 1.73, respectively and were -1.95, 8.69 and 2.04, respectively in samples for summer 2012 and indicates low contamination levels. Sobhanardakani et al (35) analyzed As, Zn, Pb, Cd and Cu content in groundwater resources of Asadabad Plain in Hamedan province in 2012 and reported that the mean values of indices in samples from spring and summer seasons were 25.61 and 27.28, respectively for HPI and were 9.29 and 8.88, respectively for HEI, and indicates low contamination levels. Nazari and Sobhanardakani (36) analyzed As and Zn concentrations in groundwater resources of Qaleh Shahin Plain in Kermanshah province and reported that the HPI values in winter 2014 vary between 1.09 to 11.4 (mean 6.11) and vary between 1.83 to 22.8 (mean 8.78) in summer 2014 and are lower than 100, the critical value for drinking water for all the sampling stations. Sobhanardakani and Nazari (37) analyzed Pb and Cd concentrations in groundwater resources of Qaleh Shahin Plain in Kermanshah province and reported that the HPI values in winter 2014 vary between 0.32 to 7.69 (mean 4.73) and vary between 8.92 to 13.90 (mean 11.74) in summer 2014 and are lower than 100, the critical value for drinking water for all the sampling stations.

Conclusion

According to the findings, heavy metal pollution was not observed in any water samples. So, groundwater samples of the Toyserkan Plain have been identified as suitable for drinking but based on the correlation matrix (Table 6), Pb has a great role in the quality of water samples. This indicates that the water quality indices proved to be a very useful tool in evaluating overall pollution of the ground water resources. Finally, it can be concluded however, that the values of these three indices in groundwater samples collected from Toyserkan Plain are totally below the critical values. Nevertheless, it is recommended that severe precautions should be taken for controlling the sources of groundwater pollution in this area.

Acknowledgments
The authors are grateful to the Hamedan Branch, Islamic Azad University for providing facilities to conduct and complete this study.

Competing interests
The authors declare that they have no competing interests.

Authors' contributions
All authors participated in the study design, literature search, writing of the manuscript, and data acquisition, analysis, and interpretation. All authors critically reviewed, refined, and approved the manuscript

References

1. Gribble GW. The natural production of chlorinated compounds. Environ Sci Technol 1994; 28(7): 310A-9A. doi: 10.1021/es00056a712.
2. Nasrabadi T. An index approach to metallic pollution in river waters. Int J Environ Res 2015; 9(1): 385-94.
3. Morillo J, Usero J, Gracia I. Partitioning of metals in sediments from the Odiel River (Spain). Environ Int 2002; 28(4): 263-71. doi: 10.1016/S0160-4120(02)00033-8.
4. Baghvand A, Nasrabadi T, Nabi Bidhendi GR, Vosoogh A, Karbassi AR, Mehrdadi N. Groundwater quality degradation of an aquifer in Iran central desert. Desalination 2010; 260(1-3): 264-75. doi: 10.1016/j.desal.2010.02.038.
5. Hosseinpour Moghaddam M, Lashkaripour GR, Dehghan P. Assessing the effect of heavy metal concentrations (Fe, Pb, Zn, Ni, Cd, As, Cu, Cr) on the quality of adjacent groundwater resources of Khorasan steel complex. Int J Plant Anim Environ Sci 2014; 4(2): 511-18.
6. Rauret G, López-Sánchez JF, Sauquillo A, Rubio R, Davidson C, Ure A, et al. Improvement of the BCR three step sequential extraction procedure prior to the certification of new sediment and soil reference materials. J Environ Monit 1999; 1(1): 57-61. doi: 10.1039/a807854h.
7. Bhaskar CV, Kumar K, Nagendrappa G. Assessment of heavy metals in water samples of certain locations situated around Tumkur, Karnataka, India. E-J Chem 2010; 7(2): 349-52. doi: 10.1155/2010/415150.
8. Rajankar PN, Gulhane SR, Tambekar DH, Ramteke DS, Wate SR. Water quality assessment of groundwater resources in Nagpur Region (India) Based on WQI. E-J Chem 2009; 6(3): 905-08. doi: 10.1155/2009/971242.
9. Rizwan R, Gurdeep S, Manish Kumar J. Application of heavy metal pollution index for ground water quality assessment in Angul District of Orassia, India. Int J Res Chem Environ 2011; 1(2): 118-22. doi: 10.1007/s12665-014-3882-5.
10. Sobhanardakani S, Tayebi L, Farmany A, Cheraghi M. Analysis of trace elements (Cu, Cd and Zn) in muscle, gill and liver tissues of some fish species using anodic stripping voltammetry. Environ Monit Assess 2012; 184(11): 6607-11. doi: 10.1007/s10661-011-2445-4.
11. Hosseini SV, Aflaki F, Sobhanardakani S, Tayebi L, Babakhani Lashkan A, Regenstein JM. Analysis of mercury, selenium and tin concentrations in canned fish marketed in Iran. Environ Monit Assess 2013; 185(8): 6407-12. doi: 10.1007/s10661-012-3033-y.
12. Carvalho ML, Santiago S, Nunes ML. Assessment of the essential element and heavy metal content of edible fish muscle. Anal Bioanal Chem 2005; 382(2): 426-32. doi: 10.1007/s00216-004-3005-3.
13. Järup L. Hazards of heavy metal contamination. Br Med Bull 2003; 68(1): 167-82. doi: 10.1093/bmb/ldg032.
14. Tahsin N, Yankov B. Research on accumulation of zinc (Zn) and cadmium (Cd) in sunflower oil. Journal of Tekirdag Agricultural Faculty 2007; 4(1): 109-12.
15. Sobhanardakani S, Jamshidi K. Assessment of metals (Co, Ni, and Zn) content in the sediments of Mighan Wetland using geo-accumulation index. Iranian Journal of Toxicology 2015; 9(30): 1386-90.
16. Abou-Arab AAK, Ayesh AM, Amra HA, Naguib K. Characteristic levels of some pesticides and heavy metals in imported fish. Food Chem 1996; 57(4): 487-92. doi: 10.1016/S0308-8146(96)00040-4.
17. Dahiya S, Karpe R, Hegde AG, Sharma RM. Lead, cadmium and nickel in chocolate and candies from suburban areas of Mumbai. India. J Food Comp Anal 2005; 18 (6): 517-22. doi: org/10.1016/j.jfca.2004.05.002.
18. Hosseini SV, Sobhanardakani S, Tahergorabi R, Delfieh P. Selected heavy metals analysis of Persian sturgeon's (*Acipenser persicus*) caviar from Southern Caspian Sea. Biol Trace Elem Res 2013; 154(3): 357-62. doi: 10.1007/s12011-013-9740-6.
19. Adekunle IM, Akinyemi MF. Lead levels of certain consumer products in Nigeria: a case study of smoked fish foods from Abeokuta. Food Chem Toxicol 2004; 42(9): 1463-8. doi: 10.1016/j.fct.2004.04.007.
20. Saracoglu S, Tuzen M, Soylak M. Evaluation of trace element contents of dried apricot samples from Turkey. J Hazard Mater 2009; 167(1-3): 647-52. doi: org/10.1016/j.jhazmat.2009.01.011.
21. Ackah M, Anim AK, Zakaria N, Osei J, Saah-Nyarko E, Gyamfi ET, et al. Determination of some heavy metal levels in soft drinks on the Ghanaian market using atomic absorption spectrometry method. Environ Monit Assess 2014; 186(12): 8499-507. doi: 10.1007/s10661-014-4019-8.
22. Prasad B, Sangita K. Hevay metal pollution index of ground water of an abandoned open cast mine filled with fly Ash: a case study. Mine Water and the Environment 2008; 27(4): 265-7. doi: 10.1007/s10230-008-0050-8.
23. Maria-Alexandra H, Roman C, Ristoiu D, Popita G, Tanaselia C. Assessing of water quality pollution Indices for hevay metal contamination. A study case from Medias City groundwaters. Agricultura 2013; 87(3-4): 25-31. doi: 10.15835/arspa.v87i3-4.9974.
24. Sobhanardakani S, Jamali M, Maànijou M. Evaluation of As, Zn, Cr and Mn concentrations in groundwater resources of Razan Plain and preparing the zoning map using GIS. Journal of Environmental Science and Technology 2014; 16(2): 25-38.
25. Sobhanardakani S, Talebiani S, Maànijou M. Evaluation of As, Zn, Pb and Cu concentrations in groundwater resources of Toyserkan Plain and preparing the zoning map using GIS. J Mazandaran Univ Med Sci 2014; 24(114): 120-9. [In Persian].

26. American Public Health Association (APHA). Standard Methods for Examination of Water and Waste water. 20th ed. Washington DC: APHA; 2005.
27. Edet AE, Offiong OE. Evaluation of water quality pollution indices for heavy metal contamination monitoring. A study case from Akpabuyo-Odukpani area, Lower Cross River Basin (southeastern Nigeria). Geo J 2002; 57(4): 295-304. doi: 10.1023/B:GEJO.0000007250.92458.de
28. Backman B, Bodis D, Lahermo P, Rapant S, Tarvainen T. Application of a groundwater contamination index in Finland and Slovakia. Environ Geol 1998; 36(1-2): 55-64. doi: 10.1007/s002540050320.
29. Prasad B, Bose JM. Evaluation of the heavy metal pollution index for surface and spring water near a Limeston mining area of the lower Himalayas. Environ Geol 2001; 41(1-2): 183-8. doi: 10.1007/s002540100380.
30. Horton RK. An index systems for rating water quality. J Water Pollut Contr Federat 1965; 37(3): 300-6.
31. Mohan SV, Nithila P, Reddy SJ. Estimation of heavy metals in drinking water and development of heavy metal pollution index. J Environ Sci Health A 1996; 31(2): 283-9. doi: 10.1080/10934529609376357.
32. Trivedy RK. Encyclopedia of Environmental Pollution and Control (vol 2). Karad, India: Environmental Media; 1995. p. 342.
33. Edet AE, Merkel BJ, Offiong OE. Trace element hydrochemical assessment of the Calabar Coastal Plain Aquifer, southeastern Nigeria using statistical methods. Environ Geol 2003; 44(2): 137-49. doi: 10.1007/s00254-002-0738-1.
34. Sobhanardakani S. Evaluation of the water quality pollution indices for groundwater resources of Ghahavand Plain, Hamedan Province, western Iran. Iran J Toxicol 2016; 10(3): 35-40.
35. Sobhanardakani S, Yari AR, Taghavi L, Tayebi L. Water quality pollution indices to assess the heavy metal contamination, case study: Groundwater resources of Asadabad Plain in 2012. Arch Hyg Sci 2016; 5(4): 221-8.
36. Nazari S, Sobhanardakani S. Assessment of pollution index of heavy metals in groundwater resources of Qaleh Shahin plain (2013-2014). J Kermanshah Univ Med Sci 2015; 19(2): 102-8. [In Persian].
37. Sobhanardakani S, Nazari S. Assessment of Pb and Cd pollution in groundwater resources of Qaleeh Shahin Plain using heavy metal pollution index in 2014. Journal of Health System Research 2016; 12(3): 300-6. [In Persian].

Use of Aloe vera shell ash supported $Ni_{0.5}Zn_{0.5}Fe_2O_4$ magnetic nanoparticles for removal of Pb (II) from aqueous solutions

Samira Namavari[1], **Farid Moeinpour**[2*]

[1]MSc in Water and Wastewater Engineering, Department Environmental Engineering, Faculty of Natural Resources, Islamic Azad University, Bandar Abbas Branch, Bandar Abbas, Iran
[2]Assistant Professor, Department of Chemistry, College of Science, Bandar Abbas Branch, Islamic Azad University, Bandar Abbas, Iran

Abstract

Background: Lead (Pb) is a heavy metal that is widely utilized in industries. It contaminates soil and groundwater. Its non-biodegradability, severe toxicity, carcinogenicity, ability to accumulate in nature and contaminate groundwater and surface water make this toxic heavy metal extremely dangerous to living beings and the environment. Therefore, technical and economic methods of removing Pb are of great importance. This study evaluated the efficiency of $Ni_{0.5}Zn_{0.5}Fe_2O_4$ magnetic nanoparticles supported by Aloe vera shell ash in removing Pb from aqueous environments.

Methods: The adsorbent was characterized by several methods, including x-ray diffraction (XRD), scanning electron microscopy (SEM), and Fourier transform infrared spectroscopy (FT-IR). Then, the potential of Aloe vera shell ash-supported $Ni_{0.5}Zn_{0.5}Fe_2O_4$ magnetic nanoparticles to adsorb Pb (II) was investigated. To determine the amount of lead absorbed by this adsorbent, different pHs (2, 4, 5, and 6), adsorbent doses (0.01-0.40 g), Pb concentrations (5, 10, 20, 30, 40, 50, 60, 80, 100, 200, 300, and 600 mg/L), and exposure times (0, 5, 10, 15, 20, 30, 40, 50, and 60 minutes until reaching equilibrium) were tested using an atomic absorption spectrometer (Varian-AA240FS). Residual concentrations of Pb were read.

Results: The results show that a time of 15 minutes, pH value of 9, and adsorbent dose of 0.2 g are the optimum conditions for Pb (II) removal by this adsorption process. Increasing the initial concentration of Pb (II) from 5 to 600 mg/L decreased removal efficiency from 98.8% to 73%. The experimental data fit well into the Freundlich isotherm model (R^2 = 0.989).

Conclusion: $Ni_{0.5}Zn_{0.5}Fe_2O_4$ magnetic nanoparticles supported by Aloe vera shell ash comprise a low-cost, simple, and environmentally benign procedure. The maximum monolayer adsorption capacity based on the Langmuir isotherm (R^2 = 0.884) is 47.2 mg g^{-1}. The prepared magnetic adsorbent can be well dispersed in aqueous solutions and easily separated from them with the aid of an external magnet after adsorption. The process for purifying water presented here is clean and safe. Therefore, this adsorbent is applicable to managing water pollution caused by Pb (II) ions.

Keywords: Adsorption, Pb (II) ions, $Ni_{0.5}Zn_{0.5}Fe_2O_4$, Aloe vera, Aqueous solutions

***Correspondence to:**
Farid Moeinpour
Email: f.moeinpour@gmail.com

Introduction

Adsorption has been established as an important and economically practical treatment technology for removing Pb (II) ions from water and wastewater. Activated carbon is the adsorbent most commonly used for removing Pb (II) ions from aqueous solution. Despite the abundance of applications for activated carbon, its uses are sometimes limited by its high cost and loss during re-formation (1). Therefore, researchers are seeking new low-cost substitute adsorbents for water pollution control, especially where cost plays an important role. Many efforts have been made toward the development of other adsorbents that are effective and inexpensive. They can be produced from a wide variety of raw materials which are abundant and have high carbon and low inorganic content. Owing to the low cost and high accessibility of these materials, it is not essential to have complex regeneration processes. Such inexpensive adsorption methods have attracted the attention of many researchers. Often, the adsorption capabilities of such adsorbents are not great; therefore, studies of more and more new adsorbents are still being developed. Some inexpensive adsorbents for the removal of Pb (II) ions that have been studied are modified agricultural waste (2), hazelnut husks (3), *Moringa oleifera* seed powder(1), olive waste (3), waste coconut buttons (4), magnetically modified tea (5), activated carbon prepared from saffron leaves (6), bael leaves (7), pomegranate peel (1), multi-walled carbon nanotube magnetic composites (8), marine green algae (9), tourmaline (1), amino-functionalized Fe_3O_4 magnetic nanoparticles (10), $Fe_3O_4@SiO_2-NH_2$ (11), waste maize bran (12), coir (13), *Caulerpa lentillifera* (14), Fraxinus tree leaves (15), lignin (16), rice husk ash (17), corn cobs (18), *Moringa oleifera* bark (19), and others. In recent years, economic problems have encouraged the creation of inexpensive, efficient alternative methods of wastewater treatment. One of the most efficient, technical, and economic methods is the use of magnetic adsorbents. These adsorbents have magnetic properties and, by using an external magnetic field, can be easily separated from solutions. In magnetic separation, the usual high costs of separation, e.g., centrifugation and filtration, are not incurred (20). Extensive research has been done in the field of magnetization of materials such as chitosan (21), silica (22), polymer (23), and activated carbon (24) for water contaminants removal. The use of this property in nanoparticles, due to their high specific surface area and adsorption capacities, is very good (25-27). Nickel-zinc ferrites have drawn noticeable consideration from researchers because of their remarkable magnetic properties, large permeability, and very high electrical resistivity (28). They have an extensive list of potential applications in such areas as high-density information storage devices, microwave devices, transformer cores, magnetic fluids, etc. (29). There are many applications for activated carbon in removing chlorine, separating gases, air pollution treatment, and recycling heavy metals from aqueous solutions. However, because of the high cost, alternative options have been suggested. Ash, due to its low cost of production, is a good alternative to activated carbon (30-34). Ash can be produced from a wide range of carbon materials, including wood, coal, shell, walnut shell, fruit stones, agricultural waste, etc (35).

This study investigated the capability of $Ni_{0.5}Zn_{0.5}Fe_2O_4$ magnetic nanoparticles that were surface-modified with ash prepared from Aloe vera shell ($Ni_{0.5}Zn_{0.5}Fe_2O_4$/ASA) as a low-cost adsorbent for the removal of Pb (II) ions from aqueous solutions and studied the adsorption mechanism of Pb (II) ions onto this adsorbent.

Methods

Analytical-grade salt Pb $(NO_3)_2$ was obtained from Merck. A 1000 mg/L stock solution of the salt was prepared in deionized water. All working solutions were made by consecutive diluting with deionized water. Deionized water was prepared using a Millipore Milli-Q (Bedford, MA) water purification system. All reagents ($Fe(NO_3)_3.9H_2O$, $Zn(NO_3)_2.6H_2O$ and $Ni(NO_3)_2.6H_2O$, NaOH and HNO_3) used in the study were of analytical grade and purchased from Aldrich. X-ray diffraction (XRD) analysis was carried out using a PAN analytical X'Pert Pro x-ray diffractometer. Surface morphology and particle size were investigated with a Hitachi S-4800 scanning electron microscopy (SEM) instrument. FT-IR spectra were determined as KBr pellets on a Bruker model 470 spectrophotometer. All the metal ion concentrations were measured with a Varian AA240FS atomic absorption spectrophotometer.

Aloe vera is grown in southern Iran. Aloe vera shell was collected from Qeshm Island, Hormozgan, Iran, and was applied as a raw material for the preparation of surface-modified adsorbent. The collected Aloe vera shells were washed and dried in an air oven at 80°C for 24 hours and then ground and sieved to the desired particle size (2-3 mm). The resultant sieved powder was carbonized in a furnace at 700°C, at a heating rate of 10°C/min for 2 hours. The method used to produce carbon materials was similar to that used in other studies (36).

Synthesis of $Ni_{0.5}Zn_{0.5}Fe_2O_4$/ASA

Ni-Zn ferrite was prepared using stoichiometric ratios of metal nitrates and freshly extracted egg white (37). The metal nitrates ($Fe(NO_3)_3.9H_2O$, $Zn(NO_3)_2.6H_2O$ and $Ni(NO_3)_2.6H_2O$) and 2 g ASA were dissolved together in a minimum amount of double-distilled water to get a clear solution. Sixty milliliters of extracted egg white dissolved in 40 mL of double-distilled water while being vigorous stirred was added to the nitrate mixture at the ambient temperature. After constant stirring for 30 minutes, the resultant sol-gel was evaporated at 80°C until the dry precursor was obtained. The dried precursors were ground and calcined in a muffle furnace at 550°C for 2 hours.

Adsorption experiments

Batch adsorption of lead ions onto the adsorbent ($Ni_{0.5}Zn_{0.5}Fe_2O_4$/ASA) was investigated in aqueous solutions under various operating conditions viz. pH 2–6, at a temperature of 298 K, for an initial Pb (II) ion concentration of 50 mg/L. About 0.20 g adsorbent was added to 50 mL of lead nitrate solution (50 mg/L). Then the mixture was agitated on a shaker at 250 rpm. The initial pH values of the lead solutions were adjusted from 2 to 6 with 0.1 mol/LHNO_3 or 0.1 mol/L NaOH solutions using a pH meter. After equilibrium was reached, the samples were centrifuged and the adsorbent ($Ni_{0.5}Zn_{0.5}Fe_2O_4$/ASA) was magnetically removed from the solution. The concentration of Pb (II) in the supernatant was measured by flame atomic absorption spectrometry. The effects of several parameters, such as contact time, initial concentration, pH, and adsorbent dose, on the extent of Pb (II) adsorption were investigated.

The Pb (II) removal percentage was calculated using Eq. (1):

$$\text{Removal }\% = \frac{C_0 - C_t}{C_0} \times 100 \qquad (1)$$

where C_0 and C_t (mg/L) are the Pb (II) concentrations in the solution at initial and equilibrium time, respectively. The amount of Pb (II) adsorbed (Q_e) was calculated using Eq. (2):

$$Q_e = \frac{(C_0 - C_e)V}{m} \qquad (2)$$

where C_0 and C_e are, respectively, the initial and equilibrium concentrations of Pb (II) (mg/L), m is the mass of adsorbent (g), and V is the volume of solution (L).

Adsorption isotherms

Adsorption isotherms were obtained using 0.20 g of adsorbent and 50 mL of Pb (II) solution with different concentrations (50-600 mg/L) at 298 K. These solutions were buffered at an optimum pH (pH = 5) for adsorption and agitated on a shaker at 250 rpm until they reached adsorption equilibrium (15 minutes). The quantity of Pb (II) adsorbed was derived from the change in concentration.

Results

In this section, the results of this study are presented in the form of diagrams and tables. $Ni_{0.5}Zn_{0.5}Fe_2O_4$/ASA nanocrystallites were characterized by FT-IR (Figure 1), XRD (Figure 2), and SEM (Figure 3).

Figure 4 shows the different sampling times, the concentration of Pb remaining, and the adsorption balance time. As shown in Figure 4, removal efficiency increased in the time range of 0 to 15 minutes and became fixed and partly decreased at 15 to 60 minutes. Thus, the adsorption balance time was determined to be 15 minutes in this test.

The effect of solution pH on Pb (II) adsorption was studied at pH 2–6 at 298 K. As shown in Figure 5, the adsorption rate of Pb (II) was enhanced by increasing pH values from 2 to 6. Figure 5 shows the removal efficiency of lead at the balance time and at 4 different pHs (2, 4, 5, and 6), equaling, respectively, 99.2%, 99.4%, 99.8%, and 99.2%. Maximum adsorption occurred at around pH 5.0, and this value was therefore selected for all adsorption experiments in this study.

The effect of change in the adsorbent amount on the process of Pb (II) adsorption was investigated with different adsorbent doses in the range of 0.01-0.40 g. The results are shown in Figure 6. The best removal of Pb (II) is at about 99.8%, using an adsorbent dosage of 0.20 g in 50 mL of 50 mg/L Pb (II) solution.

Batch adsorption experiments were performed at different initial Pb (II) concentrations (5, 10, 20, 30, 40, 50, 60, 80, 100, 200, 300, and 600 mg/L), while other experimental parameters were kept constant. Figure 7 shows that the adsorption capacity of Pb (II) increased, but the removal percent (%R) of Pb (II) decreased when initial concentra-

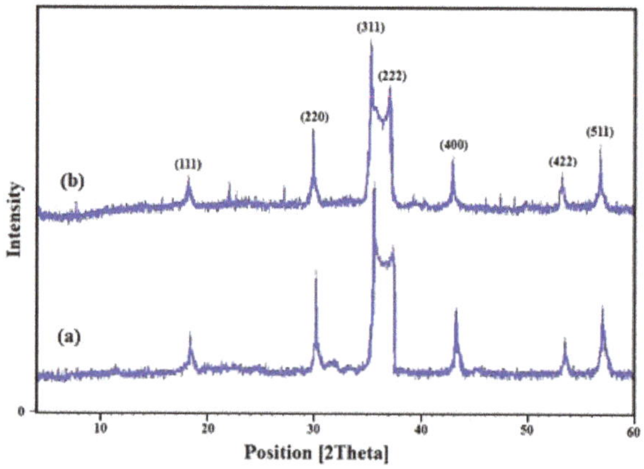

Figure 2. XRD patterns of $Ni_{0.5}Zn_{0.5}Fe_2O_4$. (a) Synthesized $Ni_{0.5}Zn_{0.5}Fe_2O_4$; (b) Standard $Ni_{0.5}Zn_{0.5}Fe_2O_4$ (JCPDS 08-0234).

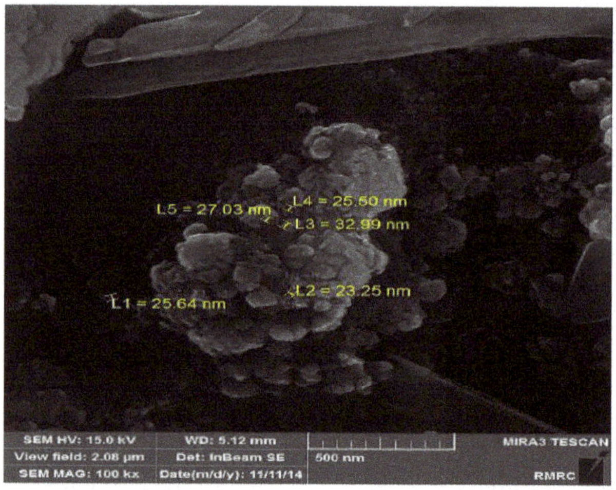

Figure 3. SEM image of $Ni_{0.5}Zn_{0.5}Fe_2O_4$/ASA nanocomposite.

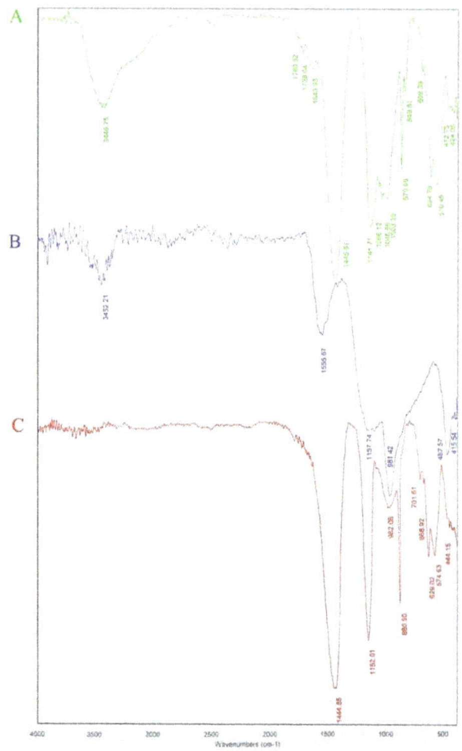

Figure 1. The FTIR spectra of: (A) $Ni_{0.5}Zn_{0.5}Fe_2O_4$/ASA; (B) ASA; (C) $Ni_{0.5}Zn_{0.5}Fe_2O_4$

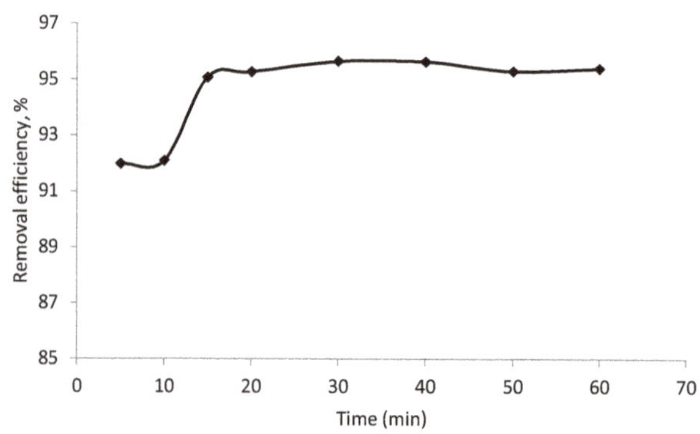

Figure 4. Effect of contact time

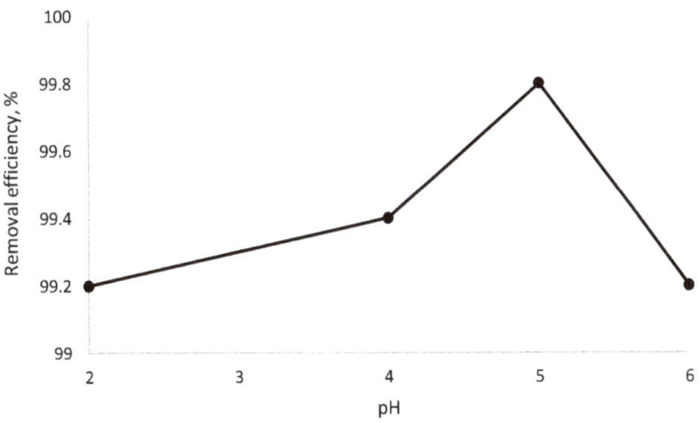

Figure 5. Effect of pH on Pb (II).

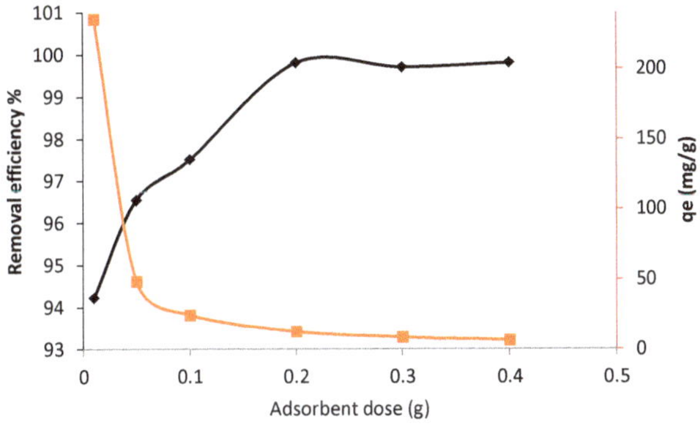

Figure 6. The effect of adsorbent dosage.

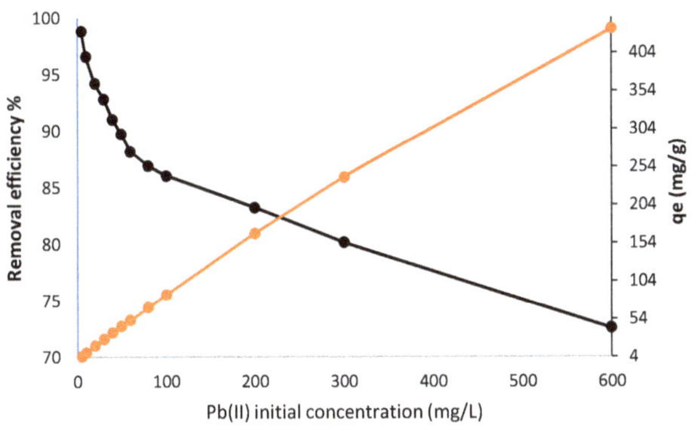

Figure 7. Effect of initial Pb (II) concentration

tion was increased, indicating that the adsorption of Pb (II) on to $Ni_{0.5}Zn_{0.5}Fe_2O_4$/ASA is highly related to initial Pb (II) concentration.
The adsorption isotherms of Pb (II) by the adsorbent are shown in Table 1. In Table 1, $R^2 = 0.884$ in the Langmuir model, $R^2 = 0.989$ in the Freundlich model, and $R^2 = 0.650$ in the Dubinin–Radushkevich model.

Discussion

$Ni_{0.5}Zn_{0.5}Fe_2O_4$ nanocrystallites were prepared according to the procedure reported by Gabal et al (39). $Ni_{0.5}Zn_{0.5}Fe_2O_4$/ASA nano-crystallites were characterized by FT-IR (Figure 1), XRD (Figure 2), and SEM (Figure 3). In the FT-IR spectrum of $Ni_{0.5}Zn_{0.5}Fe_2O_4$/ASA (Figure 1A), many of the bands of $Ni_{0.5}Zn_{0.5}Fe_2O_4$ (Figure 1C) and ASA (Figure 1B), with a slight shift for some of them, are distinct, which shows that ASA has been well supported on the $Ni_{0.5}Zn_{0.5}Fe_2O_4$. The bands in the low-frequency region (1000-500 cm^{-1}) were caused by the iron oxide framework, which is in accordance with the magnetite spectrum. The peak at 1440.85 cm^{-1} is related to the Fe–O bond (33). In Figure 1A, the presence of the – OH stretching mode is evidenced by the peak close to 3446 cm^{-1}. To verify the formation of Ni-Zn ferrite in the prepared magnetic nanoparticles, the XRD pattern of the sample was investigated. The XRD patterns (Figure 2) show that $Ni_{0.5}Zn_{0.5}Fe_2O_4$ nanoparticles have a spinal framework with all the main peaks compatible the $Ni_{0.5}Zn_{0.5}Fe_2O_4$ standard pattern (JCPDS 08-0234). The adsorbent particle size was investigated using SEM. The SEM photograph of the sample (Figure 3) shows that the average size of $Ni_{0.5}Zn_{0.5}Fe_2O_4$/ASA is slightly less than 100 nm.
The effect of contact time on the amount of Pb (II) adsorbed was studied at 50 mg/L initial concentration of lead. It can be observed from Figure 4 that the percentage of adsorptions increased with increases in contact time. Minimum adsorption was 92.0% for the time of 5 minutes, and maximum adsorption was 95.1% for the time of 15 minutes. The adsorption characteristic indicated a rapid uptake of the lead. The adsorption rate, however, diminished to a constant value as contact time was increased, because all available sites were covered and no active site was present for adsorbing.
The acidity of the aqueous solution applies a considerable effect on the adsorption process, because it can alter the solution chemistry of contaminants and the state of functional groups on the surface of adsorbents (39, 40). However, at low pH values, hydrogen ions (H^+) were likely to compete with Pb (II) and thus lower the amount of Pb (II) removed. Therefore, the great Pb (II) adsorption occurring at higher pHs could be ascribed to a decrease in competition between H^+ and Pb (II) at the same adsorption site of the adsorbent beads. At a pH value of 5.9, the

Table 1. Langmuir, Freundlich, D–R isotherm constants for the adsorption of Pb (II) ions onto $Ni_{0.5}Zn_{0.5}Fe_2O_4$/ASA

Langmuir			
q_m (mg g^{-1})	K_L	R_L	R^2
47.2	1.91	8.72×10^{-4}	0.884
Freundlich			
1/n	K_F		R^2
0.50	1.93		0.989
Dubinin–Radushkevich (D–R)			
q_m (mg g^{-1})	β (mol^2 kJ^{-2})	R^2	E (kJ mol^{-1})
6.7	7×10^{-8}	0.650	2.673

Pb (II) ions begin to hydrolyze and then form a small number of $Pb(OH)_2$ or $Pb(OH)_3^-$ varieties (41). Compared with Pb (II) ions, these species are unfavorable for adsorption, which accounts for the little reduction in adsorption capacity. Therefore, maximum adsorption occurred at around pH 5.0, and this value was therefore selected for all adsorption experiments in this study. The results are in line with the findings of Zhang et al (14) on the removal of Pb (II) by Fe_3O_4@SiO_2–NH_2 core-shell nanomaterials. From Figure 6, it can be seen that as the adsorbent dose is increased, the percentage of removal also increased until it approached a saturation point, where increases in adsorbent dose did not alter the percentage of removal. An increase in adsorption rate with adsorbent quantity can be ascribed to increased surface area and the availability of more adsorption sites.

Batch adsorption experiments were performed at different initial Pb (II) concentrations. As shown in Figure 7, these observations can be explained by considering the fact that increasing the initial Pb^{2+} concentration made more Pb^{2+} ions available, while the amount of active sites on the adsorbent remained constant which led to a lower %R.

Isotherm studies can explain how an adsorbate interacts with an adsorbent. The experimental data corresponded with the Langmuir, Freundlich and Dubinin–Radushkevich models as shown in Table 1.

The Langmuir isotherm model, which defines a monolayer adsorption, is given in Eq. (3):

$$\frac{1}{q_e} = \frac{1}{K_L q_m}\frac{1}{C_e} + \frac{1}{q_m} \tag{3}$$

where q_e = the amount of Pb (II) adsorbed per unit mass at equilibrium (mg/g);

q_m = the maximum amount of adsorbent that can be adsorbed per unit mass adsorbent (mg/g);

C_e = concentration of adsorbent (in the solution at equilibrium (mg/L);

K_L = adsorption equilibrium constant.

A plot of $1/q_e$ versus $1/C_e$ gives a straight line with a slope of $1/K_Lq_m$ and intercept $1/q_m$.

The main characteristics of the Langmuir isotherm can be expressed in terms of a dimensionless constant separation factor R_L that is given by Eq. (4) (42):

$$R_L = \frac{1}{1 + \mathrm{K_L} C_0} \tag{4}$$

Where C_0 is the highest initial concentration of adsorbate (mgL^{-1}), and K_L (Lmg^{-1}) is the Langmuir constant. The value of R_L shows the shape of the isotherm to be either unfavorable ($R_L > 1$), linear ($R_L = 1$), favorable ($0 < R_L < 1$), or irreversible ($R_L = 0$). The R_L values between 0 and 1 indicate favorable adsorption. In this study, the value of R_L is 0.116, which shows the favorable adsorption of Pb (II) by $Ni_{0.5}Zn_{0.5}Fe_2O_4$/ASA.

The Freundlich isotherm is expressed by Eq. (5). This isotherm model defines a heterogeneous adsorption with different surface energy sites and supposes the change of uptake with exponential distribution of adsorption sites and energies (43-45).

$$\log q_e = \log K_F + \frac{1}{n}\log C_e \tag{5}$$

where C_e (mg/L) and q_e (mg/g) are the equilibrium concentration of adsorbent in the solution and the amount of adsorbent adsorbed at equilibrium, respectively; K_F ($mg^{1-(1/n)}\ L^{1/n}\ g^{-1}$) and n are the Freundlich constant which indicates the adsorption capacity for the adsorbent and adsorption intensity, respectively.

A graph of $\log q_e$ versus $\log C_e$ provides a straight line with slope $1/n$ and intercept $\log K_F$. The value of 1/n identifies the adsorption intensity and type of isotherm as favorable ($0.1 < 1/n < 0.5$) or unfavorable ($1/n > 2$). The Freundlich parameter 1/n is related to the adsorption intensity of the adsorbent. When $0.1 < 1/n \le 0.5$, the adsorption of the adsorbate is easy; when $0.5 < 1/n \le 1$, the adsorption process is difficult; when $1/n > 1$, adsorption takes place with much difficulty (46, 47). In the current study, the value of 1/n (0.43) shows the favorable adsorption of Pb (II) onto $Ni_{0.5}Zn_{0.5}Fe_2O_4$/ASA.

In order to discern between physical and chemical adsorption, the sorption data was analyzed using the Dubinin–Radushkevich (D-R) equation, which is given in Eq. (6):

$$ln q_e = ln q_m - \beta\varepsilon^2 \tag{6}$$

where β is a constant related to the mean energy of adsorption ($mol^2\ kJ^{-2}$), q_m is the maximum adsorption capacity of metal ions ($mg\ g^{-1}$), and ε is the Polanyi potential given by Eq. (7):

$$\varepsilon = RTln\left(1 + \frac{1}{C_e}\right) \tag{7}$$

where R is the gas constant (8.314 J $mol^{-1}\ K^{-1}$) and T is the temperature (K). By plotting $\ln q_e$ versus ε^2 with experimental data, a straight line is obtained. From the intercept and slope, the values of q_m and β are determined. With the value of β, the mean energy E, which is the free energy transfer of one mole of solute from infinity to the surface of adsorbent, can be obtained using Eq. (8):

$$E = \frac{1}{\sqrt{2\beta}} \tag{8}$$

For E < 8 kJ mol^{-1}, the adsorption process might be performed physically, while chemical adsorption occurs when E > 8 kJ mol^{-1}(45).

All parameters are listed in Table 1. Looking at Table 1, in which the Langmuir, Freundlich, and D–R isotherm constants for the adsorption of Pb (II) are summarized, it can be derived from R^2 that the Freundlich model matched the experimental data better than either the Langmuir or D–R model. The research conducted by Senthil Kumar et al (3) on the removal of Pb (II) from aqueous solutions using chemically modified agricultural waste showed that the Freundlich model is a better fit in comparison with Langmuir. Moreover, it is clear that the adsorption of Pb (II) by $Ni_{0.5}Zn_{0.5}Fe_2O_4$/ASA may be explained as a physical

Table 2. Maximum adsorption capacity of different adsorbents for Pb (II) removal

Adsorbents	q_m (mg/g)	References
Modified agricultural waste	166.67	(46)
Hazelnut husk	109.90	(47)
Olive waste	6.57	(13)
Magnetically modified tea	44.50	(3)
Bael leaves	104.00	(4)
Marine green algae	15.62	(6)
Tourmaline	108	(48)
Amino-functionalized Fe_3O_4 magnetic nano-particles	40.10	(9)
$Fe_3O_4@SiO_2–NH_2$	243.90	(11)
Waste maize bran	142.86	(12)
Coir	48.83	(13)
Rice husk ash	91.74	(14)
Corncobs	43.40	(15)
$Ni_{0.5}Zn_{0.5}Fe_2O_4$/ASA	47.20	This study

adsorption process, for the value of E is 2.673 kJ. Adsorption capacity is a significant parameter which determines the performance of an adsorbent. Table 2 compares the maximum adsorption capacity of $Ni_{0.5}Zn_{0.5}Fe_2O_4$/ASA for Pb (II) adsorption with that of other adsorbents in the literature.

Conclusion

$Ni_{0.5}Zn_{0.5}Fe_2O_4$/ASA magnetic nanoparticles were used in the adsorption of Pb (II) ions from aqueous solutions. Maximum Pb (II) adsorption was achieved at pH 5 with a maximum adsorption capacity of 47.20 mgg^{-1} at 298 K. The adsorption isotherm fit the Freundlich model well. The prepared magnetic adsorbent can be well dispersed in the aqueous solution and easily separated from it with the aid of an external magnet after adsorption. The water treatment described here is clean and safe using magnetic nanoparticles. Thus, this adsorbent was found to be useful and valuable for controlling water pollution due to Pb (II) ions.

Acknowledgments

The authors acknowledge the Islamic Azad University-Bandar Abbas Branch for its financial support of this study.

Competing interests

The authors declare that they have no competing interests.

Authors' contributions

All authors contributed equally and were involved in designing the study, data collection, and article approval.

References

1. El-Ashtoukhy ES, Amin N, Abdelwahab O. Removal of lead (II) and copper (II) from aqueous solution using pomegranate peel as a new adsorbent. Desalination 2008; 223(1-3): 162-73.
2. Baysal Z, Cinar E, Bulut Y, Alkan H, Dogru M. Equilibrium and thermodynamic studies on biosorption of Pb (II) onto Candida albicans biomass. J Hazard Mater 2009; 161(1): 62-7.
3. Senthil Kumar P, Senthamarai C, Sai Deepthi A, Bharani R. Adsorption isotherms, kinetics and mechanism of Pb (II) ions removal from aqueous solution using chemically modified agricultural waste. Can J Chem Eng 2013; 91(12): 1950-6.
4. Imamoglu M, Şahin H, Aydın Ş, Tosunoğlu F, Yılmaz H, Yıldız SZ. Investigation of Pb (II) adsorption on a novel activated carbon prepared from hazelnut husk by K_2CO_3 activation. Desalination and Water Treatment 2016; 57(10): 4587-96.
5. Prasad M, Freitas H. Removal of toxic metals from solution by leaf, stem and root phytomass of Quercus ilex L. (holly oak). Environ Pollut 2000; 110(2): 277-83.
6. Blázquez G, Calero M, Hernáinz F, Tenorio G, Martín-Lara M. Equilibrium biosorption of lead (II) from aqueous solutions by solid waste from olive-oil production. Chem Eng J 2010; 160(2): 615-22.
7. Anirudhan T, Sreekumari S. Adsorptive removal of heavy metal ions from industrial effluents using activated carbon derived from waste coconut buttons. J Environ Sci 2011; 23(12): 1989-98.
8. Akar T, Tunali S, Kiran I. Botrytis cinerea as a new fungal biosorbent for removal of Pb (II) from aqueous solutions. Biochem Eng J 2005; 25(3): 227-35.
9. Chakravarty S, Mohanty A, Sudha TN, Upadhyay A, Konar J, Sircar J, et al. Removal of Pb (II) ions from aqueous solution by adsorption using bael leaves (Aegle marmelos). J Hazard Mater 2010; 173(1-3): 502-9.
10. Shao D, Chen C, Wang X. Application of polyaniline and multiwalled carbon nanotube magnetic composites for removal of Pb (II). Chem Eng J 2012; 185-186: 144-50.
11. Jeyakumar RS, Chandrasekaran V. Adsorption of lead (II) ions by activated carbons prepared from marine green algae: equilibrium and kinetics studies. International Journal of Industrial Chemistry 2014; 5(1): 1-9.
12. Wang C, Wu J, Sun H, Wang T, Liu H, Chang Y. Adsorption of Pb (II) ion from aqueous solutions by tourmaline as a novel adsorbent. Ind Eng Chem Res 2011; 50(14): 8515-23.
13. Tan Y, Chen M, Hao Y. High efficient removal of Pb (II) by amino-functionalized Fe_3O_4 magnetic nano-particles. Chem Eng J 2012; 191: 104-11.
14. Zhang J, Zhai S, Li S, Xiao Z, Song Y, An Q, et al. Pb (II) removal of $Fe_3O_4@SiO_2–NH_2$ core-shell nanomaterials prepared via a controllable sol-gel process. Chem Eng J 2013; 215-216: 461-71.
15. Singh K, Talat M, Hasan S. Removal of lead from aqueous solutions by agricultural waste maize bran. Bioresour Technol 2006; 97(16): 2124-30.
16. Conrad K, Hansen HC. Sorption of zinc and lead on coir. Bioresour Technol 2007; 98(1): 89-97.
17. Apiratikul R, Pavasant P. Batch and column studies of biosorption of heavy metals by Caulerpa lentillifera. Bioresour Technol 2008; 99(8): 2766-77.
18. Sangi MR, Shahmoradi A, Zolgharnein J, Azimi GH,

Ghorbandoost M. Removal and recovery of heavy metals from aqueous solution using Ulmus carpinifolia and Fraxinus excelsior tree leaves. J Hazard Mater 2008; 155(3): 513-22.
19. Guo X, Zhang S, Shan XQ. Adsorption of metal ions on lignin. J Hazard Mater 2008; 151(1): 134-42.
20. Naiya TK, Bhattacharya AK, Mandal S, Das SK. The sorption of lead (II) ions on rice husk ash. J Hazard Mater 2009; 163(2-3): 1254-64.
21. Tan G, Yuan H, Liu Y, Xiao D. Removal of lead from aqueous solution with native and chemically modified corncobs. J Hazard Mater 2010; 174(1-3): 740-5.
22. Reddy DHK, Seshaiah K, Reddy A, Rao MM, Wang M. Biosorption of Pb^{2+} from aqueous solutions by Moringa oleifera bark: equilibrium and kinetic studies. J Hazard Mater 2010; 174(1-3): 831-8.
23. Shen Y, Tang J, Nie Z, Wang Y, Ren Y, Zuo L. Preparation and application of magnetic Fe_3O_4 nanoparticles for wastewater purification. Sep Purif Technol 2009; 68(3): 312-9.
24. Cho DW, Jeon BH, Chon CM, Kim Y, Schwartz FW, Lee ES, et al. A novel chitosan/clay/magnetite composite for adsorption of Cu (II) and As (V). Chem Eng J 2012; 200-202: 654-62.
25. Shi G, Sun B, Jin Z, Liu J, Li M. Synthesis of SiO_2/Fe_3O_4 nanomaterial and its application as cataluminescence gas sensor material for ether. Sensors and Actuators B: Chemical 2012; 171-172: 699-704.
26. Dallas P, Georgakilas V, Niarchos D, Komninou P, Kehagias T, Petridis D. Synthesis, characterization and thermal properties of polymer/magnetite nanocomposites. Nanotechnology 2006; 17(8): 2046-53.
27. Mohan D, Sarswat A, Singh VK, Alexandre-Franco M, Pittman CU Jr. Development of magnetic activated carbon from almond shells for trinitrophenol removal from water. Chem Eng J 2011; 172(2-3): 1111-25.
28. Mak SY, Chen DH. Binding and sulfonation of poly (acrylic acid) on Iron oxide nanoparticles: a novel, magnetic, strong acid cation nano-adsorbent. Macromol Rapid Commun 2005; 26(19): 1567-71.
29. Nitayaphat W, Jintakosol T. Removal of silver (i) from aqueous solutions by chitosan/carbon nanotube nanocomposite beads. Advanced Materials Research 2014; 893: 166-9.
30. Nitayaphat W, Jintakosol T. Removal of silver (I) from aqueous solutions by chitosan/bamboo charcoal composite beads. Journal of Cleaner Production 2015; 87: 850-5.
31. Sharma S, Verma K, Chaubey U, Singh V, Mehta B. Influence of Zn substitution on structural, microstructural and dielectric properties of nanocrystalline nickel ferrites. Materials Science and Engineering: B 2010; 167(3): 187-92.
32. Virden A, O'Grady K. Structure and magnetic properties of Ni-Zn ferrite nanoparticles. J Magn Magn Mater 2005; 290: 868-70.
33. Mane S, Vanjara A, Sawant M. Removal of phenol from wastewater using date seed carbon. Journal of the Chinese Chemical Society 2005; 52(6): 1117-22.
34. Panday K, Prasad G, Singh V. Copper (II) removal from aqueous solutions by fly ash. Water Res 1985; 19(7): 869-73.
35. Vázquez-Rivera NI, Soto-Pérez L, St John JN, Molina-Bas OI, Hwang SS. Optimization of pervious concrete containing fly ash and iron oxide nanoparticles and its application for phosphorus removal. Constr Build Mater 2015; 93: 22-8.
36. Al. Haddabi M, Ahmed M, Al. Jebri Z, Vuthaluru H, Znad H, Al. Kindi M. Boron removal from seawater using date palm (Phoenix dactylifera) seed ash. Desalination and Water Treatment 2016; 57(11): 5130-7.
37. Agrafioti E, Kalderis D, Diamadopoulos E. Arsenic and chromium removal from water using biochars derived from rice husk, organic solid wastes and sewage sludge. J Environ Manage 2014; 133: 309-14.
38. Banat F, Al-Asheh S, Al-Makhadmeh L. Evaluation of the use of raw and activated date pits as potential adsorbents for dye containing waters. Process Biochem 2003; 39(2): 193-202.
39. Gabal M, El-Shishtawy RM, Al Angari Y. Structural and magnetic properties of nano-crystalline Ni–Zn ferrites synthesized using egg-white precursor. J Magn Magn Mater 2012; 324(14): 2258-64.
40. Pol VG, Daemen LL, Vogel S, Chertkov G. Solvent-free fabrication of ferromagnetic Fe_3O_4 octahedra. Ind Eng Chem Res 2009; 49(2): 920-4.
41. Huang M-R, Li S, Li X-G. Longan shell as novel biomacromolecular sorbent for highly selective removal of lead and mercury ions. J Phys Chem B 2010; 114(10) 3534-42.
42. Hall KR, Eagleton LC, Acrivos A, Vermeulen T. Pore-and solid-diffusion kinetics in fixed-bed adsorption under constant-pattern conditions. Industrial & Engineering Chemistry Fundamentals 1966; 5(2): 212-23.
43. Chen H, Zhao J, Wu J, Dai G. Isotherm, thermodynamic, kinetics and adsorption mechanism studies of methyl orange by surfactant modified silkworm exuviae. J Hazard Mater 2011; 192(1): 246-54.
44. Hou MF, Ma CX, Zhang WD, Tang XY, Fan YN, Wan HF. Removal of rhodamine B using iron-pillared bentonite. J Hazard Mater 2011; 186(2-3): 1118-23.
45. Kerkez Ö, Bayazit ŞS. Magnetite decorated multi-walled carbon nanotubes for removal of toxic dyes from aqueous solutions. J Nanopart Res 2014; 16(6): 1-11.
46. Luo X, Zhang L. High effective adsorption of organic dyes on magnetic cellulose beads entrapping activated carbon. J Hazard Mater 2009; 171(1-3): 340-7.
47. Samiee S, Goharshadi EK. Graphene nanosheets as efficient adsorbent for an azo dye removal: kinetic and thermodynamic studies. J Nanopart Res 2014; 16(8): 2542.
48. Lucia R, Jana S, Ivo Š, Mirka Š, Michaela C, Roman G. Magnetically Modified Tea for Lead Sorption. Adv Sci Eng Med 2014;6(4):473-6.

Assessment of chemical quality of drinking water in rural areas of Babol, Northern Iran

Zabihollah Yousefi[1], Hossein Sahebian[2*]

[1]Department of Environmental Health Engineering, School of public Health, Mazandaran University of Medical Sciences, Sari, Iran
[2]School of Health and Health Sciences Research Center, Mazandaran University of Medical Sciences, Sari, Iran

Abstract
Background: This study evaluated the quality of drinking water in rural areas of the city of Babol in terms of chemical parameters (iron, manganese, nitrite, and nitrate) and compared it with the Iranian and the World Health Organization (WHO) standards to determine the trend.
Methods: This cross-sectional study monitored chemical test results for the years 2011-2014 of samples collected from rural water and wastewater from Babol Company. A total of 375 samples from 71 drinking water wells were investigated.
Results: Results for the maximum, mean, and minimum levels of each parameter across rural sectors over 4 years were compared and analyzed against Iranian and the WHO standards. The results showed that during the four years of monitoring, the average iron concentration in Laleh Abad (0.5 ± 0.06 mg/L), Gatab (0.398 ± 0.42 mg/L), and Central (0.307 ± 0.23 mg/L) exceeded the standard concentration. Overall, the average concentrations of iron, manganese, nitrite, and nitrate in all areas during the years 2011-2014 were 0.239 ± 0.15 mg/L, 0.132 ± 0.95 mg/L, 0.008 ± 0.012 mg/L, and 2.201 ± 0.73 mg/L, respectively.
Conclusion: In general, statistical analyses showed that the average concentrations of nitrite and nitrate were below the allowable limit, which is desirable. The mean concentrations of iron and manganese (Mn^{+2}) in the water of some areas of the county during this period were higher than the allowed limit. This is an adverse finding that may be attributable to the improper locations of the water wells.
Keywords: Drinking water, Chemical quality, Babol, Northern Iran

***Correspondence to:**
Hossein Sahebian
Email: sahebianh@yahoo.com

Introduction

The chemical properties of groundwater are important and determinate factors used in domestic, irrigation, and industrial purposes with contexts (1). Water quality and its appropriateness for use can be determined by taste, smell, color, and concentration of organic or mineral substances. Importantly, pollutants in water can affect the quality and, thus, human health (2,3).

The existence of unexpected chemical compounds, even in small amounts, leads to the poisoning of humans or animals if levels are higher than amounts stipulated by international standards (4). The quality of groundwater is typically characterized through various physical and chemical features. These parameters vary widely according to different types of pollution, seasonal changes, groundwater extraction, and other changes. Based on water analyses, most drinking water resources in northern Iran are very hard, indicating poor quality. Therefore, a constant monitoring and control of groundwater pollution is required (5,6).

Drinking water can be quantified using parameters such as pH, electrical conductivity, total dissolved solids, turbidity, free residual chlorine, alkalinity, hardness, chloride, calcium, magnesium, iron nitrite, sulfate, fluoride, and phosphate. Concentrations of some parameters that are too high or too low can have adverse effects on human health (7-11). It is believed that groundwater is cleaner and freer of contamination than surface water. However, the long-term discharge of industrial wastewater, domestic sewage, and wet waste has caused groundwater to become polluted and cause health problems. Hence, the need for groundwater conservation and quality management have always existed (12). If chemical impurities in drinking water exceed recommended standard ranges over the long-term, irreversible damage and human diseases, including digestive disorders or methemoglobinemia, may occur (13).

Distinctive ecological features cause rural environments in Iran to have various spatial structures. Factors such as weather, climate, terrain, and the physical conditions of land, water resources, and other natural factors have been very effective in the formation and establishment of settlements in rural areas.

The non-homogeneous distribution of rural areas in terms of geographical distribution and the distance between residential units has created some difficulty in supply, distribution, and water quality monitoring (14). Babol is located in Mazandaran province and has six rural sectors: Gatab, Eastern Bandpei, Western Bandpie, Central, Laleh Abad, and BabolKenar. Most of the water in these parts is supplied from groundwater.

The current research evaluated the quality of drinking water in rural areas of Babol by measuring chemical parameters and comparing them with the standards of Iran and the World Health Organization (WHO).

Materials and Methods

Study population

Babol is located in the center of Mazandaran province at a latitude of 36°31'44.71"N and a longitude of 52°40'14.57"E. Babol has an area of 1578 km^2 and has more than 510 villages with a total population of over 230973 people (15).

Project method

This research evaluated the chemical quality parameters of drinking water from wells in six rural areas during the period 2011-2014. The chemical parameters iron, manganese, nitrite, and nitrate were investigated and compared with the WHO standard guidelines (the mean and standard deviation of data, etc.) in six rural areas during 2011-2014.

This cross-sectional study collected chemical test results from the years 2011 to 2014 from 71 drinking water wells in six areas (Figure 1).

Sample collection

A total of 375 water samples from groundwater drinking water wells of Babol city in Mazandaran province, northern Iran were randomly collected from six areas (Central, Western Bandpei, Eastern Bandpei, Babolkenar, Gatab, and Laleh Abad) during 2011 to 2014. All samples were collected in 1000 ml polythene bottles and carried to the laboratory. The total number of samples in 2011 2012, 2013, and 2014 were 84, 99, 93, and 99, respectively.

In Central, Western Bandpei, Eastern Bandpei, BabolKenar, Gatab, and Laleh Abad, different numbers of samples were taken, as shown in Figure 2.

Sample analyses

All samples were analyzed according to the Standard Methods for the Examination of Water and Wastewater, 22nd Edition (16). HACH-DR 2000 UV-Vis Spectrophotometer and USA colorimeter (Methods 8039 and 8507) at 500 and 507 nm were applied for the measurement of nitrite and nitrate, respectively. The standard solution method and blank preparation were prepared according to the Standard Methods for the Examination of Water and Wastewater. Measurements of iron and manganese in all samples were taken in 25 mL cells on a Hach, DR 2000 spectrophotometer.

Statistical analysis

Statistical analyses were performed on the collected data using Excel software.

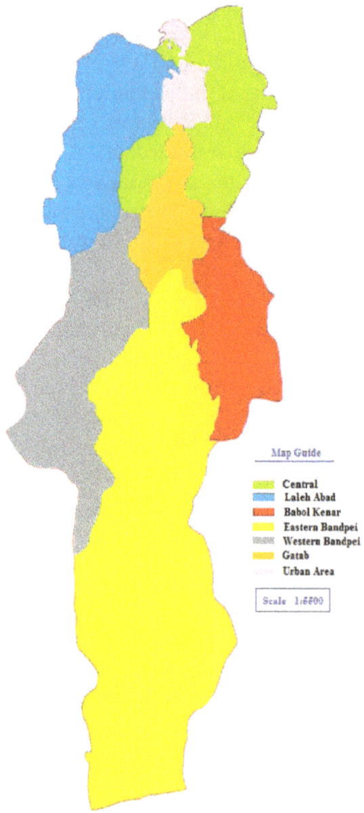

Figure 1. Geographical location of study areas.

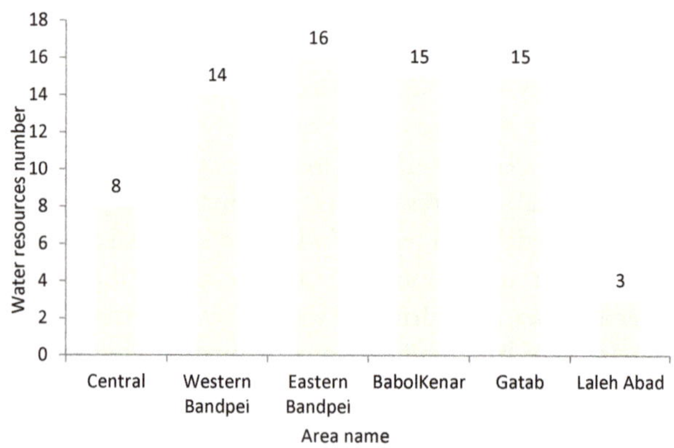

Figure 2. Number of Water Resources Evaluated in Rural Areas.

Results

For each rural area, chemical variables of water quality, i.e. iron, manganese, nitrite, and nitrate, which are all independent and quantitative variables, were studied (Table 1).

As seen in Table 1 and Figure 3, during 2011-2014, Laleh Abad, Gatab, and Central generally had maximum amounts of iron (0.5 ± 0.06 mg/L, 0.398 ± 0.42 mg/L, and 0.307 ± 0.23 mg/L, respectively), which were greater than the Iranian national and WHO standards. This may be due to the inappropriate placement of the wells. Since the standard limitations of these components were not based on consumer health, this cation poses no threat for regional water. However, it may be important to the aesthetics of the samples. In other sections, the amount of iron ions was desirable. Overall, the average concentrations of ions of iron, manganese, nitrite, and nitrate in all studied locations during 2011-2014 were within permissible levels and equal to 0.239 ± 0.15, 0.132 ± 0.95, 0.008 ± 0.012, and 2.201 ± 0.73 mg/L, respectively.

As Table 1 indicates, the results showed that the mean concentrations of water quality parameters in some cases were higher than Iran's national allowable limits. Assessing the water quality parameters in the selected cities revealed that the mean concentrations of nitrite and nitrate in all of the studied areas were within allowed limits, but the mean concentrations of iron and manganese (Mn^{+2}) in the water of Central, Gatab, and Laleh Abad exceeded the national allowable limits. Also, the mean concentration of Manganese (Mn^{+2}) in water (0.132 ± 0.95) in all of the studied areas exceeded Iran's national standard limits and the WHO guidelines.

Discussion

The results of Seth et al on drinking water of the Ashanti area in Ghana in 2014 showed that the level of chemical concentration was generally at the level allowed by the WHO guidelines, but in some aquifers, the results showed Iran high levels above the limit (0.3 mg/L) that the results of these studies were adapted to this research (16).

In a national survey, the US Environmental Protection Agency (EPA) reported about 1.2% and 2.4% of the community and rural domestic drinking water wells (respectively) had nitrate levels higher than the health advisory level. Also, this report indicated that about 1.5 million people are covered by rural domestic drinking water wells and that another 3 million people are covered by community water wells that exceed health advisory levels for nitrate (18,22).

Another research in the United States showed that in the seven states of California, Pennsylvania, New York, Illinois, Wisconsin, Minnesota, and Iowa, more than 100 000 rural residents were exposed to nitrate levels higher than the federal standard (19,22).

About 18% of rural domestic drinking water wells in Iowa had a higher level of nitrate concentration (above 10 ppm) than the allowed limit (20). In another report, pollutant resources of nitrate such as fertilizer, sewage sludge, and animal manure were the main causes of increasing nitrate in groundwater drinking water wells. Also, well waters without the influence of these pollution resources contained nitrate levels of less than 3 ppm (21,22).

Rainfall events can transfer nitrogen from lands utilized for crops into groundwater. Denitrification, sewage treatment plants, private septic systems, animal manure, legume crops, and atmospheric deposition can be important in the polluting of groundwater drinking water wells (22,23).

The findings of Mohammadi et al from their research on the quality of groundwater drinking water resources in rural areas of Babol indicate that some parameters such as nitrite, nitrate, sulfate, chloride, calcium, sodium, conductivity, and pH are at desirable levels. Other measured parameters, such as turbidity, fluoride, iron, and hardness, were not in the desirable range (14).

Generally, the results showed that the average concentrations of iron, manganese, nitrite, and nitrate ions in rural areas of Babol during 2011-2014 were within desirable ranges compared to the Iranian national and WHO standards (Table 1).

The results further showed that the number of iron ions in

Table 1. Results of water chemical analysis in rural areas of Babol, 2011-2014

Area	Fe^2 (mg/L)	Mn^2 (mg/L)	NO_2^- (mg/L)	NO_3^- (mg/L)
Central	0.317±0.307	0.220±0.180	0.015±0.009	1.807±0.647
Western Bandpei	0.112±0.105	0.122±0.090	0.014±0.008	2.475±0.950
Eastern Bandpei	0.060±0.056	0.046±0.040	0.008±0.004	2.262±0.767
Babolkenar	0.065 ± 0.043	0.055±0.041	0.003 ±0.004	2.937±1.216
Gatab	0.420±0.398	0.150±0.142	0.030±0.015	2.247±1.185
Laleh Abad	0.063±0.502	0.237±0.037	0.001 ±0.005	1.475±0.488
Mean of whole parts	0.239±0.168	0.946±0.132	0.012±0.007	2.200±0.778
Iran national standards 1053 (desirable, allowed)	0.3	0.4	3	50
WHO Guidelines	0.3	0.5	3	50

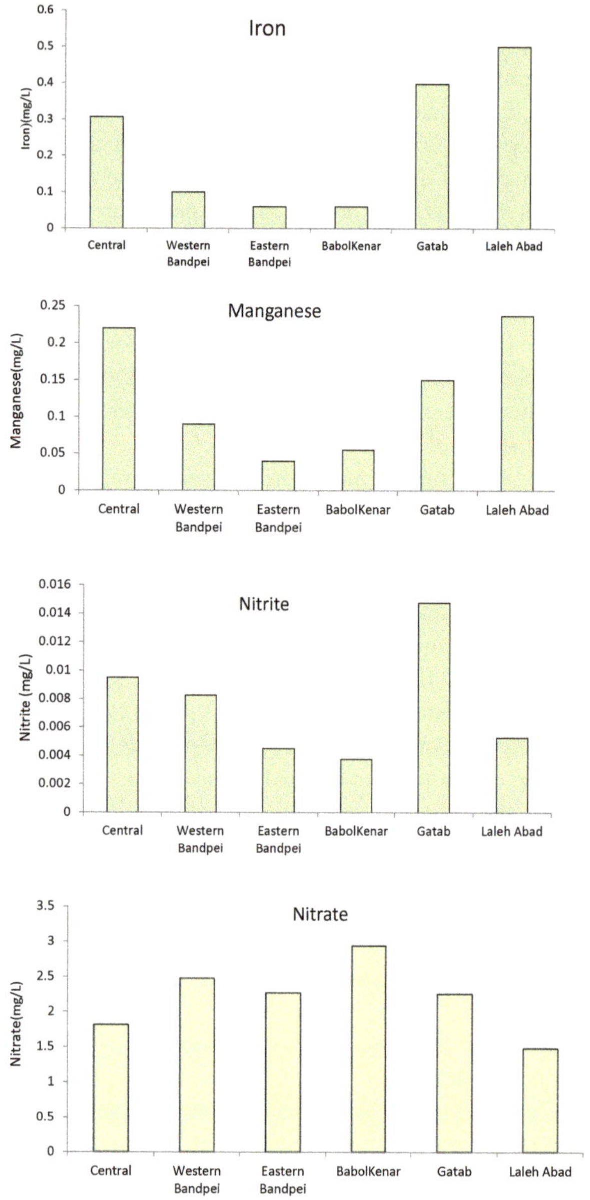

Figure 3. Average concentrations of iron, manganese, nitrite, and nitrate ions in rural parts of Babol during 2011-2014.

some parts during the 4 years were higher, possibly due to the inappropriate locations of water wells. This element is not very harmful, but in terms of the physical properties of water (color, taste, and smell), it is very good (Figure 3). In large amounts, however, it makes the water taste bitter and unpleasant. Even a small amount of iron can cause the accumulation of sediment in the water supply networks. Such a sediment is not suitable for the consumer and could cause the growth of iron bacteria, and that could explain why water quality is reduced through the production of sludge or noxious smells.

Conclusion

Statistical analysis indicated that the mean concentrations of some parameters (nitrite and nitrate) were below the Iranian national standard limits and the WHO guidelines, which is desirable. The mean concentrations of some parameters (iron and manganese) of water in some areas of the county, however, were higher than the limit, which may be due to the improper location of water wells and the improper control of discharges into the groundwater. The authors of this study recommended to the authorities of the water and wastewater corporation that a master plan for monitoring of water quality parameters and the discharges to the water resources be devised.

Acknowledgments

The authors hereby express their appreciation and gratitude to the Water and Sewage Company of Babol for the data on water quality they provided for this study.

Competing interests

The authors affirm that this article is the original work of the authors and that the authors have no conflicts of interest to declare.

Authors' contributions

All authors were involved in all stages of the study. On behalf of the co-authors, the corresponding author bears full responsibility for this submission.

References

1. Singh G, Kant Kamal R. Assessment of groundwater quality in the mining areas of Goa, India. Indian Journal of Science and Technology 2015; 8(6): 588-95. doi: 10.17485/ijst/2015/v8i6/62314.
2. Rahmanian N, Siti Hajar BA, Homayoonfard M, Ali NJ, Rehan M, Sadef Y, et al. Analysis of physiochemical parameters to evaluate the drinking water quality in the State of Perak, Malaysia. J Chem 2015; 2015: 716125. doi: 10.1155/2015/716125.
3. Mohseni Bandpei A, Yousefi Z. Status of Water Quality Parameters along Haraz River. Int J Environ Res 2013; 7(4): 1029-38. doi: 10.22059/ijer.2013.687.
4. Samaei MR, Ebrahimy A, Ehrampoosh MH, Talebi P, Khalili MH, Morovati R. A Study of the physical and chemical quality of potable water in Yazd. Tolooe Behdasht 2007; 2(20): 50-7. [In Persian].
5. Tahernezhad Z, Yousefi Z, Mousavinasab N. Asurvey on fluoride, nitrate, iron, manganese and total hardness in drinking water of Fereydoonkenar city during 2008-2013. Journal of Advances in Environmental Health Research 2016; 4(2): 102-12.
6. Srinivas Rao G, Nageswararao G. Assessment of Groundwater quality using water quality index. Arch Environ Sci 2013; 7: 1-5.
7. Shabankareh Fard E, Hayati R, Dobaradaran S. Evaluation of physical, chemical and microbial quality of distribution network drinkingwater in Bushehr, Iran. Iranian South

Medical Journal 2015; 17(6): 1223-35. [In Persian].

8. Yousefi Z, BarafrashtehPour M, Taghavi M, Mashayekh Salehi A, Sedaghat F. Survey on temporal and spatial variation of nitrate and nitrite in drinking water of Gachsaran by using geographic information system (GIS). Journal of Mazandaran University of Medical Sciences 2013; 23(Suppl-2): 158-62. [In Persian].
9. Yousefi Z, Naeej O. Study on nitrate value in rural area in Amol city. Journal of Mazandaran University of Medical Sciences 2007; 17(61): 161-5. [In Persian].
10. Yousefi Z, Taghizadeh A. Fluorine concentration in drinking-water resources in rural areas. Journal of Mazandaran University of Medical Sciences 2014; 23(109): 265-9. [In Persian].
11. Yousefii Z, Hanafi B. Fluoride level in drinking water supplies of Gonbad-e Qabus, 2008-2012. Journal of Mazandaran University of Medical Sciences 2013; 23(101): 112-6. [In Persian].
12. Vijila j, Venkatesh B, Kumaran T. Assessment of physico-chemical status of ground water samples in Tirunelveli District, Tamil Nadu. Int J Appl Res 2015; 1(8): 136-9.
13. Rajaei Q, Mehdinejad MH, Hesari Motlagh S. A survey of chemical quality of rural drinking water of Birjand and Qaen plains, Iran. Journal of Health System Research 2012; 7(6):737-45. [In Persian].
14. Raygan Shirazi A, Rezaei S, Jamshidi A, Fararoei M, Sadat A, Hashemi H. Investigating the microbial and chemical quality of drinking water. Journal of Health System Research 2012; 8: 431-7.
15. Mohammadi A, Amouei A, Tabarinia H, Faraji H. Investigating the physicochemical analysis of potable ground water resources in rural area of Babol city. Journal of Neyshabur University of Medical Sciences 2015; 3(2): 61-9. [In Persian].
16. Seth ON, Tagbor TA, Bernard O. Assessment of chemical quality of groundwater over some rock types in Ashanti Region, Ghana. American Journal of Scientific and Industrial Research 2014; 5(1): 1-6. doi: 10.5251/ajsir.2014.5.1.1.6.
17. Rice EW, Baird RB, Eaton AD, Clesceri LS. Standard methods for the examination of water and wastewater. 22nd ed. American Public Health Association, American Water Works Association, Water Environment Federation; 2012.
18. US Environmental Protection Agency. United States Environmental Protection Agency (EPA); 1992. Report No: EPA 570/9-91-020.
19. Kross BC. Nitrate toxicity and drinking water standards – A review. J Prev Med 2002; 10(1): 3-10.
20. Cohen BA, Wiles R, Campbell C. Pouring it on - Nitrate Contamination of Drinking Water. Washington DC: Environmental Working Group; 1996.
21. Kross BC, Hallberg GR, Bruner DR, Cherryholmes K, Johnson JK. The nitrate contamination of private well water in Iowa. Am J Public Health 1993; 83(2): 270-2.
22. Patrick R, Ford E, Quarles J. Groundwater Contamination in the United States. 2nd ed. Philadelphia: University of Pennsylvania Press; 1983.
23. Gholami V, Aghagoli H, Kalteh AM. Modeling sanitary boundaries of drinking water wells on the Caspian Sea southern coasts, Iran. Environ Earth Sci 2015; 74(4): 2981-90. doi: 10.1007/s12665-015-4329-3.

Evaluation of removal efficiency of phenol from synthetic aqueous solutions by *Citrullus colocynthis* seed ash

Samira Salari[1], Mojtaba Afsharnia[1], Asiyeh Moteallemi[2*], Mehdi Ghasemi[1]

[1]Department of Environmental Health Engineering, School of Health, Gonabad University of Medical Sciences, Gonabad, Iran
[2]Department of Environmental Health Engineering, School of Nursing, Torbat Jam Faculty of Medical Sciences, Torbat Jam, Iran

Abstract
Background: Phenol is a prevalent pollutant found in many industrial wastewaters, and it is paid singular attention because of its special features like high toxicity, carcinogenic properties, and vital gathering ability that affects the health of humans and the environment. One of the most important technologies for the removal of phenol is the use of adsorbents. The current study investigated the removal of phenol from synthetic aqueous solutions using *Citrullus colocynthis* seed ash.
Methods: This study is experimental and was conducted on a pilot scale. The efficiency of phenol removal by *C. colocynthis* seed ash was evaluated in a batch system, and different parameters such as initial concentration of phenol (10, 20, 50, and 80 mg/L), contact time (2, 5, 10, and 30 minutes), pH (2-12), adsorbent dose (0.5, 1, 3, 5, and 10 g/L), and temperature were studied. Excel software was used for data analysis. The adsorption process was modeled with Freundlich and Langmuir isotherms at controlled temperatures.
Results: The results showed that the highest removal rate of phenol was obtained at a pH of 2 (83.4%), initial phenol concentration of 20 ppm (66.4%), adsorbent dose of 5 g/L (86.8%), and contact time of 10 minutes. The evaluation of correlation coefficients showed that the phenol adsorbed by *C. colocynthis* seed ash was in greater accordance with the Freundlich model than the Langmuir model.
Conclusion: In general, the results of this study revealed that *C. colocynthis* seed ash has suitable potential for use in removing phenol from aqueous solutions on operation and practical scales due to its low cost and easy access.
Keywords: Phenol, Adsorption, *Citrullus colocynthis*

***Correspondence to:**
Asiyeh Moteallemi
Email: rahil_0m0@yahoo.com

Introduction

Phenol is one of the organic water pollutants with the chemical structure C_6H_5OH that occur because of industrialization (Figure 1) (1). This compound has very high solubility in water and many other organic solvents, and after dissolving in water it creates very little acidity (2). Naturally, phenol is produced from coal tar or gasoline distillation; artificially, it is created through heating sodium benzene sulfate with hydrous soda under high pressure (3). This matter and its derivatives are used in various industries such as resin and plastics, colors, pesticides, pharmaceuticals, oil refineries, and the steel, aluminum, lead, washes, synthetic textiles, and leather industries (4). Phenol is a major pollutant of industrial wastewater. Its compounds produce complexes with tangible tastes and odors in the presence of chlorine (5). The symptoms of acute phenol poisoning include xerostomia, darkening of the color of urine, and the destruction of organs and muscles. Long-term exposure to phenol vapors can cause anorexia, weakness, headache, muscle pain, and jaundice (6,7). Due to the high absorption of phenol through the skin of the hand or forehead (60%-90%), in high concentrations, it can lead to death (8,9). Diagnosis, determination, and measurement of phenol compounds in water resources and monitoring their environmental effects play key roles in the control and impact reduction of these combinations (10). According to the standard guidelines of Iran, the maximum allowed concentration of phenol compounds in superficial water, agriculture, and irrigation is 1 mg/L (11). The World Health Organization (WHO) considers the maximum allowable concentration of phenol in drinking water to be 0.002 mg/L (12). Also, according to the United States Environmental Protection Agency (USEPA), the permissible concentrations of

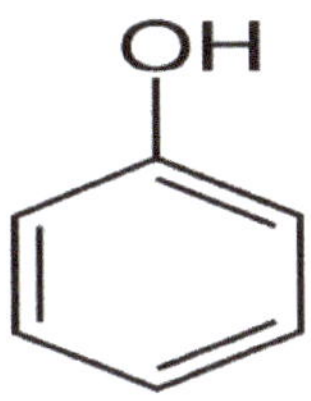

Figure 1. Chemical structure of phenol.

phenol in drinking water and industry effluent are less than 1 mg/L and 500 µg/L, respectively (13). Proper treatment based on existing standards must be carried out so as to reduce the harmful environmental effects of phenol combinations. To date, different methods have been used to treat phenol compounds, such as chemical oxidation, photochemical destruction, precipitation, ion exchange, wet oxidation, electrochemical methods, radiation, membranous filtration, enzymatic filtration, adsorption, and photocatalistic destruction (3,14). Most of these methods have disadvantages like high treatment costs, needs for additive treatment, production of hazardous by-products, high energy consumption, and low efficiency. Among different treatment methods, the adsorption process is one of the simplest and most effective methods for removing poisonous pollutants from aqueous environments (15). In the last two decades, many studies have been carried out on the adsorption of organic compounds with low degradability. In these studies, researchers used produced wastes such as activated carbon (15), coal (16), sugar cane ash (17), agricultural waste (18), and rice bran ash (19) as potential absorbents. In one study, Bazrafshan et al investigated the efficiency of *Moringa peregrina* tree shell ash as a natural adsorbent for the removal of phenol from aqueous solutions. The effect of important variables such as adsorbent dose, primary phenol concentration, and solution pH were evaluated. The highest values of removal were obtained at a pH of 6, initial concentration of 100 mg/L, and adsorbent dose of 0.4 g/L. Data experiments were better represented by the Langmuir isotherm than the Freundlich model (20). Citrullus colocynthis is a plant native to southern Iran and Gonabad city in Razavi Khorasan province. This plant has a bitter taste and is most commonly used in Iranian traditional medicine for the treatment of diabetes and the reduction of blood glucose levels. Activating the plant is very easy, and, until now, no comprehensive study has been conducted using *Citrullus colocynthis* as adsorbent. Thus, the present study applied the seed ash of *C. colocynthis* as an affordable and available adsorbent for the removal of phenol from synthetic aqueous solutions.

Methods

Preparation of adsorbent

Citrullus colocynthis seeds were collected from the vicinity of Gonabad city, Razavi Khorasan province, Iran. The seeds were separated and washed with distilled water to remove impurities like soluble, colored compounds and sand. Then they were dried in a drying oven (80°C) and electric furnaces (550°C) for 4 hours. The ash of the seeds was extracted and sieved through standard sieves of 20, 30, 40, 50, and 100 mesh to obtain particle size (21). The best results were obtained using 20 mesh.

XRD (X-ray Diffraction, D6792-PHILIPS) was used to determine the adsorbent crystalline phases (Figure 2).

Batch adsorption experiments

Synthetic solution was made by a stock solution containing phenol. Experimental solutions containing the desired concentrations of phenol were prepared by diluting the stock solution (50 mL). Then, a known mass of *C. colocynthis* seed ash (adsorbent dose) was added to the solution. Batch adsorption experiments were performed in a mechanical shaker to agitate the solutions. Finally, using Whatman grade 42 filter paper, the solutions were filtered and the phenol concentrations were analyzed (22). The amount of phenol adsorbed was calculated using the following formula:

$$q_e = \frac{(C_0 - C_e)V}{M}$$

where C_0 is the initial phenol concentration and C_e is the secondary concentration after reaching the balance, V is the volume of the solution (L), and M is the used amount of absorbent (g).

Method of measuring phenol concentration

The residual phenol concentration was measured using spectrophotometry at a wavelength of 500 nm in accordance with the standard guidelines for water and wastewater (23).

Effect of initial phenol concentration, pH, adsorbent dose, and contact time

In this study, the effects of different variables such as pH, adsorbent dosage, and initial phenol concentration were investigated (10,16,19,22). In the first stage, the initial pH of samples in the range of 2-12 was examined. In this stage, the adsorbent dose, initial phenol concentration, and contact time were 3 g/L, 50 mg/L, and 60 minutes,

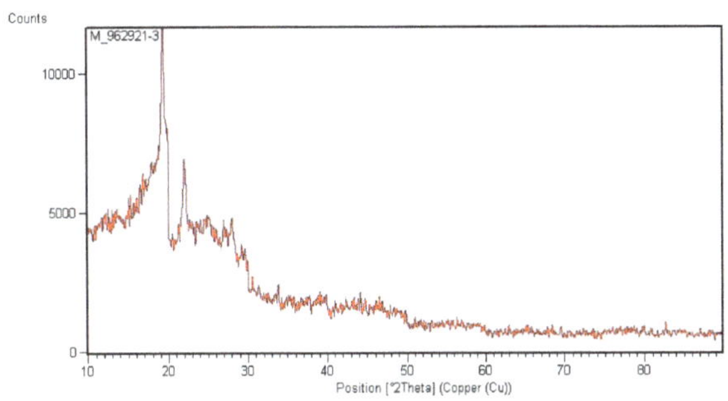

Figure 2. Characteristics of *Citrullus colocynthis* seed: XRD pattern.

respectively. After determining the optimal pH, the effects of adsorbent dose at concentrations of 0.5, 1, 3, 5, and 10 mg/L, phenol concentration of 50 mg/L, and contact time of 60 minutes were tested. Then, the effects of initial phenol concentrations (10, 20, 50, 80, and 100 mg/L) at different times (2, 5, 10, 30, and 60 minutes) were examined. The rate of phenol elimination was obtained as a percentage:

$$\% = \frac{C_0 - C_f}{C_0}.100$$

where C_0 (mg/L) is the initial concentration of phenol and C_f (mg/L) is the final concentration of phenol (after adsorption). All experimental tests were done based on the primary pilot, and all tests were repeated 2 times; the number of analyzed samples was 200.

Adsorption isotherms

Adsorption isotherms describe the interactive behavior between an adsorbent and the absorbent. Isotherms are important to optimizing the use of adsorbents and showing the relationship between the phenol concentration of the soluble and the phenol absorption rate by the solid phase, when the two phases are in balance. For this, the Langmuir and Freundlich adsorption models were used to analyze the experimental equilibrium data for the adsorption of phenol by *C. colocynthis* seed ash. To evaluate data significant to the Langmuir and Freundlich isotherms, the regression coefficient (R2) was used.

The Langmuir isotherm

The Langmuir isotherm is valid for a homogeneous system in which all sorption sites are identical. It is determined using the following formula (24):

$$\frac{C_e}{q_e} = \frac{1}{Q_m.b} + \frac{1}{Q_m} C_e$$

where C_e is the equilibrium concentration of the solution (mg/L), q_e is the value of adsorbate adsorbed per unit mass of adsorbent (mg/g), b (L/mg) is the Langmuir adsorption constant, and Q_m (mg/g) is the maximum amount adsorbed.

The Freundlich isotherm

This isotherm describes reversible adsorption in a heterogeneous system (24) and is calculated by the following formula:

$$\log q_e = \log K_f + \frac{1}{n} \log C_e$$

where K_f and n are Freundlich constants, K_f((mg/g)(L/mg)$^{1/n}$) is the adsorption capacity of the adsorbent, and n shows how favorable the adsorption process is. The values of 1/n < 1 indicate a normal Friedrich isotherm and cooperative adsorption.

Results

Chemical constituents of *Citrullus colocynthis*

The presence of the following compounds has been proven in *C. colocynthis* seeds by phytochemical analysis: proteins = 13.99 ± 0.06%, crude fibers = 46.73 ± 0.15%, moisture = 6.43 ± 0.15%, α-tocopherol = 1.90 ± 0.020 g/100 g, and fixed oil = 17%-28.5% with a high proportion of unsaturated fatty acids (79.80%), mainly linoleic acid, oleic acid, a low percentage of saturated, total saturated 20.20%, and a very low level of n-3 poly-unsaturated FA (0.5%). The seed oil of Citrullus colocynthis has the following properties: iodine value = 114.46 g I2/100 g, density at 15°C = 905.3:kg/m^3, kinematic viscosity at 40°C = 31.52 mm^2/s, saponification value = 204.44 mg KOH/g, acid value = 0.98 mg KOH/g, free fatty acid = 0.49%, caloric value = 39.37 MJ/kg, color = 5Y + 0.4R, and average molecular weight = 874 g (25).

The mineral contents of fermented and unfermented Citrullus colocynthis include: Ca 0.250 ± 0.04, Mg 0.139 ± 0.041, K 0.244 ± 0.04, Na 0.36 ± 0.02, P 0.176 ± 0.022 mg/kg; and Ca 0.341 ± 0.18, Mg 0.167 ± 0.12, K 0.327 ± 0.10, Na 0.034 ± 0.16 and P 0.097 ± 0.14 mg/kg, respectively (26).

Effect of pH

Solution pH is one of the most important parameters affecting the chemical properties of both absorbing and adsorbing. In this study, pH in the range of 2-12 was used. As presented in Figure 3, the removal efficiency was decreased when the pH value was increased. This result can be explained by the increase in OH ions and their competitive effect with the phenol anion. As seen in Figure 3, the highest and lowest removal rates were observed in the pH values of 2 (83.4%) and 12 (66.4%), respectively (Figure 3).

Effect of adsorbent dose on removal efficiency

To determine the effect of adsorbent dosage on the removal efficiency of phenol, adsorbent doses of 0.5, 1, 3, 5, and 10 g/L, initial phenol concentration of 50 mg/L, pH: 2, and 60 minutes contact time were considered. As shown in Figure 4, the phenol adsorption rate increased with

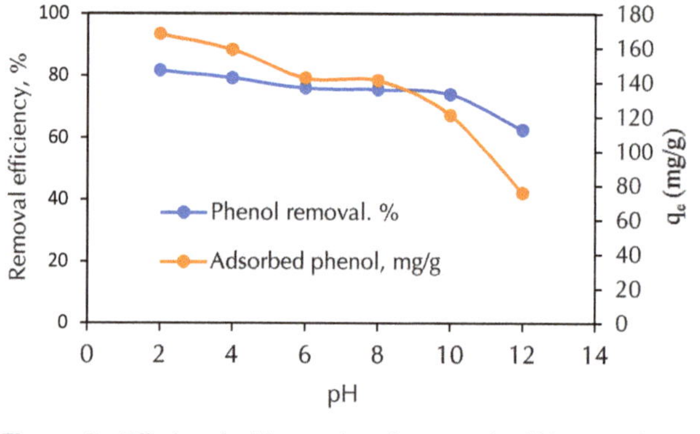

Figure 3. Effects of pH on phenol removal efficiency (phenol concentration 50 mg/L, adsorbent dose 3 g/L, contact time 60 min).

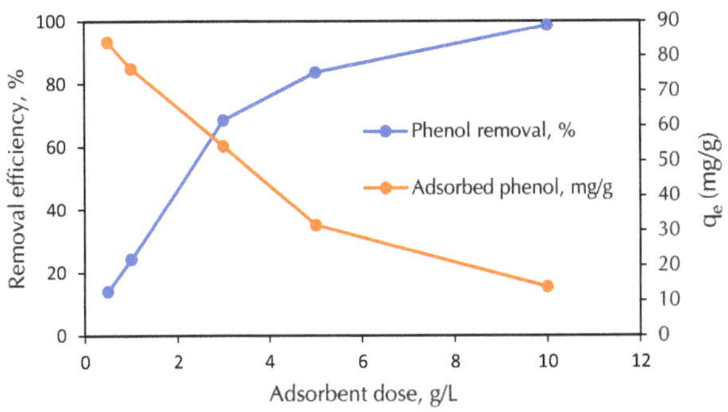

Figure 4. Effects of adsorbent dose on adsorption of phenol by *Citrullus colocynthis* seed ash (phenol concentration 50 mg/L, pH 2, contact time 60 min).

phenol concentration because of the increased number of sorption sites. Thus, the removal of phenol depends on its concentration.

Effects of initial phenol concentration on efficiency of adsorption process

The effects of primary phenol concentration (10, 20, 50, 80, and 100 mg/L) at contact times of 2, 5, 10, 30, and 60 minutes and an adsorbent dose of 3 g/L were evaluated. Figure 5 shows that by increasing the initial phenol concentration, the phenol removal efficiency was decreased. According to Figure 6, the highest removal efficiency of phenol was obtained in the first 10 minutes of the test (Figure 6), and the removal efficiency remained almost constant as it reached equilibrium values.

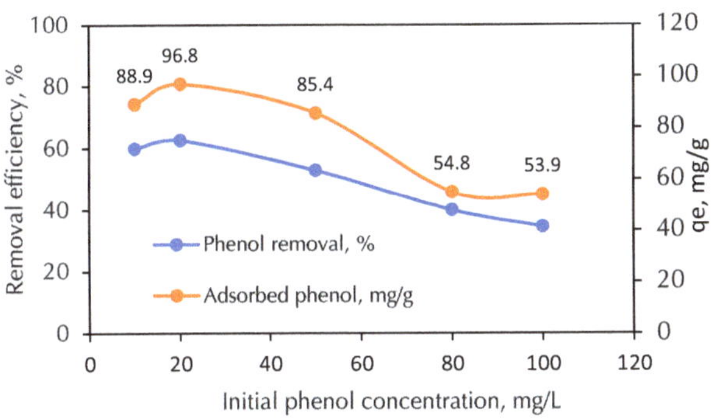

Figure 5. Effects of initial phenol concentration on adsorption process by *Citrullus colocynthis* seed ash (adsorbent dose of 3 g/L, pH 2).

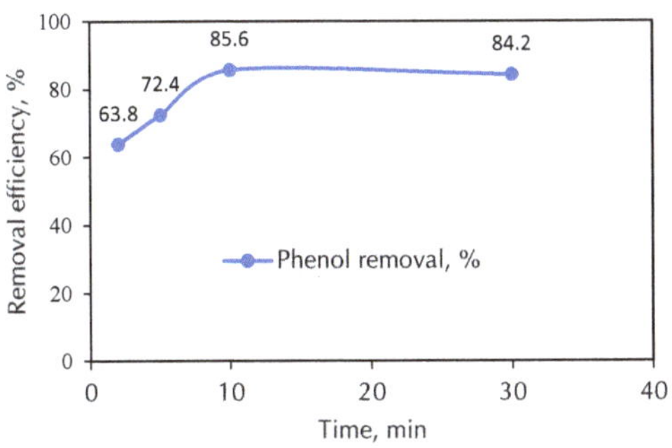

Figure 6. Effect of contact time on adsorption process by *Citrullus colocynthis* seed ash (Phenol concentration of 20 ppm, adsorbent dose of 5 g/L, pH 2).

Langmuir and Freundlich isotherms

A regression coefficient (R2) higher than 0.9 was obtained for both models. The results showed that the Freundlich model is more appropriate for concluding the balanced adsorption of phenol on the adsorbent than the Langmuir model (Figures 7 and 8). Obtained isotherm parameters for phenol adsorption are presented in Table 1.

X-ray diffraction results

One of the most powerful techniques for characterizing crystalline materials is X-ray diffraction (XRD). XRD provides information on crystal structure and texture, phase, mean grain size, crystallinity, strain, and crystal defects (27). The constructive interference of a monochromatic beam of X-rays that diffract at specific angles from each set of lattice planes produce XRD peaks. The distribution of atoms within the lattice determines the peaks' intensities. As seen in Figure 2, the two broad diffraction peaks observed at around 18° and 22° vividly show that the adsorption process made little change in the surface structure of the adsorbent (27) and included compounds such as $C_{18}H_{42}CL_3N_6O_2$ and $C_{18}H_{42}CL_3LAN_6O_{21}$.

Discussion

Effects of pH

The pH of a solution is one of the most important and

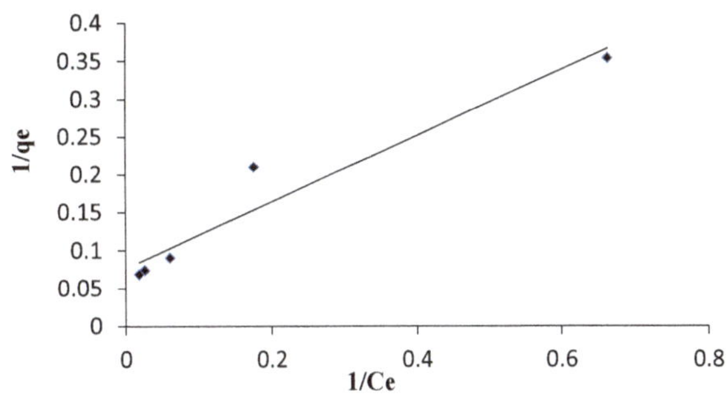

Figure 7. Appropriateness of the Langmuir model to phenol adsorption (R^2 = 0.9348).

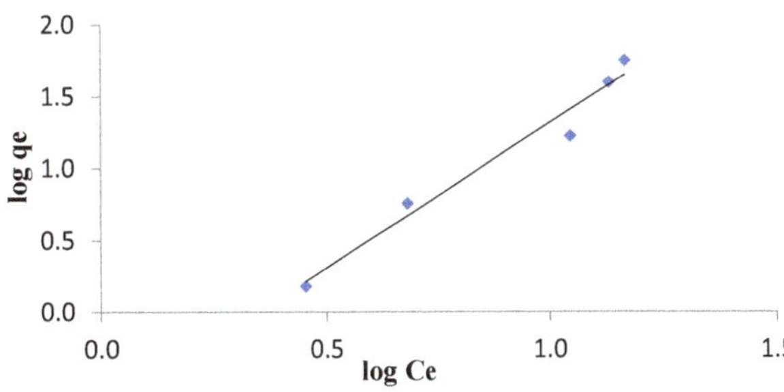

Figure 8. Appropriateness of the Freundlich model to phenol adsorption (R^2 = 0.9674).

Table 1. Obtained isotherm parameters for phenol adsorption

Langmuir isotherm			Freundlich isotherm		
q_m (mg g^{-1})	K_L (L mg^{-1})	R^2	k_f	n	R^2
13.33	0.092	0.9348	1.75	1.51	0.9674

effective parameters of the adsorption process, because it affects the electrical charge on the adsorbent surface and the ionization degree of the adsorbate (28). According to Figure 3, the removal efficiency of phenol decreased when pH was increased. The highest and lowest removal efficiency rates were 83.4% in pH=2 and 64.1% in pH = 12, respectively. These results can be due to the increase in OH ions and their competitive effects with phenol anion. The ion section of phenol is calculated by the following formula:

$$\varphi ions = \frac{1}{[1+10^{(PKa-PH)}]}$$

According to the formula above, φ_{ions} is increased by increasing pH. Phenol is a weak acid with PKa=10 that is greatly ionized in solutions. These ions are negatively charged, and because the adsorbent surface is positively charged in low values of pH, a strong attraction is created between the positive and negative ions (29). Senturk et al studied the removal rate of phenol from aqueous solutions by adsorption onto organomodified Tirebolu bentonite (30). Their results revealed that phenol adsorption was constant in pH values of 3-9, but decreased suddenly in a pH higher than 9. Similar results were presented for the adsorption of phenol by means of activated carbon, bentonite, and bagasse fly ash, by Halhouli et al (31), Banat et al. (32), and Srivastava et al, respectively (33).

Effects of adsorbent dose

According to Figure 4, the removal efficiency was increased when adsorbent dose was increased because of the increase in adsorbent surface area and the availability of more adsorption sites (31). The optimal elimination rate of phenol was 86.8 % (5 g/L). By increasing the amount of adsorbent to more than 5 g/L, phenol desorption occurred in the solution. The low adsorption rate can be explained by the lack of phenol concentration in high values of adsorbent (from 5-10 g/L). Similar results were achieved by Cengiz and Cavas (34) and SenthilKumar et al (35). However, these findings differed from the results of a study by Rahmani et al., which showed that by increasing the adsorbent dose from 0.25 g to 1 g, the residual phenol values were decreased from 40 mg/L to 1 mg/L. Moreover, the results showed that the amount of adsorbed contaminant per mass unit was decreased by increasing adsorbent values. In accordance with these results, the suitable dose of adsorbent was determined to be 0.5 g (36).

Effects of contact time and primary phenol concentration

The results of this study showed that the efficiency of phenol removal was increased by increasing contact time in the first 10 minutes of the test (Figure 6). The increase in the rate of phenol removal was probably due to the limited empty spaces of adsorbent in high phenol concentrations. Also, these results showed that phenol adsorption is a fast process, because the highest amount of the considered concentrations of phenol was absorbed in the first 10 minutes of the test. The results of a study conducted by Dehghani et al to evaluate the efficiency of multiwalled carbon nanotubes in the removal of phenol from aqueous solutions revealed that phenol removal was increased when contact time was increased. The current study results differed with these in that Dehghani obtained maximum adsorption in the first 30 minutes of the test (37). The cause of this event is the existence of many available empty sites for adsorption at first; with the passage of time, the occupation of the remained sites becomes difficult because of the repulsive force between the solute molecules between the solid phase and the solution (38). In another study conducted by Liao et al on the adsorption of chlorothenols by multiwalled carbon, an equilibrium time of 30 minutes was obtained (39).

As seen in Figure 5, the effect of initial phenol concentration on the adsorption process showed that by increasing phenol concentration, removal efficiency was decreased. The highest and lowest phenol removal rates were obtained in concentrations of 20 ppm (66.4%) and 100 ppm (33.2%), respectively, because the initial phenol concentration acts as a driving force, and the resistance of solution mass transfer increases in the adsorbent surrounding the liquid layer and adsorbent surface. Also, increasing phenol concentration decreased the ratio of available surface to the number of adsorbed moles, and that led to a reduction in removal efficiency and an increase in residual phenol in the solution. Senturk et al reported similar results (30).

Phenol adsorption isotherm (Langmuir and Freundlich)

Several different models are used to describe adsorption. One of the most important models is adsorption isotherms. In this study, the Langmuir and Freundlich models were used to describe the relationship between the amount of adsorbed phenol and its equilibrium concentration in the solution (40). Adsorption isotherm studies were conducted under optimized conditions (pH= 2, adsorbent dosage=5 g/L), phenol concentrations ranging from 10 to 80 mg/L, and a contact time of 10 minutes. The investigation of correlation coefficients showed that the obtained regression coefficient was higher than 0.9 for both models, and the phenol adsorption by *C. colocynthis* seed ash showed more suitability with the Freundlich model than the Langmuir model (Figures 7 and 8). Similar findings were reported by Feng et al. Equilibrium data was in greater accordance with the Freundlich model than the Langmuir model (41). This finding was also reported in the literature for phenol removal by other adsorbents (42,43).

Conclusion

In the current study, it was observed that Citrullus colocynthis seed ash can be used effectively for the removal of phenol from aqueous solutions. Among all parameters, pH had an important role in the adsorption process by this adsorbent. Phenol adsorption was greater in low pH values than in high values of pH. The results further revealed that the removal efficiency of phenol was increased by increasing adsorbent dose, contact time, and temperature. The evaluation of correlation coefficients showed that phenol adsorption by Citrullus colocynthis seed ash was in greater accordance with the Freundlich model than the Langmuir model. Thus, *C. colocynthis* seed ash has suitable potential for use in the removal of phenol on operation and practical scales due to its low cost and easy access.

Acknowledgments

The authors express their gratitude to the Gonabad University of Medical Sciences for the financial support of this project (Project number: 95/79).

Competing interests

The authors declare that they have no competing interests.

Authors' contributions

All authors contributed equally and were involved in the study design, data collection, and article approval.

References

1. Razeghi N, Hashemi SH, Jadidi KH, Aghapoor K. Performance of Mesopore MCM-41 with a Nanoporous Structure in Removing Phenol from Water. Journal of Water and Wastewater 2015; 25(6): 2-9. [In Persian].
2. Saeedi M, Pajooheshfar SP. Investigation of phenol adsorption from contained water by activated carbon and carbon of almonds and walnut skin. Quarterly Journal of Environmental Science and Technology 2008; 10(4): 218-31. [In Persian].
3. Afsharnia M, Saeidi M, Zarei A, Narooie MR, Biglari H. Phenol removal from aqueous environment by adsorption onto pomegranate peel carbon. Electro physic 2016; 8(11): 3248-56.
4. Nakamoto S, Machida N. Phenol removal from aqueous solutions by peroxidase-catalyzed reaction using additives. Water Res 1992; 26(1): 49-54. doi: 10.1016/0043-1354(92)90110-P.
5. Caetano M, Valderrama C, Farran A, Cortina JL. Phenol removal from aqueous solution by adsorption and ion exchange mechanisms onto polymeric resins. J Colloid Interface Sci 2009; 338(2): 402-9. doi: 10.1016/j.jcis.2009.06.062.
6. Tor A, Cengeloglu Y, Aydin ME, Ersoz M. Removal of phenol from aqueous phase by using neutralized red mud. J Colloid Interface Sci 2006; 300(2): 498-503. doi: 10.1016/j.jcis.2006.04.054.
7. Brown VK, Box VL, Simpson BJ. Decontamination procedures for skin exposed to phenolic substances. Arch Environ Health 1975; 30(1): 1-6. doi: 10.1080/00039896.1975.10666623.
8. Baker EL, Landrigan PJ, Bertozzi PE, Field PH, Basteyns BJ, Skinner HG. Phenol poisoning due to contaminated drinking water. Arch Environ Health 1978; 33(2): 89-94. doi: 10.1080/00039896.1978.10667314.
9. Babich H, Davis DL. Phenol: a review of environmental and health risks. Regul Toxicol Pharmacol 1981; 1(1): 90-109. doi: 10.1016/0273-2300(81)90071-4.
10. Maleki A, Mahvi H. Application of agricultural waste products for removal of phenol in aqueous systems. Medical Journal of Hormozgan University 2007; 10(4); 393-9. [In Persian].
11. World Health Organization. Guidelines for Drinking-Water Quality. Geneva: WHO; 2004.
12. Roostaei N, Tezel FH. Removal of phenol from aqueous solutions by adsorption. J Environ Manage 2004; 70(2): 157-64. doi: 10.1016/j.jenvman.2003.11.004.
13. Wu C, Liu X, Wei D, Fan J, Wang L. Photosonochemical degradation of phenol in water. Water Res 2001; 35(16): 3927-33. doi: 10.1016/S0043-1354(01)00133-6.
14. Mukherjee S, Kumar S, Misra AK, Fan M. Removal of phenols from water environment by activated carbon, bagasse ash and wood charcoal. Chem Eng J 2007; 129(1-3): 133-42. doi: 10.1016/j.cej.2006.10.030.
15. Potgieter JH, Bada SO, Potgieter-Vermaak SS. Adsorptive removal of various phenols from water by South African coal fly ash. Water Sa 2009; 35(1): 89-96.
16. Malakootian M, Mesdaghinia A, Rezaei S. Efficiency of ortho-chlorophenol removal from aqueous solutions using activated Fly Ash of Zarand Fossil Fuel Power Plant. Scientific Journal of School of Public Health and Institute of Public Health Research 2014; 12(2): 81-92. [In Persian].
17. Daifullah AA, Girgis BS. Removal of some substituted phenols by activated carbon obtained from agricultural waste. Water Res 1998; 32(4): 1169-77. doi: 10.1016/S0043-1354(97)00310-2.
18. Gholizadeh A, Kermani M, Gholami M, Farzadkia M, Yaghmaeian K. Removal efficiency, adsorption kinetics and isotherms of phenolic compounds from aqueous solution using rice bran ash. Asian J Chem 2013; 25(7): 3871-3878.
19. Manshouri M, Daraei H, Yazdanbakhsh AR. The feasibility study of ostrich feathers in removing phenol from aqueous solution. J Kermanshah Univ Med Sci 2012; 16(5): 359-66.
20. Bazrafshan E, Kordmostafapour F, Faridi H, Barikbin B. Application of Moringa Peregrina seed extract as a natural coagulant for Phenol removal from aqueous solution. Journal of Birjand University of Medical Sciences 2013;

19(4): 389-98. [In Persian].
21. Bazrafshan E, Biglari H, Mahvi A. Performance evaluation of electrocoagulation process for phenol removal from aqueous solutions. Fresenius Environmental Bulletin 2012; 21(2): 364-71.
22. Zarei A, Bazrafshan E, Faridi H, Khaksefidi R, Alizadeh M. The evaluation of removal efficiency of phenol from aqueous solutions using *Moringa peregrina* tree shell ash. Iranian Journal of Health Sciences 2013; 1(1): 65-74. doi: 10.18869/acadpub.jhs.1.1.65.
23. American Public Health Association (APHA). Standard Methods for examination of water and wastewater. Washington: American Public Health Association; 2012
24. Mall ID, Srivastava VC, Agarwal NK, Mishra IM. Adsorptive removal of malachite green dye from aqueous solution by bagasse fly ash and activated carbon-kinetic study and equilibrium isotherm analyses. Colloids Surf A Physicochem Eng Asp 2005; 264(1-3): 17-28. doi: 10.1016/j.colsurfa.2005.03.027.
25. Dyanati R, Yousefi Z, Yazdani Cherati J, Balarak D. Investigating phenol absorption from aqueous solution by dried azolla. Journal of Mazandaran University of Medical Sciences 2013; 22(2): 13-20. [In Persian].
26. Nath K, Panchani S, Bhakhar MS, Chatrola S. Preparation of activated carbon from dried pods of Prosopis cineraria with zinc chloride activation for the removal of phenol. Environ Sci Pollut Res Int 2013; 20(6): 4030-45. doi: 10.1007/s11356-012-1325-y.
27. Shrestha S. Chemical, structural and elemental characterization of biosorbents using FE-SEM, SEM-EDX, XRD/XRPD and ATR-FTIR Techniques. J Chem Eng Process Technol 2016; 7(3):2-11. doi: 10.4172/2157-7048.1000295.
28. Giwa S, Abdullah LC, Adam NM. Investigating "Egusi" (*Citrullus Colocynthis* L.) seed oil as potential biodiesel feedstock. Energies 2010; 3(4): 607-18. doi: 10.3390/en3040607.
29. Al-Snafi AE. Chemical constituents and pharmacological effects of *Citrullus colocynthis* - A review. IOSR J Pharm 2016; 6(3): 57-67.
30. Senturk HB, Ozdes D, Gundogdu A, Duran C, Soylak M. Removal of phenol from aqueous solutions by adsorption onto organomodified Tirebolu bentonite: equilibrium, kinetic and thermodynamic study. J Hazard Mater 2009; 172(1): 353-62. doi: 10.1016/j.jhazmat.2009.07.019.
31. Halhouli KA, Darwish NA, Al-Dhoon NM. Effects of pH and inorganic salts on the adsorption of phenol from aqueous systems on activated decolorizing charcoal. Sep Sci Technol 1995; 30(17): 3313-24. doi: 10.1080/01496399508013147.
32. Banat FA, Al-Bashir B, Al-Asheh S, Hayajneh O. Adsorption of phenol by bentonite. Environ Pollut 2000; 107(3): 391-8. doi: 10.1016/S0269-7491(99)00173-6.
33. Srivastava VC, Swamy MM, Mall ID, Prasad B, Mishra IM. Adsorptive removal of phenol by bagasse fly ash and activated carbon: equilibrium, kinetics and thermodynamics. Colloids Surf A Physicochem Eng Asp 2006; 272(1-2): 89-104. doi: 10.1016/j.colsurfa.2005.07.016.
34. Cengiz S, Cavas L. Removal of methylene blue by invasive marine seaweed: *Caulerpa racemosa* var. *cylindracea*. Bioresour Technol 2008; 99(7): 2357-63. doi: 10.1016/j.biortech.2007.05.011.
35. SenthilKumar P, Ramalingam S, Sathyaselvabala V, Dinesh Kirupha S, Sivanesan S. Removal of copper (II) ions from aqueous solution by adsorption using cashew nut shell. Desalination 2011; 266(1-3): 63-71. doi: 10.1016/j.desal.2010.08.003.
36. Rahmani A, Asgari G, Barjasteh Asgari F, Hedayati Kamran E, Alijani F. Investigation of phenol removal from aqueous solutions using copper-impregnated pumice. Avicenna Journal of Clinical Medicine 2011; 17(4): 50-6. [In Persian].
37. Dehghani MH, Alimohammadi M, Mahvi AH, Rastkari N, Mostofi M, Gholami M. Performance of multiwall carbon nanotubes for removal phenol from aqueous solutions. Iranian Journal of Health and Environment 2014; 6(4): 491-502. [In Persian].
38. Wang L, Zhang J, Zhao R, Zhang C, Li C, Li Y. Adsorption of 2,4-dichlorophenol on Mn-modified activated carbon prepared from *Polygonum orientale* Linn. Desalination 2011; 266(1-3): 175-81. doi: 10.1016/j.desal.2010.08.022.
39. Liao Q, Sun J, Gao L. Adsorption of chlorophenols by multi-walled carbon nanotubes treated with HNO3 and NH3. Carbon 2008; 46(3): 553-5. doi: 10.1016/j.carbon.2007.12.009.
40. Gholizadeh A, Kermani M, Gholami M, Farzadkia M. Kinetic and isotherm studies of adsorption and biosorption processes in the removal of phenolic compounds from aqueous solutions: comparative study. J Environ Health Sci Eng 2013; 11(1): 29. doi: 10.1186/2052-336x-11-29.
41. Feng J, Qiao K, Pei L, Lv J, Xie S. Using activated carbon prepared from *Typha orientalis* Presl to remove phenol from aqueous solutions. Ecol Eng 2015; 84: 209-17. doi: 10.1016/j.ecoleng.2015.09.028.
42. Viswanathan N, Meenakshi S. Enriched fluoride sorption using alumina/chitosan composite. J Hazard Mater 2010; 178(1-3): 226-32. doi: 10.1016/j.jhazmat.2010.01.067.
43. Dou X, Zhang Y, Wang H, Wang T, Wang Y. Performance of granular zirconium-iron oxide in the removal of fluoride from drinking water. Water Res 2011; 45(12): 3571-8. doi: 10.1016/j.watres.2011.04.002.

32

Autotrophic denitrification of synthetic nitrate-contaminated groundwater in up-flow fixed-bed bioreactor by pumice as porous media

Masoud Tourang[1], Behnoush Aminzadeh[1*], Ali Torabian[1]

[1]Department of Environmental Engineering, Graduate Faculty of Environment, University of Tehran, Iran

Abstract

Background: Background: Increasing nitrate concentrations in groundwater resources is considered a common environmental and public health problem worldwide. In this research, an autotrophic up-flow bioreactor with pumice as media was used to study the effects of the sulfur-to-nitrogen (S/N) ratio and empty bed contact time (EBCT) on nitrate removal efficiency and byproducts.

Methods: Experiments were carried out in a 3.47 L up-flow, fixed-bed reactor with 3 sampling ports. To evaluate the overall impact of S/N ratio and EBCT on the performance of the bioreactor, several phases with different S/N ratios and EBCTs were applied.

Results: At a constant S/N ratio of 3.85 g/g, as EBCT decreased from 24 hours to 2 hours, the nitrate removal efficiency decreased from 98% to 64%. On the other hand, at the desired EBCT of 4 hr, as S/N ratio decreased from 3.85 to 1.51 g/g, nitrate removal efficiency was reduced from 85% to 32%. Changing the EBCT and S/N ratio also affected the effluent nitrite and sulfate concentrations as byproducts. At the S/N ratio of 3.85 g/g and EBCT of 24 hours, effluent nitrite and sulfate concentrations were 0.1 mg NO_2^--N/L and 463 mg SO_4^{2-}/L, respectively. Decreasing the S/N ratio to 1.51 g/g and the EBCT to 4 hours caused drastic changes in effluent nitrite and sulfate concentrations.

Conclusion: The results indicated that the autotrophic denitrification with thiosulfate as electron donor and pumice as media was feasible and applicable for nitrate contaminated groundwater.

Keywords: Autotrophic Denitrification, Pumice, Nitrate, Nitrite, Thiosulfate

***Correspondence to:**
Behnoush Aminzadeh
Email: bamin@ut.ac.ir

Introduction

Inorganic nitrogen compounds, such as nitrite, nitrate, and ammonium, are abundant contaminants in ground and surface water. High NO_3–N concentrations (>10 mg/L) have been recorded in numerous aquifers in many countries (1). Nitrate that is used mainly in inorganic fertilizers leaks into the aquifer and surface waters (2, 3). It has been estimated that more than 70% of all nitrate existing in surface and groundwater comes from fertilizers and other materials used in agricultural activities. However, emissions from septic tanks and seep from sewers, atmospheric deposition, and the usage of sewage sludge as fertilizer can all be effective (4,5). In some cases, high nitrate concentrations are caused by natural sources (6). In rural and urban areas of Iran, the growth in agriculture and industry has led to groundwater nitrate concentrations over the acceptable thresholds (7).

Causing eutrophication, nitrate has adverse effects on water ecosystems (8). According to Sun et al (2) and Migeot et al (9), health implications of nitrate in humans include but are not limited to methemoglobinemia, diminished vitality, low weight in newborns, higher chance of fetal death, and formation of carcinogens in the digestive system. In livestock, nitrate contamination can cause reduced weight gain rates.

Among different physical, chemical, and biological nitrate removal processes, biological denitrification is an efficient microbial process that changes nitrate to N_2 under anoxic conditions (1,10,11). In this process, four enzymatic steps produce intermediates such as nitrite (NO_2^-), nitric oxide (NO), and nitrous oxide (N_2O) (12). Nitrate can be exploited as a terminal electron acceptor by a variety of heterotrophic and autotrophic bacteria. While the former uses organic matters, the latter relies on inorganic compounds such as molecular hydrogen, sulfur, sulfide, thiosulfate, and Fe^{++} as electron donors (7,13). Heterotrophic denitrification has been widely applied to wastewater treatment due to its high efficiency and economical aspects (14), but for natural bodies of water, it needs the organic carbon as the electron donor to reduce nitrate and metabolism of heterotrophs. The most important problem of using this process is the existence of residual organic matter such as methanol, ethanol, or acetic acid that may create several problems in the treated water (15).

Compared with heterotrophic treatment, autotrophic treatment is an attractive alternative because of 2 notable advantages. First, since autotrophic denitrification does not need any external organic carbon such as methanol and ethanol, it is very cost effective with a lower risk of secondary contamination compared to heterotrophic denitrification. Second, autotrophic denitrification has a lower cell yield of autotrophic bacteria and, therefore, less sludge production, which minimizes the handling of sludge (16). In autotrophic denitrification, inorganic carbon compounds such as carbon dioxide (CO_2) and bicarbonate (HCO_3^-) are used as carbon sources (13,17,18). The produced H^+ consumes alkalinity to balance the charge; therefore, some alkaline sources such as limestone are needed to retain the pH in the best condition for autotrophic denitrifiers (19,20).

In sulfur-based autotrophic denitrification, sulfur-oxidizing bacteria such as *Thiobacillus denitrificans* and *Thiomicrospira denitrificans* are agents for reducing nitrate to N_2 (21-24). A variety of solid materials has been applied as biofilm carriers for attached growth autotrophic denitrification processes like granular-activated carbon (25,26), polystyrene bead (27), elemental sulfur (28,29), zeolite (30), ceramsite (10), and polyurethane foam (31). The aims of this study were to determine the feasibility of using pumice as a biofilm cost-effective carrier, and the effects of S/N ratio, EBCT, and height of bioreactor on nitrate removal efficiency and obtained byproducts.

Methods

An up-flow bioreactor system, as shown in Figure 1, was used for the experiments. The column bioreactor was made of steel with an inner diameter of 8 cm, height of 80 cm, and total volume of 3.78 L. Three sampling ports were placed at intervals of 23, 46, and 69 cm from the inlet valve. The pumice particles used as biofilm carriers had diameters ranging from 2.36 mm to 4.75 mm and a porosity of 60%. Due to the high porosity of pumice, this media has a high specific surface that increases the attached growth. In addition, pumice has some other advantages such as low cost and availability in all parts of Iran. The bioreactor was fed continuously in the up-flow mode using an adjustable peristaltic pump with the minimum flow rate of 0.5 L/min. Influent groundwater was taken from one of the Tehran's *Qanats* (Iran). The chemical properties of the water of the *Qanat* are shown in Table 1. To achieve the desired influent nitrate concentration (150 mg/L), KNO_3 was added to the influent (32).

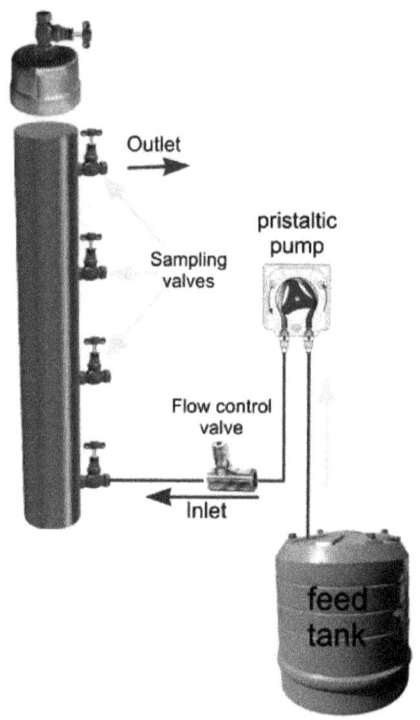

Figure 1. Experimental set-up.

The inoculum was obtained from the returned anoxic activated sludge of the A2O process treating municipal wastewater in the Shahrak-e-Ekbatan wastewater treatment plant in Tehran (Iran). To pre-enrichment the autotrophic denitrifying biomass, a synthetic medium consisting of 3.0 g KNO_3/L, 6.0 g $Na_2S_2O_3.5H_2O$/L, 1.5 g $NaHCO_3$/L, 0.3 g KH_2PO_4/L, 0.4 g $MgSO_4.7H_2O$/L, and 1 ml/L of trace element containing 5.74 g NH_4Cl/L, 5.6 g K_2HPO_4/L, 1 g $MgCl_2$/L, 1 g $FeCl_2.6H_2O$/L, 1 g $MnSO_4$. H_2O/L, and 1 g $CaCl_2$/L was added to the inoculum in a master culture reactor. After sufficient bacterial growth, the bioreactor was inoculated with one liter of the culture reactor. The rest of the bioreactor volume was filled with synthetic groundwater. After a few hours, the bioreactor was operated continuously at an empty bed contact time (EBCT) of 24 hours with synthetic groundwater as influent. The autotrophic denitrification process needs an electron donor to participate in the oxidation/reduction mechanism. Sodium thiosulfate was added to the influent as the electron donor. At first, the bioreactor was operated

Table 1. Chemical properties of the water of the *Qanat*

Chemical parameter	Concentration (mg/L)
Nitrate (No_3^-)	11±1
Nitrite (No_2^-)	1.80±0.15
Sulfate (So_4^{-2})	110±13
Alkalinity (as CaCO3)	228±24
pH	7.40±3

with the maximum electron donor concentration (500 mg/L $Na_2S_2O_3.5H_2O$) and different EBCTs (2, 4, 8, 16, and 24 hours). In this step, the desired EBCT was determined. Then, in this desired EBCT, the effects of electron donor concentration and column height on denitrification efficiency were studied. To determine the effect of the S/N ratio on the bioreactor performance, different S/N ratios (3.85, 3.05, 2.29, and 1.51 g/g) were tested. All experiments were conducted at 25±2°C.

Analysis methods

All chemical analyses were performed according to the APHA's Standard Methods for Examination of Water and Wastewater (33). Concentrations of nitrate, nitrite, and sulfate were determined using a HACH DR5000 spectrophotometer (USA). Alkalinity was measured by the titration method, and pH values were monitored using a standard digital pH meter (691 Metrohm). In each analysis, at least one in 3 samples was duplicated, and the deviation between the two samples was always less than 5%.

Results

The effects of flow rates on the denitrification of synthetic groundwater was investigated by increasing the flow rate from 3.47 L/d to 41.62 L/d with a corresponding EBCT of 24 to 2 hours at the S/N ratio of 3.85 g/g. Figure 2 shows the variations of nitrate removal efficiency and effluent nitrate concentration during the operation. The maximum nitrate removal efficiency of 98% was obtained with an EBCT of 24 hours. As EBCT was decreased from 24 to 2 hours, the nitrate removal efficiency decreased from 98% to 64% and the effluent nitrate concentration increased from 3 to 54 mg/L. According to Figure 2, changes in nitrate removal efficiency in different EBCTs are not the same. In lower EBCTs, these changes were severe. However, an increase in the nitrate-nitrogen removal rate was observed during EBCT reduction; the nitrate-nitrogen removal rate increased from 33 mg/L/d NO_3^--N in an EBCT of 24 hours to 260 mg/L/d NO_3^--N in an EBCT of 2 hours. In an EBCT of 4 hours, since the effluent nitrate concentration of 23 mg/L was lesser than the regulatory maximum contaminant level of 50 mg/L for drinking water (according to Iranian standards), this EBCT was selected as the desired EBCT for subsequent experiments (34).

The effects of EBCT on the effluent nitrite and produced sulfate concentrations are shown in Figure 3. Nitrite concentration in the effluent represented the incomplete denitrification.

The optimal range of pH for autotrophic denitrifiers is 6.8-8.2 (35), and maximum denitrification activity is observed at pH 7.5-8.0 (36). During the experiments, pH was controlled by adding buffer.

Autotrophic denitrifiers consume alkalinity. According to the current results, the ratio of alkalinity (as $CaCO_3$) consumed per NO_3^--N removed (g/g) was 2.92, which was similar to the theoretical ratio (37), and the results found by Chung et al (38) indicated that autotrophic denitrification consumed 3-4 mg $CaCO_3$/mg NO_3^--N.

Effects of S/N ratio on nitrate removal and byproducts

After determining the EBCT of 4 hours as the desired EBCT, the effect of S/N ratio on nitrate removal efficiency was investigated. S/N ratios of 3.85, 3.05, 2.29, and 1.51 (g/g) were tested. The stoichiometric S/N ratio for complete denitrification is 3.84. Previous studies have shown that in S/N ratios between 3.70 and 6.67, no significant changes in denitrification rate were observed (3).

In Figure 4 the effect of S/N ratio on nitrate removal efficiency is shown. At an EBCT of 4 hours, decreasing

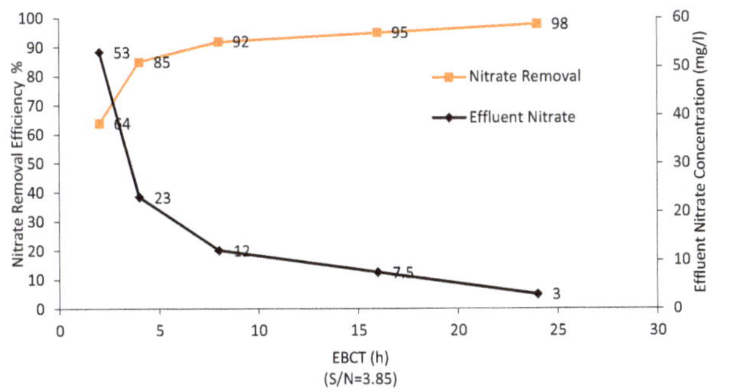

Figure 2. Effects of EBCT on nitrate removal efficiency and effluent nitrate concentration.

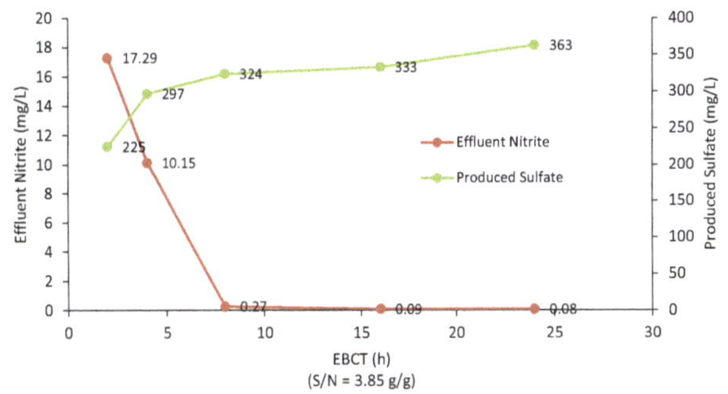

Figure 3. Effects of EBCT on effluent nitrite, and produced sulfate.

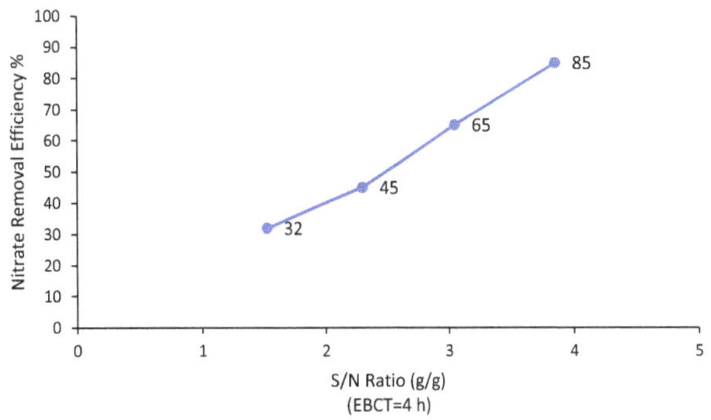

Figure 4. Effect of S/N ratio on nitrate removal efficiency.

the S/N ratio from 3.85 to 1.51 (g/g) reduced nitrate removal efficiency from 85% to 32% and increased nitrite accumulation.

Figure 5 illustrates the effect of S/N ratio on nitrate, nitrite, and sulfate concentrations. As the S/N ratio was reduced, a decrease in the amount of produced sulfate was observed. At the S/N ratio of 3.85, the nitrite concentration in the effluent was 3.95 mg NO_2^-/L. By decreasing the S/N ratio, nitrite accumulation was increased so that at the S/N ratio of 1.51, about 25 mg/L nitrite was observed in the effluent.

Figures 6, 7 and 8 show the profile of effluent nitrate and nitrite concentrations and sulfate production versus reactor height (23, 46, and 69 cm), at different EBCTs and an S/N ratio of 3.85 g/g, respectively. High rates of nitrate removal and sulfate production at the lower sampling port (23 cm) were observed in all cases. Generally, the 46 cm and 69 cm sampling ports were associated with lower removal and production rates. Increased cell growth at the lower sampling port was anticipated considering the high nitrate and thiosulfate concentration in the feed.

It was observed that at an EBCT of 4 hours, more than 90% of nitrate removal efficiency occurred in the first 23 cm of the column. The reduction of EBCT increased the effect of higher parts on the removal of nitrate removal, so that at an EBCT of 2 hours, 50% of total nitrate removal occurred in the mentioned part of the column (Figure 6).

As shown in Figure 7, a rapid accumulation of nitrite was observed in the first 23 cm of the column at EBCTs of 24, 16, and 8 hours and an S/N ratio of 3.85 g/g. However, upon completion of denitrification at the height of 46 cm, an intensive nitrite decrease was observed.

Figure 8 shows the effect of column height on effluent sulfate concentration at different EBCTs. The sulfate production profile was very similar to that of nitrate; therefore, a fast increase in sulfate concentration was observed in the first 23 cm of the column. In upper ports, the increase in sulfate concentration occurred at a slower rate.

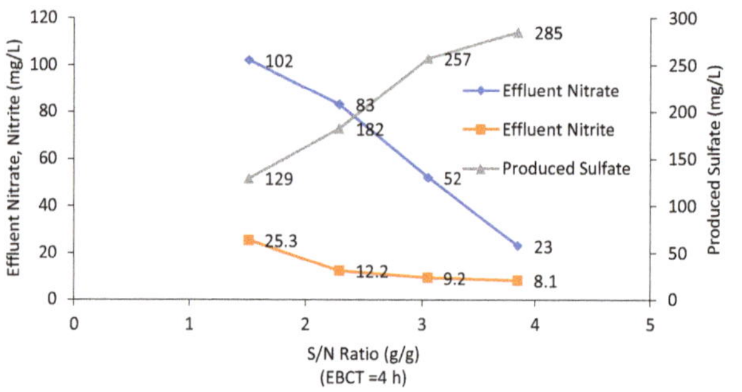

Figure 5. Effects of S/N ratio on effluent nitrate, nitrite, and produced sulfate concentrations.

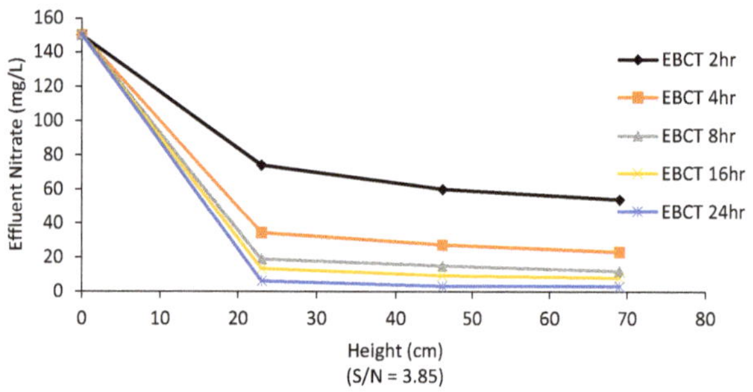

Figure 6. Effects of height and EBCT on effluent nitrate concentration.

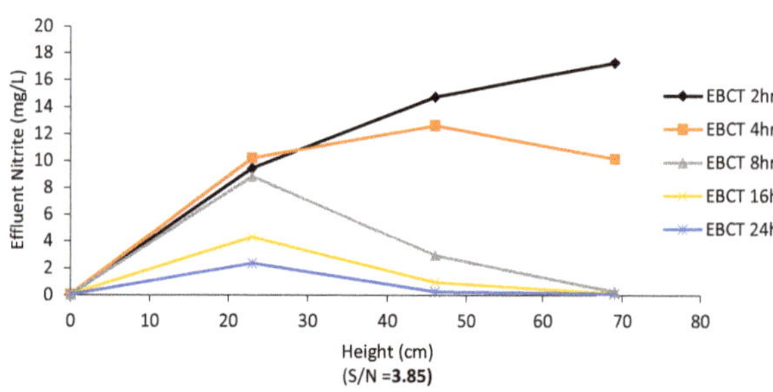

Figure 7. Effects of height and EBCT on effluent nitrite concentration.

Discussion

Effects of EBCT on nitrate removal and byproducts

According to Zhou et al (16), low influent concentration and longer EBCTs improve nitrate removal efficiency. In addition, they reported that the nitrogen in groundwater and effluent from municipal WWT can obtain removal rates of up to 90% with a four-hour EBCT. However, longer EBCTs are required for water with a nitrate contamination over 70 mg/L. According to this study, reducing the EBCT from 24 hours to 2 hours led to the reduction of nitrate removal efficiency from 98% to 64% (with a constant S/N ratio of 3.85 g/g).

As shown in Figure 3, at EBCTs of 24, 16, and 8 hours, the effluent nitrite concentrations were less than the maximum allowable level of 3 mg/L. As EBCT decreased to 4 and 2 hours, a drastic nitrite accumulation occurred in the effluent due to incomplete denitrification. Although molecular analysis associates the same microorganisms to denitrifications of nitrate and nitrite into nitrogen (16), the nitrite accumulation in autotrophic denitrification is associated with different causes. High specific utilization rate of nitrate, the reduction in induction time of NO_2^- reducing enzymes caused by nitrate, the variety in

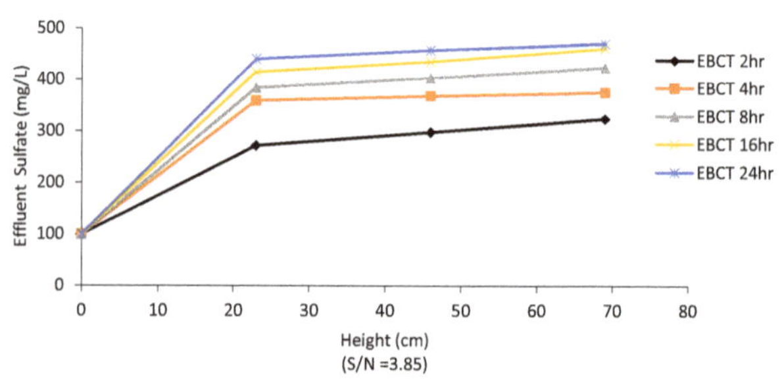

Figure 8. Effects of height and EBCT on effluent sulfate concentration.

saturation rates and affinities of electron acceptors, and the type of bacteria are all considered to be contributors to nitrate accumulation in this process (39).

When thiosulfate is used as the electron donor, sulfate ion is produced as a byproduct (31). Figure 9 shows the relationship between the produced sulfate and removed nitrate during the steady state condition for the influent S/N ratio of 3.85 g/g at different EBCTs. It was found that for each mg NO_3^- removed, 2.43 mg/L SO_4^{2-} was produced, which corresponds closely with the stoichiometric ratios of 2.62 and 2.24 derived from Equations (1) and (2), respectively (37,40). The difference between these values is caused by several factors such as feed composition, microbial population, pure culture, temperature, pH, etc.

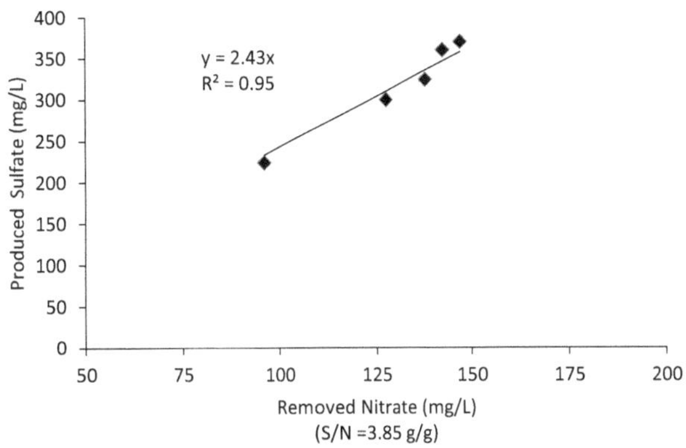

Figure 9. Relationship between removed nitrate and produced sulfate.

$$0.844S_2O_3^{2-} + NO_3^- + 0.434H_2O + 0.347CO_2 + 0.0865HCO_3^- + 0.0865NH_4^+ \rightarrow 0.0865C_5H_7NO_2 + 0.5N_2 + 1.689SO_4^{2-} + 0.697H^+ \quad (1)$$

$$NO_3^- + 0.844S_2O_3^{2-} + 0.347CO_2 + 0.0865HCO_3^- + 0.0865NH_4^+ \rightarrow 0.036C_5H_7O_2N + 0.48N_2 + 1.45SO_4^{2-} + 0.045H^+ \quad (2)$$

As shown in Figure 9, the produced sulfate had a linear relationship with removed nitrate, and the EBCT had no significant effect on the stoichiometric ratio.

Effect of S/N ratio on Nitrate removal and byproducts

As shown in Figure 4, decreasing the S/N ratio to 1.51 caused a notable decrease in the nitrate removal efficiency. This reduction occurred because of the electron donor shortage. In this case, increasing EBCT did not cause any significant changes in the nitrate removal efficiency; so, it seems that the concentration of electron donor is the controller agent. Campos et al (39) showed that when S/N ratio was less than 2.44, an increase in nitrite concentration occurred during denitrification.

It can be observed from Figure 5 that, in the denitrification process, a rapid initial decrease in thiosulfate concentration resulting in sudden nitrite accumulation occurred (38). In this case (low S/N ratio), the electron donor shortage caused incomplete denitrification and led to the ultimate nitrite accumulation. The results of this study correspond with those of previous studies and show that at S/N ratios lower than 4.35 (g/g), the accumulation of nitrite is inevitable (39). In this condition, even in greater EBCTs, nitrite accumulation occurred. Also, 2.43 mg of SO_4^{2-} was produced for each milligram of NO_3^--N removed, and this ratio did not change with different S/N ratios.

Campos et al (39) also showed that sulfate concentrations greater than 500 mg/L induced inhibitory effects on the nitrate removal efficiency. According to their study, autotrophic denitrifying activity was inhibited to 85% of that in the control test at a concentration of 5000 mg SO_4^{2-}-S/L. Oh et alreported that sulfate inhibition began at concentrations above 2000 mg SO_4^{2-}-S/L using a mixed culture (41), while Claus and Kutzner found that this compound started to inhibit at 1600 mg SO_4^{2-}-S/L (36). However, in this study, according to the low concentrations of nitrite and sulfate, the inhibitory effects were negligible.

Effects of column height on nitrate removal and byproducts

Moon et al (42) showed that a larger portion of nitrate removal took place in the bottom part of the column. Their findings are very similar to the results of this study that show more than 90% of nitrate removal occurred in the first 23 cm of the column (Figure 6). They observed that almost all the nitrate removal occurred in the first 25 cm of the column, and the effect of S/N ratio was negligible. Figure 7 shows the effluent nitrite concentrations at different EBCTs and heights. The results showed a great nitrite accumulation in the first 23 cm, similar to the findings of Moon et al (42) which showed nitrite accumulation at the height of 17 cm. Incomplete denitrification caused primary nitrite accumulation, but in upper heights, nitrite concentration decreased because of complete denitrification. In shorter EBCTs, such as 2 hours and 4 hours, nitrite accumulation continued, so that a great nitrite concentration in the effluent was observed. The behavior of the sulfate concentration was very similar to revers trend nitrate removal (Figure 8). The results showed a notable accumulation of sulfate concentration in the first 23 cm due to the high rate of nitrate removal in this section of the bioreactor.

Conclusion

Continuous bioreactor tests showed that pumice granules can be used as neutral media in autotrophic bio-filters and have a significant performance in reducing the nitrate concentration of groundwater. However, parameters such as EBCT, S/N ratio, and column height play important roles in nitrate removal efficiency. While using autotrophic microorganisms for denitrification, byproducts such as nitrite and sulfate (if sulfur-oxidizing bacteria are used) should always be monitored. In fact, parameters that control the efficiency of nitrate removal can also control the byproducts.

Acknowledgments

The authors thank the Graduate Faculty of Environment at the University of Tehran for providing laboratory facilities.

Competing interests

The authors declare that they have no competing interests.

Authors' contributions

All authors contributed equally and participated in the collection, analysis, and interpretation of the data. All authors critically reviewed, refined, and approved the manuscript.

References

1. Huang G, Huang Y, Hu H, Liu F, Zhang Y, Deng R. Remediation of nitrate-nitrogen contaminated groundwater using a pilot-scale two-layer heterotrophic-autotrophic denitrification permeable reactive barrier with spongy iron/pine bark. Chemosphere 2015; 130: 8-16. doi: 10.1016/j.chemosphere.2015.02.029.
2. Sun Y, Nemati M. Evaluation of sulfur-based autotrophic denitrification and denitritation for biological removal of nitrate and nitrite from contaminated waters. Bioresour Technol 2012; 114: 207-16. doi: 10.1016/j.biortech.2012.03.061.
3. Fajardo C, Mosquera-Corral A, Campos JL, Méndez R. Autotrophic denitrification with sulphide in a sequencing batch reactor. J Environ Manage 2012; 113: 552-6. doi: 10.1016/j.jenvman.2012.03.018.
4. Rivett MO, Buss SR, Morgan P, Smith JW, Bemment CD. Nitrate attenuation in groundwater: a review of biogeochemical controlling processes. Water Res 2008; 42(16): 4215-32. doi: 10.1016/j.watres.2008.07.020.
5. Breisha GZ, Winter J. Bio-removal of nitrogen from wastewaters - a review. J Am Sci 2010; 6(12): 508-28.
6. Darbi A, Viraraghavan T, Butler R, Corkal D. Pilot-scale evaluation of select nitrate removal technologies. J Environ Sci Health A Tox Hazard Subst Environ Eng 2003; 38(9): 1703-15.
7. Aminzadeh B, Torabian A, Azimi AA, Nabi Bidhendi GR, Mehrdadi N. Salt Inhibition Effects on Simultaneous Heterotrophic/Autotrophic Denitrification of High Nitrate Wastewater. Int J Environ Res 2010; 4(2): 255-62. doi: 10.22059/ijer.2010.17.
8. Wilkinson GM. Eutrophication of Freshwater and Coastal Ecosystems A2 - Abraham, Martin A. Encyclopedia of Sustainable Technologies. Oxford: Elsevier; 2017. p. 145-52. doi: 10.1016/B978-0-12-409548-9.10160-5
9. Migeot V, Albouy-Llaty M, Carles C, Limousi F, Strezlec S, Dupuis A, et al. Drinking-water exposure to a mixture of nitrate and low-dose atrazine metabolites and small-for-gestational age (SGA) babies: a historic cohort study. Environ Res 2013; 122: 58-64. doi: 10.1016/j.envres.2012.12.007.
10. Zhou W, Liu X, Dong X, Wang Z, Yuan Y, Wang H, et al. Sulfur-based autotrophic denitrification from the micro-polluted water. J Environ Sci (China) 2016; 44: 180-8. doi: 10.1016/j.jes.2016.01.002.
11. Moraes BS, Souza TS, Foresti E. Characterization and kinetics of sulfide-oxidizing autotrophic denitrification in batch reactors containing suspended and immobilized cells. Water Sci Technol 2011; 64(3): 731-8.
12. Sierra-Alvarez R, Beristain-Cardoso R, Salazar M, Gomez J, Razo-Flores E, Field JA. Chemolithotrophic denitrification with elemental sulfur for groundwater treatment. Water Res 2007; 41(6): 1253-62. doi: 10.1016/j.watres.2006.12.039.
13. Cardoso RB, Sierra-Alvarez R, Rowlette P, Flores ER, Gomez J, Field JA. Sulfide oxidation under chemolithoautotrophic denitrifying conditions. Biotechnol Bioeng 2006; 95(6): 1148-57. doi: 10.1002/bit.21084.
14. Park JY, Yoo YJ. Biological nitrate removal in industrial wastewater treatment: which electron donor we can choose. Appl Microbiol Biotechnol 2009; 82(3): 415-29. doi: 10.1007/s00253-008-1799-1.
15. Wan D, Liu H, Qu J, Lei P, Xiao S, Hou Y. Using the combined bioelectrochemical and sulfur autotrophic denitrification system for groundwater denitrification. Bioresour Technol 2009; 100(1): 142-8. doi: 10.1016/j.biortech.2008.05.042.
16. Zhou W, Sun Y, Wu B, Zhang Y, Huang M, Miyanaga T, et al. Autotrophic denitrification for nitrate and nitrite removal using sulfur-limestone. J Environ Sci (China) 2011; 23(11): 1761-9. doi: 10.1016/S1001-0742(10)60635-3.
17. Zhu I, Getting T. A review of nitrate reduction using inorganic materials. Environ Technol Rev 2012; 1(1): 46-58. doi: 10.1080/09593330.2012.706646.
18. Cecconet D, Devecseri M, Callegari A, Capodaglio AG. Effects of process operating conditions on the autotrophic denitrification of nitrate-contaminated groundwater using bioelectrochemical systems. Sci Total Environ 2018; 613-614: 663-71. doi: 10.1016/j.scitotenv.2017.09.149.
19. Sahinkaya E, Dursun N. Sulfur-oxidizing autotrophic and mixotrophic denitrification processes for drinking water treatment: elimination of excess sulfate production and alkalinity requirement. Chemosphere 2012; 89(2): 144-9. doi: 10.1016/j.chemosphere.2012.05.029.
20. Park S, Seon J, Byun I, Cho S, Park T, Lee T. Comparison of nitrogen removal and microbial distribution in wastewater treatment process under different electron donor conditions. Bioresour Technol 2010; 101(9): 2988-95. doi: 10.1016/j.biortech.2009.12.012.
21. Shao MF, Zhang T, Fang HH. Sulfur-driven autotrophic denitrification: diversity, biochemistry, and engineering applications. Appl Microbiol Biotechnol 2010; 88(5): 1027-42. doi: 10.1007/s00253-010-2847-1.
22. Trouve C, Chazal PM, Gueroux B, Sauvaitre N. Denitrification by New Strains of Thiobacillus Denitrificans under Non-Standard Physicochemical Conditions. Effect of Temperature, pH, and Sulphur Source. Environ Technol 1998; 19(6): 601-10. doi: 10.1080/09593331908616716.
23. Trouve C, Ghazal PM. Autotrophic Denitrification by Thiobacillus Denitrificans with Thiosulphate as Sole Energy Source. Pilot Scale Experiments at Low Temperature. Environ Technol 1999; 20(4): 387-95. doi:

10.1080/09593332008616831.
24. Bruckner CG, Mammitzsch K, Jost G, Wendt J, Labrenz M, Jurgens K. Chemolithoautotrophic denitrification of epsilonproteobacteria in marine pelagic redox gradients. Environ Microbiol 2013; 15(5): 1505-13. doi: 10.1111/j.1462-2920.2012.02880.x.
25. Di Capua F, Milone I, Lakaniemi AM, Hullebusch EDV, Lens PN, Esposito G. Effects of different nickel species on autotrophic denitrification driven by thiosulfate in batch tests and a fluidized-bed reactor. Bioresour Technol 2017; 238: 534-41. doi: 10.1016/j.biortech.2017.04.082.
26. Zou G, Papirio S, Lakaniemi AM, Ahoranta SH, Puhakka JA. High rate autotrophic denitrification in fluidized-bed biofilm reactors. Chem Eng J 2016; 284: 1287-94. doi: 10.1016/j.cej.2015.09.074.
27. Wang Z, Fei X, He S, Huang J, Zhou W. Comparison of heterotrophic and autotrophic denitrification processes for treating nitrate-contaminated surface water. Sci Total Environ 2017; 579: 1706-14. doi: 10.1016/j.scitotenv.2016.11.194.
28. Christianson L, Summerfelt S. Fluidization velocity assessment of commercially available sulfur particles for use in autotrophic denitrification biofilters. Aquacultural Engineering 2014; 60: 1-5. doi: 10.1016/j.aquaeng.2014.03.004.
29. Kostrytsia A, Papirio S, Frunzo L, Mattei MR, Porca E, Collins G, et al. Elemental sulfur-based autotrophic denitrification and denitritation: microbially catalyzed sulfur hydrolysis and nitrogen conversions. J Environ Manage 2018; 211: 313-22. doi: 10.1016/j.jenvman.2018.01.064.
30. Montalvo S, Huilinir C, Galvez D, Roca N, Guerrero L. Autotrophic denitrification with sulfide as electron donor: Effect of zeolite, organic matter and temperature in batch and continuous UASB reactors. Int Biodeterior Biodegradation 2016; 108: 158-65. doi: 10.1016/j.ibiod.2015.12.022.
31. Moraes BS, Souza TS, Foresti E. Effect of sulfide concentration on autotrophic denitrification from nitrate and nitrite in vertical fixed-bed reactors. Process Biochem 2012; 47(9): 1395-401. doi: 10.1016/j.procbio.2012.05.008.
32. Torabian A, Aminzadeh B, Hashemi SH. A study of nitrate in Tehran ground water and a method of its removal. Int J Environ Stud 2000; 57(6): 725-34. doi: 10.1080/00207230008711307.
33. Association, American Water Works Association, Water Environment Federation. Standard Methods for the Examination of Water and Wastewater. Washington DC: APHA, AWWA, WPCF; 2012.
34. Institute of Standards and Industrial Research of Iran. Chemical Specifications of Drinking Water. 5th ed. Tehran: ISIRI; 2009. [Persian].
35. Vidal S, Rocha C, Galvao H. A comparison of organic and inorganic carbon controls over biological denitrification in aquaria. Chemosphere 2002; 48(4): 445-51.
36. Claus G, Kutzner HJ. Autotrophic denitrification by Thiobacillus denitrificans in a packed bed reactor. Appl Microbiol Biotechnol 1985; 22(4): 289-96. doi: 10.1007/bf00252032.
37. Hashimoto S, Furukawa K, Shioyama M. Autotrophic denitrification using elemental sulfur. J Ferment Technol 1987; 65(6): 683-92. doi: 10.1016/0385-6380(87)90011-2.
38. Chung J, Amin K, Kim S, Yoon S, Kwon K, Bae W. Autotrophic denitrification of nitrate and nitrite using thiosulfate as an electron donor. Water Res 2014; 58: 169-78. doi: 10.1016/j.watres.2014.03.071.
39. Campos JL, Carvalho S, Portela R, Mosquera-Corral A, Mendez R. Kinetics of denitrification using sulphur compounds: Effects of S/N ratio, endogenous and exogenous compounds. Bioresour Technol 2008; 99(5): 1293-9. doi: 10.1016/j.biortech.2007.02.007.
40. Byun IG, Ko JH, Jung YR, Lee TH, Kim CW, Park TJ. The feasibility of using spent sulfidic caustic as alternative sulfur and alkalinity sources in autotrophic denitrification. Korean J Chem Eng 2005; 22(6): 910-6. doi: 10.1007/bf02705674.
41. Oh SE, Kim KS, Choi HC, Cho J, Kim IS. Kinetics and physiological characteristics of autotrophic dentrification by denitrifying sulfur bacteria. Water Sci Technol 2000; 42(3-4): 59-68.
42. Moon HS, Ahn KH, Lee S, Nam K, Kim JY. Use of autotrophic sulfur-oxidizers to remove nitrate from bank filtrate in a permeable reactive barrier system. Environ Pollut 2004; 129(3): 499-507. doi: 10.1016/j.envpol.2003.11.004.

Predicting arsenic and heavy metals contamination in groundwater resources of Ghahavand plain based on an artificial neural network optimized by imperialist competitive algorithm

Meysam Alizamir[1*], Soheil Sobhanardakani[2]

[1]Young Researchers & Elite Club, Hamedan Branch, Islamic Azad University, Hamedan, Iran
[2]Department of the Environment, School of Basic Sciences, Hamedan Branch, Islamic Azad University, Hamedan, Iran

Abstract
Background: The effects of trace elements on human health and the environment gives importance to the analysis of heavy metals contamination in environmental samples and, more particularly, human food sources. Therefore, the current study aimed to predict arsenic and heavy metals (Cu, Pb, and Zn) contamination in the groundwater resources of Ghahavand Plain based on an artificial neural network (ANN) optimized by imperialist competitive algorithm (ICA).
Methods: This study presents a new method for predicting heavy metal concentrations in the groundwater resources of Ghahavand plain based on ANN and ICA. The developed approaches were trained using 75% of the data to obtain the optimum coefficients and then tested using 25% of the data. Two statistical indicators, the coefficient of determination (R^2) and the root-mean-square error (RMSE), were employed to evaluate model performance. A comparison of the performances of the ICA-ANN and ANN models revealed the superiority of the new model. Results of this study demonstrate that heavy metal concentrations can be reliably predicted by applying the new approach.
Results: Results from different statistical indicators during the training and validation periods indicate that the best performance can be obtained with the ANN-ICA model.
Conclusion: This method can be employed effectively to predict heavy metal concentrations in the groundwater resources of Ghahavand plain.
Keywords: Neural networks (computer), Groundwater, Models, Algorithms, Trace elements

***Correspondence to:**
Meysam Alizamir
Email: meysamalizamir@gmail.com

Introduction

Environmental pollutants, especially toxic heavy metals, can discharge into natural cycles, (e.g., soil, water, and air) through urbanization, industrialization, agriculture, mining, and exploitation of natural resources (1). Many elements such as Cu, Fe, Mn, Ni, and Zn are essential for human life and play major roles in health in low concentrations, but they can be toxic at high levels. Others, including As, Cd, Cr, Hg, Pb, and Sn, have no known essential function in living organisms and are toxic even at low concentrations. Heavy metals can cause serious adverse health effects in humans; thus, they are known as the most dangerous pollutants (2-6).

Surface and groundwater resources are important for human life and for economic development (1). More than 50% of the world's population depends on groundwater resources for drinking, agriculture, and for general survival (7,8). Therefore, the contamination of groundwater by toxic heavy metals is a serious global environmental problem.

Arsenic is a widely distributed metalloid that is also a carcinogen for humans, even at low levels of exposure (9). The combustion of fossil fuels, smelting of non-ferrous metals, and use of arsenical pesticides in agriculture are the main sources of this element in the environment (10). Some foods, such as vegetables, fruits, nuts, red meat, and shellfish, are known as sources of copper. Although Cu can play a critical role in various biochemical processes (11), a constant diet of this element results in the dissolution of the barrier that keeps undesirable toxins from entering the brain. Critical doses of this element can cause adverse health effects such as fatigue, hair loss, inflammation of brain tissues, panic attacks, premenstrual syndrome, anorexia, allergies, liver and kidney dysfunction, and also cancer (12).

Poor reproductive capacity, blood pressure, impaired organ function, tumors, and hepatic abnormalities are known symptoms of chronic exposure to lead (13). Lead can also affect brain activity by interfering with synaptogenesis and neurotransmitter release. It has been proven that the consumption of Pb-contaminated food can cause adverse effects on human health, such as a reduction in IQ, learning disabilities, kidney failure, hyperactivity, slow growth, impaired hearing, and antisocial behaviors (9,14). Zinc is known as an essential element in biological systems because of its role in catalyzing reactions and the reversible changes in the oxidation state of metal ions. It should be noted that exposure to high levels of Zn can cause disruptions in some physiological activities, particularly breathing (15,16).

In recent years, different artificial neural network (ANN) approaches have been successfully applied in a large number of studies on forecasting water resources problems because of their ability to model nonlinear systems (17-20). Nor et al (21) developed ANN-based models for estimating nitrate and sulfate in water sources. Their results showed the good accuracy of ANN models. Mandal et al (22) presented an ANN model based on a backpropagation (BP) training algorithm (ANN-BP) for predicting removal efficiency. Their results showed that the ANN-BP can predict adsorption efficiency with acceptable accuracy. Keskin et al (23) investigated the applicability of ANN models for predicting water pollution sources in several areas of Turkey. They found that the ANN model can yield acceptable results. Hossain and Piantanakulchai (24) proposed a model based on GIS and the classification tree method to predict groundwater arsenic contamination risk. They demonstrated that the proposed model can effectively forecast the degree of As accumulation in groundwater with acceptable accuracy. Alizamir and Sobhanardakani (19) applied ANNs to forecast As, Pb and Zn concentrations in the groundwater resources of Asadabad plain. Their results showed the feasibility of ANNs in modeling the concentrations of heavy metals. Alizamir et al (20) applied two ANNs (MLP and RBF) to estimate heavy metals concentrations in the Asadabad plain. As demonstrated in their study, the MLP model offered better results than the single RBF model. In the current study, an ANN with two training algorithms was proposed for the prediction of heavy metal concentrations in the groundwater resources of Ghahavand plain. Artificial intelligence models are modeling tools that can identify statistical relationships between the input and output parameters of a complex system. This study introduces a model for predicting heavy metal concentrations using an ANN with imperialist competitive algorithm (ICA) and Levenberg-Marquardt (LM).

Methods

Study area

Ghahavand plain with an area of 2360 km^2 is located in Hamadan province, western Iran. Drinking water for residents of this plain is supplied by 1788 wells, 104 springs, and 96 aqueducts (25,26).

Sample collection

Based on Cochran's formula, a total of 60 groundwater samples were collected from 20 different wells under exploitation in the study area, including agricultural and residential regions. The locations of groundwater sampling stations are presented in Figure 1.

Sample preparation and analysis

In the current study, groundwater samples were taken according to the method introduced by Sobhanardakani et al (1). Then they were filtered with Whatman No. 42, preserved with 65% nitric acid (Merck, Germany), and kept at 4°C for further analysis (1,27). Finally, the concentrations of arsenic and heavy metals (Cu, Pb, and Zn) in groundwater samples were determined using an inductively coupled plasma-optical emission spectrometer at the wavelengths of 188.980 nm for As, 324.754 nm for Cu, 220.353 nm for Pb, and 206.200 nm for Zn (710-ES, Varian, Australia).

Artificial neural network

ANNs can describe nonlinear and complex relationships using a part of the input and output training patterns from the dataset. These approaches establish a non-linear relationship between inputs and outputs (28). An ANN can be demonstrated based on architecture that shows the

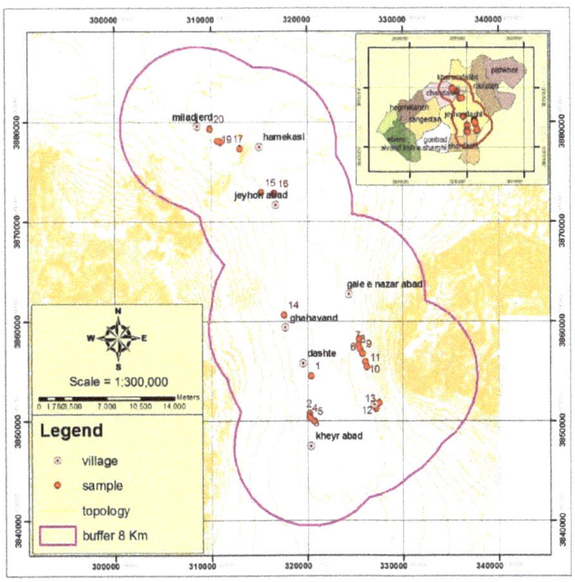

Figure 1. Map of sampling stations.

connection pattern between nodes, connection weights method determination, and the activation function (29). Because of their ability to learn a system's dynamics from data, ANNs are able to solve large-scale complex problems (30,31). The most commonly used neural network architecture is the feed-forward neural network (FFNN). The structure of a three-layered FFNN is based on some neurons in each layer and elements which link them (30). The training of a network is based on the optimization process for weights to obtain the appropriate weights to minimize errors; this process continues until the values of the output layer are as close as possible to the actual outputs (28). In this study, the LM training algorithm was utilized to tune the weights (29,32). Figure 2 shows the feed-forward network for this study, having one hidden layer with several nodes between the input and output layers.

Imperialist competitive algorithm

The ICA was proposed by Atashpaz-Gargari and Lucas (33) as a novel optimization algorithm. This algorithm was inspired by imperialistic competition. Like other evolutionary algorithms, it starts with an initial population. In concept of this approach, population individuals are called countries and are in two types: colonies and imperialists. All together, they form empires (33). In competition with each other, powerful empires obtain new colonies, and weak ones collapse. At the end of the algorithm, only the most powerful imperialist exists, and all the countries are colonies of the strongest empire. These colonies have the same position and cost as the imperialist. The ICA was applied in several benchmark problems and it revealed reliability in the optimization of different cost functions. A flowchart of the ICA is presented in Figure 3.

Model performance evaluation

The following statistical indicators were selected in the performance evaluation ANN models:

1) root-mean-square error (RMSE) (Eq. 1)

$$RMSE = \sqrt{\frac{\sum_{i=1}^{n}(P_i - O_i)^2}{n}} \tag{1}$$

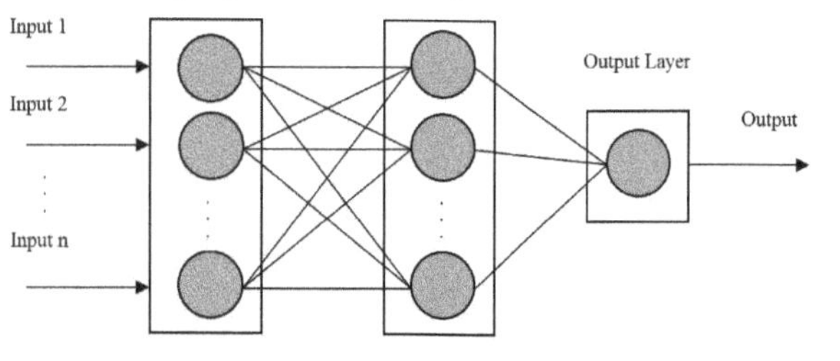

Figure 2. The neural network model for estimating heavy metals concentrations in groundwater resources of Ghahavand plain.

2) Pearson correlation coefficient (r) (Eq. 2)

$$r = \frac{n\left(\sum_{i=1}^{n} O_i \cdot P_i\right) - \left(\sum_{i=1}^{n} O_i\right)\cdot\left(\sum_{i=1}^{n} P_i\right)}{\sqrt{\left(n\sum_{i=1}^{n} O_i^2 - \left(\sum_{i=1}^{n} O_i\right)^2\right)\cdot\left(n\sum_{i=1}^{n} P_i^2 - \left(\sum_{i=1}^{n} P_i\right)^2\right)}} \tag{2}$$

3) Coefficient of determination (R^2) (Eq. 3)

$$R^2 = \frac{\left[\sum_{i=1}^{n}\left(O_i - \overline{O_i}\right)\cdot\left(P_i - \overline{P_i}\right)\right]^2}{\sum_{i=1}^{n}\left(O_i - \overline{O_i}\right)\cdot\sum_{i=1}^{n}\left(P_i - \overline{P_i}\right)} \tag{3}$$

where n is the total number of data, and P_i and O_i are the heavy metal concentrations predicted by the ANN methods and measured values, respectively.

Results

Descriptive statistics of elements content (µg/L) in groundwater samples collected from Ghahavand plain are indicated in Table 1. The average levels of As, Cu, Pb, and Zn in groundwater samples were 8.26 ± 1.09 µg/L, 9.25 ± 0.06 µg/L, 2.57 ± 0.30 µg/L, and 10.41 ± 4.68 µg/L, respectively. The results of statistical analysis (one sample *t* test) showed that the mean concentrations of analyzed elements were lower than the maximum permissible limits (µg/L) (100.0, 200.0, 100.0, and 2000.0 for As, Cu, Pb, and Zn, respectively) established by the World Health Organization (WHO) (25).

Neural networks have been successfully applied in different fields for environmental problems. In the present study, the same training and testing data sets were employed for the development of ANN-ICA and ANN-LM models. The collected data was divided into training and testing parts (80% and 20%, respectively). Since there is no criteria in ANN modeling to tell how many hidden nodes are needed, selecting the optimum number of hidden nodes is a difficult task. Here, a three-layer MLP with one hidden layer and the trial and error procedure were applied to select the number of hidden nodes (32,34). Sigmoid and linear functions were employed for the hidden and output node activation functions, respectively. For all heavy metal concentrations, the ANN models were first trained using the data in the training sets to obtain the optimized set of learning coefficients and then tested. RMSE, determination coefficients (R^2), and Pearson

Table 1. Descriptive statistics of metals contents (µg/L) in groundwater resources of Ghahavand plain

Element	Min.	Max.	Mean	SD
As	2.25	17.16	8.26	1.09
Cu	1.10	20.08	9.25	0.06
Pb	0.05	13.68	2.57	0.30
Zn	0.74	32.50	10.41	4.68

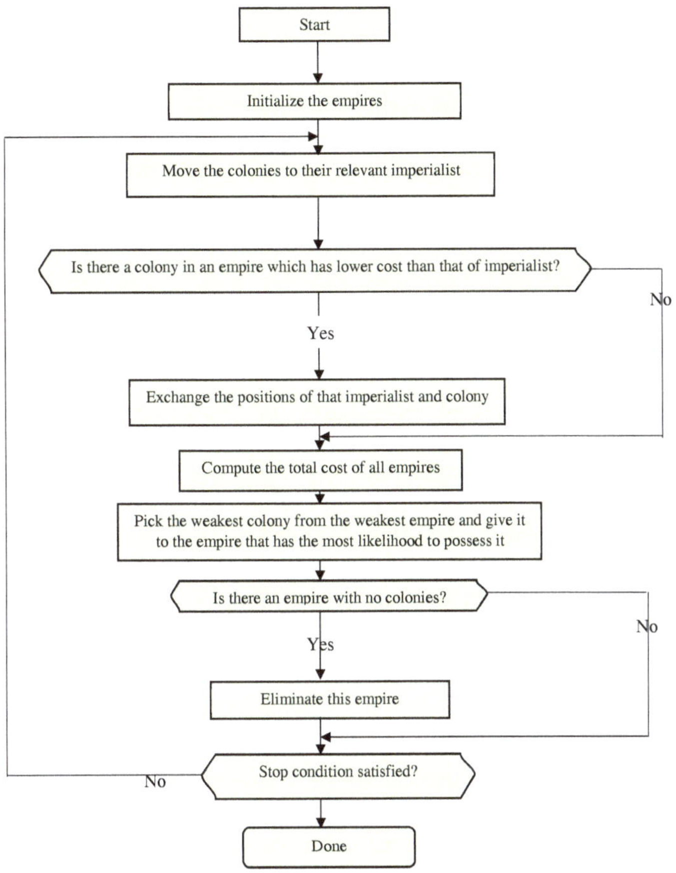

Figure 3. Flowchart of the ICA algorithm (33).

correlation coefficients (r) were used as evaluation criteria. For the ANN simulations, program codes were written in MATLAB software.

Discussion

To demonstrate the merits of the proposed ANN-ICA approach, the prediction accuracy of the model was compared to the prediction accuracy of the ANN-LM method, which was used as the benchmark. Table 2 presents a numerical comparison of the ANN-ICA and the ANN-LM models in terms of three statistical indicators: the RMSE, the model coefficient of determination (R^2), and the Pearson correlation coefficient (r). In terms of RMSE, the ANN-ICA model showed itself to be better than the ANN-LM model in both the training and the testing periods as shown in Table 2. Comparisons of the model efficiency statistic (R^2) between the ANN-ICA model and the ANN-LM model, presented in Table 2, revealed that the ANN-ICA model outperformed the ANN-LM model in both the training period and the testing period. In the testing period, the ANN-LM model efficiency was less than 90%, while the ANN-ICA model efficiency was 90%, which is a significant improvement over the ANN-LM model results.

A graphical performance comparison of the ANN-ICA and ANN-LM models is presented in Figures 4-11 as scatterplots of simulated versus observed As, Cu, Pb, and Zn concentrations. In Figures 4-11, the left columns

Table 2. Comparative performance of ANNs for As, Cu, Pb, and Zn concentrations

Heavy metal concentration	Methods	Training			Training		
		RMSE	r	R^2	RMSE	r	R^2
As	ANN-ICA	0.263	0.995	0.992	0.972	0.964	0.930
	ANN-LM	0.970	0.944	0.892	1.810	0.872	0.762
Cu	ANN-ICA	0.899	0.973	0.947	2.071	0.949	0.901
	ANN-LM	1.640	0.908	0.825	3.074	0.896	0.804
Pb	ANN-ICA	0.780	0.961	0.924	0.192	0.950	0.903
	ANN-LM	1.002	0.933	0.872	0.441	0.860	0.741
Zn	ANN-ICA	2.358	0.961	0.925	3.798	0.953	0.909
	ANN-LM	3.621	0.907	0.823	4.167	0.928	0.862

represent the models' results for the model training period, and the right columns show the corresponding results for the testing period. For the heavy metal As, the estimations of the two models are presented in Figures 4 and 5 in the form of scatterplots. It is seen from the scatterplots that the ANN-ICA estimations are closer to the corresponding observed heavy metal concentrations than those of the ANN-LM approach. As seen in the figures, the ANN-ICA model had a higher R^2 value (0.93) than the ANN-LM approach. The Cu and Pb concentrations observed and estimated using the ANN-ICA model are shown in Figs. 6 and 8, respectively. As can be seen from Table 1, the ANN-ICA model had a lower RMSE (2.0712) and higher R^2 (0.901) in the testing period for Cu concentration. For Pb concentration, the RMSE (0.1928) was lowest with the ANN-ICA model. Here also, the ANN-ICA model performed better than the ANN-LM model. Both the ANN-ICA and ANN-LM models estimated the Zn concentration in the testing period very closely. The

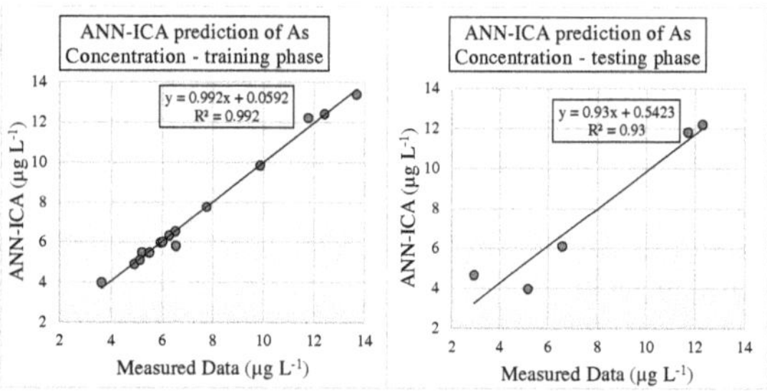

Figure 4. Observed and simulated As concentrations by ANN-ICA model during training and testing phases.

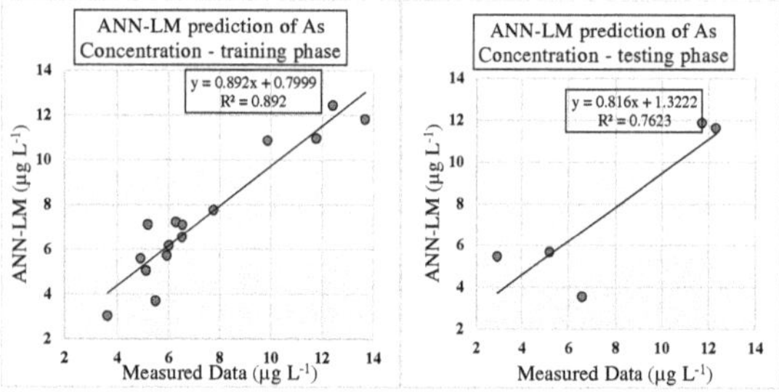

Figure 5. Observed and simulated As concentrations by ANN-LM model during training and testing phases.

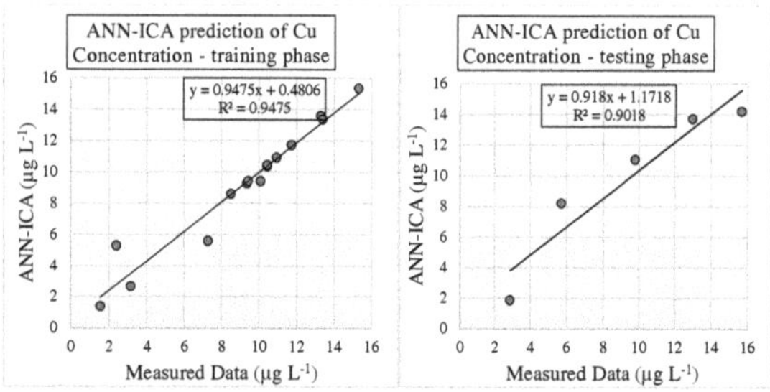

Figure 6. Observed and simulated Cu concentrations by ANN-ICA model during training and testing phases.

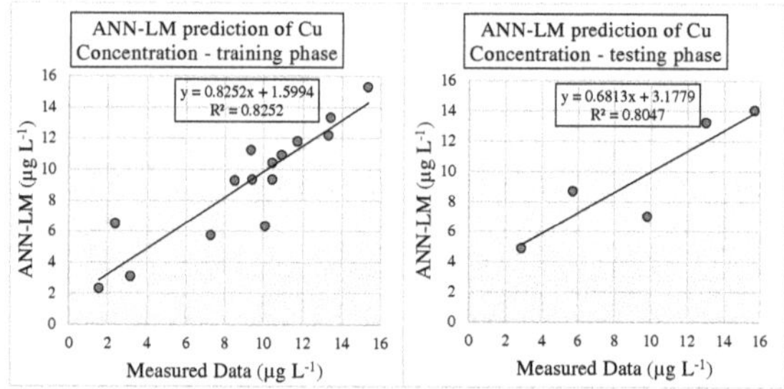

Figure 7. Observed and simulated Cu concentrations by ANN-LM model during training and testing phases.

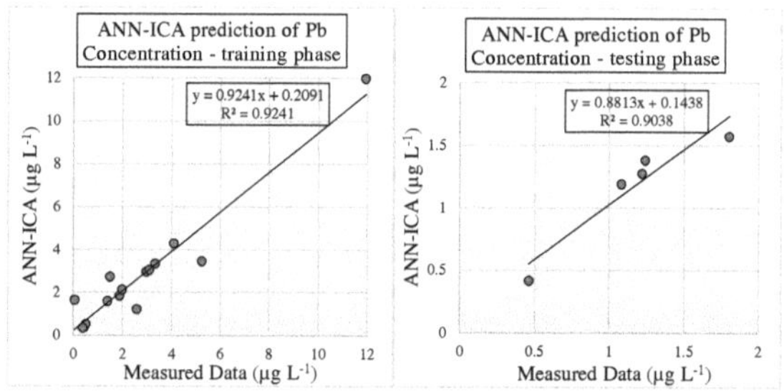

Figure 8. Observed and simulated Pb concentrations by ANN-ICA model during training and testing phases.

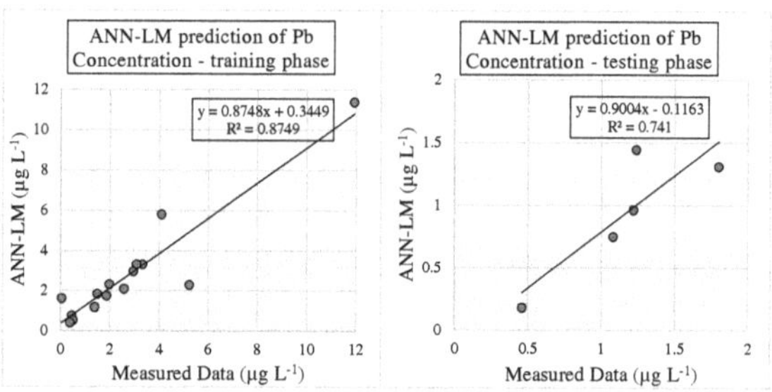

Figure 9. Observed and simulated Pb concentrations by ANN-LM model during training and testing phases.

ANN-ICA model gave a lower RMSE (3.7986) and higher R^2 (0.909) for the Zn concentration.

Overall, Figures 10 and 11 further emphasize the better performance of the ANN-ICA model over the ANN-LM model. It can be seen from Table 2 that the r values obtained while training and testing the ANN-ICA model were 0.995 and 0.964, respectively, which shows the acceptable forecasting performance of the ANN-ICA model. On the other hand, the r computed when testing the ANN-LM model was 0.872, which supports the higher capability of the ANN-ICA model for forecasting As concentrations. Table 2 also indicates that the ANN-ICA model had the highest r values for other heavy metals.

It can be seen from the scatterplots that in both the model training phase and the model testing phase, the ANN-ICA simulated data show more agreement with the observed data than the ANN-LM simulated data for all heavy metal concentrations. The ANN-ICA model scatterplots have less spread than the scatterplots for the ANN-LM

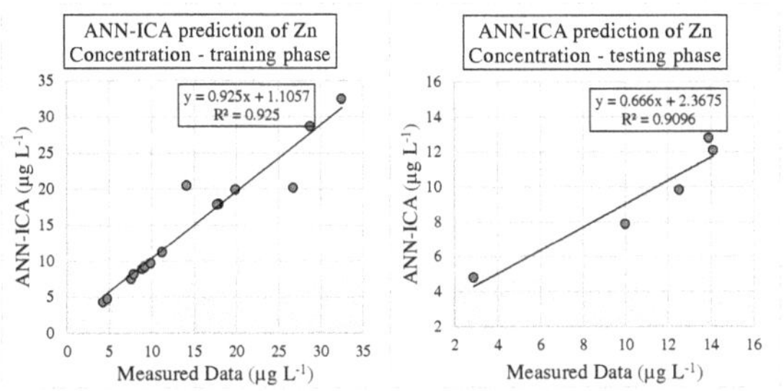

Figure 10. Observed and simulated Zn concentrations by ANN-ICA model during training and testing phases.

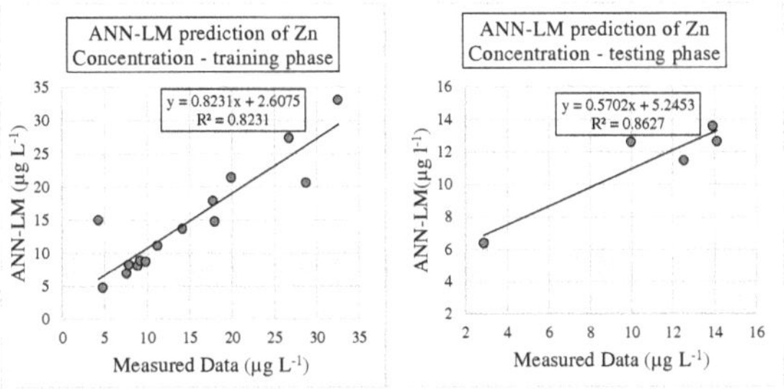

Figure 11. Observed and simulated Zn concentrations by ANN-LM model during training and testing phases.

model. The results prove the effectiveness, robustness, and compatibility of the ICA-ANN model.

The application of ANN models has been investigated by other researchers. For example, Hosseini and Mahjouri (35) applied fuzzy neural network-based support vector regression (FNN-SVR) and ANN (ANN) models to predict nitrate concentrations in the groundwater of Karaj aquifer. They obtained an R^2 of 0.71 from the FNN-SVR. Gholami et al (36) estimated groundwater quality using ANN and GIS at the Mazandaran plain of Iran. The optimal ANN model provided an R^2 of 0.73.

Conclusion

In the current study, the long-term changes in trends of heavy metal levels (As, Cu, Pb, and Zn) in groundwater resources of Ghahavand plain were estimated using an ANN with ICA and LM. Observations collected in the Ghahavand plain were used for model training and testing. Four predictive models for As, Pb, Cu, and Zn were created using the ANN-ICA approach. The ANN-ICA and ANN-LM methods were compared to assess prediction accuracy. The results, measured in terms of RMSE, *r*, and R^2, revealed that the ANN-ICA model was superior to the ANN-LM model. Heavy metal concentrations can be estimated from easily available data using the ANN-ICA technique. The proposed ANN-ICA approach can be implemented for forecasting heavy metal concentrations in groundwater resources data in environmental modelling studies.

Acknowledgments

The authors are grateful to the Hamedan Branch, Islamic Azad University for providing the facilities necessary to conduct and complete this study.

Competing interests

The authors affirm that this article is their original work, and they have no conflicts of interest to declare.

Authors' contributions

All authors were involved in all stages of the study.

References

1. Sobhanardakani S, Taghavi L, Shahmoradi B, Jahangard A. Groundwater quality assessment using the water quality pollution indices in Toyserkan Plain. Environmental Health Engineering and Management Journal 2017; 4(1): 21-7. doi: 10.15171/ehem.2017.04.
2. Tuzen M, Soylak M. Trace heavy metal levels in microwave digested honey samples from Middle Anatolia, Turkey. J Food Drug Anal 2005; 13(4): 342-7.
3. Yalcin MG, Aydin O, Elhatip H. Heavy metal contents and the water quality of Karasu Creek in Nigde, Turkey. Environ Monit Assess 2008; 137(1-3): 169-78. doi: 10.1007/s10661-007-9737-8.
4. Hague T, Petroczi A, Andrews PL, Barker J, Naughton DP. Determination of metal ion content of beverages and estimation of target hazard quotients: a comparative study. Chem Cent J 2008; 2: 13. doi: 10.1186/1752-153x-2-13.
5. Hosseini SV, Aflaki F, Sobhanardakani S, Tayebi L, Lashkan AB, Regenstein JM. Analysis of mercury, selenium, and tin concentrations in canned fish marketed in Iran. Environ Monit Assess 2013; 185(8): 6407-12. doi: 10.1007/s10661-012-3033-y.
6. Iwegbue CMA, Nwozo SO, Overah CL, Ossai EK, Mkpado CI, Osazuwa O, et al. Concentrations of selected metals in chicken eggs from commercial farms in Southern Nigeria. Toxicol Environ Chem 2012; 94(6): 1152-63. doi: 10.1080/02772248.2012.693492.
7. Rajankar PN, Gulhane SR, Tambekar DH, Ramteke DS, Wate SR. Water quality assessment of groundwater resources in Nagpur Region (India) Based on WQI. E-Journal of Chemistry 2009; 6(3): 905-8. doi: 10.1155/2009/971242.
8. Reza R, Singh G, Jain MK. Application of heavy metal pollution index for ground water quality assessment in Angul District of Orassia, India. Int J Res Chem Environ 2011; 1(2): 118-22.
9. Sobhanardakani S, Talebiani S, Maanijou M. Evaluation of As, Zn, Pb and Cu concentrations in groundwater resources of Toyserkan Plain and preparing the zoning map using GIS. Journal of Mazandaran University of Medical Sciences 2014; 24(114): 120-9. [In Persian].
10. Jarup L. Hazards of heavy metal contamination. Br Med Bull 2003; 68(1): 167-82. doi: 10.1093/bmb/ldg032.
11. Saracoglu S, Tuzen M, Soylak M. Evaluation of trace element contents of dried apricot samples from Turkey. J Hazard Mater 2009; 167(1-3): 647-52. doi: 10.1016/j.jhazmat.2009.01.011.
12. Ackah M, Anim AK, Zakaria N, Osei J, Saah-Nyarko E, Gyamfi ET, et al. Determination of some heavy metal

levels in soft drinks on the Ghanaian market using atomic absorption spectrometry method. Environ Monit Assess 2014; 186(12): 8499-507. doi: 10.1007/s10661-014-4019-8.
13. Abou-Arab AA, Ayesh AM, Amra HA, Naguib K. Characteristic levels of some pesticides and heavy metals in imported fish. Food Chem 1996; 57(4): 487-92. doi: 10.1016/S0308-8146(96)00040-4.
14. Hosseini SV, Sobhanardakani S, Tahergorabi R, Delfieh P. Selected heavy metals analysis of Persian sturgeon's (Acipenser persicus) caviar from Southern Caspian Sea. Biol Trace Elem Res 2013; 154(3): 357-62. doi: 10.1007/s12011-013-9740-6.
15. Tahsin N, Yankov B. Research on accumulation of zinc (Zn) and cadmium (Cd) in sunflower oil. Journal of Tekirdag Agricultural Faculty 2007; 4(1): 109-12.
16. Sobhanardakani S, Jamshidi K. Assessment of Metals (Co, Ni, and Zn) Content in the Sediments of Mighan Wetland Using Geo-Accumulation Index. Iranian Jornal of Toxicology 2015; 9(30): 1386-90.
17. Tapoglou E, Trichakis IC, Dokou Z, Nikolos IK, Karatzas GP. Groundwater-level forecasting under climate change scenarios using an artificial neural network trained with particle swarm optimization. Hydrological Sciences Journal 2014; 59(6): 1225-39. doi: 10.1080/02626667.2013.838005.
18. Chang J, Wang G, Mao T. Simulation and prediction of suprapermafrost groundwater level variation in response to climate change using a neural network model. J Hydrol 2015; 529(Part 3): 1211-20. doi: 10.1016/j.jhydrol.2015.09.038.
19. Alizamir M, Sobhanardakani S. Forecasting of heavy metals concentration in groundwater resources of Asadabad plain using artificial neural network approach. Journal of Advances in Environmental Health Research 2016; 4(2): 68-77. doi: 10.22103/jaehr.2016.40223.
20. Alizamir M, Sobhanardakani S, Taghavi L. Modeling of groundwater resources heavy metals concentration using soft computing methods: Application of different types of artificial neural networks. Journal of Chemical Health Risks 2017;7(3): 207-16.
21. Nor ASM, Faramarzi M, Yunus M, Ibrahim S. Nitrate and sulfate estimations in water sources using a planar electromagnetic sensor array and artificial neural network method. IEEE Sens J 2015; 15(1): 497-504. doi: 10.1109/JSEN.2014.2347996.
22. Mandal S, Mahapatra SS, Sahu MK, Patel RK. Artificial neural network modelling of As(III) removal from water by novel hybrid material. Process Saf Environ Prot 2015; 93: 249-64. doi: 10.1016/j.psep.2014.02.016.
23. Keskin TE, Düğenci M, Kaçaroğlu F. Prediction of water pollution sources using artificial neural networks in the study areas of Sivas, Karabük and Bartın (Turkey). Environ Earth Sci 2015; 73(9): 5333-47. doi: 10.1007/s12665-014-3784-6.
24. Hossain MM, Piantanakulchai M. Groundwater arsenic contamination risk prediction using GIS and classification tree method. Eng Geol 2013; 156: 37-45. doi: 10.1016/j.enggeo.2013.01.007.
25. Sobhanardakani S, Razban SS, Maànijou M. Evaluation of concentration of some heavy metals in ground water resources of Qahavand Plain-Hamedan. J Kermanshah Univ Med Sci 2014; 18(6): 339-348. [Persian].
26. Sobhanardakani S. Evaluation of the Water Quality Pollution Indices for Groundwater Resources of Ghahavand Plain, Hamadan Province, Western Iran. Iranian Jornal of Toxicology 2016; 10(3): 35-40.
27. Edet AE, Offiong OE. Evaluation of water quality pollution indices for heavy metal contamination monitoring. A study case from Akpabuyo-Odukpani area, Lower Cross River Basin (southeastern Nigeria). GeoJournal 2002; 57(4): 295-304. doi: 10.1023/B:GEJO.0000007250.92458.de.
28. Hornik K, Stinchcombe M, White H. Universal approximation of an unknown mapping and its derivatives using multilayer feedforward networks. Neural Netw 1990; 3(5): 551-60. doi: 10.1016/0893-6080(90)90005-6.
29. Kisi Ö. Constructing neural network sediment estimation models using a data-driven algorithm. Math Comput Simul 2008; 79(1): 94-103. doi: 10.1016/j.matcom.2007.10.005.
30. Haykin S. Neural Network: A Comprehensive Foundation. Englewood Cliffs, NJ: Prentice-Hall; 1999.
31. ASCE Task Committee on Application of Artificial Neural Networks in Hydrology. Artificial neural networks in hydrology. I: preliminary concepts. J Hydrol Eng 2000; 5(2): 115-23. doi: 10.1061/(ASCE)1084-0699(2000)5:2(115).
32. Kişi Ö. Daily pan evaporation modelling using multi-layer perceptrons and radial basis neural networks. Hydrol Process 2009; 23(2): 213-23. doi: 10.1002/hyp.7126.
33. Atashpaz-Gargari E, Lucas C. Imperialist competitive algorithm: an algorithm for optimization inspired by imperialistic competition. IEEE Congress on; 2007 Sept 25-28; IEEE, Singapore; 2007. doi: 10.1109/CEC.2007.4425083.
34. Hecht-Nielsen R. Kolmogorov's mapping neural network existence theorem. Proceedings of the International Symposium on Neural Networks. New York, USA; 1987.
35. Hosseini SM, Mahjouri N. Developing a fuzzy neural network-based support vector regression (FNN-SVR) for regionalizing nitrate concentration in groundwater. Environ Monit Assess 2014; 186(6): 3685-99. doi: 10.1007/s10661-014-3650-8.
36. Gholami V, Sebghati M, Yousefi Z. Integration of artificial neural network and geographic information system applications in simulating groundwater quality. Environmental Health Engineering and Management Journal 2016; 3(4): 173-182. doi: 10.15171/ehem.2016.17.

Sonochemical degradation of azithromycin in aqueous solution

Atiyeh Yazdani[1], Mohammad Hossein Sayadi[1*]

[1]Department of Environmental Sciences, Faculty of Natural Resources and Environment, University of Birjand, Birjand, Iran

Abstract

Background: The presence of pharmaceutical substances and their residual in water resources is an important environmental concern. Azithromycin, an antibiotic that is used for the treatment of infectious diseases, is a pollutant agent in the wastewater. The aim of this study was to investigate azithromycin degradation in aqueous solution through ultrasonic process in the presence of zinc oxide nanoparticles as catalysts.

Methods: Sonocatalytic experiments were performed at variable conditions including pH (3-8), temperature (20-60°C), time (3-21 minutes), catalyst dosage (0.25-2 g/L), hydrogen peroxide concentration (15-100 mg/L) and initial azithromycin concentration (10-100 mg/L).

Results: The optimum values for pH (3), temperature (40°C), time (15 minutes), catalyst dosage (1 g/L), H_2O_2 concentration (50 mg/L) and initial azithromycin concentration (20 mg/L) were determined. The highest degradation efficiency of 98.4% was achieved after 15 minutes of ultrasonic irradiation under optimum condition.

Conclusion: According to the results, ultrasonic irradiation is able to degrade azithromycin. In addition, ZnO and hydroxyl radical can successfully accelerate the reaction process in the shortest possible time.

Keywords: Azithromycin, Catalysis, Wastewater, Ultrasonics

***Correspondence to:**
Mohammad Hossein Sayadi
Email: Mh_sayadi@birjand.ac.ir

Introduction

One of the important pollutants in the environment is drug substance. More than 90% of the drugs are disposed and discharged into urban wastewater, and since the conventional treatment system is not efficient to remove these materials, hence, they enter the environment as water resources (1-3). Moreover, pharmacies and unprocessed drugs are also the main sources of drug entry into the environment. Antibiotics account for about 15% of total drug usage and are considered as one of the major pollutants of aquatic systems such as wastewater treatment plants besides surface, drinking and ground waters (4,5). The presence of antibiotics and pharmaceuticals in water has become a significant concern in recent years.

azithromycin as one of the most important macrolide antibiotics in medicine, is used to treat infection in humans and animals (6). Usually, these substances are not removed in the usual primary and secondary purification processes of a refinery, therefore, they can enter into the water supplies (7,8). The presence of these compounds in the aquatic environment can lead to toxicity, development of antibiotic-resistant strains, and subsequently, treat the human health and ecosystem function. Therefore, an efficient purification system is necessary to remove these compounds (9). In recent years, the chemical oxidation processes such as ozonation, oxidation, and photolysis have been introduced to remove such contaminants. Baran et al studied the removal of four antibiotics (sulfadiazine, sulfamethoxazole, sulfathiazole and sulfacetamide) by photocatalytic method and reported that the produced intermediates had less toxicity than the primary compounds (10). Uslu and Balcioglu reported almost total removal of antibiotics i.e. oxytetracycline and sulfamethazine, but the main concern with the ozonation process is the possibility of converting toxic compounds into undegradable compounds (11). The ultrasonic method is a sample of advanced oxidation processes (AOPs) that decomposes and degrades organic pollutants. The ultrasonic method has some unique benefits such as non-use of chemicals, ease of use and high efficiency compared to other technologies (12). Recently, the use of ultrasonic method alone or combined with other materials or processes has been considered as the most efficient method. Using the ultrasonic process alone will require a long time and high energy consumption to remove pollutants. Therefore, in order to reduce the time and energy consumption, semiconductor nanoparticles such as titanium dioxide (TiO2), zinc oxide (ZnO) and copper oxide (Cu2O) are used (13-15). The ultrasonic process using catalysts, is a relatively new process in water treatment compared to the other AOPs that can make antibiotics less toxic or less toxic byproducts, therefore, it is used as an effective method for the decomposition of pollutants with low biodegradability, such as antibiotics (16). Villaroel et al examined ultrasonic degradation of acetaminophen, and reported that ultrasound has shown remarkable ability to degrade acetaminophen, and in comparison to other AOPs (Fenton, photo-Fenton, UV/H_2O_2), it showed no harmful effects (17). Ultrasonic method can improve the catalytic oxidation process by activating the catalyst surface. The combination of a catalyst with ultrasound has a synergistic effect on the pollutant degradation. In addition, the ultrasonic process consumes considerable energy and time, which makes it a costly affair. To solve this problem, catalysts are added to the ultrasonic reaction system to reduce the activation energy and accelerate the response (18). Hartmann et al showed that the degradation of diclofenac in water by ultrasonic process in the presence of a catalyst is a useful method for the degradation of pharmaceutical compounds in water. Using this combined method in the presence of titanium dioxide (TiO_2), 84% of diclofenac was removed from water in 30 minutes (19). Zinc oxide nanoparticles are cheap and its catalytic properties will be appeared if the input energy be more than the energy required for its stimulation (20). Accordingly, the ultrasonic process with ZnO as a catalyst, can be used in the process of sonocatalysis (ultrasonically assisted catalysis) to purify water resources. This catalyst is a non-toxic, corrosion resistant and strong oxidizer (21). The purpose of this study was to investigate the azithromycin degradation through ultrasonic process in water using a ZnO catalyst considering the effect of different parameters such as pH, azithromycin concentration, catalyst concentration, time and temperature in order to find the optimal values.

Materials and Methods

Materials

Azithromycin powder was prepared by Sigma (St. Louis, MO, USA), ZnO nanoparticles as a catalyst, with size ranging from 20 to 60 nm and purity of 99.5%, was taken from Rouinsazan company, Iran. The SEM of ZnO nanoparticles is shown in Figure 1. Methanol and H_2O_2 were purchased from Merck (Darmstadt, Germany). In addition, pH meter (Istek, 915PDC), ultrasonic device (PARSONIC 7500 S, 220 VAC, Iran), UV-Vis spectrophotometer (Shimadzu Corporation, Japan), were also used.

Preparation of azithromycin solution

Azithromycin is an erythromycin derivative that its chemical formula is $C_{38}H_{72}N_2O_{12}$. Its solvent is generally methanol. Figure 2 shows its molecular structure.

The concentrations of azithromycin solution were determined by dissolving different amounts of azithromycin powder in 100 mL of methanol (20, 40, 70, and 100 mg/L). The solution was adjusted at various pHs (3, 4, 5, 6, 7 and 8) using HCl and 0.01 M NaOH. At each stage, different parameters such as ultrasonic time (3, 6, 12, 15, 18 and 21 minutes), temperature (20, 30, 40, 50 and 60°C), catalyst concentration (0, 0.25, 0.5, 0.75, 1, and 2 g/L) and H_2O_2 at different concentration (15, 20, 40, 50, 75, and 100 mg/L) were tested using an ultrasonic device at a frequency of 35 kHz. Samples were taken at selected times, filtered through Whatman no. 0.45 filter paper and prepared for reading.

Analysis

To determine the concentration of azithromycin, UV analysis was performed by UV-Vis spectrophotometer at a wavelength of 547 nm, and the optimal value of each parameter and removal rate were determined.

Results

Effect of pH solution on azithromycin degradation

Several studies have shown that pH of the solution plays an important role in the ultrasonic removal of organic pollutants. This can be due to its effects on the distribution of electric charge on the catalyst surface and oxidation potential. Studies have also shown that pH has a significant effect on the degradation and removal of antibiotics. In

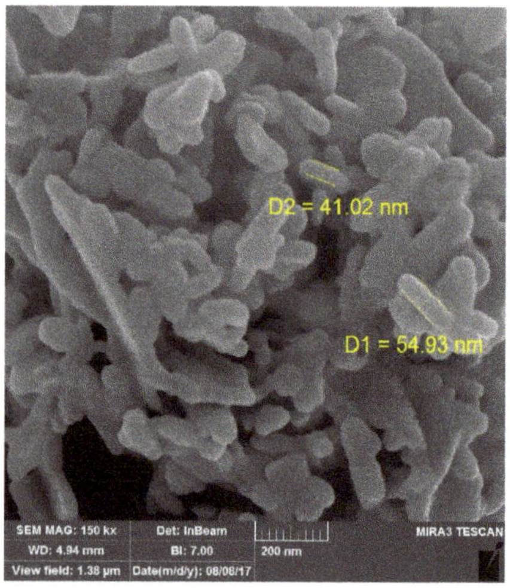

Figure 1. SEM of ZnO nanoparticles.

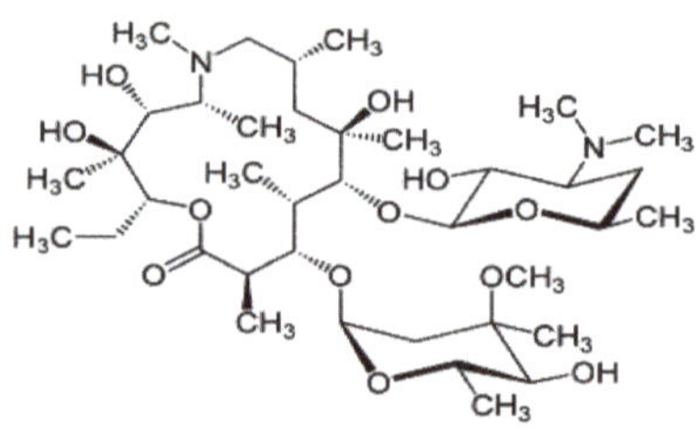

Figure 2. Molecular structure of azithromycin.

this study, the pH values from 3 to 8 were tested under test conditions (concentration of azithromycin = 40 mg/L, catalyst concentration = 1 g/L and temperature = 30°C) for 15 minutes in the presence of ZnO catalyst (Figure 3).

Effect of reaction temperature on azithromycin degradation

To determine the effect of temperature on azithromycin degradation, experiments were performed at 5 different temperatures ranging from 20 to 60°C and constant values of other parameters (contaminant concentration = 40 mg/L, ZnO catalyst concentration = 1 g/L and time = 15 minutes) and optimized pH value. As shown in Figure 4, the removal rate has increased with increasing temperature, especially from 20 to 40°C. But at temperatures from 40 to 60°C, there was a steady decrease in the removal rate.

Effect of reaction time on azithromycin degradation level

To investigate the effect of reaction time on the azithromycin degradation, under optimized conditions (pH = 3 and temperature = 40°C) and constant parameters (azithromycin concentration = 40 mg/L, and catalyst concentration = 1 g/L), sampling was carried out at intervals of 3 to 21 minutes after experiment using low-frequency ultrasound (35 kHz). Figure 5 shows the removal rate of azithromycin over time. The results demonstrated that there was a significant increase in the percentage of removal in the first 15 minutes, then, no significant change was observed.

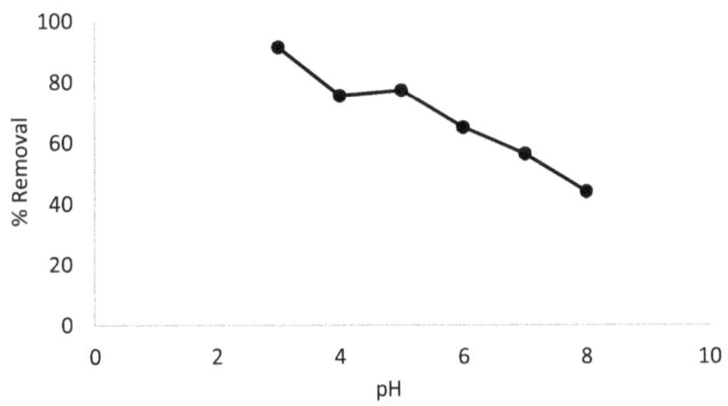

Figure 3. Effect of pH on azithromycin degradation at azithromycin concentration=40 mg/L, catalyst concentration=1 g/L, temperature=30°C, and time=15 min.

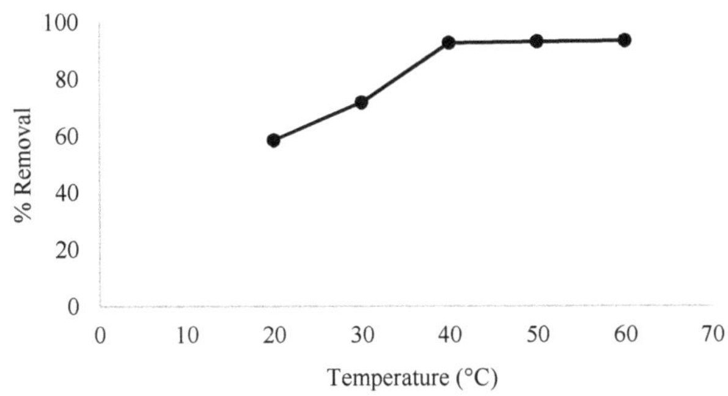

Figure 4. Effect of temperature on azithromycin degradation at contaminant concentration=40 mg/L, catalyst concentration=1 g/L, time=15 min, and pH= 3.

Effect of zinc oxide catalyst concentration on azithromycin degradation

Degradation of azithromycin in water was performed in the presence of ZnO catalyst. The presence of catalysts accelerates the degradation of pollutants in the reaction. In a solution of 40 mg/L azithromycin at pH = 3 and 40°C, the optimum catalyst level was obtained by ultrasonication of 1 g/L ZnO catalyst after 15 min of degradation. According to Figure 6, the increase in catalyst content in the solution from 0.25 to 2 g/L leads to an increase in the removal efficiency of azithromycin.

Effect of H_2O_2 on azithromycin degradation rate

According to researchers, the addition of an oxidizing agent, such as H_2O_2, as a hydroxyl radical producer, significantly increases the efficiency of organic pollutant degradation in AOPs. To maximize efficiency, H_2O_2 concentration should be selected according to the type and concentration of pollutants. To investigate the effect

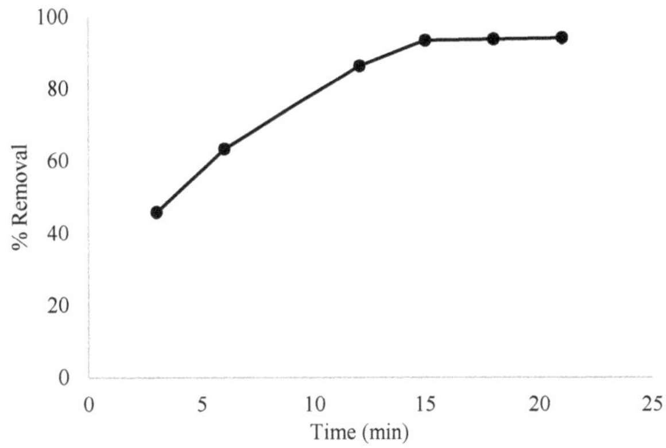

Figure 5. Effect of time on azithromycin degradation level at pH=3, temperature=40°C, azithromycin concentration=40 mg/L, and catalyst concentration=1 g/L.

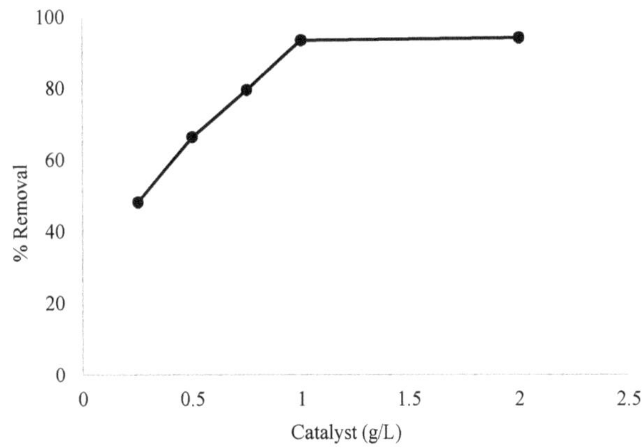

Figure 6. Effect of catalyst concentration on azithromycin degradation at pH=3, temperature=40°C, azithromycin concentrations=40 mg/L, and time=15 min.

of H_2O_2 at different concentrations (15, 20, 40, 50, 75, and 100 mg/L), azithromycin concentration = 40 mg/L, ZnO catalyst concentration = 1g/L, temperature = 40°C, pH = 3 in 15 minutes are essential. The percentage of azithromycin removal at different concentrations of H_2O_2 is shown in Figure 7. The results reveal that the use of H_2O_2 with a catalyst has a direct effect on the azithromycin removal rate. So that the sonocatalysts solely increased from 81.78% to 98.4% at 40 mg/L azithromycin in a 15-minute period using 50 mg/L H_2O_2 (optimal amount).

Effect of azithromycin concentration on degradation rate
By optimizing the effective parameters of degradation efficiency, the azithromycin concentration that had the highest degradation efficiency in optimal conditions, was selected. For this purpose, different concentrations (10, 20, 40, 70, and 100 mg/L) of azithromycin solution were prepared under optimal conditions (pH = 3, temperature = 40°C, catalyst concentration = 1 g/L and H_2O_2 concentration = 50 mg/L) was sonolyzed for 15 minutes. Figure 8 shows the azithromycin degradation percentage at 5 different concentrations. Based on the results, the decrease in azithromycin concentration increased the removal efficiency. Decreasing contaminant concentration had the same effect as increasing the catalyst concentration. By decreasing azithromycin concentration from 100 mg/L to 10 mg/L, the degradation rate increased from 30% to 99.8% after 15 minutes.

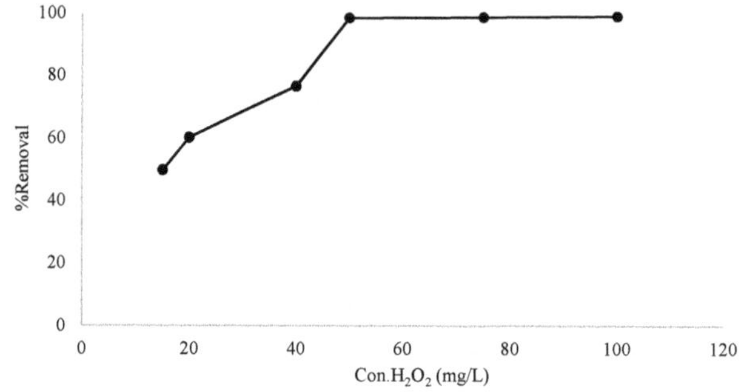

Figure 7. Effect of H_2O_2 on azithromycin removal percentage at azithromycin concentration=40 mg/L, ZnO catalyst=1 g/L, temperature=40°C, pH=3, and time=15 min.

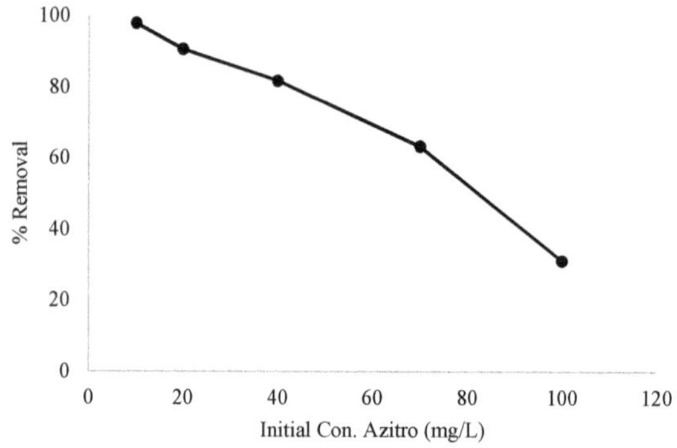

Figure 8. Effect of initial azithromycin concentration on removal percentage at pH=3, temperature=40°C, catalyst concentration=1 g/L, H2O2 concentration=50 mg/L, and time=15 min.

Discussion

According to the results obtained in Figure 2, with increasing pH, the degradation process of azithromycin decreased and acidic conditions helped degrade azithromycin. As pH decreased due to the reduction of hydroxyl radicals, the efficiency of the process also decreased. The effect of acidic environment to increase the efficiency can be shown as follows:

$2HO_{20} \longrightarrow O^{\circ}_2 + H_2O_2$ (Eq. 1)

$H_2O_2 + O^{\circ}_2 \longrightarrow OH^0 + OH^- + O_2$ (Eq. 2)

$e^- + O_2 \longrightarrow O^{\circ-}_2$ (Eq. 3)

$O^{\circ-}_2 + H^+ \longrightarrow 2HO^{\circ}_2$ (Eq. 4)

The basis of AOPs is hydroxyl radical production. But at higher pHs, H_2O_2 decomposes quickly and reduces radical formation (7,22). The effect of pH on azithromycin degradation depends on the structure and properties of the agent and also on its PKa value. The pKa value of azithromycin is 8.74. Accordingly, in acidic solutions, it will usually be molecular. The molecular forms mainly directed to the liquid bubble transfer region, where there is a high OH concentration. According to the studies, the accumulation of catalyst particles in the acidic solution decreases, resulting in an increase in the effective surface of the catalyst, which leads to an increase in the sonocatalytic degradation in acidic conditions (15,23-25). Villaroel et al reported that in the degradation of acetaminophen by ultrasonic method, acidic medium (pH = 3-5.6) is more suitable, thus acetaminophen (PKa 9.5) in acidic solutions in molecular form could has a higher degradation (17). Im et al also performed experiments at three different pHs (3,7 and 10) to investigate acetaminophen and naproxen degradation by oxidation processes (Sonocatalytic and Fenton). The results showed that when pH of the solution is low, the decomposition of compounds increases that in turn, increases the hydrophobic property of the acidic medium (25).

Ultrasonic tests are usually performed in the temperature control system to ensure that the isothermal conditions are maintained (26). The increase in temperature significantly increases the cavitation intensity and leads to azithromycin reduction. As the temperature increases, the vapor pressure increases and leads to the generation of more cavitation bubbles. Decomposition at a lower temperature reduces the concentration of radical hydroxyl, and subsequently decreases the degradation of contaminant (27-29). Braeutigam et al have shown that the rate constant depends on the reaction temperature, and the increase of the temperature has a positive effect on the reaction process. Reactions at temperatures below 25°C lead to the lower cavitation bubbles formation and concentration, and for substances with low solubility, the reaction is improved at higher temperatures (30). The removal rate of organic compounds is directly proportional to the temperature, because organic molecules migrate from the solution to the region where the hydroxyl radical concentration is high (31). Su et al investigated the sonocatalytic degradation of amoxicillin using sulfate radicals to determine the effect of temperature. They performed the experiments at a temperature ranging from 24 to 70°C and reported that the temperature increase, particularly from 24 to 40°C, increased the degradation efficiency (32).

The reaction time is one of the variables that influences the performance of oxidation process. Optimizing time in the removal reactions will save the cost of utilization and energy consumption (33). Increased reaction time results in the production of hydroxyl radicals, further contact of these radicals with antibiotics, and ultimately, further degradation of azithromycin. Over time, active sites will change for antibiotic absorption and the number of products produced by the catalyst reaction increases in the aqueous medium, and subsequently the degradation efficiency increases (34,35). The continuation of reaction for a long time leads to high energy consumption. Hence, achieving 93% removal in 15 minutes will save energy (36).

The direct effect of increasing the catalyst on degradation efficiency is due to an increase in the level of access or active catalyst positions. As the amount of catalyst increases, the amount of hydroxyl radical produced also increases. In other words, the presence of catalysts in the sonocatalyst process leads to the production of additional nuclei, which increases the number of active bubbles and radicals (37,38). In an ultrasonic process, the water is pyrolyzed in a portion of hydrogen radical, hydroxyl radical and oxygen radical, and then reacts with the pollutant. The reactions that occur are as follows (39,40):

$H_2O \rightarrow OH^0 + H^0$ (Eq. 5)

$OH^0 + OH^0 \rightarrow H_2O_2$ (Eq. 6)

$2\, OH^0 \rightarrow H_2O + O^0$ (Eq. 7)

$ZnO + H^0 \rightarrow Zn^{2+} + H_2O$ (Eq. 8)

$Zn^{2+} + 2\, OH \rightarrow ZnO + H_2O$ (Eq. 9)

$Zn^{+} + 2\, OH^0 \rightarrow Zn(OH)_2$ (Eq. 10)

$Zn(OH)_2 \rightarrow ZnO + H_2O$ (Eq. 11)

$Zn^{2+} + 2O_2{}^0 \rightarrow ZnO + 3/2\, O_2$ (Eq. 12)

According to Figure 5, azithromycin removal efficiency increased with increase of catalyst concentration up to 1 g/L and then remained almost constant. The addition of more amounts of catalyst to the reaction system results in the interactions between catalyst particles, thus, ultrasonic energy cannot reach the catalyst surface, resulting in the less active production of active radicals. This can be explained by the fact that when all of the antibiotic molecules settle on the nanoparticles, adding more amounts of catalyst does not affect the efficiency due to the absence of antibiotic molecules (41,42). Actually, ultrasonic device is used as an energy source to activate catalyst nanoparticles. Due to the high energy levels, nanoparticles tend to accumulate, but ultrasonic waves lead to their dispersion and non-accumulation. When the concentration of the catalyst nanoparticles exceeds a certain limit, ultrasonic energy will not be enough to disperse them, therefore, the removal efficiency remains constant despite the addition of more catalysts (43). These results were consistent with the results of other studies that investigated the effects of various concentrations of ZnO nanoparticles on the organic pollutants decomposition in AOPs (44). Excessive catalyticity in the sonolysis due to the turbulence occurring in the process, leads to increased darkness in the solution, and consequently its incremental effect will not continue (45).

The formation of hydroxyl radicals may be due to the breakdown of O-O bond in H_2O_2 by ultrasonic waves (14,46,47). The effect of H_2O_2 (the production of most hydroxyl radicals) is defined in the following formulas:

$H_2O + US \rightarrow \bullet H + HO\bullet$ (Eq. 13)

$H_2O_2 + \bullet H \rightarrow H_2O + HO\bullet$ (Eq. 14)

$H_2O_2 + \bullet O \rightarrow HO_2 + HO\bullet$ (Eq. 15)

$H_2O + \bullet O \rightarrow 2HO\bullet$ (Eq. 16)

$H_2O_2 + US \rightarrow 2HO\bullet$ (Eq. 17)

The hydrogen formed in reaction 13 reacts with H_2O_2 to produce hydroxyl radicals (Eq. 14). On the other hand, the oxygen atoms produced by the dissolution of molecular oxygen in the reaction with H_2O_2 (Eq. 15) and water (Eq. 16), generate more radicals. In addition, the sonolysis can directly act as a source of hydroxyl radical production in response to H_2O_2 (Eq17) (48). H_2O_2 is broken down by ultrasound (sonolysis) into OH, in two following ways (49,50):

$\bullet H + H_2O_2 \rightarrow \bullet OH + H_2O$ (Eq. 18)

$H_2O_2 \rightarrow 2\, \bullet OH$ (Eq. 19)

In the study on sonochemical degradation of ofloxacin in the presence of 100 mM H_2O_2 by Hapeshi et al, degradation efficiency increased from 28% to 50% (51). The reason for increase in the degradation efficiency along with the reduction of the initial concentration can be explained by the fact that, under the same conditions, other parameters (catalyst concentration, contact time and pH) increased the density of OH-radicals in the lower concentrations, which increased the decomposition of azithromycin released by radicals (51). On the other hand, at higher initial concentration of azithromycin, due to the limited amount of adsorption at the catalyst level, the removal percentage decreases. By increasing the concentration of azithromycin, the probability of a reaction between the pollutant and the reactive species is decreased. In addition, intermediates may be formed during side reactions (32,52,53). Serna-Galvis et al examined the sonochemical degradation of fluoxetine (FLX) with various parameters, including the initial concentration of the pollutant. The results showed that high concentration of FLX leads to an increase in degradation rates, and stated that the amount of H_2O_2 accumulation decreased with increasing the concentrations of pollutants. On the other hand, at higher concentration of the pollutant, the FLX molecules

Table 1. The degradation of pharmaceutical compounds by ultrasonic methods using various catalysts

Pollutant	Catalyst		Time (min)	Removal (%)	Reference
	Type	Concentration			
Amoxicillin	Co^{2+}	0.045 mmol/L	60	98.7	32
Levofloxacin	Ccl_4	0.02 mL	35	80	37
Tetracycline	Fe_3O_4	1 g/L	90	89	52
Acetaminophen, Naproxen	TiO_2	-	30	AAP 60.1, NPX 72.4	25
Fluoxetine	-	-	60	80	54
Sulfamethoxazole	Carbon nanotubes	10 g/L	60	88	55
Ibuprofen	Carbon nanotubes	10 g/L	60	94	55
Ibuprofen	-	-	30	98	56
Azithromycin	ZnO	1 g/L	15	98.4	Present study

are closer to the cavitation bubbles, therefore, the reaction with OH^0 is performed (54).

Table 1 shows the degradation of pharmaceutical compounds by ultrasonic methods in the optimum conditions. As shown in Table 1, various catalysts increased the degradation efficiency. However, the use of ZnO as a catalyst has some advantages including high efficiency, short reaction time, and easy workup.

Conclusion

In this study, degradation of azithromycin from aqueous sources was investigated by ultrasonic method using ZnO catalyst. Studies have shown that parameters such as pH, temperature, initial concentration of pollutant, catalyst, and time, affect the degradation process. According to the results, a high percentage of azithromycin removal was obtained in optimal conditions at pH = 3, temperature = 40°C, initial concentration of azithromycin = 20 mg/L and ZnO dosage = 1 g/L after 15 min. In these conditions, the azithromycin removal was 90.59%. Addition of H_2O_2 as an oxidizing agent, significantly increased the removal percentage (98.4%). Recently, researchers have explored the effective and cost-effective methods for wastewater treatment. Therefore, based on the previous studies and the results of the present study, it is concluded that the AOPs are suitable methods for the treatment of wastewaters containing pharmaceuticals.

Acknowledgments

This study, conducted as a research project (ID:1396/24355) in 2016, was funded by the Research Council of University of Birjand, South Khorasan. Authors genuinely appreciate the authorities of the Research Council and Faculty of Natural Resources and Environment, University of Birjand, for their sincere co-operation.

Competing interests

The authors declared no competing interests.

Authors' contributions

All authors were participated in all stages of this article, like data collection, analysis and interpretation and the manuscript approval.

References

1. Sayadi MH, Trivedy RK, Pathak RK. Pollution of pharmaceuticals in environment. I Control Pollution 2010; 26(1): 89-94.
2. Gadipelly C, Perez-Gonzalez A, Yadav GD, Ortiz I, Ibanez R, Rathod VK, et al. Pharmaceutical Industry wastewater: review of the technologies for water treatment and reuse. Ind Eng Chem Res 2014; 53(29): 11571-92. doi: 10.1021/ie501210j.
3. Lin AY, Lin CF, Chiou JM, Hong PK. O(3) and O(3)/H(2) O(2) treatment of sulfonamide and macrolide antibiotics in wastewater. J Hazard Mater 2009; 171(1-3): 452-8. doi: 10.1016/j.jhazmat.2009.06.031.
4. Grujic S, Vasiljevic T, Lausevic M. Determination of multiple pharmaceutical classes in surface and ground waters by liquid chromatography-ion trap-tandem mass spectrometry. J Chromatogr A 2009; 1216(25): 4989-5000. doi: 10.1016/j.chroma.2009.04.059.
5. Bhandari A, Close LI, Kim W, Hunter RP, Koch DE, Surampalli RY. Occurrence of ciprofloxacin, sulfamethoxazole, and azithromycin in municipal wastewater treatment plants. Practice Periodical of Hazardous, Toxic, and Radioactive Waste Management 2008; 12(4): 275-81. doi: 10.1061/(ASCE)1090-025X(2008)12:4(275).
6. Bahal N, Nahata MC. The new macrolide antibiotics: azithromycin, clarithromycin, dirithromycin, and roxithromycin. Ann Pharmacother 1992; 26(1): 46-55. doi: 10.1177/106002809202600112.
7. Homem V, Santos L. Degradation and removal methods of antibiotics from aqueous matrices--a review. J Environ Manage 2011; 92(10): 2304-47. doi: 10.1016/j.jenvman.2011.05.023.
8. Rivera-Utrilla J, Sanchez-Polo M, Ferro-Garcia MA, Prados-Joya G, Ocampo-Perez R. Pharmaceuticals as emerging contaminants and their removal from water. A

review. Chemosphere 2013; 93(7): 1268-87. doi: 10.1016/j.chemosphere.2013.07.059.
9. Sayadi MH, Ahmadpour N. The ultrasonic of drug removal using catalysts from aqueous solutions. Int J Environ Sci Nat Res 2017; 5(4): 555668. doi: 10.19080/IJESNR.2017.05.555668
10. Baran W, Sochacka J, Wardas W. Toxicity and biodegradability of sulfonamides and products of their photocatalytic degradation in aqueous solutions. Chemosphere 2006; 65(8): 1295-9. doi: 10.1016/j.chemosphere.2006.04.040.
11. Uslu MO, Balcioglu IA. Simultaneous removal of oxytetracycline and sulfamethazine antibacterials from animal waste by chemical oxidation processes. J Agric Food Chem 2009; 57(23): 11284-91. doi: 10.1021/jf902188j.
12. Bremner DH, Burgess AE, Chand R. The chemistry of ultrasonic degradation of organic compounds. Curr Org Chem 2011; 15(2): 168-77. doi: 10.2174/138527211793979862.
13. Abdelsalam ME, Birkin PR. A study investigating the sonoelectrochemical degradation of an organic compound employing Fenton's reagent. Phys Chem Chem Phys 2002; 4(21): 5340-5. doi: 10.1039/B205987H.
14. Pang YL, Abdullah AZ, Bhatia S. Review on sonochemical methods in the presence of catalysts and chemical additives for treatment of organic pollutants in wastewater. Desalination 2011; 277(1-3): 1-14. doi: 10.1016/j.desal.2011.04.049.
15. Hoseini M, Safari GH, Kamani H, Jaafari J, Ghanbarain M, Mahvi AH. Sonocatalytic degradation of tetracycline antibiotic in aqueous solution by sonocatalysis. Toxicol Environ Chem 2013; 95(10): 1680-9. doi: 10.1080/02772248.2014.901328.
16. Michael I, Rizzo L, McArdell CS, Manaia CM, Merlin C, Schwartz T, et al. Urban wastewater treatment plants as hotspots for the release of antibiotics in the environment: A review. Water Res 2013; 47(3): 957-95. doi: 10.1016/j.watres.2012.11.027.
17. Villaroel E, Silva-Agredo J, Petrier C, Taborda G, Torres-Palma RA. Ultrasonic degradation of acetaminophen in water: effect of sonochemical parameters and water matrix. Ultrason Sonochem 2014; 21(5): 1763-9. doi: 10.1016/j.ultsonch.2014.04.002.
18. Lifka J, Ondruschka B, Hofmann J. The use of ultrasound for the degradation of pollutants in water: aquasonolysis – a review. Eng Life Sci 2003; 3(6): 253-62. doi: 10.1002/elsc.200390040.
19. Hartmann J, Bartels P, Mau U, Witter M, Tumpling WV, Hofmann J, et al. Degradation of the drug diclofenac in water by sonolysis in presence of catalysts. Chemosphere 2008; 70(3): 453-61. doi: 10.1016/j.chemosphere.2007.06.063.
20. Hu B, Zhou J, Wu XM. Decoloring methyl orange under sunlight by a photocatalytic membrane reactor based on ZnO nanoparticles and polypropylene macroporous membrane. Int J Polym Sci 2013; 2013: 451398. doi: 10.1155/2013/451398.
21. Wang J, Lv Y, Zhang L, Liu B, Jiang R, Han G, et al. Sonocatalytic degradation of organic dyes and comparison of catalytic activities of CeO2/TiO2, SnO2/TiO2 and ZrO2/TiO2 composites under ultrasonic irradiation. Ultrason Sonochem 2010; 17(4): 642-8. doi: 10.1016/j.ultsonch.2009.12.016.
22. Elmolla ES, Chaudhuri M. Photocatalytic degradation of amoxicillin, ampicillin and cloxacillin antibiotics in aqueous solution using UV/TiO2 and UV/H2O2/TiO2 photocatalysis. Desalination 2010; 252(1-3): 46-52. doi: 10.1016/j.desal.2009.11.003.
23. Goel M, Hongqiang H, Mujumdar AS, Ray MB. Sonochemical decomposition of volatile and non-volatile organic compounds--a comparative study. Water Res 2004; 38(19): 4247-61. doi: 10.1016/j.watres.2004.08.008.
24. De Bel E, Dewulf J, Witte BD, Van Langenhove H, Janssen C. Influence of pH on the sonolysis of ciprofloxacin: Biodegradability, ecotoxicity and antibiotic activity of its degradation products. Chemosphere 2009; 77(2): 291-5. doi: 10.1016/j.chemosphere.2009.07.033.
25. Im JK, Yoon J, Her N, Han J, Zoh KD, Yoon Y. Sonocatalytic-TiO2 nanotube, Fenton, and CCl4 reactions for enhanced oxidation, and their applications to acetaminophen and naproxen degradation. Sep Purif Technol 2015; 141: 1-9. doi: 10.1016/j.seppur.2014.11.021.
26. Mason TJ, Lorimer JP. Sonochemistry in Environmental Protection and Remediation. In: Applied Sonochemistry: Uses of Power Ultrasound in Chemistry and Processing. New York: Wiley; 2003. p. 131-156. doi: 10.1002/352760054X.ch4.
27. Leong T, Ashokkumar M, Kentish S. The fundamentals of power ultrasound-A review. Acoust Aust 2011; 39(2): 54-63.
28. Thokchom B, Kim K, Park J, Khim J. Ultrasonically enhanced electrochemical oxidation of ibuprofen. Ultrason Sonochem 2015; 22: 429-36. doi: 10.1016/j.ultsonch.2014.04.019.
29. Xu M, Du H, Gu X, Lu S, Qiu Z, Sui Q. Generation and intensity of active oxygen species in thermally activated persulfate systems for the degradation of trichloroethylene. RSC Adv 2014; 4(76): 40511-7. doi: 10.1039/C4RA04942J.
30. Braeutigam P, Franke M, Schneider RJ, Lehmann A, Stolle A, Ondruschka B. Degradation of carbamazepine in environmentally relevant concentrations in water by Hydrodynamic-Acoustic-Cavitation (HAC). Water Res 2012; 46(7): 2469-77. doi: 10.1016/j.watres.2012.02.013.
31. Neis U. The use of power ultrasound for wastewater and biomass treatment. In: Gallego-Juárez JA, Graff KF, eds. Power Ultrasonics: Applications of High-Intensity Ultrasound. Oxford: Woodhead Publishing; 2015: 973-96. doi: 10.1016/B978-1-78242-028-6.00032-6.
32. Su S, Guo W, Yi C, Leng Y, Ma Z. Degradation of amoxicillin in aqueous solution using sulphate radicals under ultrasound irradiation. Ultrason Sonochem 2012; 19(3): 469-74. doi: 10.1016/j.ultsonch.2011.10.005.
33. Bazrafshan E, Mohammadi L, Kord Mostafapour F, Zazouli MA. Adsorption of Methylene blue from aqueous solutions onto low-cost ZnCl 2 treated Pistachio-nut shell ash. Wulfenia 2013; 20(11): 149-63.
34. Meskin PE, Ivanov VK, Barantchikov AE, Churagulov BR, Tretyakov YD. Ultrasonically assisted hydrothermal synthesis of nanocrystalline ZrO2, TiO2, NiFe2O4 and Ni0.5Zn0.5Fe2O4 powders. Ultrason Sonochem 2006; 13(1): 47-53. doi: 10.1016/j.ultsonch.2004.12.002.
35. Billik P, Plesch G. Mechanochemical synthesis of nanocrystalline TiO2 from liquid TiCl4. Scr Mater 2007; 56(11): 979-82. doi: 10.1016/j.scriptamat.2007.01.048.

36. Villegas-Guzman P, Silva-Agredo J, Giraldo-Aguirre AL, Florez-Acosta O, Petrier C, Torres-Palma RA. Enhancement and inhibition effects of water matrices during the sonochemical degradation of the antibiotic dicloxacillin. Ultrason Sonochem 2015; 22: 211-9. doi: 10.1016/j.ultsonch.2014.07.006.
37. Guo W, Shi Y, Wang H, Yang H, Zhang G. Intensification of sonochemical degradation of antibiotics levofloxacin using carbon tetrachloride. Ultrason Sonochem 2010; 17(4): 680-4. doi: 10.1016/j.ultsonch.2010.01.004.
38. Wang J, Jiang Z, Zhang Z, Xie Y, Wang X, Xing Z, et al. Sonocatalytic degradation of acid red B and rhodamine B catalyzed by nano-sized ZnO powder under ultrasonic irradiation. Ultrason Sonochem 2008; 15(5): 768-74. doi: 10.1016/j.ultsonch.2008.02.002.
39. Kummerer K. Antibiotics in the aquatic environment--a review--part I. Chemosphere 2009; 75(4): 417-34. doi: 10.1016/j.chemosphere.2008.11.086.
40. Gao XD, Li XM, Yu WD, Li L, Peng F, Zhang CY. Microstructure analysis and formation mechanism of ZnO nanoporous film via the ultrasonic irradiation mediated SILAR method. J Cryst Growth 2006; 291(1): 175-82. doi: 10.1016/j.jcrysgro.2006.02.008.
41. Jia X, Fan H, Zhang F, Qin L. Using sonochemistry for the fabrication of hollow ZnO microspheres. Ultrason Sonochem 2010; 17(2): 284-7. doi: 10.1016/j.ultsonch.2009.09.010.
42. Jamalluddin NA, Abdullah AZ. Reactive dye degradation by combined Fe(III)/TiO2 catalyst and ultrasonic irradiation: Effect of Fe(III) loading and calcination temperature. Ultrason Sonochem 2011; 18(2): 669-78. doi: 10.1016/j.ultsonch.2010.09.004.
43. Salehi M, Hashemipour H, Mirzaee M. Experimental study of influencing factors and kinetics in catalytic removal of methylene blue with TiO2 nanopowder. Am J Environ Eng 2012; 2(1): 1-7. doi: 10.5923/j.ajee.20120201.01.
44. Sanchez-Prado L, Barro R, Garcia-Jares C, Llompart M, Lores M, Petrakis C, et al. Sonochemical degradation of triclosan in water and wastewater. Ultrason Sonochem 2008; 15(5): 689-94. doi: 10.1016/j.ultsonch.2008.01.007.
45. Ertugay N, Acar FN. The degradation of Direct Blue 71 by sono, photo and sonophotocatalytic oxidation in the presence of ZnO nanocatalyst. Appl Surf Sci 2014; 318: 121-6. doi: 10.1016/j.apsusc.2014.01.178.
46. Bhavani R, Sivasamy A. Sonocatalytic degradation of malachite green oxalate by a semiconductor metal oxide nanocatalyst. Ecotoxicol Environ Saf 2016; 134(Pt 2): 403-11. doi: 10.1016/j.ecoenv.2015.10.029.
47. Voncina DB, Le Marechal AM. Reactive Dye Decolorization Using Combined Ultrasound/H2O2. Dyes Pigm 2003; 59(2): 173-9. doi: 10.1016/S0143-7208(03)00101-3.
48. Petrier C, Torres-Palma R, Combet E, Sarantakos G, Baup S, Pulgarin C. Enhanced sonochemical degradation of bisphenol-A by bicarbonate ions. Ultrason Sonochem 2010; 17(1): 111-5. doi: 10.1016/j.ultsonch.2009.05.010.
49. Suri RPS, Kamrajapuram A, Fu H. Ultrasound destruction of aqueous 2-chlorophenol in presence of silica and peroxide. Environ Eng Sci 2008; 25(10): 1447-54. doi: 10.1089/ees.2007.0145.
50. Ku Y, Tu YH, Ma CM. Effect of hydrogen peroxide on the decomposition of monochlorophenols by sonolysis in aqueous solution. Water Res 2005 39(6): 1093-8. doi: 10.1016/j.watres.2004.11.036.
51. Hapeshi E, Achilleos A, Papaioannou A, Valanidou L, Xekoukoulotakis NP, Mantzavinos D, et al. Sonochemical degradation of ofloxacin in aqueous solutions. Water Sci Technol 2010; 61(12): 3141-6. doi: 10.2166/wst.2010.921.
52. Hou L, Zhang H, Xue X. Ultrasound enhanced heterogeneous activation of peroxydisulfate by magnetite catalyst for the degradation of tetracycline in water. Sep Purif Technol 2012; 84: 147-52. doi: 10.1016/j.seppur.2011.06.023.
53. Worch E. Adsorption technology in water treatment: fundamentals, processes, and modeling. Berlin, Boston: De Gruyter; 2012.
54. Serna-Galvis EA, Silva-Agredo J, Giraldo-Aguirre AL, Torres-Palma RA. Sonochemical degradation of the pharmaceutical fluoxetine: Effect of parameters, organic and inorganic additives and combination with a biological system. Sci Total Environ 2015; 524-525: 354-60. doi: 10.1016/j.scitotenv.2015.04.053.
55. Al-Hamadani YAJ, Chu KH, Flora JRV, Kim DH, Jang M, Sohn J, et al. Sonocatalytical degradation enhancement for ibuprofen and sulfamethoxazole in the presence of glass beads and single-walled carbon nanotubes. Ultrason Sonochem 2016; 32: 440-8. doi: 10.1016/j.ultsonch.2016.03.030.
56. Mendez-Arriaga F, Torres-Palma RA, Petrier C, Esplugas S, Gimenez J, Pulgarin C. Ultrasonic treatment of water contaminated with ibuprofen. Water Res 2008; 42(16): 4243-8. doi: 10.1016/j.watres.2008.05.033.

Permissions

All chapters in this book were first published in EHEMJ, by Kerman University of Medical Sciences; hereby published with permission under the Creative Commons Attribution License or equivalent. Every chapter published in this book has been scrutinized by our experts. Their significance has been extensively debated. The topics covered herein carry significant findings which will fuel the growth of the discipline. They may even be implemented as practical applications or may be referred to as a beginning point for another development.

The contributors of this book come from diverse backgrounds, making this book a truly international effort. This book will bring forth new frontiers with its revolutionizing research information and detailed analysis of the nascent developments around the world.

We would like to thank all the contributing authors for lending their expertise to make the book truly unique. They have played a crucial role in the development of this book. Without their invaluable contributions this book wouldn't have been possible. They have made vital efforts to compile up to date information on the varied aspects of this subject to make this book a valuable addition to the collection of many professionals and students.

This book was conceptualized with the vision of imparting up-to-date information and advanced data in this field. To ensure the same, a matchless editorial board was set up. Every individual on the board went through rigorous rounds of assessment to prove their worth. After which they invested a large part of their time researching and compiling the most relevant data for our readers.

The editorial board has been involved in producing this book since its inception. They have spent rigorous hours researching and exploring the diverse topics which have resulted in the successful publishing of this book. They have passed on their knowledge of decades through this book. To expedite this challenging task, the publisher supported the team at every step. A small team of assistant editors was also appointed to further simplify the editing procedure and attain best results for the readers.

Apart from the editorial board, the designing team has also invested a significant amount of their time in understanding the subject and creating the most relevant covers. They scrutinized every image to scout for the most suitable representation of the subject and create an appropriate cover for the book.

The publishing team has been an ardent support to the editorial, designing and production team. Their endless efforts to recruit the best for this project, has resulted in the accomplishment of this book. They are a veteran in the field of academics and their pool of knowledge is as vast as their experience in printing. Their expertise and guidance has proved useful at every step. Their uncompromising quality standards have made this book an exceptional effort. Their encouragement from time to time has been an inspiration for everyone.

The publisher and the editorial board hope that this book will prove to be a valuable piece of knowledge for researchers, students, practitioners and scholars across the globe.

List of Contributors

Mohammad Malakootian
Environmental Health Engineering Research Center, Department of Environmental Health, School of Public Health, Kerman University of Medical Sciences, Kerman, Iran

Alireza Mesdaghinia
Department of Environmental Health Engineering, School of Public Health, Water Quality Research Centre, Institute for Environmental Research, Tehran University of Medical Sciences, Tehran, Iran

Shima Rezaei
Department of Environmental Health, School of Public Health, Kurdistan University of Medical Sciences, Sanandaj, Iran

Mojtaba Pourakbar
MSc of environmental health engineering, Department of Environmental Health Engineering, Faculty of Health, Tabriz University of Medical Sciences, Tabriz, Iran

Mohammad Mosaferi
Tabriz Health Services Management Research Center, Tabriz University of Medical Sciences, Tabriz, Iran

Saeed Lak
Department of Environmental Health Engineering, Faculty of Health, Ardabil University of Medical Sciences, Ardabil, Iran

Hossein Jafari Mansoorian
Environmental Health Engineering Research Center, Department of Environmental Health, Kerman University of Medical Sciences, Kerman, Iran

Salman Zarei
MSc of Environmental Health Engineering, Chief of Environmental Protection, Ilam Gas Company, Ilam, Iran

Narges Khanjani
Department of Environmental Health Engineering, School of Public Health, Kerman University of Medical Sciences, Kerman, Iran

Mohammad Hossien Salmani Nodoushan
PhD in Analytical Chemistry, Department of Environment Health, Shahid Sadoughi University of Medical Sciences, Yazd, Iran

Mohammad Hassan Ehrampoush
Department of Environment Health, Shahid Sadoughi University of Medical Sciences, Yazd, Iran

Lida Rafati
Department of Environment Health, Shahid Sadoughi University of Medical Sciences, Yazd, Iran

Maryam Farzi
Department of Environment Health, Shahid Sadoughi University of Medical Sciences, Yazd, Iran

Kazem Barzegar
Head of the English Language Department, School of Medicine, Shahid Sadoughi University of Medical Science, Yazd, Iran

Yaser Sharifi
Young Researchers and Elite Club, Qom Branch, Islamic Azad University, Qom, Iran

Azadeh Abedzadeh and Atieh Salighe
Department of Microbiology, Qom Branch, Islamic Azad University, Qom, Iran

Naser Kalhor
Academic Center for Education Culture and Research, Qom Branch, Qom, Iran

Mohammad Khodadad Motlagh
Faculty of Veterinary Medicine, Shiraz University, Shiraz, Iran

Ali Javadi
Students' Scientific Research Center, Tehran University of Medical Sciences, Tehran, Iran

Abdolmajid Gholizadeh
Department of Environmental Health Engineering, School of Public Health, Shahid Sadoughi University of Medical Sciences, Yazd, Iran

Mitra Gholami
Department of Environmental Health Engineering, School of Public Health, Iran University of Medical Sciences, Tehran, Iran

Reza Davoudi
MSc of Environmental Health Engineering, Department of Environmental Health Engineering, School of Public Health, Iran University of Medical Sciences, Tehran, Iran

Ayoob Rastegar
Department of Environmental Health Engineering School of Public Health, Sabzevar University of Medical Sciences, Sabzevar, Iran

Mohammad Miri
School of Public Health, Shahid Sadoughi University of Medical Sciences, Yazd, Iran

Nammam Ali Azadi
Kurdistan Research Center for Social Determinants of Health, Medical School, Kurdistan University of Medical Sciences, Sanandaj, Iran
Department of Epidemiology and Biostatistics, Medical School, Kurdistan University of Medical Sciences, Sanandaj, Iran

Reza Ali Falahzadeh
Department of Environmental Health Engineering, School of Public Health, Shahid Sadoughi University of Medical Sciences, Yazd, Iran

Shahram Sadeghi
Student Research Committee, Kurdistan University of Medical Sciences, Sanandaj, Iran

A'zam Golzari
Department of Environmental Engineering, Estahban Branch, Islamic Azad University, Estahban, Iran

Mohammad Mehdi Taghizadeh
Department of Environmental Engineering, Estahban Branch, Islamic Azad University, Estahban, Iran

Hamzeh Ali Jamali
Department of Environmental Health Engineering School of Health, Qazvin University of Medical Sciences, Qazvin, Iran
Department of Environmental Health Engineering, School of Health Isfahan University of Medical Sciences, Isfahan, Iran

Maryam Moradnia
Health Network of Divandareh, Kurdistan University of Medical Sciences, Sanandaj, Iran

Mohsen Pourkhosravani
Department of Geography Shahid Bahonar University of Kerman, Kerman, Iran

Zeynab Akbarlou
MSc of Environmental Health Engineering, Student Research Committee, Department of Environmental Health Engineering School of Health, Hormozgan University of Medical Sciences, Bandar Abbas, Iran

Vali Alipour and Mohsen Heidari
Department of Environmental Health Engineering, School of Health, Hormozgan University of Medical Sciences, Bandar Abbas, Iran

Kavoos Dindarloo
Department of Environmental Health Engineering, School of Health, Hormozgan University of Medical Sciences, Bandar Abbas, Iran

Dipankar Ghosh
PhD of Ecological Studies, Department of Ecological Studies, University of Kalyani, Kalyani, West Bengal, India

Jayanta Kumar Biswas
Department of Ecological Studies and Coordinator of International Centre for Ecological Engineering, University of Kalyani, Kalyani, West Bengal, India

Sohrab Golmohammadi
MSc of Water and Wastewater, Environmental Engineering (Trends Water and Wastewater), Kurdistan Rural Water & Wastewater CO (Kamyaran), Sanandaj, Iran

Mohammad Ahmadpour
Department of Environmental Health Engineering, School of Public Health, Maragheh University of Medical Science, Maragheh, Iran

Aliakbar Mohammadi
Department of Environmental Health Engineering, School of Public Health, Neyshabur University of Medical Sciences, Neyshabur, Iran

Azim Alinejad
Ph.D of Environmental Health Engineering, Social Determinants in Health Promotion Research Center, Hormozgan University of Medical Sciences, Bandar Abbas, Iran

Nezam Mirzaei
Environmental Health Research Center, Kurdistan University of Medical Sciences, Sanandaj, Iran

Mansour Ghaderpoori
Ph.D of Environmental Health Engineering, Students Research Committee, Shahid Beheshti University of Medical Sciences, Tehran, Iran

Afshin Ghaderpoori
Department of Environmental Health Engineering, Shahid Beheshti University of Medical Sciences, Tehran, Iran

Mohsen Dehghani
M.Sc. in Water and Wastewater Engineering, Islamic Azad University Bandar Abbas Branch, Bandar Abbas, Iran

Mohammad Hossein Alizadeh
Environmental Engineering and Science, Department of the Environmental Engineering, Islamic Azad University Bandar Abbas Branch, Bandar Abbas, Iran

Ali Asghar Najafpoor
Health Sciences Research Center, Department of Environmental Health Engineering, School of Health, Mashhad University of medical Sciences, Mashhad, Iran

Sina Dousti
MSc in Environmental Health Engineering, Student Research Committee, School of Health, Mashhad University of medical Science, Mashhad, Iran

Ahmad Joneidi Jafari
Department of Environmental Health Engineering, Center for Environmental Health Technology, School of Health, Iran University of Medical Sciences, Tehran, Iran

Ahmad Hosseinzadeh
MSc Student of Environmental Health Engineering, Student Research Committee, Department of Environmental Health Engineering, School of Health, Mashhad University of Medical Sciences, Mashhad, Iran

Amirreza Talaiekhozani
Civil Engineering Department, Jami Institute of Technology, Isfahan, Iran

Zeinab Eskandari
Chemical Engineering Department, Jami Institute of Technology, Isfahan, Iran

Marzieh Bagheri
Chemical Engineering Department, Jami Institute of Technology, Isfahan, Iran

Mohammad Reza Talaei
Chemical Engineering Department, Shiraz University, Shiraz, Iran

Malihe Salari
Civil Engineering Department, Jami Institute of Technology, Isfahan, Iran

Oshin Rezaei Raja
Department of the Environment, College of Basic Sciences, Hamedan Branch, Islamic Azad University, Hamedan, Iran

Soheil Sobhanardakani and Mehrdad Cheraghi
Department of the Environment, College of Basic Sciences, Hamedan Branch, Islamic Azad University, Hamedan, Iran

Fariba Sherri and Amir Hessam Hassani
Department of Environmental Engineering, Faculty of Natural Resources and Environment, Science and Research Branch, Islamic Azad University, Tehran, Iran

Amir Hossein Mahvi
Department of Environmental Health Engineering, School of Public Health, Tehran University of Medical Sciences, Tehran, Iran

Abbas Toloie Eshlaghy
Department of Industrial Management, Faculty of Management and Economic, Science and Research Branch, Islamic Azad University, Tehran, Iran

Mohammad Mehdi Amin
Department of Environmental Health Engineering, School of Health, Isfahan University of Medical Sciences, Isfahan, Iran

Sepideh Sadeghi, Ghazal Raki, Asrin Amini and Nezamaddin Mengelizadeh
Department of Environmental Health Engineering, School of Health, Isfahan University of Medical Sciences, Isfahan, Iran
Student Research Committee, Isfahan University of Medical Sciences, Isfahan, Iran

Majid Hashemi
Department of Environmental Health Engineering, School of Health, Isfahan University of Medical Sciences, Isfahan, Iran
Student Research Committee, Isfahan University of Medical Sciences, Isfahan, Iran
Environmental Health Engineering Research Center, School of Health, Kerman University of Medical Sciences, Kerman, Iran

Fatemeh Khademi
Islamic Azad University, Bandar Abbas Branch, Bandar Abbas, Iran

Kamyar Yaghmaeian
Department of Environmental Health Engineering, School of Public Health, Tehran University of Medical Sciences, Tehran, Iran

Mohammad Taheri and Mohammad Ali Hayatabadi
Kerman Water and Sewage Company, Kerman, Iran

Alireza Nasiri
Environmental Health Engineering Research Center, Kerman University of Medical Sciences, Kerman, Iran

Mohammad Malakootian
Environmental Health Engineering Research Center, Kerman University of Medical Sciences, Kerman, Iran
Department of Environmental Health, School of Public Health, Kerman University of Medical Sciences, Kerman, Iran

Mehdi Bibak and Javid Imanpour Namin
Department of Fisheries, School of Natural Resources, University of Guilan, Sowmehsara, Iran

Masoud Sattari
Department of Fisheries, School of Natural Resources, University of Guilan, Sowmehsara, Iran
Department of Marine Sciences, Caspian Sea Basin Research Center, University of Guilan, Sowmehsara, Iran

Ali Agharokh
Research Center for Persian Gulf, School of Persian Gulf, University of Persian Gulf, Bushehr, Iran

Saeid Tahmasebi
Department of Statistics, School of Basic Sciences, University of Persian Gulf, Bushehr, Iran

Soheil Sobhanardakani
Associate Professor, Department of the Environment, College of Basic Sciences, Hamedan Branch, Islamic Azad University, Hamedan, Iran

Raziyeh Zandipak
Msc, Young Researchers & Elite Club, Hamedan Branch, Islamic Azad University, Hamedan, Iran

Lobat Taghavi
Department of the Environmental Pollution, College of Environment and Energy, Science and Research Branch, Islamic Azad University, Tehran, Iran

Nematollah Jafarzadeh
Department of Environmental Health Engineering, School of Public Health, Ahvaz Jundishapur University of Medical Sciences, Ahvaz, Iran
Environmental Technologies Research Center, Ahvaz Jundishapur University of Medical Sciences, Ahvaz, Iran

Darioush Ranjbar Vakil Abadi
Environmental Health Engineering, Department of Environmental Health Engineering, School of Public Health, Bushehr University of Medical Sciences, Bushehr, Iran

Maryam Ravanbakhsh
Environmental Health Engineering, Student Research Committee, Ahvaz Jundishapur University of Medical Sciences, Ahvaz, Iran
Environmental Health Engineering, Department of Environmental Health Engineering, School of Public Health, Ahvaz Jundishapur University of Medical Sciences, Ahvaz, Iran

Kambis Ahmadi Angali
Department of Biostatistics, Ahvaz Jundishapur University of Medical Sciences, Ahvaz, Iran

Ahmad Zare Javid
Nutrition and Metabolic Diseases Research Center, Ahvaz Jundishapur University of Medical Sciences, Ahvaz, Iran
Hyperlipidemia Research Center, Ahvaz Jundishapur University of Medical Sciences, Ahvaz, Iran

Shirin Ardeshirzadeh
Observer at Cleveland Clinic Foundation, Family Medicine Department, Ohio, USA

Mohammad Taghi Samadi
Department of Environmental Health Engineering, School of Public Health and Center for Health Research, Hamadan University of Medical Sciences, Hamadan, Iran

Shahram Sadeghi
Environmental Health Engineering, Student Research Committee, Department of Environmental Health Engineering, School of Public Health, Kurdistan University of Medical Sciences, Sanandaj, Iran

Alireza Rahmani
Department of Environmental Health Engineering, School of Public Health and Center for Health Research, Hamadan University of Medical Sciences, Hamadan, Iran

Mohammad Hossien Saghi
Environmental Health Engineering, Department of Environmental Health Engineering, School of Public Health, Shahid Beheshti University of Medical Sciences, Tehran, Iran

Ahmad Reza Yazdanbakhsh
Environmental Health Engineering, Department of Environmental Health Engineering, School of Public Health, Shahid Beheshti University of Medical Sciences, Tehran, Iran

Mohammad Hasan Abasi
BSc of Environmental Health, Ilam University of Medical Sciences, Ilam, Iran

Zeinab Gholami
MSc of Environmental Health Engineering, Student Research Committee, School of Public Health, Tehran University of Medical Sciences, Tehran, Iran

Moayed Avazpour
Research Office of Environmental Health, Department of Environmental Health Engineering, School of Public Health, Shahid Beheshti University of Medical Sciences, Tehran, Iran
Environmental Health, Department of Environmental Health Engineering, Ilam University of Medical Science, Ilam, Iran

Hamed Fathi Dokht
Department of Soil Sciences, Gorgan University of Agricultural Sciences and Natural Resources, Gorgan, Iran

Seyed Alireza Movahedi Naeini and Esmaeil Dordipour
Department of Soil Sciences, Gorgan University of Agricultural Sciences and Natural Resources, Gorgan, Iran

Liz Wollsene De Jong
Department Agroecology, Aarhus University, Aarhus, Denmark

Ebrahim Hezarjaribi
Research Instructor of Plant Breeding, Golestan Agricultural and Natural Resources Research Center, Gorgan, Iran

Samad Gholami Yengejeh and Gharib Majidi
MSc of Environmental Health Engineering, Department of Environmental Health Engineering, School of Public Health, Qom University of Medical Sciences, Qom, Iran

Hossein Jafari Mansoorian
Environmental Health Engineering Research Center, Department of Environmental Health, School of Public Health, Kerman University of Medical Science, Kerman, Iran
Hamedan Branch, Islamic Azad University, Hamedan, Iran

Ahmad Reza Yari
Research Center for Environmental Pollutants, Qom University of Medical Sciences, Qom, Iran

Narges Khanjani
Environmental Health Engineering Research Center, Department of Environmental Health Engineering, School of Public Health, Kerman University of Medical Sciences, Kerman, Iran

Soheil Sobhanardakani
Department of the Environment, College of Basic Sciences, Hamedan Branch, Islamic Azad University, Hamedan, Iran

Lobat Taghavi
Department of the Environmental Pollution, College of Environment and Energy, Science and Research Branch, Islamic Azad University, Tehran, Iran

Behzad Shahmoradi
Kurdistan Environmental Health Research Center, Kurdistan University of Medical Sciences, Sanandaj, Iran

Amin Jahangard
MSc, Young Researchers & Elite Club, Hamedan Branch, Islamic Azad University, Hamedan, Iran

Samira Namavari
MSc in Water and Wastewater Engineering, Department Environmental Engineering, Faculty of Natural Resources, Islamic Azad University, Bandar Abbas Branch, Bandar Abbas, Iran

Farid Moeinpour
Department of Chemistry, College of Science, Bandar Abbas Branch, Islamic Azad University, Bandar Abbas, Iran

Zabihollah Yousefi
Department of Environmental Health Engineering, School of public Health, Mazandaran University of Medical Sciences, Sari, Iran

Hossein Sahebian
School of Health and Health Sciences Research Center, Mazandaran University of Medical Sciences, Sari, Iran

Samira Salari, Mojtaba Afsharnia and Mehdi Ghasemi
Department of Environmental Health Engineering, School of Health, Gonabad University of Medical Sciences, Gonabad, Iran

Asiyeh Moteallemi
Department of Environmental Health Engineering, School of Nursing, Torbat Jam Faculty of Medical Sciences, Torbat Jam, Iran

Masoud Tourang, Behnoush Aminzadeh and Ali Torabian
Department of Environmental Engineering, Graduate Faculty of Environment, University of Tehran, Iran

Meysam Alizamir
Young Researchers & Elite Club, Hamedan Branch, Islamic Azad University, Hamedan, Iran

Atiyeh Yazdani and Mohammad Hossein Sayadi
Department of Environmental Sciences, Faculty of Natural Resources and Environment, University of Birjand, Birjand, Iran

Ali Asghar Najafpoor
Health Sciences Research Center, Department of Environmental Health Engineering, School of Health, Mashhad University of Medical Sciences, Mashhad, Iran

Ghorban Soleimani
MS of Environmental Health Engineering, Shahid Sadoughi University of Medical Sciences, Yazd, Iran

Mohammad Hasan Ehrampoush
Environmental Health Engineering Department, Shahid Sadoughi University of Medical Sciences, Yazd, Iran

Mohammad Taghi Ghaneian
Environmental Health Engineering Department, Shahid Sadoughi University of Medical Sciences, Yazd, Iran

Elham Rahmanpour Salmani and Maryam Dolatabadi Takabi
Environmental Health Engineering, Student Research Committee, Department of Environmental Health, School of Health, Mashhad University of Medical Sciences, Mashhad, Iran

Amir Hussein Javid
Associate Professor of Environmental Engineering, Graduate School of Marine Science and Technology, Science and Research Branch, Islamic Azad University, Tehran, Iran

Mohammad Malakootian
Professor of Environmental Health Engineering Research Center and Department of Environmental Health, Kerman University of Medical Sciences, Kerman, Iran

Mohsen Mehdipour
Environmental Health in SBMU, Member of Environmental Health Engineering Research Center, Kerman University of Medical Sciences, Kerman, Iran

Mozhdeh Murkani and Mohammad Nasrollahi
Department of Environmental Health Engineering, School of Public Health, Ahvaz Jundishapur University of Medical Sciences, Ahvaz, Iran

Maryam Ravanbakhsh and Parvaneh Bahrami
MSc, Department of Environmental Health Engineering, School of Public Health, Ahvaz Jundishapur University of Medical Sciences, Ahvaz, Iran

Neamat Jaafarzadeh Haghighi Fard
Environmental Technology Research Center, Ahvaz Jundishapur University of Medical Sciences, Ahvaz, Iran and Department of Environmental Health Engineering, School of Public Health, Ahvaz Jundishapur University of Medical Sciences, Ahvaz, Iran

Mahdi Farzadkia
Department of Environmental Health Engineering, School of Public Health, Iran University of Medical Sciences, Tehran, Iran

Mohammad Hassan Ehrampush
Department of Environmental Health Engineering, School of Public Health, Shahid Sadoughi University of Medical Sciences, Yazd, Iran

Ehsan Abouee Mehrizi
Environment Research Center, Department of Environmental Health Engineering, School of Public Health, Iran University of Medical Sciences, and Department of Environmental Health Engineering, Faculty of Public Health, North Khorasan University of Medical Sciences, Bojnurd, Iran

Shahram Sadeghi
Department of Environmental Health Engineering, School of Public Health, Student Research Committee, Kurdistan University of Medical Sciences, Sanandaj, Iran

Parvaneh Talebi and Akbar Salehi
Department of Environmental Health Engineering, School of Public Health, Islamic Azad University, Maybod Branch, Maybod, Iran

Majid Kermani
Department of Environmental Health Engineering, School of Public Health, Iran University of Medical Sciences, Tehran, Iran

Index

CPSIA information can be obtained
at www.ICGtesting.com
Printed in the USA
LVHW060853030122
707669LV00002B/123

9 781639 895663